Primate Communities

Although the behavior and ecology of primates has been more thoroughly studied than that of any other group of mammals, there have been very few attempts to compare the communities of living primates found in different parts of the world. In *Primate Communities*, an international group of experts compares the composition, behavior and ecology of primate communities in Africa, Asia, Madagascar and South America. They examine the factors underlying the similarities and differences between these communities, including their phylogenetic history, climate, rainfall, soil type, forest composition, competition with other vertebrates and human activities. As it brings together information about primate communities from around the world for the very first time, it will quickly become an important source book for researchers in anthropology, ecology and conservation, and a readable and informative text for undergraduate and graduate students studying primate ecology, primate conservation or primate behavior.

JOHN G. FLEAGLE is Professor of Anatomical Sciences at the State University of New York at Stony Brook. He is the author of *Primate Adaptation and Evolution* (1988, 1999), and co-editor of several other books including *The Human Evolution Sourcebook* (1993), *Anthropoid Origins* (1994) and *Primate Locomotion: Recent Advances* (1998), and is founding editor of the journal *Evolutionary Anthropology*.

CHARLES JANSON is Professor of Ecology and Evolution at the State University of New York at Stony Brook. His major research interests are the evolutionary ecology of primate social behavior and the evolution of seed dispersal.

KAYE E. REED is Assistant Professor of Anthropology at Arizona State University, Research Associate at the Institute of Human Origins, and Co-Director of the IHO/University of the Witwatersrand Summer Paleoanthropological Field School. Her primary research interests are in the evolutionary paleoecology of fossil primates and hominids, and the structure and dynamics of mammalian and primate communities. Current paleontological and paleoecological research is focused in Ethiopia and South Africa.

Primate Communities

EDITED BY   JOHN G. FLEAGLE
State University of New York, Stony Brook

CHARLES H. JANSON
State University of New York, Stony Brook

KAYE E. REED
Arizona State University, Tempe

CAMBRIDGE
UNIVERSITY PRESS

PUBLISHED BY THE PRESS SYNDICATE OF THE UNIVERSITY OF CAMBRIDGE
The Pitt Building, Trumpington Street, Cambridge, United Kingdom

CAMBRIDGE UNIVERSITY PRESS
The Edinburgh Building, Cambridge CB2 2RU, UK   www.cup.cam.ac.uk
40 West 20th Street, New York, NY 10011–4211, USA   www.cup.org
10 Stamford Road, Oakleigh, Melbourne 3166, Australia
Ruiz de Alarćon 13, 28014 Madrid, Spain

First published 1999

Printed in the United Kingdom at the University Press, Cambridge

Typeset in Ehrhardt 9/12 [wv]

A catalogue record for this book is available from the British Library

Library of Congress Cataloguing in Publication data

Primate communities / edited by John G. Fleagle, Charles H. Janson,
and Kaye E. Reed.
     p.   cm.
1. Primates. 2. Animal communities. I. Fleagle, John G.
II. Janson, Charles Helmar. III. Reed, Kaye E., 1951–   .
QL737.P9P67225 1999
599.8′1782–dc21   98-54414 CIP

ISBN 0 521 62044 9 hardback
ISBN 0 521 62967 5 paperback

Contents

Contributors

COLIN A. CHAPMAN
Department of Zoology, University of Florida, Gainesville, FL 32611, USA

DAVID J. CHIVERS
Wildlife Research Group, Department of Anatomy, University of Cambridge, Downing Site, Cambridge CB2 3DY, UK

HARRIET A. C. EELEY
The Forest Biodiversity Programme, Department of Zoology and Entomology, University of Natal, Pietermaritzburg, Private Bag X01, Scottsville 3209, South Africa

LOUISE H. EMMONS
Smithsonian Institution, Division of Mammals, MRC 108, Washington, DC 20560, USA

JOHN G. FLEAGLE
Department of Anatomical Sciences, Health Sciences Center, State University of New York, Stony Brook, Stony Brook NY 11794-8081, USA

JÖRG U. GANZHORN
Zoologisches Institut, Martin-Luther-King-Platz 3, 20146 Hamburg, Germany

ANNIE GAUTIER-HION
CNRS, Université de Rennes I, Station Biologique, 35380 Paimport, France

A. K. GUPTA
Wildlife Research Group, Department of Anatomy, University of Cambridge, Downing Site, Cambridge CB2 3DY, UK

CHARLES H. JANSON
Department of Ecology and Evolution, State University of New York, Stony Brook, Stony Brook, NY 11794-5245, USA

JUKKA JERNVALL
Institute of Biotechnology, University of Helsinki, PO Box 56 (Viikinkaari 9), FIN 00014, Finland

PETER M. KAPPELER
ABt Verhaltensforschung/Ökologie, Deutsches Primatenzentrum, Kellnerweg 4, 37077 Göttingen, Germany

MICHAEL J. LAWES
The Forest Biodiversity Programme, Department of Zoology and Entomology, University of Natal, Pietermaritzburg, Private Bag X01, Scottsville 3209, South Africa

JOHN F. OATES
Department of Anthropology, Hunter College, 695 Park Avenue, New York, NY 10021, USA

DAPHNE A. ONDERDONK
Department of Zoology, University of Florida, Gainesville, FL 32611, USA

CARLOS A. PERES
School of Environmental Sciences, University of East Anglia, Norwich NR4 7JT, UK

JONAH RATSIMBAZAFY
Department of Anthropology, State University of New York, Stony Brook, Stony Brook, NY 11794, USA

KAYE E. REED
Institute of Human Origins, Arizona State University, Box 874101, Tempe, AZ 85287-4101, USA

THOMAS T STRUHSAKER
Department of Biological Anthropology and Anatomy, Duke University, Box 90383, Durham, NC 27708-0383, USA

CAROLINE TUTIN
S.E.G.C., BP 7847, Libreville, Gabon

LEE WHITE
Wildlife Conservation Society, New York, and Institute of Cell, Animal and Population Biology, University of Edinburgh, UK

PATRICIA C. WRIGHT
Department of Anthropology, State University of New York, Stony Brook, Stony Brook, NY 11794, USA

Preface

During the last four decades, the primates of Africa, Asia, Madagascar and South America have been the subject of hundreds of field studies involving millions of hours of observation. Despite this remarkable effort, there have been only a handful of attempts to undertake broad comparisons of the primate faunas in different biogeographical regions in order to document and understand their similarities and differences. This volume is an effort to make a start in addressing that major gap in our understanding of primate evolution. By bringing together a group of researchers with many decades of combined experience in all the major regions of the world inhabited by primates today, we hoped to summarize our current understanding of the factors determining primate community biology, highlight the many lacunae in our knowledge, and provide a baseline for future research in the area.

Like many projects of this nature, this one has a long history and has only been possible through the efforts and generosity of many people and organizations, especially the citizens and governments of countries inhabited by non-human primates today who have permitted and supported the research by primate field workers that ultimately formed the basis of the studies summarized here. The Wenner–Gren Foundation provided funds, and the Department of Anthropology at the University of Wisconsin provided the space, for a workshop on "Primate Communities" in 1996 that enabled many of the authors to discuss this topic face-to-face over three intense days in Madison, Wisconsin. We especially thank Dr Sydel Silverman, President of the Wenner–Gren Foundation, and Drs Karen Strier and Margaret Schoeninger for their support of the workshop. Joan Kelly was indispensable in organizing the workshop.

The task of transforming a collection of manuscripts from all over the world into a single coherent volume greatly benefitted from the talents of many individuals. In addition to those acknowledged in the individual papers, we thank Frances White, Diane Doran, Carel van Schaik, Scott McGraw, Charles Lockwood, Nancy Stevens, Lea Ann Jolley, Margaret Hall, Jay Norejko, Chris Heesy, Richard Kay, and Michael Alvard for many helpful comments and suggestions as well as seemingly endless editorial assistance. Mary Maas and Lea Ann Jolley compiled the indices. Luci Betti, Stephen Nash, Russ Mittermeier, Andy Young, Richard Tenaza, and Frances White generously provided illustrations. At Cambridge University Press, Tracey Sanderson, Lesley Bennun, and Sue Tuck shepherded the volume through the many steps associated with publication and printing.

John G. Fleagle
Charles H. Janson
Kaye E. Reed

1 • African primate communities: Determinants of structure and threats to survival

COLIN A. CHAPMAN, ANNIE GAUTIER-HION, JOHN F. OATES
AND DAPHNE A. ONDERDONK

INTRODUCTION

Africa is an immense continent covering approximately 30 million km^2 and encompassing 49 countries. For decades Africa has been considered a continent of great mystery, partly stemming from the fact that when Europeans first traveled to Africa, they found large expanses of seemingly impenetrable forests. Further, the first explorers often pressed inland by following rivers and thus often encountered only long stretches of riverine forest. In reality, the majority of the rainforest in Africa is situated in a belt that extends less than 10° north and south of the equator, and it is frequently broken by savanna or dry forest (e.g., the 300 km wide Dahomey Gap in West Africa). Thus, unlike the initial impressions of continuous homogenous forest, Africa actually contains a myriad of habitats from multi-strata tropical forest, to dry deciduous forest, woodland, savanna, and desert. Along with the variety of habitats found within the continent, Africa harbors a great diversity of primate communities: at least 64 species of primates are found in Africa (15 prosimians, 46 monkeys, and 3 apes, Oates, 1996a; Fig. 1.1).

The objective of this chapter is to provide a template with which to begin to understand the diversity of primate communities in Africa's tropical forests. To do this, we first review the nature of the forested habitats in which primates occur, describing general habitat characteristics and, when possible, providing detailed contrasts of rainfall and forest structure. Subsequently, since the majority of primate research taking a community level approach is derived from only a handful of sites, we describe each of these field locations and the key studies conducted at each site. This descriptive information is then used to evaluate how data from Africa can provide insights into determinants of

primate community structure. Finally, we describe the major threats faced by primate communities in Africa. It is clear that African tropical forests and the primate communities they support are seriously threatened by accelerating rates of forest conversion and degradation and by subsistence and commercial hunting. The impact of these threats is so great that some primate communities have been lost altogether, while others have been irrevocably changed before they can even be described.

ENVIRONMENTS OF AFRICAN PRIMATES

Tropical Africa covers a larger area than either tropical Asia or America, but the climate is generally drier. Only a few regions in Africa receive more than 2500 mm of rain a year, and many areas experience a pronounced dry season. Many central and eastern African forests lie close to the equator and therefore have two rainy seasons and two dry seasons each year. Any particular region receives rain when the rainbelt passes over the area as it moves to its southern maximum in January and once again when it moves to its northern-most extent in July (Fig. 1.2). In some regions in West Africa, winds blow parallel to the coast and not towards it from the ocean, producing a single peak in rainfall (Fig. 1.3).

Partially as a result of the continent's drier climate, the forests of Africa are fragmented by what was likely dry forest or woodland prior to human activities, but is now savanna and cultivated land (Fig. 1.2). In addition, major rivers further divide sections of forest, often isolating primate populations on either side. For example, before the central block of forest in Africa reaches Cameroon, it is cut by the Congo River, which can be over 15 km across

(a)

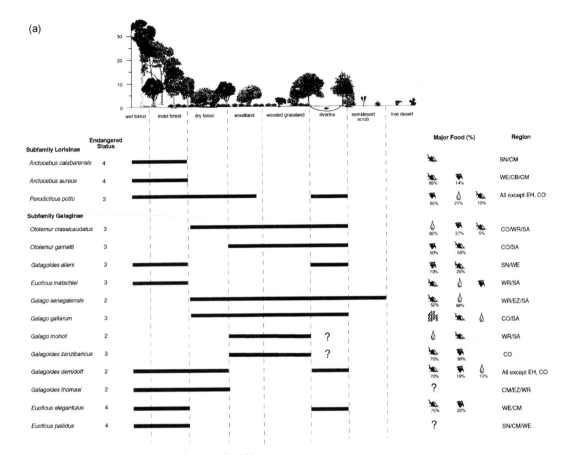

Fig. 1.1. A description of the primates found on the African
continent, their endangered status, habitat preference, diet, and
the region in which they are endemic.

(Beadle, 1981). The forest continues westward following
the coast. However, from the western frontier of Nigeria
to just west of the Volta River in Ghana, the forest belt is
broken by the Dahomey Gap, a 300 km wide stretch of
what is now savanna and farmland, but was probably once
dry forest. After the Dahomey Gap, forest continues
westward until it eventually gives way to dry forest and
grassland in Sierra Leone and Guinea-Bissau at about
10° north of the equator. There is evidence that even more
arid conditions affected much of Africa in the past and
further divided and isolated blocks of forest (Booth, 1958;
Struhsaker, 1981; Chapman, 1983; Colyn et al., 1991). For
example, layers of Kalahari sand underlie some parts of the
central basin, indicating that much of this forest area was
severely reduced in the Pleistocene (Moreau, 1966).

The richest primate communities in Africa are found in
forest habitats. Relative to drier savanna and desert habitats, these areas usually are rich in plant species, harbor
luxuriant plant growth, and are structurally complex. However, corresponding to Africa's generally dry climate, there
are very few areas that are classified as true wet-evergreen
forests. Rather, the majority of Africa's forests are considered moist forests, which do not receive as much rainfall as
wet-evergreen forests, and as little as 50–100 mm of rain
may fall in some months. Interestingly, original descriptions considered genuine rainforest to occur only in areas
that receive more than 2000 mm of rain a year (Richards,
1996). However, a number of areas in Africa that receive
as little as 1250 mm of rain a year have tall closed-canopy
forest. This has led a number of researchers to re-consider

(b)

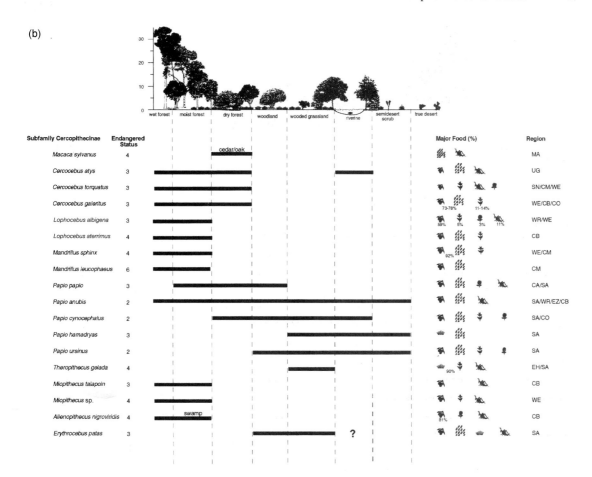

the minimum rainfall level for true rainforest (Lawson, 1986). One reason some African areas can maintain closed canopy forests while receiving little rain has to do with patterns of cloud cover. For example, during the four month dry season in Gabon and Congo, a quasi-permanent and non-precipitating stratiform cloud cover extends at least 800 km inland from the ocean. This cloud cover dramatically reduces temperature and evaporation. Maley (in press) speculates that without this atmospheric humidity, the severity of the dry season would result in the replacement of forest by savanna.

The tall closed-canopy forests of Africa are by no means homogenous, and the richness of plant species found in different forested areas can vary greatly. Hall (1977) and Hall & Swaine (1976, 1981) documented patterns of plant diversity in Nigeria and Ghana and found that in areas where annual precipitation is at least 1750 mm, one can find up to 200 plant species in a 25 m by 25 m plot. As areas become drier, the species richness declines, so that in areas that receive between 1500 and 1750 mm of rain a year, there are often less than 170 species in a 25 m by 25 m plot, and areas receiving between 1250 and 1500 mm of rain per year harbor between 40–100 plant species in the same sized plots.

Even within a single forested area receiving similar levels of rainfall, areas are by no means homogenous and the primate communities that occupy different areas can vary dramatically. For example, in the Ituri Forest in the Democratic Republic of the Congo, there are at least five distinct habitat types found in this area: mbau forest,

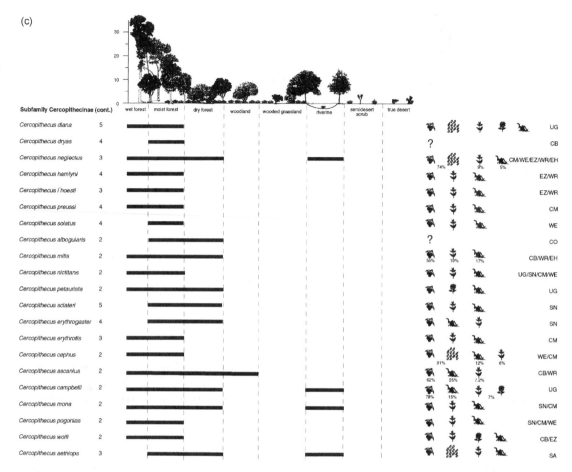

Fig. 1.1. (*cont.*)

which is dominated by *Gilbertiodendron dewevrei* (this one tree species comprises over 70–90% of the canopy level trees in these stands; Hart *et al.*, 1989; Hart, 1995), mixed forest, riparian forests, flooded forests, and secondary forests, which occur in areas of abandoned agricultural clearings and blow-downs. Because of the dominance of *Gilbertiodendron dewevrei*, the mbau forest exhibits a low diversity of trees (an average of 18 tree species greater than 10 cm DBH (diameter at breast height) in 0.5 ha plots; Hart, 1985). In contrast, the mixed forest has higher levels of tree diversity (an average of 65 tree species greater than 10 cm DBH in 0.5 ha plots). The richness and density of anthropoid primates varies between these habitat types, but can be fairly high (13 anthropoid species, 112 primates/km² or 710 kg/km²). However, in monodominant stands

of *Gilbertiodendron* all species are encountered at a low rate, and in the largest monodominant stand no primates were seen in nine samples of a 2.15 km transect. In spite of this generality such monodominant stands may be important to some species; observations of the owl-faced monkey (*Cercopithecus hamlyni*) suggest that, while they are still rare and hard to see in stands of *Gilbertiodendron*, they are more common here than elsewhere (Oates, pers. obs.).

Freshwater swamp forests are extremely extensive in Africa. Conservative accounts estimate that there are over 60 000 km² of permanent swamps and 400 000 km² of seasonally inundated swamps (Thompson & Hamilton, 1983). In the central basin of the Congo River alone, over 80 000 km² of forested land is permanently or seasonally inundated. Many of the plant species common in such

(d)

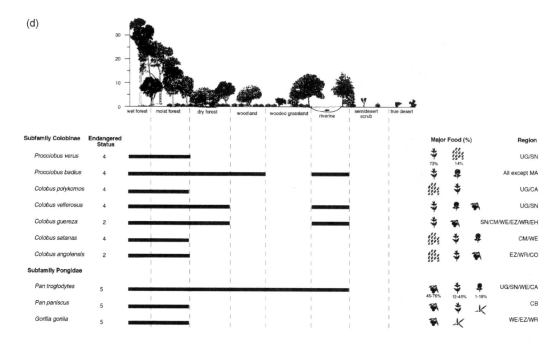

Conservation priorities follow Oates 1996 – higher numbers are more endangered. Distinct Regional Communities following Oates 1996 (see Figure 1) – MA = Maghreb, CA Casanance, UG = Upper Guinea, SN = Southern Nigeria, CM = Cameroon, WE = Western Equatorial, CB = Congo Basin, EZ = Eastern Zaire, WR = Western Rift, EH = Ethiopian Highlands, CO = Coastal East Africa. SA is used to depict species that are common in savanna regions. Dietary values are from Rowe 1996. Fruit, Gum, Insect, Seeds, Flowers, Leaves, Grass, Terrestrial herbs. Distribution based on Kingdon 1984, Rowe 1996 (personal communication), and distribution maps – Typically this should be viewed as an estimate. 'Riverine' is meant as forested habitats along rivers in a savanna area.

swamps (e.g., *Ficus congensis, Raphia farinifera, Phoenix reclinata, Uapaca* sp.) are eaten by a variety of primate species. Considering the extent of inundated swamp forest, it is not surprising that there is a primate species whose name reflects its dependence on swamp habitat: Allen's Swamp Monkey (*Allenopithecus nigroviridis*) is described as being restricted to regularly inundated forests (Verheyen, 1963). Similarly, species such as *Cercocebus galeritus, C. torquatus, Miopithecus* sp., and to some extent *Colobus guereza* can be dependent on inundated forests in certain areas.

Since the majority of the primate research taking a community-level approach is derived from studies conducted in forested habitats, we will only briefly mention other habitats. Savannas, which range from humid woodlands to dry grasslands, cover approximately 60% of Africa. Here, annual rainfall is often less that 1000 mm, and while trees and shrubs are often present, there is

always extensive grass cover (Deshmukh, 1986). Compared to forested habitats, the primate communities in savanna areas are relatively depauperate. These communities are typically composed of only a few hardy species such as patas monkeys (*Erythrocebus patas*), vervet monkeys (*Cercopithecus aethiops*), baboons (*Papio* spp.), some bushbabies (e.g., *Galago gallarum, Galago moholi, Galago senegalensis, Otolemur crassidaudatus*), and possibly chimpanzees (*Pan troglodytes*). However, in such savanna regions, riverine forests play particularly important roles (Fig. 1.1). At Tana River, Kenya, an area receiving less than 500 mm of rain annually, one finds a riverine habitat that supports six primate species (*Procolobus badius, Cercocebus galeritus, Papio cynocephalus, Cercopithecus albogularis, Otolemur crassicaudatus, Galago senegalensis*; Marsh, 1979). The area surrounding Tana River is thorn scrubland, but there are a number of tall forest trees found near the river (e.g., *Ficus sycomorus, Diospyros mespiliformis*).

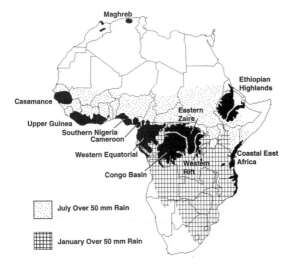

Fig. 1.2. Distinct regional communities of African primates (adapted from Oates, 1996a). The gridded area indicates the region where over 50 mm of rain falls in January, while the dotted area indicates the region over which 50 mm of rain falls in July.

A FOCUS ON FOREST STUDY SITES

By far the most diverse primate communities found in Africa are encountered in forest habitats. However, there is only a handful of study sites for which detailed data are available on the richness and abundance of the members of these communities, and even fewer that describe the behavior of the members and the ecological context in which they interact (Table 1.1).

In this section, we review the ecology and history of disturbance of these sites (reviewed from west to east) and outline the investigations that have taken place at each location. At these nine sites, there are on average nine anthropoid primate species (range 7 to 11; Table 1.2), and between three to six prosimians. Primate biomass is typically impressive, averaging 982 kg/km^2 and reaching values as high as 2710 kg/km^2. From these figures, it is clear that primates constitute a major component of the frugivore/folivore community in tropical forests (see Bourlière, 1985; Terborgh, 1986; Waser, 1987; Davies, 1994 for comparative data).

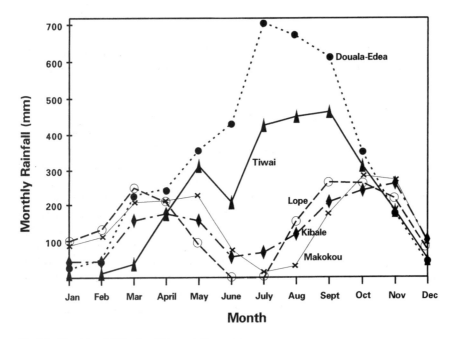

Fig. 1.3. Monthly rainfall at five African rainforest sites where detailed studies of primate communities have been conducted.

Table 1.1. *Descriptive characteristics of the major field sites where forest primates have been studied in Africa*

Site	Country	Vegetation classification	Rainfall	Elevation (m)	Coordinates	Source
Tiwai	Sierra Leone	Lowland moist forest	2708	85–112	7°31′N 11°20′W	Oates *et al.*, 1990 Oates, 1994; Fimbel, 1994*a,b*
Taï	Côte d'Ivoire	Lowland moist forest	1800	90	6°10′N 4°20′W	Boesch & Boesch 1983 Merz, 1986
Douala-Edéa	Cameroon	Lowland wet forest	4000	Sea level	3°29′N 9°50′E	McKey *et al.*, 1981
Lopé	Gabon	Lowland moist forest	1506	100–700	0°10′S 11°35′E	White, 1994*a,b* Tutin *et al.*, 1994
Forêt des Abeilles	Gabon	Lowland moist forest	1755	200–600	0°20′S 11°45′E	Gautier-Hion *et al.*, 1997
Makokou	Gabon	Lowland moist forest	1755	430	0°34′N 12°52′E	Gautier-Hion *et al.*, 1981 Gautier-Hion *et al.*, 1985
Botsima	DR Congo	Seasonally flooded forest	1774	350–700	1°15′S 20°00′E	Gautier-Hion & Maisels, 1994
Ituri	DR Congo	Lowland moist forest	1802	750	1°25′N 28°35′E	Hart, 1995
Kibale	Uganda	Mid-altitude moist forest	1662	1500	0°13′N 30°19′E	Struhsaker, 1975 Chapman *et al.*, 1997
Budongo	Uganda	Medium altitude moist forest	1495	1050	1°35′N 31°18′E	Eggeling, 1947 Plumptre *et al.*, 1994

Studies conducted at each of these major sites have had slightly different objectives, making it difficult to compare quantitative data from all sites. At all sites, however, estimates of annual rainfall have been made, which vary from 1495 to 4000 mm (Fig. 1.3). Rainfall strongly influences forest structure and productivity and, in doing so, could influence primate community structure.

Between them, the nine communities contain 40 of the 51 species of moist forest primates that we recognize in this chapter (see Fig. 1.1). More than half of the species not included are endemic to southern Nigeria and western Cameroon, a region in which primates are heavily hunted for their meat and where habituating animals for behavioral observation is difficult; as a consequence, reliable biomass and dietary data are not available from primate communities in this area (see chapter 17).

Tiwai Island, Sierra Leone

Tiwai is a 12 km² island located in southern Sierra Leone, 30 km from the Liberian border and approximately 60 km from the coastline at the mouth of the Moa River. The island harbors nine anthropoids and two prosimians (Table 1.2). Like many forests in Africa, the vegetation on the island has a history of human-induced disturbance. At the time that the majority of the primate research was conducted, there were only a few active farms on the island. However, approximately 30% of the island was regenerating forest growing on formerly agricultural lands that had been abandoned less than 20 years earlier (Fimbel, 1994*a, b*). Palm swamps and riverine forest fringe the island, and the remaining 60% of the island is old secondary high forest (Oates *et al.*, 1990). In this high forest, there are gaps resulting from felling of trees by local people for timber and canoes. Hunting in the area is believed to be relatively light, both because of the difficulty of crossing the river and because of antipathy towards Liberians who were the chief hunters of monkeys in southern Sierra Leone (Oates, 1994).

Primate studies were initiated on the island in 1982, and studies were conducted at various intervals until 1991, when regional unrest ended research. Studies were conducted

Table 1.2. *Descriptions of the primate community found at the major field sites where forest primates have been studied in Africa*

	Rainfall	Density	Biomass	Frugivore biomass	Folivore biomass	Group size	Diet		
Tiwai Island, Sierra Leone[a]									
	2708		1379	785.5	599.5		Leaves	Fruit	Flowers
Perodicticus potto						1			
Galagoides demidoff						1			
Cercocebus atys			196			35			
Cercopithecus campbelli			88			14			
Cercopithecus petaurista			120			14			
Cercopithecus diana			165			24	12.6	42.8	16.1 (27.7 Insect)
Cercopithecus aethiops			rare						
Colobus polykomos			382			9	58 (30Young)	35 (32seed)	3
Procolobus badius			380			35	52 (32Young)	32 (25seed)	16
Procolobus verus			23			6	70 (59Young)	19 (14seed)	11
Pan troglodytes			25						
Taï, Côte d'Ivoire[b]									
	1800	~183.6	802	244	558		Leaves	Fruit	Insects
Perodicticus potto									
Galagoides demidoff									
Cercocebus atys		10	58				1.3	93.5	1.3
Cercopithecus petaurista		29.3	52.7				5.5	77.2	7.3
Cercopithecus nictitans									
Cercopithecus diana		17.5	75.3				7.7	76.3	4.7
Cercopithecus campbelli		15	61.5				0	78.3	15.2
Procolobus badius		66	455.4			83.5	10.4	0	
Colobus polykomos		23.5	166.9				53.1	32.5	0
Colobus verus		21	81.9				94.1	0	0
Pan troglodytes		1.3	58.5						
Douala-Edéa, Cameroon[c]									
	4000	–	409	198	217				
Lophocebus albigena			88						
Cercopithecus mona			14						
Cercopithecus pogonias			15						
Cercopithecus nictitans			67						
Cercopithecus erythrotis			55						
Colobus satanas			198				43.1 (leaves)	53.2 (fruit and seeds)	
Pan troglodytes			25						
Gorilla gorilla			rare/ extinct(?)						
Forêt des Abeilles, Gabon[d]							Leaves + flowers	Ripe fruit	Unripe fruit + seeds
	1755								
Arctocebus aureus									

Table 1.2. (*cont.*)

	Rainfall	Density	Biomass	Frugivore biomass	Folivore biomass	Group size	Diet			
Perodicticus potto										
Euoticus elegantulus										
Galagoides alleni										
Galagoides thomasi										
Galagoides demidoff										
Mandrillus sphinx										
Lophocebus albigena		8.4					57	26	57	
Cercopithecus nictitans		31.8					14.5	35.5	50	
Cercopithecus pogonias		5.3					6.5	25.5	56	
Cercopithecus cephus		5.8					–	–	–	
Cercopithecus solatus		27.9					–	–	–	
Colobus satanas		20.5					49.5	9	40.5	
Pan troglodytes										
Gorilla gorilla										
Lopé Reserve, Gabon[e]							Fruit	Seeds	Leaves	Insects
	1506		318.6	227.9	90.7					
Cercopithecus nictitans		19.2	62.8			13.5				
Cercopithecus pogonias		4.6	10.1			12.6				
Cercopithecus cephus		5.1	10.2			9.6				
Lophocebus albigena		8.1	33.7			18.9	30	36	4	28
Colobus satanas		10.8	90.7			12.1	4	60	26	–
Gorilla gorilla		0.6	45.3			5.3				
Pan troglodytes		0.6	22.5			–				
Mandrillus sphinx		3.8	43.9			150				
Makokou, Gabon[f]							Leaves	Ripe fruit	Unripe fruit	Insect
	1755									
Arctocebus aureus						1	0 gum	14	0	85
Perodicticus potto						1	21 gum	65	0	10
Euoticus elegantulus						1	75 gum	5	0	20
Galagoides alleni						1	0 gum	73	0	25
Galagoides demidoff						1	10 gum	19	0	70
Galagoides thomasi						1				
Lophocebus albigena							6	81	5	6
Cercocebus galeritus		10.1				10	14	73	0.5	3
Mandrillus sphinx		~5								
Cercopithecus neglectus		33	110			4.5	9	74	3	5
Cercopithecus nictitans		30	100			20	28	61	1	8
Cercopithecus pogonias		22.5	60			15	2	84	0	14
Cercopithecus cephus		25	80			10	8	79	0	10
Miopithecus sp.		65	60			85	2	43	1.5	36
Colobus guereza							52	49		
Pan troglodytes										
Gorilla gorilla										

Table 1.2. (*cont.*)

	Rainfall	Density	Biomass	Frugivore biomass	Folivore biomass	Group size	Diet			
Botsima, Salonga National Park[g]							Leaves + flowers	Ripe fruit	Unripe fruit + seeds	
	1774									
Perodicticus potto										
Galagoides alleni										
Galagoides demidovii										
Lophocebus aterrimus							34	36	30	
Cercopithecus wolfi							43	39	18	
Cercopithecus ascanius							32	44	24	
Allenopithecus nigroviridis										
Colobus angolensis							33	17	50	
Procolobus badius							62	7	31	
Pan paniscus										
Ituri Forest, Democratic Republic of the Congo[h]							Leaves	Seed	Fruit	Flower
	1802	~112.4	709.6	401.9	307.7					
Lophocebus albigena		6.9	53.1			10				
Cercocebus galeritus		~2.0	15.4							
Papio anubis		?	?							
Cercopithecus ascanius		18.9	68.0			7				
Cercopithecus mitis		24.2	145.2			10				
Cercopithecus wolfi		23.1	87.8			14				
Cercopithecus neglectus		~.4	2.1							
Cercopithecus lhoesti		~.4	2.5							
Cercopithecus hamlyni		~.1	0.8							
Procolobus badius		26.7	218.9			28				
Colobus guereza		~1.2	12.6				57.9	22.0	24.6	2.9
Colobus angolensis		7.7	76.2			11	51.1	22.1	27.5	7.2
Pan troglodytes		~.8	27.7							
Kibale National Park, Uganda[i]							Leaves	Ripe fruit	Unripe fruit	Insect
	1662	~656.8	2710	633.5	2077					
Perodicticus potto		17.7	1.9			1				
Galagoides thomasi										
+Euoticus matschiei		79.5	12.6			1				
Lophocebus albigena		9.2	60			15	5.4	58.8	3.4	26
Papio anubis		–	–			–				
Cercopithecus ascanius		140	328			35	16.1	43.7	15.3	21.8
Cercopithecus mitis		41.8	133			25	20.9	45.1	12.5	19.8
Cercopithecus l'hoesti		8	13			–				
Cercopithecus aethiops		rare	rare			–				
Procolobus badius		300	1760			50	74.8	5.6	15.9	2.6

Table 1.2. (*cont.*)

	Rainfall	Density	Biomass	Frugivore biomass	Folivore biomass	Group size	Diet			
Colobus guereza		58.1	317			9	76.3	13.2	2.1	0
Pan troglodytes		2.5	85			7.1	8	80	0	0 (12% THV)
Budongo Forest Reserve, Uganda[j]	1495	67.4	545	261.5	283.5		Leaves	Ripe fruit	Unripe fruit	Insect
Perodicticus potto										
Galagoides demidoff										
Papio anubis		14	231							
Cercopithecus mitis		15.6	93.6				39	39	6	10
Cercopithecus ascanius		8.3	29.5							
Cercopithecus aethiops		rare								
Colobus guereza		27	284							
Pan troglodytes		2.5	89.2							

Note: Annual rainfall in mm, biomass in kg/km², density in individuals/km². Prosimians are only listed if they are described for the community, thus many lists are incomplete. Forêt des Abeilles and Lopé are neighboring sites, but since rainfall and diet differ they are presented separately.

[a]Oates *et al.* (1990) (mid-point in range used), Diet – Oates (1994), Oates & Whitesides (1990).

[b]Galat & Galat-Luong (1985) with *Pan* added from Bourlière (1985), Terborgh & van Schaik (1987).

[c]Oates *et al.* (1990).

[d]Gautier *et al.* (1997), Brugiere & Gautier unpublished data, Gautier-Hion & Gautier (in prep.).

[e]Primate density White (1994a, mean of 5 neighboring sites), diet data for *L. albigena* from Ham (1994), diet data for *Colobus satanas* from Harrison (1986).

[f]Makokou – from Waser (1987) – midpoint of pop density: diet – Charles-Dominique (1977). Gautier-Hion (1978, 1980, 1983), Gautier-Hion *et al.* (1980).

[g]Maisel *et al.* (1994), Maisel & Gautier-Hion (1994) – data only concerned plant diet.

[h]Thomas 1991, Group size was taken as the midpoint of estimated social group sizes which was a range corresponding to minimum and maximum values for "good" counts. Rainfall Hart (1985), *Colobus* diets from Bocian (1997)

[i]Struhsaker (1975, 1978, 1981), Struhsaker & Leland (1979), Weisenseal *et al.* (1993), Chapman and Wrangham (1993), Chapman unpublished data, Wrangham *et al.* (1996), Chapman *et al.* (1995)

[j]Plumptre *et al.* (1994), Plumptre & Reynolds (1994) (density – unlogged forest only), Harvey *et al.* (1987) (weights assuming 50/50 sex ratio and 1/2 group immature weighing 1/2 adult), Eggeling (1947) (rainfall), *C. mitis* diet (unlogged; Fairgrieve, 1995)

on *Procolobus verus* between 1983 and 1986 (Oates, 1988; Oates *et al.*, 1990), on *Colobus polykomos* (Dasilva, 1992; 1994) and *Procolobus badius* between 1984 and 1986 (Oates, 1994), on Diana monkeys (*Cercopithecus diana;* Whitesides, 1989; Oates & Whitesides, 1990; Hill, 1994), and on the success of the different primate species in the various modified habitats on the island (Fimbel, 1994a,b). Primate richness is comparable to Kibale, but biomass is approximately one half that of Kibale. Much of this difference in

biomass is simply a reflection of the fact that red colobus (*Procolobus badius*) is extremely abundant in Kibale, but not at Tiwai (Table 1.2).

Taï National Park, Côte d'Ivoire

Taï National Park (3500 km²) in Côte d'Ivoire received its first protected status in 1926 when it was declared a forest and wildlife refuge, and it received its park status in 1972

(IUCN, 1987). The perceived importance of this forest has been steadily increasing. It became a Biosphere Reserve in 1978, a World Heritage Site in 1982, and was included in the IUCN's list of eleven most threatened areas in 1984 (Galat & Galat-Luong, 1985; Merz, 1986). As with other parts of the Upper Guinea region of West Africa, the plant community found in Taï shows a high level of endemism; 54% of the 1300 plant species identified are endemic to the region (Martin, 1991). Possibly related to the richness of the plant communities, the area harbors a rich and abundant primate community. In Taï there are 11 species of primates (including two prosimians), and primate biomass has been estimated to exceed 800 kg/km^2 (Galat & Galat-Luong, 1985).

Certain areas in Taï have come under heavy pressure from new settlers. In 1972, the human population density surrounding Taï was estimated to be 1.3 individuals per km^2 (Martin, 1991). Since that time, timber exploitation has opened up large areas neighboring the park for agriculture, and farmers from the Sahel region have immigrated to the area. In eight years, the human population density increased sixfold. This led to increasing encroachment and slash-burn activities in the park. At times, rice was cultivated in swampy areas, and several hundred gold washers were active in the park (Martin, 1991). Further, timber companies have overstepped their concessions and cut extensively in northern areas of the park.

The majority of the primate research, which has focused on chimpanzees, has been conducted in the western part of the park, approximately 20 km from the nearest village and the Liberian border (Boesch & Boesch, 1983, 1989; Galat & Galat-Loung, 1985; Boesch, 1994; Zuberbuhler et al., 1997). This region of the park is effectively protected against logging, and hunting pressure on primates is low (Boesch & Boesch, 1983).

Douala-Edéa Forest Reserve, Cameroon

The Douala-Edéa Forest Reserve (1300 km^2) is located in Cameroon at the mouth of the Sanaga River (McKey, 1978). The sandy soils of the reserve are impoverished and acidic with a pH as low as 2.7. Soils of this acidity are toxic to the roots of many plants, cause phosphorus to become unavailable, limit nitrogen fixation, and inhibit decay and nutrient cycling (Gartlan et al., 1978; Newbery et al., 1986). As a result, tree species diversity in the forest is relatively low; there was a mean of only 39 tree species

($\geq$ 30 cm GBH) in each of 104 80 × 80 m plots (Newbery et al., 1986). The primate community of the reserve is not very rich or abundant. Originally, the low primate densities were ascribed to the abundance of plants that, as a result of poor soils, have high levels of chemical defenses (McKey et al., 1978). However, Tiwai was recently documented to have sandy soils with markedly lower levels of all mineral nutrients than Douala-Edéa and a similarly low pH, yet it supports one of the highest estimates of primate biomass (Oates et al., 1990). By contrasting results of studies from Douala-Edéa, Tiwai, and Kibale, Oates et al. (1990) argued that nutrient-poor soils and high tannin levels in tree foliage do not necessarily correlate with a low primate biomass. They did, however, find a positive relationship between the ratio of protein to fiber in the foliage and colobine biomass.

There has been only a handful of primate studies conducted in this region (Gartlan et al., 1978; McKey et al., 1981, McKey & Waterman, 1982). When studies were first initiated in the reserve in 1973, there had been little interference with the vegetation of the reserve, probably due to the poor soils and the scarcity of commercially valuable timber trees (Oates, 1994). Hunting in the area was probably light (Oates, 1994).

Lopé Forest Reserve, Gabon and Forêt des Abeilles – Makandé, Gabon

The Lopé Reserve of Central Gabon covers a 5000 km^2 area that is primarily mature semi-evergreen forest, but 300 km^2 along its northern and eastern border is a savanna and forest-savanna mosaic (Harrison & Hladik, 1986; White et al., 1993; White, 1994a,b). Some areas of the reserve were logged in the 1960s (White, 1994b); however, with the recent construction of a railroad, it has become much more economically feasible to extract timber from the area, and thus logging has intensified (White, 1994b). Hunting is forbidden within the reserve, but with the increased presence of workers at forestry camps, it is a concern.

Research began at the Station d'Etudes des Gorillas et Chimpanzés (SEGC) in 1983 and is on-going. The main focus of the project has been studies of the ecology of gorillas and chimpanzees (Rogers et al., 1990; Tutin et al., 1994), but research interests have expanded to include forest elephants (Loxodonta africana; White et al., 1993;

White, 1995) and forest ecology (Tutin & Fernandez, 1993; White, 1994a,b; White et al., 1995). The majority of the primate studies have been conducted in the 50 km² SEGC study area, an area that was selectively logged for *Aucoumea klaineana* between 1960 and 1970 at an average density of 1.5 trees per hectare. Primates in this area are not overly abundant, at least in comparison to sites like Kibale and Tiwai, but this site presently offers unique opportunities for primate studies. Lopé is presently one of the few well-established sites where one can study sympatric gorillas and chimpanzees, as well as mandrills (*Mandrillus sphinx*), a species that the scientific community knows little about (Rogers et al., 1996).

Makandé is a site that neighbors the Lopé Reserve and is located in the Forêt des Abeilles (approximately 10 000 km²) in Central Gabon. It is mainly covered by primary dense forest dominated by Caesalpiniaceae (38% of the trees > 10 cm DBH). As with the neighboring Lopé Reserve, the area is home to a rich primate community with a total of 15 species including the endemic guenon, *Cercopithecus solatus*. Studies on primate populations began in 1993 and continued until 1996 (Gautier-Hion et al., 1997). Unfortunately, the area has no protected status and is in the process of being logged.

M'passa Reserve – Makokou study site, Gabon

This study site was first located at Makokou, where a station was founded in 1962 under the name of the Mission Biologique au Gabon (Charles-Dominique, 1977). In 1968 a research station was built by the Centre National de la Recherche Scientifique (C. N. R. S.) (IUCN, 1987). Subsequently, a reserve was created in the forest of M'passa, 10 km from Makokou. Studies were undertaken in the reserve as well as in areas up to 100 km around Makokou. In the area, mature forest is interspersed with areas of riparian and swamp forest (C. M. Hladik, 1973; Quris, 1976; A. Hladik, 1978). Where the banks are not steep, the river often floods extensive areas, resulting in characteristic flooded forest communities that include *Uapaca* sp. and various palms. Some of the riparian areas contain sections of secondary forest where there were formerly villages and agricultural activity.

In the M'passa reserve and the area near Makokou, there are six species of nocturnal primates and 11 species of monkeys and apes. Except for the gorilla, all have been

the subject of study (Gautier-Hion, 1978). Many of the projects conducted here were made in a comparative manner to provide information on the entire community (Gautier-Hion & Gautier, 1979; Emmons et al., 1983; Gautier-Hion et al., 1985). As a result, studies at this site remain some of the most comprehensive community-wide studies available.

As with Kibale, there is a great deal of small scale variation in the abundance and even presence of species between particular localities, making it difficult to rigidly define the primate community. For example, the mandrill (*Mandrillus sphinx*) only occurs on the right bank of the Ivindo River, and *Cercocebus galeritus* is only found on the left bank of the river. Furthermore, several species of this community are typically only observed in riverine forests: *Cercopithecus neglectus, Cercocebus galeritus, Colobus guereza,* and *Miopithecus talapoin*. These species consequently have a patchy distribution.

At the time when many of the primate studies were conducted, clearing for agriculture was conducted only on a very small scale and was restricted to areas near roads (Charles-Dominique, 1977). Hunting in the area was only intense near villages.

Salonga National Park, Democratic Republic of the Congo

The Salonga National Park, created in 1970, is located within the Congo Basin and covers 36 000 km² in two different blocks separated by a corridor about 45 km long. Research on the primate community in the area was conducted at Botsima, a site located within a meander of the Lomela River. Information on primate diets was obtained during 12 months in an inundated rainforest area largely dominated by Caesalpiniaceae (Gautier-Hion & Maisels, 1994). Studies in the area were initiated in 1989 but ended in 1991 following political troubles.

Ituri Forest, Democratic Republic of the Congo

The east of the Congo Basin is home to one of the world's richest primate communities: at least 17 species living sympatrically. There are four major protected areas in the region covering 38 300 km² (Parc National de Kahuzi-Biega, Parc National de la Maiko, Parc National des Virungas, Réserve de Faune à Okapis (Ituri Forest); Hart

& Hall, 1996). However, the only community-level primate research has been conducted in the Réserve de Faune à Okapis (13 000 km²), part of the Ituri Forest. Studies in this region were initiated in 1981 and have been conducted more or less continuously since then. However, information on the primates in the area is limited to a study by Thomas (1991), in which he estimated total anthropoid primate density in several areas using line transect censuses, and a study by Bocian (1997) on niche separation of two sympatric species of black-and-white colobus.

Human populations in the reserve were estimated in 1993 at approximately 10 000–11 000, including about 5 000 Mbuti and Efe hunter-gatherers and approximately 5 000–6 000 shifting cultivators (J. Hart, unpublished data). There are currently no regulations controlling hunting in the reserve. However, human activities including hunting are typically concentrated within 10 km of the roads, with decreasing activity out to 35 km, and only sporadic human presence beyond 35 km (Hart & Hall, 1996). Evidence of larger scale market hunting has been found in remote areas of the forest, though. One of the greatest threats to the forest in this region is immigration. In a survey of villages in the area, 86% exhibited some level of immigration among the Bantu population (Stephenson & Newby, 1997). Of course, with the current political instability in the region, the long-term fate of all reserves in the area remains to be determined (Hart et al., 1996; Hart & Hart, 1997).

Kibale National Park, Uganda

The Kibale National Park, located in western Uganda (0° 13′–0° 41′ N and 30° 19′–30° 32′ E) near the base of the Ruwenzori Mountains, is a moist, evergreen forest, transitional between lowland rainforest and montane forest (Wing & Buss, 1970; Struhsaker, 1975; Skorupa, 1988; Butynski, 1990; Chapman et al., 1997). Kibale obtained its first legal status when it was gazetted a crown forest reserve in 1932. The reserve was established to provide sustained production of hardwood timber and for production of softwoods from plantations established in the grasslands (Osmaston, 1959; Kingston, 1967). During the 1960s, different areas within the reserve received different management treatments; some areas were left undisturbed, some were lightly logged, while others were more heavily harvested and "refined" (poisoning of unwanted species and cutting of vines). The diverse management of the area

has provided opportunities to examine the effects of logging on primate communities (Skorupa, 1988; Struhsaker, 1997; Chapman & Chapman, 1997). In November 1993, Kibale was declared a National Park, providing long-term protection for the forest and the primate communities. The Batoro, the predominate people living around the park, do not eat monkeys (Struhsaker, 1975), and as a result there has been virtually no hunting of primates in the area.

Long-term studies began in the area in 1970, and research has been continuously conducted in Kibale since that date. The majority of the primate research conducted to date has been based at two research sites: Kanyawara and Ngogo. Recently three additional sites have been established in the park (Sebatoli, Dura River, and Mainaro). The placement of these five sites takes advantage of the north-south gradient in forest structure (Parinari forest in the north, Cynometra forest in the south) to examine how primate communities respond to small-scale variation in ecological conditions. Because of this extensive history of continuous primate research, a wealth of data on primate and plant communities is available over a 28 year period.

Kibale has one of the richest and most abundant primate communities in the world; it is home to 12 species of non-human primates (including three prosimians), and the primate biomass of the area is estimated to reach 2710 kg/km² (Table 1.2). However, all of these species rarely, if ever, inhabit the same area. For example, while vervet monkeys (Cercopithecus aethiops) and baboons (Papio anubis) are in the national park and adjacent areas, vervet monkeys have only been seen at the main study site at Kanyawara twice in 28 years (Struhsaker, pers. comm.), and baboons use this area typically only when a large fruiting crop is available. Furthermore, blue monkeys (Cercopithecus mitis) are one of the most abundant monkeys in the north of the park, but their densities decline as one travels south so that approximately a third of the way down the park on a north/south axis, blue monkeys are rare (Butynski, 1990), and approximately half-way down the park they drop out altogether (Chapman & Chapman, unpublished data). Such small scale variation in primate community structure is intriguing (see below).

Budongo Forest Reserve, Uganda

The Budongo Forest Reserve (793 km², 428 km² of which is forest) has a long history of exploitation. The reserve was gazetted in 1932, but commercial extraction of timber began as early as 1915, and rubber tapping began in 1905 (Paterson, 1991; Reynolds, 1993; Plumptre *et al.*, 1994). The first major sawmill was established in 1925 (Eggeling, 1947; Fairgrieve, 1995), and harvesting has occurred in approximately 77% of the reserve (Howard, 1991; Plumptre, 1996). The majority of this harvest has involved the extraction of mahogany (*Khaya* and *Entandrophragma*), but other species have been extracted as well. In addition to harvesting mahogany, the management of the reserve has also involved extensive mahogany replanting and a variety of silvicultural treatments. For example, poisoning of unwanted trees was carried out extensively in the 1950s and 60s, and by mid-1966 a total of approximately 4800 ha had been treated (Paterson, 1991). Part of the management involved setting aside a Nature Reserve, where timber harvest was prohibited. Despite the reserve's protected status, illegal timber extraction by pit-sawyers is widespread (Reynolds, 1993; Fairgrieve, 1995). Unlike other areas of Uganda, illegal hunting of primates is another problem in Budongo. It has been estimated that 40% of communities neighboring the reserve hunt, and some of the cultural groups in the area regularly eat primate meat (Johnson, 1996). Pressures such as these have been increasing as a result of both the growth of the local Banyoro population and immigration. Thus, areas adjacent to the park that were grasslands in the 1960s are now agricultural settlements (Reynolds, 1993).

A few primate studies were conducted in the reserve in the 1960s and 1970s (Reynolds & Reynolds, 1965; Sugiyama, 1968; 1969; Aldrich-Blake, 1970; Suzuki, 1979). However, there was a research hiatus until 1991, at which time the Budongo Forest Project was initiated. The main objective of this project was to investigate the responses of wildlife (including primates) to logging and the role of fruit eating primates in forest regeneration (Plumptre *et al.*, 1994).

PATTERNS AND CAUSES OF VARIATION IN PRIMATE COMMUNITY STRUCTURE AND BIOMASS

These sites display considerable variation both in the numbers and kinds of primates present, and in the total and relative abundances of these species. Here, we consider some possible explanations for these variations, both in terms of proximate ecological factors and long-term evolutionary processes.

The species structure of communities

Table 1.3 shows the species structure of the nine forest primate communities that we have described with the species grouped in six eco-taxonomic categories, or guilds: pottos, galagos, terrestrial cercopithecines, arboreal cercopithecines, colobines, and apes. Members of these categories vary in body size, activity pattern, locomotor behavior, and/or diet. The terrestrial cercopithecines are monkeys that often travel on the ground, but usually obtain at least part of their diet in the forest canopy.

The communities cluster into geographical sets, within which many species are shared, and between which there are considerable differences. These sets correspond to some of the large regional communities recognized in the IUCN/SSC Status Survey and Conservation Action Plan for African Primates (Oates, 1996a), and the "Primate Zones" of regional endemism described by Grubb (1990) (Fig. 1.2). Tiwai and Taï are part of the Upper Guinea region, or west African zone; Douala-Edéa, Lopé, and Makokou are part of a western equatorial region, or west-central zone; Salonga is in the Congo Basin region, or south-central zone; Ituri is in the eastern region of the Democratic Republic of the Congo, or east-central zone; and Kibale and Budongo are within the western Rift regional community.

Comparing species across communities, it can be seen that only two species are ubiquitous, occurring in every moist-forest community: the potto (*Perodicticus potto*) and the dwarf galago (*Galagoides demidoff*). The common chimpanzee (*Pan troglodytes*) is very widespread, occurring in all communities except Salonga, where it is replaced by the pygmy chimpanzee (*Pan paniscus*). A few species such as *Lophocebus albigena* and *Procolobus badius* occur widely, but patchily. Most species, however, are localized and occur in only two or three of the communities we have described. In some cases, these localized forms have no close relatives outside the limited geographical area in which they occur (e.g., *Euoticus matschiei*, *Cercopithecus diana*, *Cercopithecus hamlyni*, and *Procolobus verus*), but others are members of species groups or superspecies which have one form in all or most of the communities (such as *Colobus polykomos* and other black-and-white

Table 1.3. *The species structure of the nine forest primate communities that we have described, with the species grouped in six eco-taxonomic categories*

Species	Site								
	Tiwai	Taï	D-Edéa	Abeilles/ Lopé	Makokou	Salonga	Ituri	Kibale	Budongo
Pottos									
Arctocebus aureus			+?	+	+				
Perodicticus potto	+	+	+?	+	+	+?	+	+	+
Galagos									
Galagoides alleni			+?	+	+				
Euoticus matschiei							+?	+	
Galagoides demidoff	+	+	+?	+	+	+?	+	+	+
Galagoides thomasi				+	+		+?	+	
Euoticus elegantulus			+?	+	+				
Terrestrial cercopithecines									
Cercocebus atys	+	+							
Cercocebus galeritus					+		+		
Mandrillus sphinx				+	+				
Papio anubis							+	(+)	+
Cercopithecus hamlyni							+		
Cercopithecus lhoesti							+	+	
Cercopithecus solatus				+					
Cercopithecus aethiops	(+)							(+)	(+)
Arboreal cercopithecines									
Lophocebus albigena			+	+	+		+	+	
Lophocebus aterrimus						+			
Cercopithecus diana	+	+							
Cercopithecus neglectus					+		+		
Cercopithecus mitis						+?	+	+	+
Cercopithecus nictitans		(+)	+	+	+				
Cercopithecus petaurista	+	+							
Cercopithecus erythrotis			+						
Cercopithecus cephus				+	+				
Cercopithecus ascanius						+	+	+	+
Cercopithecus campbelli	+	+							
Cercopithecus mona			+						
Cercopithecus pogonias			+	+	+				
Cercopithecus wolfi						+	+		
Miopithecus sp.			+		+				
Colobines									
Procolobus verus	+	+							
Procolobus badius	+	+				+	+	+	
Colobus polykomos	+	+							
Colobus guereza					+		+	+	+
Colobus satanas			+	+					
Colobus angolensis						+	+		

Table 1.3 (*cont.*)

Species	Site								
	Tiwai	Tai	D-Edéa	Abeilles/ Lopé	Makokou	Salonga	Ituri	Kibale	Budongo
Apes									
Pan troglodytes	+	+	+	+	+		+	+	+
Pan paniscus						(+)			
Gorilla gorilla			(+)	+	+				

Note: +, rare and/or localized; +?, presence suspected but not confirmed.

Table 1.4. *Composition of African forest primate communities according to eco-taxonomic groups, based on data in Table 1.3*

Site	Annual rainfall (mm)	Primate biomass (kg/km²)	Number of species per group						Total species number
			Pottos	Galagos	Terrestr. cercop.	Arboreal cercop.	Colobines	Apes	
Tiwai	2708	1379	1	1	1(–2)	3	3	1	11
Taï	1800	802	1	1	1	3(–4)	3	1	11
Douala-Edéa	4000	409	2?	3?	0	6	1	1(–2)	13–14
Lopé	1505	319	2	4	2	4	1	2	15
Makokou	1755	—	2	4	2	6	1	2	17
Salonga	1774	—	1?	1?	0	4	2	(1)	8–9
Ituri	1802	710	1	1(–3)	4	5	3	1	17
Kibale	1662	2710	1	3(?)	1(–3)	3	2	1	11–13
Budongo	1495	545	1	1	1(–2)	2	1	1	8

Note: Terrestr. cercop., terrestrial cercopithecines

colobus, and *Cercopithecus cephus* and closely related small guenons).

The eco-taxonomic composition of the different communities is summarized in Table 1.4. Except for Douala-Edéa and Salonga, which lack a terrestrial cercopithecine, each community contains one or more species in each eco-taxonomic category. However, the communities show considerable variation in their total number of species (from only eight in Budongo, to 17 in Makokou and the Ituri) and in the distribution of these species across eco-taxonomic categories. For example, communities in western-central Africa (Cameroon and Gabon) are relatively much richer than other sites in nocturnal prosimians (pottos and galagos), communities in Central Africa (Cameroon to the Democratic Republic of the Congo) are relatively rich in arboreal cercopithecines, and west Africa

(Sierra Leone and Côte d'Ivoire) is relatively rich in colobines.

What has produced these patterns of variation in forest primate community structure, and in particular what have been the relative contributions of proximate ecological factors and more distant evolutionary events?

Habitat and community structure

Proximate ecology may play a significant role in producing the high species-richness at Makokou and Ituri. Oates *et al.* (1990) argued that habitat heterogeneity tends to increase both the species richness and biomass of primate communities, and both Makokou and Ituri display high habitat heterogeneity. The M'passa forest, where Makokou is located, is a mosaic of dry-land, riparian, and swamp forest,

Table 1.5. *The species of primates found in mainland forest and riverine forest at three sites in central Africa. These lists illustrate the increased diversity that is associated with the increased habitat heterogeneity of having riverine forest in the region*

Species	Mainland forests			Riverine forests		
	Makokou Gabon	Ngotto CAR	Odzala Congo	Makokou Gabon	Ngotto CAR	Odzala Congo
Subfamily Colobinae						
Procolobus badius	—	—	—	—	x	x
Colobus guereza	—	—	—	x	x	x
Subfamily Cercopithecinae						
Lophocebus albigena	x	x	x	x	x	x
Cercocebus galeritus	—	—	—	x	x	x
Cercopithecus cephus	x	x	x	x	x	x
Cercopithecus pogonias	x	x	x	x	x	x
Cercopithecus neglectus	—	—	—	x	x	x
Cercopithecus nictitans	x	x	x	x	x	x
Miopithecus talapoin	—	—	—	x	x	—
Subfamily Pongidae						
Gorilla gorilla	x	x	x	x	x	x
Pan troglodytes	x	x	x	x	x	x
Total	6	6	6	10	10	10

Table 1.6. *A description of the communities of primates found in mainland, swamp forest, and riparian forests at Odzala, Congo*

Species	Mainland		Swamp forest		Riparian forest	
	n	%	*n*	%	*n*	%
Colobus guereza	2	0.7	16	10.4	19	10.6
Cercocebus galeritus	0		1	0.6	10	5.6
Cercopithecus cephus	48	15.7	56	36.4	43	23.9
Cercopithecus neglectus	0		1	0.6	42	23.3
Cercopithecus nictitans	133	43.6	50	32.5	35	19.4
Cercopithecus pogonias	53	17.4	15	9.7	8	4.4
Miopithecus sp.	0		0		14	7.8
Number of species	7		9		10	
Diversity index	3.37		3.74		5.74	
Equitability	0.34		0.46		0.68	

Source: Gautier-Hion, 1996.

n, number of troops.

while the Ituri is a mosaic of mixed and monodominant *Gilbertiodendron* forest on dry land, veined with a network of rivers and streams which are often bordered by swamps. *Cercopithecus neglectus*, *Cercocebus galeritus*, and *Colobus guereza* are particularly associated with riverine forest in both the Makokou and Ituri communities; *Miopithecus* is restricted to these forests at Makokou; and in the Ituri, *Cercopithecus hamlyni* is most commonly encountered in *Gilbertiodendron* forest. Because of the habitat preferences of these primates, habitat heterogeneity contributes to high primate species-richness at these sites. A similar pattern is found in several other central African forests (Tables 1.5 and 1.6, Gautier-Hion, 1996).

Some of the patchiness in the distribution of the red colobus (*Procolobus badius*) may also be a reflection of habitat variation. Red colobus appear to select for diets that are diverse and contain relatively high proportions of young leaves, flowers, and buds (Struhsaker, 1975; Oates, 1994), and it may be that some habitats cannot provide such diets. However, it is far from certain that sites such as Makokou are incapable of supporting red colobus, and the great differences in pelage and other features among red colobus populations in different parts of the African forest zone (leading them to be classified in different subspecies, or even different species) suggest that the patchy distribution of these populations has a strong historical component. Indeed, many of the differences in the species composition among African forest primate communities can be best understood as resulting from long-term historical processes of evolution, dispersal, and extinction.

History and community structure

The moist and wet forests of tropical Africa have no doubt been greatly affected (like rainforests in other parts of the tropics) by the climatic vicissitudes of the last few million years. During the arid phases of the glacial cycle which began in the Pliocene, these forests would have greatly contracted, becoming restricted to distinct "refuges". During interglacial warming, the area of moist forest would have expanded. Between the peaks and troughs of the cycle, climate has changed erratically (deMenocal, 1995), so that forest vegetation rarely would have been stable for long periods. Such a pattern of environmental change would have led to a variety of evolutionary forces operating on primates living in African forests. As forests

alternately contracted and expanded, and changed in their climate and species-composition, primate populations must have sometimes been fragmented into long-term isolates, have sometimes expanded, and sometimes gone extinct. During isolation, evolutionary processes such as genetic drift and adaptation to changing local conditions would often have caused populations to differentiate to the extent that taxonomists would regard them as distinct subspecies or species (Haffer, 1969; Hamilton, 1988). Further evolutionary change would have occurred during population dispersal at times of forest expansion (Grubb, 1978), and if populations of similar animals then met, hybridization may have occurred, or further adaptive change may have resulted as a consequence of competition. Different species no doubt responded differently to these events and opportunities, depending on their attributes. For example, at times of forest contraction, species able to maintain viable populations in small areas may have been most likely to persist in small refuges, while at times of forest expansion, differing dispersal abilities and ecological tolerances probably influenced the likelihood with which various species spread out and crossed ecogeographic boundaries. Such ecogeographic boundaries might be obvious barriers, such as large river or mountain ranges, or more subtle features such as forest types not providing appropriate food items at certain times of year.

Such historical events and adaptive features of these kinds would have played major roles in producing the patterns of variation seen in present-day primate communities. For example, the angwantibos (*Arctocebus aureus* and *A. calabarensis*), the western needle-clawed galagos (*Euoticus elegantulus* and *E. pallidus*), Allen's galago (*Galagoides alleni*), and mandrills (*Mandrillus leucophaeus* and *M. sphinx*) are all restricted to west-central forests. It may be hypothesized that the ancestors of these primates were isolated in this area during one or more glacial maxima, and that if any close relatives were living elsewhere they became extinct. During interglacial forest expansion, dispersal of these primates out of the west-central forests was inhibited, perhaps by rivers. For example, in the west, *A. calabarensis*, *E. pallidus*, and *G. alleni* have all been found living close to the east bank of the Niger River in southern Nigeria, but none of them occurs west of the Niger (Oates & Jewell, 1967). As we noted above, *M. sphinx* occurs only on one side of the Ivindo River in Gabon. That rivers have some important function in limiting the distribution of these taxa is shown also by the fact that the lower Sanaga River in Cameroon separates the members of the species pairs *A. aureus/calabarensis*, *E. elegantulus/pallidus*, and *M. leucophaeus/sphinx*. Indeed, a majority of primate taxa are represented by different forms on either side of the Sanaga, and similar differences are found on the left and right banks of the Congo River, accounting for some of the species differences between the Ituri and Salonga communities that we have described.

Major rivers also separate many primate taxa in other parts of the world, including Amazonia (Hershkovitz, 1977; Kinzey, 1982; Peres *et al.*, 1996), but there has been debate about the contribution of rivers to patterns of diversification. The present consensus view seems to be that other factors, such as forest fragmentation into refugia, are mainly responsible for producing initial population differentiation, with rivers acting later to limit dispersal, and thus to reinforce differentiation. Wide rivers with strong currents are more likely to limit dispersal (especially if they are fringed by special vegetation types such as flooded forest) than are narrow headwater channels or slow meandering rivers that frequently change their course.

We have described one historical process that could account for some part of the complex pattern of similarities and differences in species-structure among the different African forest communities we have described. The prosimians and mandrills of the west-central forest area may have differentiated there in a past period of forest retraction, and been subsequently limited in their spread by rivers or river-associated ecosystems. A similar but slightly different process may account for the distribution of species such as *Procolobus verus* (restricted to the forests of Upper Guinea, including Nigeria) and *Cercopithecus hamlyni* (in eastern Democratic Republic of the Congo). These two taxa appear to have originated, or been isolated, in a restricted area (perhaps with related forms going extinct elsewhere) and have subsequently not spread very widely, but their distributions are not obviously limited by rivers. Perhaps their dispersal has been limited by some other habitat feature. Each appears to be a habitat specialist, with *P. verus* being most common near water at low altitude, and *C. hamlyni* being most frequent in or near *Gilbertiodendron* forest in the Ituri.

Similarly, the various distinct populations of *Procolobus badius* across the African forest zone may have originally differentiated during episodes of forest fragmentation, and

then been limited in their subsequent dispersal into some areas by habitat features. The absence of any red colobus monkeys between western Cameroon and the eastern part of the Congo Republic is not easily explained by any particular river barrier, but exactly what features have limited dispersal into this area remain unclear. There may be general habitat features of the west-central African forests (such as the seasonal pattern of food production and/or the chemistry of tree foliage) that make it a relatively unfavorable habitat for colobus monkeys, with the result that colobines have both few species and a low biomass in these forests.

The species groups which have different members in many of the forest communities (such as the *cephus* and *mona* groups of guenons, and the black-and-white colobus) are probably the result of yet other historical processes, where forest fragmentation has produced local differentiation, but with this differentiation accompanied either by the absence of widespread extinctions, or by good dispersal ability. Each of these attributes could be associated with broad ecological tolerance (particularly the ability to survive in a range of forest types), and could therefore explain why these groups are so widely distributed. Yet even in these cases, rivers have sometimes been found to form boundaries between taxa, such as the Cross River separating *Cercopithecus sclateri* and *C. erythrotis* in the *cephus* group, and the Sanaga separating *C. pogonias pogonias* and *C. p. grayi* in the *mona* group. This could be because rivers, even when they do not completely limit dispersal, may slow the spread of a population, allowing two different, but related, taxa to spread out from centers of differentiation to opposite river banks. In this case, occasional migration across a river might have little effect on the maintenance of population differences; the migrants might have low survivorship or breeding success in competition with members of a closely related taxon, or their genes might be swamped should they interbreed with members of the other taxon.

Whatever the validity of the particular evolutionary scenarios described here, long-term historical processes of some kind have surely contributed strongly to the present pattern of variation in the species structure of African forest primate communities. These processes have interacted with contemporary patterns of habitat diversity and forest size. Struhsaker (1981) has presented an argument for floristic diversity, proximity to a glacial refuge, and forest size acting together to determine the number of anthropoid primate species that occur in a range of Ugandan forests, including Budongo. Of these factors, proximity to a glacial refuge (a factor which presumably interacts with dispersal abilities) seems to have the strongest correlation with species number. The Budongo Forest is larger than Kibale, but contains three fewer anthropoids and possibly one less galago; it is 200 km further than Kibale from the nearest presumed refuge just west of the Ruwenzori Mountains. However, Budongo lies at a lower altitude than large parts of Kibale, and its vegetational differences (even before recent felling operations) may have played a role in producing its different primate community; the two forests are likely also to have had different ecological histories since the last glacial maximum.

Consequences of variation in species structure

The presence and absence of different species can have fundamental influences on other aspects of community structure and function. For example, more than half of the biomass of the Kibale primate community near Kanyawara is made up of red colobus monkeys (Table 1.2). The presence of red colobus at Kibale is one of the chief contributors to the difference in primate biomass between Kibale (2710 kg/km^2) and Budongo (545 kg/km^2). Despite the absence of red colobus, black-and-white colobus occur at approximately the same density in Budongo as they do at Kibale. The biomass of nocturnal prosimians is almost certainly higher at Makokou, where five or six species occur, than at the numerous sites with only two species; the three species restricted to the forests of west central Africa have little niche overlap with the much more widespread *Perodicticus potto* and *Galagoides demidoff*. Furthermore, where talapoins, mangabeys, or gorillas are present in a community, they add an ecological component (in terms of features such as body size, diet, and/or feeding level) not closely matched by any other species.

Population densities and community biomass

While historical and biogeographical factors have played important roles in determining the presence or absence of species at a given site, ecological factors are mainly responsible for determining the total and relative abundances of those species within a community. There is some evidence,

however, (e.g., the decline and disappearance of blue monkeys along a north–south gradient in Kibale – described earlier) that ecological factors can determine the presence or absence of a species. Here, we address ecological determinants of primate communities and discuss possible mechanisms responsible for between-site differences in total and relative abundances of primate species, using African examples wherever possible. The factors we consider are food supply (quality and quantity), disease/parasites, and predation (Table 1.7).

Food supply

Available studies suggest that, in general, habitats with more rainfall have higher levels of plant productivity than areas with less rainfall. This is illustrated by the fact that the annual net primary productivity for dry forest averages 50–75% that of wet forest (Hartshorn, 1983; Murphy & Lugo, 1986). Similarly, in dry forests the total plant biomass ranges between 78 and 320 tons/ha, while in wet forest, biomass ranges between 269 and 1186 tons/ha (Murphy & Lugo, 1986). While it may be a general trend for rainfall to be positively correlated with plant productivity, this need not imply a correlation between rainfall and primate biomass. For example, the biomass of primates at the site with the greatest rainfall, Douala-Edéa (4000 mm/year), is approximately six times less than Kibale, which has an annual rainfall of 1662 mm.

In addition to the total amount of rainfall, the seasonal distribution of rain could influence plant communities and in turn influence primate communities. The general pattern of rainfall at five sites is depicted in Fig. 1.3. While Kibale (1662 mm) and Lopé (1506 mm) receive less rainfall each year than Taï (1800 mm) and Douala-Edéa (4000 mm), rainfall at these two sites is less than or equal to 50 mm for only 2 months of the year. In contrast, at Taï and Douala-Edéa there are 4 and 3 months, respectively, each year when rainfall is less that 50 mm.

Differences in the floristic composition and richness of forests could likely influence the food supply and therefore primate community structure and biomass. For example, if one forest had a higher tree species richness than another forest, the primates in the richer habitat would likely have more foraging options and be less likely to experience periods of food scarcity than the animals at the more depauperate forest, since there would be more types of resources

Table 1.7. *A summary of studies examining possible mechanisms that could regulate primate population size, and the effect that was observed*

Ecological factor	Reference	Community	Effect
Food Supply			
Direct quantification & contrasts between sites and times			
quantity	Chapman & Onderdonk unpub	Community	No
	Terborgh & van Schaik, 1987	Community	?
Quality	Oates *et al.*, 1990	African colobine community	Yes
	Ganzhorn, 1992	Folivore community	Yes
Both	Peres, 1997	S. American community	Yes
	Davies, 1994	Asian colobine community	Yes
Ecological change			
Natural	Milton, 1982, 1990	*Alouatta palliata*	No
	Dittus, 1977, 1985	*Semnopithecus entellus, Macaca sinica*	Yes
Human induced	Skorupa, 1986,1988	African community	Yes
	Johns, 1988, 1992	Asian community	Yes
	Bennett & Dahaban, 1995	Asian community	Yes
	Plumptre & Reynolds, 1994	African community	Yes
Both	Marsh, 1981, 1986	*Procolobus badius*	Yes
	Decker, 1989	*Procolobus badius*	Yes
Ecological release[a]	Struhsaker, 1978	African community	Yes
	Davies, 1994	Asian community	Yes
Key foods	Oates, 1978	*Colobus guereza*	Inferred
Disease			
Die offs	Collias & Southwick, 1952	*Alouatta palliata*	Yes
	Work *et al.*, 1957	*Presbytis entellus, Macaca radiata*	Yes
Parasite load	Stuart *et al.*, 1993	*Brachyteles arachnoides*	No
	Stoner, 1996	*Alouatta palliata*	No
	Stuart *et al.*, 1990	*Alouatta palliata*	No
Predation	Struhsaker & Leakey, 1990	African community	No
	Stanford *et al.*, 1994	*Procolobus badius*	Inferred

[a]Ecological release – if interactions with other primate species make critical resources unavailable to particular species, one would expect to observe both changes in behavior and compensatory changes in density when the composition of the primate community changes.

to fall back upon. Gautier-Hion (1983) proposed that the higher consumption of fruit by *Lophocebus albigena* and *Colobus guereza* in Makokou relative to Kibale, and the generally more frugivorous diet of the entire community at Makokou, were a result of the fact that fruit is available in greater quantities and there are more fruiting species in central Africa than east Africa. At Makokou, 95 species

(≥ 5 cm DBH) were identified in a 0.4 ha plot (A. Hladik, 1978), while at Kibale, Struhsaker (1975) describes only approximately 34 species in a 1 ha sampling area.

One ecological factor that has received little attention is the heterogeneity of habitats available to the primates in a particular region (Oates *et al.*, 1990). A mosaic of forest of different ages can increase vegetation diversity and the availability of palatable leaves, fleshy fruits, and thus lead to high carrying capacities for some species (Fimbel 1994*b*). For example, members of the *Cercopithecus cephus* superspecies often frequent secondary growth areas with dense vegetation and tangles of lianas, but occur at relatively low density in primary forest with little undergrowth.

While food supply is often assumed to be the chief critical factor determining overall and relative abundances of primates in a community, few tests of this proposition have been conducted. One method of testing the relationship between food supply and primate populations is to correlate measures of each across several sites. One study found a positive correlation between abundance of leguminous trees, which are important colobine food trees, and colobine biomass in Asia (Davies, 1994). However, this relationship was not upheld for African colobines (Davies, 1994).

We attempted to examine the relationship between food availability and primate biomass by obtaining published and unpublished data for both of these variables. Primate biomass data collected in conjunction with fruit abundance data estimated using phenology transects were found for five sites (Fig. 1.4) and as estimated from fruit traps were found for four sites (Fig. 1.5). While a statistical comparison is not possible given the limited sample size, no relationship is apparent between fruit abundance and frugivore biomass, and if there is a trend, it is for biomass to decrease with increasing fruit availability (Figs 1.4 and 1.5).

While the logic behind considering the quantity of food is clear, it would also seem important to consider the quality of food. Several studies have found a relationship between food quality and folivorous primate biomass. A positive correlation exists between Asian and African colobine biomass and an index of leaf quality, the ratio of protein to fiber (Davies, 1994; Oates *et al.*, 1990; Waterman *et al.*, 1988). A similar relationship was found between the quality of leaves (ratio of protein to fiber) and the biomass of folivorous lemurs of Madagascar (Ganzhorn, 1992). Further support for this relationship comes from South America. In his comparison of 20 Amazonian sites, Peres (1997) found a significantly higher primate biomass in annually flooded forests than in terre firme forests. He attributes this to the high fertility of alluvial soils, which may in turn lead to higher fruit production (quantity) and/or to lower levels of secondary compounds in the foliage (quality). However, as soil fertility was not a predictor of primate biomass for three African sites (Oates

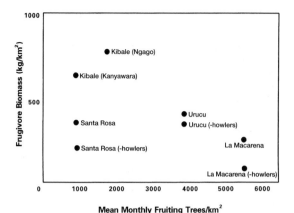

Fig. 1.4. Monthly average number of fruiting trees per km² versus the biomass of frugivorous primates. New World sites are shown with and without howler monkeys (*Alouatta* sp.) as there is debate with respect to their tropic categorization. Sources for the phenology data: Chapman *et al.* (unpublished data), Peres (1994), Stevenson *et al.* (1994), Chapman & Chapman (1990), and Chapman (unpublished data). All phenology studies monitored trees > 10 cm DBH, except Stevenson *et al.* (1994) who monitored trees > 5 cm DBH.

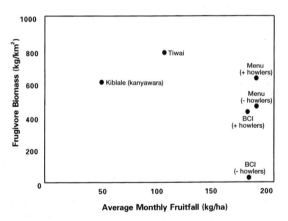

Fig. 1.5. Average monthly fruitfall from fruit traps versus the biomass of frugivorous primates. Fruitfall data are taken from all forest types. New World sites are shown with and without howler monkeys (*Alouatta* sp.) as there is debate with respect to their tropic categorization. Sources for fruit trap data: Chapman *et al.* (1994), Terborgh (1983), Janson (1984), Goldizen *et al.* (1988), Goldizen (unpublished data), Fimbel (unpublished data), Smythe (1970).

et al., 1990), the mechanisms controlling these observed differences in biomass need further investigation.

It is also possible to assess the relationship between primate populations and food supply by examining single sites at which the food supply has changed over time, either through natural transformations or through human disturbance. Few examples exist of changes in food supply due to entirely natural causes. One example is the failed fruit crop of 1970 on Barro Colorado Island, Panama, which led to a widespread famine and decline of frugivore populations (Foster, 1982). However, the population of howler monkeys (*Alouatta palliata*) soon recovered (Milton, 1982), indicating that a failed fruit crop in a single year does not regulate populations in the long term. This example also indicates that superannual periods of fruit scarcity ("ecological crunches", as described by Weins (1977) and Cant (1980)) may not be long-term regulating factors.

Studies at Lopé, however, indicate that fruit availability may be limiting to the primate community (Tutin *et al.*, 1997). For the eight species of diurnal primates at Lopé, there is a negative correlation between biomass and degree of frugivory, suggesting that species are limited by the amount of available fruit. Fruit production at Lopé varies seasonally within years, as well as between years, and so intra- and inter-annual periods of fruit scarcity may keep frugivorous primate biomass low (Tutin *et al.*, 1997). This relationship between degree of frugivory and biomass does not account for the low overall primate biomass at Lopé, relative to other African sites (Table 1.4). Tutin *et al.* (1997) suggest that an "ecological catastrophe" in the recent past would explain the low overall biomass; however, differences in overall food abundance could result in the differences in overall primate biomass, and so quantitative comparisons of overall food abundance should be made between Lopé and other African sites.

Unfortunately, examples of human-induced changes that affect primate food supply are abundant. One of the most common forms of human-induced change is selective logging, which has been studied at several African sites (Kibale – Skorupa, 1986; Lopé – White, 1992; Budongo – Plumptre & Reynolds, 1994). Due at least partially to differences in patterns and intensity of logging regimes, the effects of logging on primate populations are quite variable. Different species at the same site are affected in diverse ways, and the same species can respond differently to logging at different sites. An example involves the effect of logging on blue monkeys (*Cercopithecus mitis*) in Uganda. Blue monkeys have been classed as an extreme generalist based on studies conducted in Kibale (Johns & Skorupa, 1987; Butynski, 1990). However, at this same site, logging had a severe negative effect on this species. Fifteen years after logging in Kibale, logged areas had 20–30% fewer blue monkeys than unlogged areas (Skorupa, 1988). In contrast, blue monkeys in Budongo are 3.7 times more abundant in logged areas than in unlogged areas (Plumptre & Reynolds, 1994). This positive response may be due to the fact that a reduction of timber trees does not always lead to a reduction of food supply. In fact, selective logging of the monodominant *Cynometra* forest in Budongo has led to a diversification of the canopy-level tree community. In some cases, though, logging has clearly led to a reduction of food available for certain primate species, and thus to a decline in population density, such as with red colobus (*Procolobus badius*) in Kibale (Skorupa, 1986). Better quantification of the changes in food supply after logging is necessary to fully understand the mechanisms by which logging affects primate populations.

A combination of natural and human-induced disturbance has been implicated in the decline of red colobus (*Procolobus badius rufomitratus*) of Tana River, Kenya (Decker, 1994). In the period between 1975 and 1985, the total population of Tana River red colobus declined by approximately 80% (Marsh, 1986). Decker (1994) attributes this decline to loss and fragmentation of habitat in the 1960s due mostly to clearcutting of forest for agriculture, but also to changes in the course of the Tana River. She suggests that the high population densities reported in the 1975 census reflect a concentrated population well above its carrying capacity, while the subsequent censuses in the 1980s reflect a population adjusted to the new constraints of its habitat. The observation that red colobus in the 1975 study were eating a much higher proportion of mature leaves in their diet than the red colobus a decade later suggests that loss of food resources was a mechanism regulating population density (Decker, 1994).

A relationship between food supply and primate abundance can also be demonstrated in competitive situations. Competition is of course extremely difficult to demonstrate without experimentation (Simberloff & Connor, 1981), but certain lines of circumstantial evidence can indicate its existence (Waser, 1987). One line of evidence suggesting competition is density compensation, the response of one species' density to a change in the density of a competitor. Density compensation has been suggested to explain the differences in relative abundances of blue monkeys (*Cercopithecus mitis*) and grey-cheeked mangabeys (*Lophocebus albigena*) at two sites within Kibale National Park (Struhsaker, 1978). At the Ngogo study site, mangabey density is much higher than that of blue monkeys, but at the Kanyawara study site, the situation is reversed. Struhsaker suggests that the lower density of mangabeys at Kanyawara allows for a competitive release of blue monkeys by providing them access to more food resources, thereby allowing the blue monkeys to attain a much higher density at this site than at Ngogo. Without experimental manipulation, however, it is impossible to determine which species is limiting which, or if they are directly competing at all. Competition, when it does exist, is likely to affect the relative abundances of species within a community or guild, rather than the overall abundance of all primates in the community or guild.

The availability of certain critical food resources, rather than the overall abundance of food, could also limit primate populations. This has been suggested to explain variations in black-and-white colobus (*Colobus guereza*) densities in different areas of Kibale National Park (Oates, 1978). Oates suggested that the presence of certain sodium-rich aquatic plants in swampy areas of the Kanyawara study area could be responsible for the relatively high population density of black-and-white colobus in that area. All known guereza groups in Kanyawara included the swamp in their home ranges, and the estimated guereza biomass in Kanyawara was approximately seven times higher than in a neighboring area with no known swamps. However, without knowing the dietary requirements of this species, it is difficult to discern if these sodium-rich plants are required by the guerezas or merely preferred.

Disease and parasites

Disease and parasites are other factors potentially limiting primate populations (Freeland, 1977; Anderson & May, 1979; Scott, 1988). Data from Africa are very limited, but disease has been implicated in a major population decrease of howler monkeys on BCI, Panama between 1933 and 1951 (Collias & Southwick, 1952). A yellow fever epidemic spreading through Panama at the time is thought to have cut the howler population size almost in half. However,

within eight years the population reached a size higher than that before the yellow fever outbreak, and so disease cannot be said to be a long-term limiting factor. The effects of parasite load have also been studied on the howler monkeys of BCI (Milton, 1996). Milton (1996) found that the relative density of bot-fly infestations on the howlers was positively correlated with howler mortality. She concludes that howler mortality is often due to the combined factors of physical condition, degree of bot-fly infestation, and dietary stress. In a study of gelada baboons in Ethiopia, Dunbar (1980) found that parasitic infestation of a larval stage of tapeworm was a major cause of mortality, along with old age and exposure to the harsh wet season climate. In two different years of observation, 9.8% and 10.6% of the population showed swellings indicating the infestation of a larval stage of tapeworm, and several animals died from or had their movement severely impeded by the rupturing of these swellings.

Predation

A final factor potentially involved in regulating primate populations is predation (Isbell, 1994; Boinski & Chapman, 1995). Struhsaker & Leakey (1990) observed 41 cases of crowned hawk-eagle predation (*Stephanoaetus coronatus*) on primates in Kibale over three and a quarter years. This mortality represented only a small fraction of the primate population at Kibale (approximately 0.3%), and so cannot be considered a strong factor in limiting primate populations. However, crowned hawk-eagle predation could potentially limit black-and-white colobus populations, since they were selected most frequently by the eagles relative to their abundance. Stanford *et al.* (1994) found that at least 20% of the red colobus population at Gombe National Park, Tanzania was lost to chimpanzee predation in 1992. They suggest that chimpanzee predation is probably the main factor limiting red colobus population size. Isbell (1990) documented a substantial increase in the rate of predation by leopards on vervet monkeys (*Cercopithecus aethiops*) in Amboseli National Park, Kenya. Between 1977–1986, the average vervet monkey predation rate was at least 11%. This rate increased to at least 45% in 1987, possibly because of an increase in the leopard population. Clearly, predation on vervets, even at baseline levels, could be an important factor limiting their population density.

Table 1.8. *Descriptive statistics of African countries that are known to have at least 10 primate species*

Country	Size	Number of primate species	Original wildlife habitat (1000 ha)	Amount remaining (1000 ha)	Habitat loss	Popu. (millions)	Growth
DR Congo	2 645 000	33–34	233 590	105 116	55	41.8	3.25
Cameroon	475 000	31	49 940	19 245	59	14.0	3.48
Nigeria	394 000	26	91 980	22 995	75	127.7	3.17
Equatorial Guinea	28 000	22–23	2 500	920	63	0.4	2.60
Congo	342 000	22	34 200	17 422	49	2.7	3.35
Uganda	236 000	20	19 370	4261	78	22.7	3.47
CAR	623 000	19–20	62 300	27 412	56	3.5	2.98
Gabon	265 000	19–20	26 700	17 355	35	1.4	3.09
Tanzania	945 000	18–19	88 620	50 513	43	32.8	3.68
Angola	1 247 000	18–21	124 670	76 085	39	11.5	2.85
Kenya	583 000	18–19	56 950	29 614	48	29.0	3.81
Côte d'Ivoire	322 000	16	31 800	6678	79	14.5	3.83
Rwanda	26 000	15–18	2510	326	87	8.6	3.41
Ghana	239 000	15	23 000	4600	80	17.6	3.1
Sierra Leone	72 000	15	7170	1076	85	4.7	2.75
Guinea	246 000	14	24 590	7377	70	6.7	3.12
Liberia	111 000	12	11 140	1448	87	3.0	3.30
Burundi	28 000	11–14	2750	359	86	6.4	2.91
Sudan	2 505 700	11	170 370	51 090	70	29.1	2.87
Ethiopia	1 222 000	10–11	110 100	3030	70	57.1	2.99
Guinea-Bissau	36 000	10–11	3610	794	78	1.1	2.20
Togo	57 000	9–11	5600	1904	66	4.1	3.22
Zambia	752 957	9–10	75 260	53 435	29	10.2	3.65
Senegal	197 000	8–11	19 260	3532	82	8.4	2.9

Source: Species numbers from Oates (1996a), population size and projected growth rates are from World Resources Institute (1994).

CONSERVATION

There is no way to estimate how drastically primate populations in Africa have been reduced in the recent past as a result of human activities, or even how many unique primate communities have been lost. Too little research is currently being conducted to quantify the present situation. Furthermore, even if the scientific community could increase research efforts, we do not have the needed baseline data. Regardless of the lack of quantitative data, it is clear that primate populations have been drastically reduced and are being assaulted from a number of directions, including timber extraction, agricultural clearing, and subsistence and commercial hunting. It is impossible to attribute the responsibility to the different factors, since each is inevitably linked. For example, while timber operations may harvest relatively few trees (typically <10% of the canopy level trees), they construct roads and open areas up for agriculturists and provide economic incentives for immigration to regions, which increases human population, leading to increased hunting pressure.

It is estimated that the rainforests of Africa originally covered 3 620 000 km^2 prior to agricultural clearing and habitat alterations by people. Seventy-four percent of this was found in central Africa, 19% in west Africa, and 7% in east Africa (Martin, 1991; Table 1.8; Fig. 1.6). While estimates from different sources vary considerably, it is clear that the area of forest remaining today is drastically

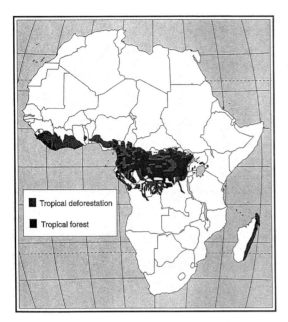

Fig. 1.6. Map of the African continent illustrating the major regions of moist and wet forest and the extent of deforestation in these areas (adapted from National Geographic Atlas of the World 1992, and Chapman in prep.).

reduced. One estimate suggests that the amount of forest remaining is approximately 1 490 000 km² or 55% of the original area in central Africa, 190 000 km² or 28% of the area in west Africa, and 70 000 km² or 28% of the original area of east Africa (Sommer in Martin, 1991). FAO statistics estimate that in 1985 west African forests encompassed 143 260 km², while central Africa forests included some 1 717 450 km² and east African forests encompass only 30 000 km². While these estimates show discrepancies and are open to debate, it is clear that African forests have been drastically reduced, particularly in west and east Africa. These estimates suggest that a great effort must be placed into protecting and properly managing remaining forests, particularly in east and west Africa, and that great potential remains within central Africa.

The level of protection for primate communities within different regions is clearly limited and constantly changing. Within the countries of Africa with closed canopy forest, an average of 3.2% of each country's area has been protected in national parks or similar protected areas (IUCN, 1985; Table 1.9). The investment of different countries in

national parks is very dynamic, making it difficult to interpret the significance of the values given in Table 1.9. New parks are being created in some countries, while in other countries parks are being heavily degraded or even degazetted. For example, in the northern part of Taï National Park, 770 km² (21% of the total park area) was temporarily ceded for exploitation and is now heavily impacted (IUCN, 1987). Similarly, Bia National Park in Ghana was gazetted in 1974 to include 306 km², it was reduced in size to 230 km² in 1979, and further reduced to 77.7 km² in 1980. The area excised from the park has been classified as a game production reserve (IUCN, 1987; now called a resource reserve) and has been largely opened up to timber exploitation. In contrast, the amount of protected land in Uganda has increased from 7698 km² in the 1980s to 11 145 km² as of 1995 (Table 1.9). In Gabon, the status of area around Lopé is debated. It is now called "Aire d'Exploitation Rationelle de la Faune". A recent decree recognized a central core where hunting and logging are banned and a peripheral zone which is open to logging. The M'passa reserve has no protection at all and a logging concession has recently been given for the south of the reserve.

It should be cautioned that values such as those presented in Table 1.9 represent the theoretical maximum protected area. Given the current lack of funding in many regions for the enforcement and education of existing wildlife laws, many national parks are suffering serious encroachment. Furthermore, given the current situation of political unrest in many areas, particularly in the Democratic Republic of the Congo, Rwanda, and Burundi, it is difficult to know the current status of protected areas in these regions (Hart & Hall, 1996).

Clearly, within Africa, tropical forests and the animals they support are increasingly threatened by accelerating rates of forest conversion and degradation and by commercial and subsistence hunting (Lanly et al., 1991; FAO, 1993; Oates, 1996a). Traditionally, conservation efforts have concentrated on protecting plant and animal populations through the establishment of national parks in pristine or semi-pristine habitats, but this is only one strategy to protect these communities. The majority of tropical forests are outside of parks and have experienced or will soon experience major anthropogenic pressure. It seems somewhat inevitable that African countries will turn to timber exploitation as a means of raising foreign income. High levels of foreign debt, with a mean of 58% of the

Table 1.9. *Tropical African countries containing closed canopy forest and description of their park system, based on IUCN (1985, 1987) (Categories 1 and 2 from the IUCN 1985, 1987). Note that this listing includes all national parks, many of which will not be forested. Recent changes involve parks added to a country's parks system after the IUCN (1985, 1987) publications: this listing is not complete*

Country	Size	Number of national parks	Size of national parks (km²)	% of the country protected	Recent changes
Senegal	197 000	6	10 094	5.12	
Guinea-Bissau	36 000	No areas listed	0		
Guinea	246 000	1	130	0.05	
Sierra Leone	72 000	1	980	1.36	(A)
Liberia	111 000	1	1307	1.18	
Côte d'Ivoire	322 000	9	17 920	5.57	
Ghana	239 000	6	11 626	4.86	
Togo	57 000	No areas listed	0		
Benin	112 620	2	8435	0	
Nigeria	394 000	1	9800	2.50	(B)
Cameroon	475 000	7	12 079	2.54	(C)
Equatorial Guinea	28 000	No areas listed	0		(D)
Gabon	265 000	No areas listed	0		
Congo	342 000	1	1266	0.37	(E)
DR Congo	2 645 000	8	87 940	3.32	
Angola	1 247 000	1	9960	0.80	
CAR	623 000	3	28 960	4.65	
Rwanda	26 000	2	2620	10.08	
Burundi	28 000	No areas listed	0		
Uganda	236 000	4	7698	3.26	(F)
Tanzania	945 000	10	170 382	18.03	
Kenya	583 000	23	30 509	5.23	
Sudan	2 505 700	3	19 157	0.76	
Ethiopia	1 222 000	8	8795	0.72	

Note: (A) Proposed but not fully gazetted.
(B) Cross River 4320 km², Gashaka 6363 km², Chad Basin 2358 km², Oyo 2550 km², Yankari 2250 km².
(C) Korup 1259 km².
(D) Parc national de Monte Alen 800 km².
(E) Odzala 2830 km², Nouabalé-Ndoki 3866 km².
(F) Kibale 766 km², Mt. Elgon 1145 km², Bwindi 321 km², Semliki 220 km², Mgahinga 29 km², Rwenzori 996 km². From Howard (1991), USAID (1992).

sub-Saharan countries' gross national product (as high as 241%), places strong pressures on governments to encourage timber harvesting (Stuart *et al.*, 1990). This calls for investigations into the long-term effects of different patterns and intensities of forest conversion on forest regeneration and primate populations. Considerations of habitat heterogeneity are important from a conservation perspec- tive, as areas where there is a mix of inundated and non-inundated forests often support a more abundant and diverse primate community. Furthermore, inundated forests may be easier to protect because they may be less prone to human pressures (logging and agriculture) than non-inundated forest (Gautier-Hion, 1996).

With the reduction in forested area in west Africa and

the decline in timber production in these countries, timber companies are increasingly turning their attention to central Africa (Martin, 1991). The lessons learned in west Africa have a great deal to offer planners when considering various conservation activities in central Africa. For example, in 1945, the Tropical Shelterwood System (TSS), which involves the cutting of vines and the poisoning of unwanted species of medium size, was introduced to manage the forest reserves of Nigeria and Ghana (between 1958 and 1970, 188 tons of sodium arsenite was used in Ghana to poison trees – Martin, 1991). This system was abandoned when it was realized that natural regeneration under this system did not meet expectations (Martin, 1991; Chapman & Chapman, in press). In the taungya system, which was first attempted in Nigeria in 1945, local farmers were allocated land after logging, subsequently they were to plant timber species, tend the trees, then move on (Oates, 1995). The social consequences of this approach, such as the increased immigration into areas, are just now becoming clear (Oates, 1995). Such experiments need to be carefully evaluated before planning extraction regimes in central Africa.

More must be known about the consequences of logging on primate communities (White, 1994b; Chapman & Fimbel, in press), as comparable data on the effects of logging are generally lacking. Evidence from west Africa suggests that timber trees can contribute disproportionately to the diets of some primate species, indicating that logging could have severe impacts on these species, unless they have extremely flexible diets. In Bia National Park, Ghana 43% of the plant species found in the diet of red colobus are from economically valuable timber species. Diana monkeys and black-and-white colobus also fed heavily on timber trees (20% and 25% respectively, Martin, 1991). However, as discussed earlier, the effects of logging on primates vary within and between sites, and the mechanisms contributing to these mixed results are poorly understood.

The negative effects of logging on primate communities are typically exacerbated by intensive hunting that typically follows logging operations. Subsistence and commercial hunting are having devastating impacts on primate populations in many areas (Martin, 1983; Peres, 1990; Wilkie et al., 1992; Redford, 1992; Bodmer et al., 1994). From case studies, it is clear that wildlife hunting provides a major source of food for many local communities in Africa and that primates are often the target of such hunting

activities (Martin, 1983; Fa et al., 1995; Fitzgibbon et al., 1995; Johnson, 1996). For example, a market survey in two cities in Equatorial Guinea with a combined population size of 107 000, recorded 4222 primate carcasses being brought to market in 424 days (Fa et al., 1995). In Arabuko-Sokoke Forest, Kenya (372 km^2), 1202 Cercopithecus mitis and 683 Papio cynocephalus were reported to have been killed by subsistence hunters in a year, a number far above that which would represent a sustainable harvest (Fitzgibbon et al., 1995). Martin (1983) found that, in Nigeria, 50% of the human population ate bushmeat regularly, and bushmeat was popular with all income groups. A study conducted in Congo, Central African Republic, and Cameroon documented that primates represent 16%, 11%, and 4.5%, respectively of the bushmeat taken in villages (Gally & Jeanmart, 1996).

As dramatic as many of these figures are, it is likely that the majority of these data represent an underestimate of the actual extent of the harvest (Colell et al., 1994). For example, harvest estimates from market surveys do not include primates that are consumed in villages. In the Democratic Republic of the Congo, 57.1% of the primates harvested are eaten in the villages and do not make it to the market, and in Liberia primates are more valuable in rural than urban areas (Lahm, 1993). Also, interview results will be biased since hunting is officially prohibited in many areas where it occurs (Johnson, 1996). Regardless of any potential bias, it is clear that primates are being harvested in great numbers in unsustainable fashions in many areas of Africa.

CONCLUSIONS

Uncovering the processes determining primate communities is a difficult task for primate ecologists, as experimental manipulations used by community ecologists studying other taxa are usually not possible with primates. Therefore, patterns must be sought by comparing a diverse array of primate communities. The numerous and varied primate communities of Africa provide an excellent opportunity for examining these patterns. The wide range of biomass and species richness and the array of evolutionary and ecological histories of African primate communities provide the variation necessary for uncovering these patterns.

So far, the data on African primate communities is patchy and somewhat inconsistent. In future studies, it is crucial that comparable methods be used at different sites to allow comparisons of various parameters. This is particularly important in measuring food availability. Phenological cycles have been quantified at several African sites, but often the methods of measuring fruit and leaf abundance are different, preventing meaningful comparisons from being made. Furthermore, plots of the same size must be used for quantifying floristic diversity so that this parameter can be compared across sites.

Scale is an important consideration in the comparison of primate communities. It is valuable to understand on what spatial scale large differences in forest composition occur, and thus on what scale we might expect to see differences in primate communities. The premise for making broad geographical contrasts is that enough variation will exist in the ecology of these areas to permit detection of differences in response variables, such as primate community composition. However, if significant differences in primate community structure exist on smaller scales, fine scale contrasts may be more sensitive to detecting ecological determinants of community structure and thus could be the more appropriate contrasts to make.

Human disturbance is rapidly destroying, or at least altering, many of Africa's primate communities. It is critical that conservation and research activities accelerate so that the future of the remaining populations are ensured. For effective protection and management of the remaining communities, an understanding of factors influencing community composition will be crucial. As each community studied adds to the variation necessary for identifying determinants of community structure, it is crucial that as many communities as possible be studied before they are lost. When disturbance has already occurred or is inevitable, the disturbance can be used as an "experiment" to help determine the requirements of a community.

ACKNOWLEDGMENTS

We thank John Fleagle, Charlie Janson, and Kaye Reed for inviting us to participate in this project and Wenner-Gren for making the initial workshop possible. The work described here from Kibale National Park was supported by the Wildlife Conservation Society, National Geographic Grants, Lindbergh Foundation, an NSF Grant, and USAID PSTC Funding. We thank the Government of Uganda, the Forest Department, and Makerere University for permission to work in Kibale Forest. Recent work undertaken by A. Gautier-Hion at Salonga and Makande was supported by the CEE (DGVII and DGXII). John Oates thanks the National Science Foundation, the Wildlife Conservation Society and Research Foundation of CUNY for support of his field work in west Africa. Lauren Chapman, and Amy Zanne provided helpful comments on this work.

REFERENCES

Aldrich-Blake, F. P. G. (1968). A fertile hybrid between two *Cercopithecus* spp. in the Budongo Forest, Uganda. *Folia Primatologica*, **9**, 15–21.

Aldrich-Blake, F. P. G. (1970). Problems of social structure in forest monkeys. In *Social Behavior in Birds and Mammals*, ed. J. H. Crook, pp. 79–101. London: Academic Press.

Anderson, R. M. & May, R. M. (1979). Population biology of infectious diseases. *Nature*, **271**, 361–6.

Beadle, L. C. (1981). *The Inland Waters of Tropical Africa: An Introduction to Tropical Limnology*. Longman, London.

Bennett, E. L. & Dahaban, Z. (1995). Wildlife responses to disturbances in Sarawak and their implications for forest management. In *Ecology, Conservation, and Management of Southeast Asian Rainforests*, eds. R. B. Primack & T. E. Lovejoy, pp. 66–86. New Haven: Yale University Press.

Bocian, C. M. (1997). *Niche separation of black-and-white colobus monkeys* (Colobus angolensis *and* Colobus guereza) *in the Ituri Forest*. Ph.D. Dissertation, City University of New York, New York.

Bodmer, R. E., Fang, T. G., Moya, L. & Gill, R. (1994). Managing wildlife to conserve Amazonian forests: Population biology and economic consideration of game hunting. *Biological Conservation*, **67**, 29–35.

Boesch, C. (1994). Chimpanzee-red colobus monkeys: a predator-prey system. *Animal Behaviour*, **47**, 1135–48.

Boesch, C. & Boesch, H. (1983). Optimization of nut-cracking with natural hammers by wild chimpanzees. *Behaviour*, **83**, 265–86.

Boesch, C. & Boesch, H. (1989). Hunting behavior of wild chimpanzees in the Taï National Park. *American Journal of Physical Anthropology*, **78**, 547–73.

Boinski, S. & Chapman, C. A. (1995). Predation on primates: Where are we and what's next? *Evolutionary Anthropology*, **4**, 1–3.

Booth, A. H. (1958). The zoogeography of West African primates: A review. *Bulletin de l'l F. A. N.,* **20**, 587–622.

Bourlière, F. (1985). Primate communities: their structure and role in tropical ecosystems. *International Journal of Primatology,* **6**, 1–26.

Butynski, T. M. (1990). Comparative ecology of blue monkeys (*Cercopithecus mitis*) in high- and low-density subpopulations. *Ecological Monograph,* **60**, 1–26.

Cant, J. G. H. 1980). What limits primates? *Primates,* **21**, 538–44.

Chapman, C. A. (1983). Speciation of tropical rainforest primates of Africa: insular biogeography. *African Journal of Ecology,* **21**, 297–308.

Chapman, C. A. & Fimbel, R. (in press). An evolutionary perspective on natural disturbance and logging: Implications for forest management. In *Conserving Wildlife in Managed Tropical Forests,* ed. A Grajal, J. Robinson & R. Fimbel. New York: Columbia University Press.

Chapman, C. A. & Chapman, L. J. (1990). Density and growth rate of some tropical dry forest trees: comparisons between successional forest types. *Bulletin of the Torrey Botanical Club,* **117**, 226–31.

Chapman, C. A. & Chapman, L. J. (1996). Mid-elevation forests: A history of disturbance and regeneration. In: *Ecosystems and Their Conservation in East Africa,* eds. T. R. McClanahan & T. P. Young, pp. 385–400. London: Longman.

Chapman, C. A. & Chapman, L. J. (1997). Forest regeneration in logged and unlogged forests of Kibale National Park, Uganda. *Biotropica,* **29**, 396–412.

Chapman, C. A., Chapman, L. J., Wrangham, R. W., Isabirye-Basuta, G. & Ben-David, K. (1997). Spatial and temporal variability in the structure of a tropical forest. *African Journal of Ecology,* **35**, 287–302.

Chapman, C. A. & Fedigan, L. M. (1990). Dietary differences between neighboring *Cebus capucinus* groups: local traditions, food availability or responses to food profitability? *Folia Primatologica,* **54**, 177–86.

Chapman, C. A. & Wrangham, R. W. (1993). Range use of the forest chimpanzees of Kibale: implications for the evolution of chimpanzee social organization. *American Journal of Primatology,* **31**, 263–73.

Chapman, C. A., Wrangham, R. W. & Chapman, L. J. (1994). Indices of habitat-wide fruit abundance in tropical forests. *Biotropica,* **26**, 160–71.

Chapman, C. A., Wrangham, R. W. & Chapman, L. J. (1995). Ecological constraints on group size: An analysis of spider monkey and chimpanzee subgroups. *Behavioral Ecology and Sociobiology,* **36**, 59–70.

Chapman, L. J. (in press). Fishes of African rain forests: Diverse adaptation to environmental challenges.

Charles-Dominique, P. (1977). *Ecology and Behaviour of Nocturnal Primates: Prosimians of Equatorial West Africa.* New York: Columbia University Press.

Cheney, D. L. & Wrangham, R. W. (1987). Predation. In *Primate Societies,* ed. B. B. Smuts, D. L. Cheney, R. M. Seyfarth, R. W. Wrangham & T. T. Struhsaker, pp. 227–39. Chicago: University of Chicago Press.

Colell, M., Maté, C. & Fa, J. E. (1995). Hunting among Moka Bubis: Dynamics of faunal exploitation at the village level. *Biodiversity and Conservation,* **3**, 939–50.

Collias, N. & Southwick, C. (1952). A field study of population density and social organization in howling monkeys. *Proceedings of the American Philosophical Society,* **96**, 143–56.

Colyn, M., Gautier-Hion, A. & Verheyen, W. (1991). A re-appraisal of palaeoenvironmental history in Central Africa: evidence for a major fluvial refuge in the Zaire Basin. *Journal of Biogeography,* **18**, 403–7.

Dasilva, G. A. (1992). The western black-and-white colobus as a low-energy strategist: activity budgets, energy expenditure, and energy intake. *Journal of Animal Ecology,* **61**, 79–91.

Dasilva, G. A. (1994). Diet of *Colobus polykomos* on Tiwai Island: selection of food in relation to its seasonal abundance and nutritional quality. *International Journal of Primatology,* **15**, 1–26.

Davies, G. A. (1994). Colobine populations. In *Colobine Monkeys: Their Ecology, Behaviour and Evolution,* ed. A. G. Davies & J. F. Oates, pp. 285–310. Cambridge: Cambridge University Press.

Decker, B. S. (1989). *Effects of habitat disturbance on the behavioral ecology and demographics of the Tana River colobus* (Colobus badius rufomitratus). Ph.D. Dissertation Emory University, Atlanta.

Decker, B. S. (1994). Effects of habitat disturbance on the behavioral ecology and demographics of the Tana River red colobus (*Colobus badius rufomitratus*). *International Journal of Primatology,* **15**, 703–737

deMenocal, P. B. (1995). Plio-Pleistocene African climate. *Science,* **270**, 53–59.

Deshmukh, I. (1986). *Ecology and Tropical Biology.* London: Blackwell Scientific Publications.

Dittus, W. P. J. (1975). Population dynamics of the toque monkey, *Macaca sinica.* In. *Socioecology and Psychology of Primates,* ed. R. H. Tuttle, pp. 125–51. Hague: Mouton.

Dittus, W. P. J. (1977). The social regulation of population density and age-sex distribution in the toque monkey. *Behaviour*, **63**, 281–322.

Dittus, W. P. J. (1979). The evolution of behaviour regulating density and age-specific sex ratios in a primate population. *Behaviour*, **69**, 265–302.

Dittus, W. P. J. (1985). The influence of cyclones on the dry evergreen forest of Sri Lanka. *Biotropica*, **17**, 1–14.

Dunbar, R. I. M. (1980). Demographic and life history variables of a population of gelada baboons (*Theropithecus gelada*). *Journal of Animal Ecology*, **49**, 485–506.

Eggeling, W. (1947). Observations on the ecology of the Budongo rainforest, Uganda. *Journal of Ecology*, **34**, 20–87.

Emmons, L. Gautier-Hion, A. & Dubost, G. (1983). Community structure of the frugivorous–folivorous forest mammals of Gabon. *Journal of Zoology (London)*, **199**, 209–22.

Fa. J. E., Juste, J., del Val, J. P. & Castroviejo, J. (1995). Impact of market hunting on mammal species in Equatorial Guinea. *Conservation Biology*, **9**, 1107–15.

Fairgrieve, C. (1995). *The Comparative Ecology of Blue Monkeys* (Cercopithecus mitis stuhlmanni) *in Logged and Unlogged Forest, Budongo Forest Reserve, Uganda. The Effects of Logging on Habitat and Population Density*. Ph.D. Dissertation, University of Edinburgh.

FAO (Food and Agriculture Organizations of the United Nations). 1993. Forest resource assessment, 1990: Tropical Countries. FAO Forestry paper 112, Rome.

Fimbel, C. (1994a). Ecological correlates of species success in modified habitats may be disturbance- and site-specific: the primates of Tiwai Island. *Conservation Biology*, **8**, 106–13.

Fimbel, C. (1994b). The relative use of abandoned farm clearings and old forest habitats by primates and a forest antelope at Tiwai, Sierra Leone, West Africa. *Biological Conservation*, **70**, 277–86.

Fitzgibbon, C. D., Mogaka, H. & Fanshawe, J. H. (1995). Subsistence hunting in Arabuko-Sokoke Forest, Kenya, and its effects on mammal populations. *Conservation Biology*, **9**, 1116–26.

Foster, R. B. (1982). A seasonal rhythm of fruitfall on Barro Colorado Island. In *The Ecology of Tropical Forests*, ed. E. G. Leigh, A. S. Rand & D. M. Windsor, pp. 151–72. Washington, DC: Smithsonian Institution Press.

Freeland. W. (1977). *The Dynamics of Primate Parasites*. Ph.D. Thesis. University of Michigan,. Ann Arbor.

Galat, G. & Galat-Luong, A. (1985). La communaute de primates diurnes de la forêt de Taï, Côte d'Ivoire. *Revue d'Ecologie (Terre et Vie)*, **40**, 3–32.

Gally, M. & Jeanmart, P. (1996). Etude de la chasse villageoise en forêt dense humide d'Afrique Centrale. Memoire de Fin d'Etudes. Faculte des Sciences Agronomiques de Gembloux (projet ECOFAC, CEE DGVIII).

Ganzhorn, J. U. (1992). Leaf chemistry and the biomass of folivorous primates in tropical forests: test of a hypothesis. *Oecologia*, **91**, 540–7.

Gartlan, J. S., McKey, D. B. & Waterman, P. G. (1978). Soils, forest structure and feeding behaviour of primates in a Cameroon coastal rain-forest. In: *Recent Advances in Primatology, Vol. 1*. ed. D. J. Chivers & J. Herbert, pp. 259–67. New York: Academic Press.

Gautier-Hion, A. (1978). Food niches and coexistence in sympatric primates in Gabon. In *Recent Advances in Primatology, Vol. 1*. ed. D. J. Chivers & J. Herbert, pp. 269–86. New York: Academic Press.

Gautier-Hion, A. (1980). Seasonal variation of diet related to species and sex in a community of *Cercopithecus* monkeys. *Journal of Animal Ecology*, **49**, 237–69.

Gautier-Hion, A. (1983). Leaf consumption by monkeys in western and eastern Africa: A comparison. *African Journal of Ecology*, **21**, 107–13.

Gautier-Hion, A. (1996). In *Statut des Populations de Primates au Sein du Bloc Forestier d'Afrique Centrale*. Rapport Project ECOFAC, CEE DGVIII (AGRECO-CTFT).

Gautier-Hion A., Duplantier, J. M., Quris, R., Feer, F., Sourd, C., Decous, J. P., Doubost, G., Emmons, L., Erard, C., Hecketsweiler, P., Moungazi, A., Roussilhon, C. & Thiollay, J. M. (1985). Fruit characters as a basis of fruit choice and seed dispersal in a tropical forest vertebrate community. *Oecologia*, **65**, 324–37.

Gautier-Hion A., Duplantier, J. M., Emmons, Feer, F, Hecketsweiler, P., Moungazi, A., Quris, R. & Sourd, C. (1985). Coadaption entre rythmes de fructification et frugivorie en forêt tropicale humide du Gabon: Mythe ou realité. *Revue d'Ecologie (Terre et Vie)*, **40**, 405–34.

Gautier-Hion, A., Emmons, L. H. & Dubost, G. (1980). A comparison of the diets of three major groups of primary consumers of Gabon (primates, squirrels and ruminants). *Oecologia*, **45**, 182–9.

Gautier-Hion, A. & Gautier, J. -P. (1979). Niche écologique et diversité des espèces sympatriques dans le genre *Cercopithecus*. *Revue d'Ecologie (Terre et Vie)*, **33**, 493–508.

Gautier-Hion, A., Gautier, J-P. & Moungazi, A. (1997). Do black colobus in mixed troops benefit from increased foraging efficiency. *C. R. Acad. Sci., Paris*, **320**, 67–71.

Gautier-Hion, A., Gautier, J. P. & Quris, R. (1981). Forest structure and fruit availability as complementary factors influencing habitat use by a troop of monkeys (*Cercopithecus cephus*). *Revue d'Ecologie (Terre et Vie)*, **35**, 511–36.

Gautier-Hion, A. & Maisels, F. (1994). Mutualism between leguminous trees and large African monkeys as pollinators. *Behavioural Ecology & Sociobiology*, **34**, 203–10.

Goldizen, A. W., Terborgh, J., Cornejo, F., Porras, D. T. & Evans, R. (1988). Seasonal food shortages, weight loss, and the timing of births in saddle-backed tamarins (*Saguinus fuscicollis*). *Journal of Animal Ecology*, **57**, 893–902.

Grubb, P. (1978). Patterns of speciation in African mammals. *Bulletin of the Carnegie Museum of Natural History*, **6**, 152–67.

Grubb, P. (1990). Primate geography in the Afro-tropical forest biome. In *Vertebrates in the Tropics*, ed. G. Peters & R. Hutterer, pp. 187–214. Bonn: Museum Alexander Koenig.

Haffer, J. (1969). Speciation in Amazonian forest birds. *Science*, **165**, 131–7.

Hall, J. B. (1977). Forest types in Nigeria: an analysis of pre-exploitation forest enumeration. *Journal of Ecology*, **65**, 187–99.

Hall, J. B. & Swaine, M. D. (1976). Classification and ecology of closed-canopy forest in Ghana. *Journal of Ecology*, **64**, 913–51.

Hall, J. B. & Swaine, M. D. (1981). *Distribution and Ecology of Vascular Plants in a Tropical Rain Forest*. Hague: Dr. W. Junk Publishers.

Ham, R. M. (1994). *Behaviour and Ecology of Grey-cheeked Mangabeys* (Cercocebus albigena) *in the Lopé Reserve, Gabon*. Ph.D. Thesis, Stirling University, Scotland.

Hamilton, A. C. (1988). Guenon evolution and forest history. In *A Primate Radiation: Evolutionary Biology of the African Guenons*. ed. A. Gautier-Hion, F. Bourlière, J.-P. Gautier & J. Kingdon, pp. 13–34. Cambridge: Cambridge University Press.

Harrison, M. J. S. (1986). Feeding ecology of black colobus, Colobus satanas, in Gabon. In *Primate Ecology and Conservation*, ed. L. Else & P. C. Lee, pp. 31–7. Cambridge: Cambridge University Press.

Harrison, M. J. S. & Hladik, C. M. (1986). Un primate granivore: le colobe noir dans la forêt du Gabon; potentialité d'evolution du comportement alimentaire. *Revue d'Ecologie (Terre et Vie)*, **41**, 281–98.

Hart, J. A. (1985). *Comparative Dietary Ecology of a Community of Frugivorous Forest Ungulates in Zaire*. Ph.D. Dissertation. Michigan State University, Lansing, Michigan.

Hart, J. A. & Hall, J. S. (1996). Status of Eastern Zaire's forest parks and reserves. *Conservation Biology*, **10**, 316–27.

Hart, T. B. (1995). Seed, seedling, and sub-canopy survival in monodominant and mixed forests of the Ituri Forest, Africa. *Journal of Tropical Ecology*, **11**, 443–59.

Hart, T. B., & Hart, J. A. (1997). Zaire: New models for an emerging state. *Conservation Biology*, **11**, 309.

Hart, T. B., Hart, J. A. & Hall, J. (1996). Conservation in the declining nation state: A view from Eastern Zaire. *Conservation Biology*, **10**, 685–6.

Hart, T. B., Hart, J. A. & Murphy, P. G. (1989). Monodominant and species-rich forests of the humid tropics: causes for their co-occurrence. *American Naturalist*, **133**, 613–33.

Hartshorn, G. S. (1983). Plants. In *Costa Rican Natural History*, ed. D. H. Janzen, pp. 118–57. Chicago: University of Chicago Press.

Harvey, P. H., Martin, R. D. & Clutton-Brock T. H. (1987). Life history in comparative perspective. In *Primate Societies*, ed. B. B. Smuts, D. L. Cheney, R. M. Seyfarth, R. W. Wrangham, & T. T. Struhsaker, pp. 181–96. Chicago: Chicago University Press.

Hershkovitz, P. (1977). *Living New World Monkeys (Platyrrhini)*. Volume 1. Chicago: University of Chicago Press.

Hill, C. M. (1994). The role of female diana monkeys, *Cercopithecus diana*, in territorial defense. *Animal Behaviour*, **47**, 425–31.

Hladik, A. (1978). Phenology and leaf production in rain forests of Gabon: distribution and composition of food for arboreal folivores. In *The Ecology of Arboreal Folivores*, ed. G. G. Montgomery, pp. 51–71. Washington, DC: Smithsonian Institution Press.

Hladik, C. (1973). Alimentation et activité d'un groupe de chimpanzes reintroduits en forêt Gabonaise. *Revue d'Ecologie (Terre et Vie)*, **27**, 343–413.

Howard, P. C. (1991). *Nature Conservation in Uganda's Tropical Forest Reserves*. Gland, Switzerland: IUCN.

Isbell, L. A. (1990). Sudden short-term increase in mortality of vervet monkeys (*Cercopithecus aethiops*) due to leopard predation in Amboseli National Park, Kenya. *American Journal of Primatology*, **21**, 41–52.

Isbell, L. A. (1994). Predation on primates: ecological patterns and evolutionary consequences. *Evolutionary Anthropology*, **3**, 61–71.

IUCN (1985). *United Nations List of National Parks and Protected Areas.* Gland, Switzerland: IUCN.

IUCN (1987). *IUCN Directory of Afrotropical Protected Areas.* Gland, Switzerland: IUCN.

Janson, C. H. (1984). Female choice and mating system of the brown capuchin monkey *Cebus apella* (Primates: Cebidae). *Zeitschrift für Tierpsychologie,* **65**, 177–200.

Johns, A. D. (1988). Effects of "selective" timber extraction on rain forest structure and composition and some consequences for frugivores and folivores. *Biotropica,* **20**, 31–7.

Johns, A. D. (1992). Vertebrate responses to selective logging: Implications for the design of logging systems. *Philosophical Transactions of the Royal Society London B,* **335**, 437–42.

Johns, A. D. & Skorupa, J. P. (1987). Responses of rain-forest primates to habitat disturbance: A review. *International Journal of Primatology,* **8**, 157–91.

Johnson, K. (1996). Hunting in the Budongo Forest, Uganda. *Swara,* Jan–Feb 1996.

Kingston, B. (1967). *Working Plan for the Kibale and Itwara Central Forest Reserves.* Entebbe: Ugandan Forest Department, Government of Uganda Printer.

Kinzey, W. G. (1982). Distribution of primates and forest refuges. In *Biological Diversification in the Tropics,* ed. G. T. Prance, pp. 455–82. New York: Columbia University Press.

Lahm, S. A. (1993). Utilization of forest resources and local variation of wildlife populations in Northeastern Gabon. In *Tropical Forest, People, and Food,* ed. C. M. Hladik, A. Hladik, O. F. Linarea, H. Pagezy, A. Semple & M. Hadley, pp 213–26. Paris: Parthenon Publishing.

Lanly, J. P., Singh, K. D. & Janz, K. (1991). FAO's 1990 reassessment of tropical forest cover. *Nature and Resources,* **27**, 21–6.

Lawson, G. W. (1986). Vegetation and environment in West Africa. In *Plant Ecology in West Africa.* ed. G. W. Lawson, pp. 1–11. New York: John Wiley and Sons.

Maisel, F., Gautier-Hion, A. & Gautier, J-P. (1994). Diets of two sympatric colobus monkeys in Zaire: More evidence on seed eating in forests on poor soils. *International Journal of Primatology,* **15**, 681–701.

Maisel, F. & Gautier-Hion, A. (1994). Why are Caesalpinioideae so important for monkeys in hydromorphic rainforests of the Zaire Basin? In *Advances in Legume Systematics 5: The Nitrogen Factor.* ed. J. I. Sprent & D. McKey, pp. 189–204. Royal Botanical Gardens, Kew.

Maley, J. (in press). Late quaternary climatic changes in the African rain forest: Forest refugia and the major role of sea surface temperature variations. In *Paleoclimatology and Paleometeorology: Modern and Past Patterns of Global Atmospheric Transport,* ed. M. Leinen & M. Sarnthein, NATO Atmospheric Sciences Series. Dordrecht: Kluwer.

Marsh, C. W. (1979). Comparative aspects of social organization in the Tana River red colobus (*Colobus badius rufomitratus*). *Zeitschrift für Tierpsychologie,* **51**, 337–62.

Marsh, C. W. (1981). Ranging behaviour and its relation to diet selection in Tana River red colobus (*Colobus badius rufomitratus*). *Journal of Zoology (London),* **195**, 473–92.

Marsh, C. W. (1986). A resurvey of Tana primates and their forest habitat. *Primate Conservation,* **7**, 72–81.

Martin, C. (1991). *The Rainforests of West Africa: Ecology, Threats, Conservation.* Basel: Borkhauser Verlag.

Martin, G. H. G. (1983). Bushmeat in Nigeria as a natural resource with environmental implications. *Environmental Conservation,* **10**, 125–32.

McKey, D. B. (1978). Soils, vegetation, and seed-eating by black colobus monkeys. In: *The Ecology of Arboreal Folivores.* ed. G. G Montgomery, pp. 423–38. Washington, DC: Smithsonian Institution Press.

McKey, D. B., Gartlan, J. S., Waterman, P. G. & Choo, C. M. (1981). Food selection by black colobus monkeys (*Colobus satanas*) in relation to plant chemistry. *Biological Journal of the Linnean Society,* **16**, 115–146.

McKey, D. B. & Waterman, P. G. (1982). Ranging behaviour of a group of black colobus (*Colobus satanas*) in Douala-Edéa Reserve, Cameroon. *Folia Primatologica,* **39**, 264–304.

McKey, D. B., Waterman, P. G., Mbi, C. N., Gartlan, J. S. & Struhsaker, T. T. (1978). Phenolic content of vegetation in two African rain forests: ecological implications. *Science,* **202**, 61–4.

Merz, G. (1986). Counting elephants (*Loxodonta africana cyclotis*) in tropical rain forests with particular reference to the Taï National Park, Ivory Coast. *African Journal of Ecology,* **24**, 61–8.

Milton, K. (1982). Dietary quality and demographic regulation in a howler monkey population. In *The Ecology of a Tropical Forest,* ed. E. G. Leigh, A. S. Rand & D. M. Windsor, pp. 273–89. Washington, DC: Smithsonian Institution Press.

Milton, K. (1990). Annual mortality patterns of a mammal community in central Panama. *Journal of Tropical Ecology,* **6**, 493–9.

Milton, K. (1996). Effects of bot fly (*Alouattamyia baeri*) parasitism on a free-ranging howler (*Alouatta palliata*)

Population in Panama. *Journal of Zoology (London)*, **239**, 39–63.

Mitani, M. (1992). Preliminary results of studies on wild western lowland gorillas and other sympatric diurnal primates in the Ndoki Forest, Northern Congo. In *Topics in Primatology*, ed. N. Itoigawa, Y. Sugiyama, G. P. Sackett, & R. K. R. Thompson, pp. 215–224. Tokyo: University of Tokyo Press.

Moreau, R. E. (1966). *The Bird Faunas of Africa and its Islands*. London: Academic Press.

Murphy, P. G. & Lugo, A. E. (1986). Ecology of tropical dry forest. *Annual Review of Ecology and Systematics*, **17**, 67–88.

Newbery, D. M., Gartlan, J. S., McKey, D. B. & Waterman, P. G. (1986). The influence of drainage and soil phosphorus on the vegetation of Douala-Edéa Forest Reserve, Cameroon. *Vegetation*, **65**, 149–62.

Oates, J. F. (1978). Water-plant and soil consumption by guereza monkeys (*Colobus guereza*): A relationship with minerals and toxins in the diet. *Biotropica*, **10**, 241–53.

Oates, J. F. (1988). The diet of the olive colobus monkey, *Procolobus verus* in Sierre Leone. *International Journal of Primatology*, **9**, 457–78.

Oates, J. F. (1994). The natural history of African colobines. In *Colobine Monkeys: Their Ecology, Behaviour and Evolution*, ed. A. G. Davies & J. F. Oates. pp. 75–128. Cambridge: Cambridge University Press.

Oates, J. F. (1995). The dangers of conservation by rural development – a case-study from the forests of Nigeria. *Oryx*, **29**, 115–22.

Oates, J. F. (1996*a*). *African Primates: Status Survey and Conservation Action Plan. Revised Edition*. Gland, Switzerland: IUCN.

Oates, J. F. (1996*b*). Habitat alteration, hunting and the conservation of folivorous primates in African forests. *Australian Journal of Ecology*, **21**, 1–9.

Oates, J. F. & Jewell, P. A. (1967). Westerly extent of the range of three African lorisoid primates. *Nature*, **215**, 778–9.

Oates, J. F. & Whitesides, G. H. (1990). Association between Olive colobus (*Procolobus verus*), Diana guenons (*Cercopithecus diana*), and other forest monkeys in Sierra Leone. *American Journal of Primatology*, **21**, 129–46.

Oates, J. F., Whitesides, G. H., Davies, A. G., Waterman, P. G., Green, S. M., Dasilva, G. L. & Mole, S. (1990). Determinants of variation in tropical forest primate biomass: new evidence from West Africa. *Ecology*, **71**, 328–43.

Osmaston, H. A. (1959). *Working plan for the Kibale and Itwara Forests*. Entebbe, Ugandan Forest Department, Government of Uganda Printer.

Paterson, J. D. (1991). The ecology and history of Uganda's Budongo Forest. *Forest and Conservation History*, **35**, 179–87.

Peres, C. A. (1990). Effects of hunting on western Amazonian primate communities. *Biological Conservation*, **54**, 47–59.

Peres, C. A. (1994). Primate responses to phenological change in an Amazonian terra firme forest. *Biotropica* **26**, 98–112.

Peres, C. A. (1997). Primate community structure at twenty western Amazonian flooded and unflooded forests. *Journal of Tropical Ecology*, **13**, 381–405.

Peres, C. A., Patton, J. L. & da Silva, M. N. (1996). Riverine barriers and gene flow in Amazonian saddle-back tamarins. *Folia Primatologica*, **67**, 113–24.

Plumptre, A. J. (1996). Changes following 60 years of selective timber harvesting in the Budongo Forest Reserve, Uganda. *Forest Ecology and Management*, **89**, 101–13.

Plumptre, A. J. & Reynolds, V. (1994). The effect of selective logging on the primate populations in the Budongo Forest Reserve, Uganda. *Journal of Applied Ecology*, **31**, 631–41.

Plumptre, A. J., Reynolds, V. & Bakuneeta, C. (1994). *The Contribution of Fruit Eating Primates to Seed Dispersal and Natural Regeneration after Selective Logging*. ODA Report.

Quris, R. (1976). Données comparatives sur la socio-écologie de huit espèces de Cercopithecidae vivant dans une même zone de forêt primative periodiquement inondée (Nord-Est du Gabon). *Revue d'Ecologie (Terre et Vie)*, **30**, 193–209.

Redford, K. H. (1992). The empty forest. *Bioscience*, **42**, 412–22.

Reynolds, V. (1993). Sustainable forestry: The case of the Budongo Forest, Uganda. *Swara* July-August 13–17.

Reynolds, V. (1993). Conservation of chimpanzees in the Budongo Forest Reserve. *Primate Conservation*, **11**, 41–3.

Reynolds, V. & Reynolds, F. (1965). Chimpanzees of the Budongo Forest. In *Primate Behavior: Field Studies of Monkeys and Apes*, ed. I. DeVore, pp. 368–424. New York: Holt, Rinehart, and Winston.

Richards, P. W. (1996). *The Tropical Rainforest*, 2nd edn. Cambridge: Cambridge University Press.

Rogers, M. E., Abernethy, K. A., Fontaine, B., Wickings, J. E., White, L. J. T. & Tutin, C. E. G. (1996). Ten days in the life of a mandrill horde in the Lopé

Reserve, Gabon. *American Journal of Primatology*, **40**, 297–313.

Rogers, M. E., Maisels, F., Williamson, E. A., Fernandez, M. & Tutin, C. E. G. (1990). Gorilla diet in the Lopé Reserve, Gabon: A nutritional analysis. *Oecologia*, **84**, 326–39.

Rowe, N. (1996). *The Pictorial Guide to Living Primates*. East Hampton, New York: Pogonias Press.

Scott, M. E. (1988). The impact of infection and disease on animal populations: implications for conservation biology. *Conservation Biology*, **2**, 40–56.

Simberloff, D. & Connor, E. F. (1981). Missing species combinations. *American Naturalist*, **118**, 215–39.

Skorupa, J. P. (1986). Responses of rainforest primates to selective logging in Kibale Forest, Uganda: A summary report. In *Primates: The Road to Self-Sustaining Populations*, ed. K. Benirschke, pp. 57–70. New York: Springer-Verlag.

Skorupa, J. P. (1988). *The Effect of Selective Timber Harvesting on Rain-forest Primates in Kibale Forest, Uganda*. Ph.D. Dissertation, University of California, Davis.

Smythe, N. (1970). Relationships between fruiting seasons and seed dispersal method in a neotropical forest. *American Naturalist*, **104**, 25–35.

Stanford, C. B., Wallis, J., Matama, H. & Goodall, J. (1994). Patterns of predation by chimpanzees on red colobus monkeys in Gombe National Park, 1982–1991. *American Journal of Physical Anthropology*, **94**, 213–28.

Stoner, K. E. (1996). Prevalence and intensity of intestinal parasites in mantled howling monkeys (*Alouatta palliata*) in Northeastern Costa Rica: implications for conservation biology. *Conservation Biology*, **10**, 539–46.

Stephenson, P. J. & Newby, J. E. (1997). Conservation of the Okapi Wildlife Reserve, Zaire. *Oryx*, **31**, 49–58.

Stevenson, P. R., Quinones, M. J. & Ahumada, J. A. (1994). Ecological strategies of woolly monkeys (*Lagothrix lagotricha*) at Tinigua National Park, Colombia. *American Journal of Primatology*, **32**, 123–40.

Struhsaker T. T. (1975). *The Red Colobus Monkey*. Chicago: University of Chicago Press.

Struhsaker, T. T. (1978). Food habits of five monkey species in the Kibale Forest, Uganda. In *Recent Advances in Primatology*, ed. D. J. Chivers & J. Herbert, pp. 225–48. New York: Academic Press.

Struhsaker, T. T. (1981). Forest and primate conservation in East Africa. *African Journal of Ecology*, **19**, 99–114.

Struhsaker, T. T. (1997). *Ecology of an African Rain Forest: Logging in Kibale and the Conflict between Conservation and Exploitation*. Gainesville, Florida: University Presses of Florida.

Struhsaker, T. T. & Leakey, M. (1990). Prey selectivity by crowned hawk-eagles on monkeys in the Kibale Forest, Uganda. *Behavioural Ecology and Sociobiology*, **26**, 435–43.

Struhsaker, T. T. & Leland, L. (1979). Socioecology of five sympatric monkey species in the Kibale Forest, Uganda. *Advances in the Study of Behavior*, **9**, 159–228.

Stuart, M. D., Greenspan, L. L., Glander, K. E. & Clarke, M. (1990). A coprological survey of parasites of wild mantled howling monkeys *Alouatta palliata palliata*. *Journal of Wildlife Disease*, **26**, 547–9.

Stuart, M. D., Strier, K. B. & Pierberg, S. M. (1993). A coprological survey of parasites of wild muriquis, *Brachyteles arachnoides*, and brown howling monkeys, *Alouatta fusca*. *Journal of the Helminthological Society*, **60**, 111–15.

Sugiyama, Y. (1968). Social organization of chimpanzees in the Budongo Forest, Uganda. *Primates*, **9**, 225–58.

Sugiyama, Y. (1969). Social behaviour of chimpanzees in the Budongo Forest, Uganda. *Primates*, **10**, 197–225.

Suzuki, A. (1979). The variation and adaptation of social groups of chimpanzees and black and white colobus monkeys. In: *Primate Ecology and Human Origins*. ed. I. S. Bernstein & E. O. Smith, pp. 153–74. New York: Garland STPM Press.

Terborgh, J. (1986). Community aspects of frugivory in tropical forests. In *Frugivores and Seed Dispersal*, ed. A. Estrada & T. H. Fleming, pp. 371–84. Dordrecht: D. W. Junk Publishers.

Terborgh, J. (1983). *Five New World Primates*. Princeton: Princeton University Press.

Terborgh, J. & van Schaik, C. P. (1987). Convergence vs. nonconvergence in primate communities. In *Organization of Communities, Past and Present*, ed. J. H. R. Gee & P. S. Giller, pp. 205–26. Oxford: Blackwell Scientific.

Thomas, S. C. (1991). Population densities and patterns of habitat use among anthropoid primates of the Ituri Forest, Zaire. *Biotropica*, **23**, 68–83.

Thompson, K. & Hamilton, A. C. (1983). Peatlands and swamps of the African continent. In *Ecosystems of the World*, ed. A. J. P. Gore, pp. 331–70. Amsterdam: Elsevier Scientific.

Tutin, C. E. G. (1996). Ranging and social structure of lowland gorillas in the Lopé Reserve, Gabon. In *Great Ape Societies*, ed. W. C. McGrew, L. Marchant & T. Nishida, pp. 58–70. Cambridge: Cambridge University Press.

Tutin, C. E. G. & Fernandez, M. (1993). Relationships between minimum temperature and fruit production in some tropical forest trees in Gabon. *Journal of Tropical Ecology*, **9**, 241–8.

Tutin, C. E. G., Ham, R. M., White, L. J. T. & Harrison, M. J. S. (1997). The primate community of the Lopé Reserve, Gabon: Diets, responses to fruit scarcity, and effects on biomass. *American Journal of Primatology*, **42**, 1–24.

Tutin, C. E. G., White, L. J. T., Williamson, E. A., Fernandez, M. & McPherson, G. (1994). List of plant species identified in the northern part of the Lopé Reserve, Gabon. *Tropics*, **3**, 249–76.

USAID (1992). *USAID Country Program Strategic Plan – Uganda 1992–1997*. Washington, DC: USAID.

Verheyen, W. N. (1963). New data on the geographical distribution of *Cercopithecus* (*Allenopithecus*) *nigroviridis* Pocock 1907. *Revue de Zoologie et de Botanique Africaines*, **68**, 393–6.

Waser, P. M. (1987). Interactions among primate species. In *Primate Societies*, ed. B. B. Smuts, D. L. Cheney, R. M. Seyfarth, R. W. Wrangham & T. T. Struhsaker, pp. 210–26. Chicago: Chicago University Press.

Waterman, P. G., Ross, J. A. M., Bennett, E. L. & Davies, A. G. (1988). A comparison of the floristics and leaf chemistry of the tree flora in two Malaysian rain forest and the influence of leaf chemistry on populations of colobine monkeys in the Old World. *Biological Journal of the Linnean Society*, **34**, 1–32.

Weins, J. A. (1977). On competition and variable environments. *American Scientist*, **65**, 590–7.

Weisenseel, K., Chapman, C. A. & Chapman, L. J. (1993). Nocturnal primates of Kibale Forest: Effects of selective logging on prosimian densities. *Primates*, **34**, 445–50.

White, L. J. T. (1992). *Vegetation History and Logging Disturbance: Effects on Rain Forest Mammals in the Lopé Reserve, Gabon (with Special Emphasis on Elephants and Apes)*. Ph.D. Thesis, University of Edinburgh.

White, L. J. T. (1994*a*). Biomass of rain forest mammals in the Lopé Reserve, Gabon. *Journal of Animal Ecology*, **63**, 499–512.

White, L. J. T. (1994*b*). The effects of commercial mechanized selective logging on a transect in lowland

rainforest in the Lopé Reserve, Gabon. *Journal of Tropical Ecology*, **10**, 313–22.

White, L. J. T. (1995). Factors affecting the duration of elephant dung piles in rain forest in the Lopé Reserve, Gabon. *African Journal of Ecology*, **33**, 142–50.

White, L. J. T., Tutin, C. E. G. & Fernandez, M. (1993). Group composition and diet of forest elephants, *Loxodonta africana cyclotis* Matschie 1900, in Lopé Reserve, Gabon. *African Journal of Ecology*, **31**, 181–99.

White, L. J. T., Rogers, M. E. R., Tutin, C. E. G., Williamson, E. A. & Fernandez, M. (1995). Herbaceous vegetation in different forest types in the Lopé Reserve, Gabon: Implications for keystone food availability. *African Journal of Ecology*, **33**, 124–41.

Whitesides, G. H. (1989). Interspecific association of Diana monkeys, *Cercopithecus diana*, in Sierra Leone, West Africa: Biological significance of chance? *Animal Behaviour*, **37**, 760–76.

Wilkie, D. S., Sidle, J. G. & Boundzanga, G. C. (1992). Mechanized logging, market hunting and a bank loan in Congo. *Conservation Biology*, **6**, 570–80.

Wing, L. D. & Buss, I. O. (1970). Elephants and forest. *Wildlife Monographs*, Number 19.

Work, T. H., Trapido, H., Narasimba Murthy, D. P., Laxmana Rao, R., Bhatt, R. N. & Kulkarni, K. G. (1957). Kyasanur forest disease. III. A preliminary report on the nature of the infection and clinical manifestations in human being. *Indian Journal of Medical Science*, **11**, 619–45.

World Resources Institute (1994). World Resources 1994–1995. The World Resources Institute, New York.

Wrangham, R. W., Chapman, C. A. Clark-Arcadi, A. P. & Isabirye-Basuta, G. (1996). Social ecology of Kanyawara chimpanzees: Implications for understanding the costs of great ape groups. In *Great Ape Societies*. ed. W. C. McGrew, L. F. Marchant & T. Nishida, pp. 45–57. Cambridge: Cambridge University Press.

Zuberbuhler, K., Noe, R. & Seyfarth, R. M. (1977). Diana monkey long-distance calls: messages for conspecifics and predators. *Animal Behaviour*, **53**, 589–604.

2 · Biomass and use of resources in south and south-east Asian primate communities

A. K. GUPTA AND DAVID J. CHIVERS

INTRODUCTION

Primate habitats in south-east Asia span the two main forest formations of this part of the Oriental region – the deciduous monsoon rainforests of mainland Asia north of the isthmus of Kra and the evergreen rain forests of the "islands" of the Sunda Shelf (Whitmore, 1975). The evergreen rainforests of the Sunda Shelf comprise the main primate habitat, but significant numbers of taxa and individuals occur in the more seasonal forests of mainland Asia, mostly concentrated in pockets of evergreen forest surviving in the moister areas under maritime influence, e.g., Indo-China, Thailand, south India and Sri Lanka, Burma, and Bangladesh (Table 2.1). Intermittent volcanic activity over the last 100 million years has thrown up several arcs of mountains, creating the Sunda Shelf only about 12 million years ago; they run mainly from the north-west to the south-east, with the largest down the west of Sumatra.

The Sunda Shelf owes its uniquely rich fauna (and flora) to an admixture of immigrants. First came the Siva-Malayan fauna from the Indian subcontinent, nearly two million years ago during the Villafranchian. Then, nearly one million years ago, the Sino-Malayan fauna came from southern China, while the spread of tropical rainforest and/or the spread of the cold influence of the Himalayas blocked the Indian route.

It is important to note that during periods of lowest sea level, the center of the Shelf dried out. The key rainforest relicts, into which primates and other forest animals retreated and out of which they spread when sea level rose, were in eastern Indo-China and southern China, north-eastern Borneo, west Java, north Sumatra and southern Burma, as well as the Mentawai Islands. The Lower Pleistocene was

moist, with the evolution of mixed dipterocarp forest, and there was considerable volcanic activity. The Middle Pleistocene was driest, coinciding with the major glaciations in temperate regions. Since the central part of the Shelf was savanna, such animals as hippopotami, rhinoceri, elephants, and cattle were common from one-and-a-half to half a million years ago (Medway, 1972). In some cases, these species adapted to the spread of forest, but many other large or savanna mammals became extinct during the upper Pleistocene, when major pluvials led to changes in sea level. Faunal elements mixed together when sea level was low and the whole Sunda Shelf was exposed.

Five climatic forest formations have been recognized in the Malay Peninsula (Wyatt-Smith, 1953). The forest canopy up to about 1220 m (4000 ft) above sea level is dominated by trees of the Dipterocarpaceae. Within this altitudinal range there is zonation into lowland, hill, and upper dipterocarp forests, with the main transition zones at 305 m (1000 ft) and 760 m (2500 ft) respectively. The first two belong to the tropical lowland evergreen rainforest formation (Whitmore, 1975) and the last, along with oak-laurel forest up to 1525 m (5000 ft), belongs to the world-wide formation of lower montane forest. Above 1524 m (5000 ft) belongs to the world-wide formation of upper montane forest. Above 1524 m the upper montane forests are characterized as ericaceous, because of the predominance of species of Ericaceae (and Coniferae and Myrtaceae).

Such altitudinal zonation is approximate, and varies from place to place, being more compressed on isolated massifs. Across the Straits of Melaka in Sumatra, for example, the transition from lower to upper montane forest occurs around 1830 m (6000 ft) – the highest mountains in Sumatra are around 3000–3800 m (10000–12500 ft), compared with 2817 m (7184 ft) in the Malay Peninsula

Table 2.1. *Size, activity, diet and habitat for Asian primates*

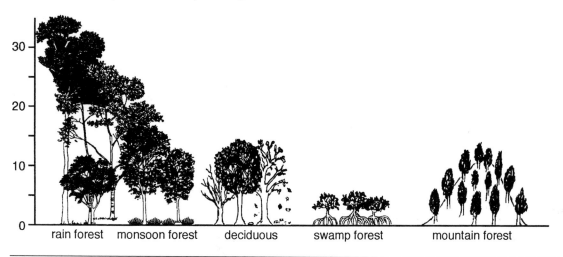

Species	Body mass (kg)	Activity	Diet	Rainforest	Monsoon	Deciduous	Swamp	Montane	Geographic region
Loris tardigradus	0.1–0.28	N	OM	+	+	+	+	+	I
Nycticebus coucang	0.23–0.61	N	OM	+	+	+			B,SEA,PM,SS
Nycticebus pygmaeus	0.372–0.462	N	GUM	+				+	SEA
Tarsius bancanus	0.12	N	FAUN	+					SS
Tarsius dianae	0.10	N	FAUN						SS
Tarsius pumilus	?	N	FAUN					+	SS
Tarsius spectrum	0.1	N	FAUN	+				+	SS
Tarsius syrichta	0.12	N	FAUN						Phil.
Macaca arctoides	7.5–10.0	D	OM	+	+	+			SEA,PM
Macaca assamensis	4.9–15.0	D	OM	+	+	+		+	I,B,SEA
Macaca cyclopis	4.9–6.0	D	OM						Taiwan
Macaca fascicularis	2.5–8.3	D	OM	+	+		+		B,SEA.PM.SS
Macaca fuscata	8.3–18.0	D	OM	+		+			Japan
Macaca maura	?	D	OM						SS
Macaca mulatta	4.4–10.9	D	OM	+	+	+	+		I,B,SEA
Macaca nemestrina	4.7–15.4	D	OM	+	+		+	+	B,SEA,PM,SS
Macaca nigra	?	D	OM	+					SS
Macaca ochreata	?	D	OM	+					SS
Macaca radiata	3.9–6.8	D	OM	+	+	+			I
Macaca silenus	3.0–10.0	D	OM	+					I
Macaca sinica	3.4–8.4	D	OM			+			Sri Lanka
Macaca thibetana	?	D	?	+	+				SEA
Presbytis comata	6.45	D	FO	+				+	SS
Presbytis femoralis	5.8–8.8	D	FR	+			+		PM,SS
Presbytis frontata	?	D	?						SS
Presbytis hosei	5.5–6.2	D	FO	+					SS
Presbytis melalophos	5.8	D	FO	+					PM,SS
Presbytis potenziani	6.4	D	FO/FR	+			+		SS
Presbytis rubicunda	5.7–6.2	D	FO	+					SS
Presbytis thomasi	?	D	FR	+					SS
Semnopithecus entellus	11.2–18.3	D	OM	+	+	+			I
Trachypithecus auratus	7.1	D	FO	+					SS
Trachypithecus cristatus	5.7–6.6	D	FO	+			+		PM,SS
Trachypithecus delacouri	?	D	FO						SEA
Trachypithecus francoisi	5.9	D	?	+					SEA

Table 2.1. (*cont.*)

Species	Body mass (kg)	Activity	Diet	Rainforest	Monsoon	Deciduous	Swamp	Montane	Geographic region
Trachypithecus geei	9.5–10.9	D	FR/FO	+		+			B
Kasi johnii	10.8–12.7	D	FO/FR	+		+			I
Trachypithecus obscurus	6.6–7.3	D	FO/FR	+			+	+	PM
Trachypithecus phayrei	6.9–7.9	D	FO/FR	+	+				B,SEA
Trachypithecus pileatus	10.0–12.8	D	FR/FO	+		+			B
Kasi vetulus	5.1–9.4	D	FO/FR			+		+	Sri Lanka
Pygathrix nemaeus	8.2–10.9	D	FO	+					PM
Pygathrix nigripes	?	D	FO	+					SEA
Rhinopithecus avunculus	8.5–14.0	D	FR/FO	+	+				SEA
Rhinopithecus bieti	9.2–15.0	D	FO/FR	+		+		+	SEA
Rhinopithecus brelichi	?	D	FO	+	+	+			SEA
Rhinopithecus roxellana	6.5–39.0	D	FO	+	+	+			SEA
Nasalis larvatus	10.0–21.2	D	FO	+			+		SS
Simias concolor	7.1–8.7	D	FO/FR	+			+		SS
Hylobates hoolock	6.5	D	FR	+	+	+			B,SEA
Hylobates agilis	5.5–6.4	D	FR/FO	+			+		PM,SS
Hylobates klossii	5.8	D	FR	+					SS
Hylobates lar	4.4–7.6	D	FR	+	+	+		+	SEA,PM, SS
Hylobates moloch	5.7	D	FR	+					SS
Hylobates muelleri	5.0–6.4	D	FR	+					SS
Hylobates pileatus	6.3–10.4	D	FR	+		+		+	SEA
Hylobates gabriellae	5.7	D	?	+					SEA
Hylobates leucogenys	5.7	D	?	+					SEA
Hylobates syndactylus	10.0–14.7	D	FO	+				+	PM
Pongo abelii	?	D	FR/OM				+	+	SS
Pongo pygmaeus	37.0–77.5	D	FR	+			+		SS

Note: I, India; PM, Peninsular Malaysia; Phil, Philippines; B, Bangladesh; SEA, South East Asia; SS, Sunda Shelf

(West Malaysia). The Kota Kinabalu massif in the north-east of Borneo rises to more than 4000 m (13 000 ft).

The ever-wet humid tropics, the evergreen rainforest zone, is characterized by the main north-east monsoon at the turn of the year, with a milder south-west monsoon in April or May, with an annual rainfall of 5000mm (200 inches), as in West Malaysia. In areas of rain shadow the rainfall may be less than 2000mm (78 inches); rarely are there months with no rain, most often in February, and the mean annual temperature is 29 °C (Chivers, 1980). In north Sumatra, the annual mean rainfall at Ketambe in the Gunung Leuser National Park is 3329 mm (125 inches). In Kalimantan, over six years, the average rainfall was 3722 mm (147 inches) (Chivers, 1997). On the more seasonal mainland such as Bangladesh, the monsoon rain, from May to October, amounts to 1200–3500 mm (47–138 inches), with temperatures rising from May to October, sometimes to 35 °C by day; there is virtually no rain from November to February, inclusive, with temperatures dropping as low as 0 °C at night (Ahsan, 1994).

Between 24 and 50 tree families have been documented from Asian primate habitats, averaging 37, with about 400 trees/hectare (Ahsan, 1994). Trees of the family Dipterocarpaceae are typical of most Asian primate habitats, ranging from 1–43% of forest composition (Mather, 1992), averaging 16% (Ahsan, 1994). Moraceae and Euphorbiaceae are the commonest tree families used as food sources in gibbon habitats, followed by Leguminosae, Myrtaceae, Annonaceae, Rubiaceae, Guttiferae, and Anacardiaceae. These families present 161 species – 45% of all known gibbon foods (Mather, 1992).

Forests in Bangladesh and much of India do not fall within the recognized major forest types; they are not markedly evergreen. The majority of small trees are evergreen, and most of the tall trees are deciduous. Some of the deciduous trees shed their leaves in the cold season, others in the hot season, thereby giving the forest the appearance and character of an evergreen forest. Thus, most of the area has mixed evergreen forest with a large proportion of deciduous species. Further west is a more open, drier, more seasonal Sal forest. Thus, the major vegetation type, where forest persists, is tropical wet semi-evergreen forest, sub-divided into tropical evergreen (dominated by *Artocarpus*, and species of Meliaceae and Lauraceae), mixed evergreen (dominated by species of *Dipterocarpus, Artocarpus, Protium, Prunus,* and *Ficus*), and low evergreen (stunted and poorer). Deciduous forests are either loha (*Xylia kerii*), teak (*Tectona grandis*), jarul (*Lagerstroemia speciosa,* with lianus), kadam (*Anthocephalus chinesis*), garjan (*Dipterocarpus* spp.), sal (*Shorea robusta*), and the open deciduous on the drier hill tops. There are patches of bamboo jungle and savanna, almost devoid of trees and covered with sun grass.

Thus, rich vegetation, in terms of biomass and species diversity, results in a comparable richness in animal species. Primates, birds, and bats converge dramatically on the frugivorous niche, playing varying roles in seed dispersal (and predation), hence in the natural regeneration of rainforests.

Biomass and primate communities

Biomass of the primate community at a given place refers to the total weight of primate populations of all different species per unit area (kg/km²). It depends on both the species and total population (as a function of number of groups and group size) of primates at a given site. Thus, it might be presumed that the biomass of primate communities at different sites with the same species composition would be relatively similar, but the presence or absence of a particular species and its group size may differ between locations, even if located only a few tens of kilometres apart (Bourlière, 1985).

The few available estimates of forest primate biomass show high site-to-site variation (Bourlière, 1985; Oates *et al.,* 1990; Laidlaw, 1994). Oates *et al.* (1990), quoting Davies (1984) and Whitesides *et al.* (1988), considered that variations in primate community biomass at different sites cannot be explained alone by sampling errors, differences in the intensity of hunting pressure or in sampling and analytical techniques. Bourlière (1985) was the first to list three main categories of factors responsible for differences in primate community structure (and hence biomass): history and paleoecology, current spatial heterogeneity of the vegetation, and competition with other taxonomic groups. In addition, community-level chemical composition of the

vegetation (Janzen, 1974; McKey, 1978; Waterman *et al.,* 1988) and soil fertility (Emmons, 1984; Oates *et al.,* 1990) have been identified as important factors affecting primate biomass. Chivers (1974) concluded that biomass may be related to animals' tolerance to poor-quality food during the period of scarcity.

Over the past few years, human disturbance in varying forms and intensity (logging, fragmentation, hunting, shifting and settled agriculture, production forestry, and so on) have hardly left any primate community habitat unaffected (Tutin & White, chapter 13; Peres, chapter 15; Struhsaker, chapter 17, this volume). In these human-induced habitats, the biomass of primate community at any given site may not necessarily be related to high primate-species diversity, because certain species may be favored at the cost of others. For example, the population density of *Macaca mulatta* in many sites in China is inflated due to human provisioning (John MacKinnon, pers. comm.). Generalist colobines and macaques are favored in habitats affected by shifting cultivation at the cost of hylobatid species (hoolock gibbon) in Tripura, north-east India (Gupta & Kumar, 1994). Some populations of *Macaca* sp. and colobines are high because of their proximity to certain favored food-plant cultivation fields (Caldecott, 1986).

In all these cases, although primate species diversity has been affected adversely, overall primate biomass is even higher than in those sites with more diverse primate communities. Thus, a higher value of primate community biomass at any given site may not always be related to a diverse primate community. We hypothesize that various human-induced disturbances work at the microhabitat level, so much so, that even geographically adjacent habitats, although appearing similar in their vegetation and primate community compositions, may differ in resource use by the primate communities and hence in their biomass. As a result, primate community biomass in Asia may not be a simple function of species diversity nor easily related to variations in the chemical composition of the vegetation and soil fertility.

In this chapter we examine the relationships between the biomass of primate communities and various habitat parameters (different habitat types, rainfall, tree density, productivity, seasonality, and forage quality) at different south and south-east Asian sites (Figs 2.1 and 2.2). We also propose a new model, based at the microhabitat level, for a better explanation of site-to-site variations in the structure

(a)

(b)

(c)

(d)

Fig 2.1. Forest and some primates of east India/Bangladesh.
(a) hoolock gibbon, (b) pig-tailed macaque, (c) capped langur
(courtesy of Md. M. Feeroz), (d) forest.

(biomass) and function (resource use) of different primate communities.

STUDY SITES

Data were used from available studies at different sites in south and south-east Asia. In total, data were collated from 33 sites from four countries (Table 2.1): in south Asia are Bangladesh (four sites) and India (eight sites), and in south-east Asia are Malaysia (13 sites) and Indonesia (eight sites). The eight sites in India further differed in their geographical locations (four from north-east India, and two each from north and south Indian regions). Of the 13 sites in Malaysia, 11 were from West (Peninsular) Malaysia, and the remaining two were from East Malaysia (on the island of Borneo). The eight sites in Indonesia comprised four in Kalimantan (Borneo), two in Java and two in Sumatra. All these sites are fairly representative of prevailing different habitat types, but most are influenced by human activity.

METHODS

Data on various habitat parameters were not equally available for all the sites. Tree species diversity was not available for most sites. Only a few sites lacked data on rainfall (one site), tree density (four sites), productivity of fruits (seven sites), productivity of young leaves and flowers (eight sites each), and number of dry months (eight sites). Seasonality of production is expressed as the number of dry months, in which the monthly rainfall does not exceed 100 mm. Productivity is expressed as the production of young leaves, flowers and fruit, represented here as the number of months of peak production of these three plant parts. Forage quality refers to the vegetation composition and is represented here as the three commonest plant families in the habitat of the given primate community. The ratio between protein (P) and acid detergent fibre (ADF) used in this study represents an average value of P/ADF for three dominant families at each given site. The P/ADF ratio for each dominant family was calculated from the values available for different species in various publications (Primack, 1985; Davies *et al.*, 1988; Waterman *et al.*, 1988; Oates *et al.*, 1990; Dasilva, 1994; Gupta & Kumar, 1994).

All simple relationships between variables are tested using nonparametric correlation (Spearman's rho), but parametric methods were used for more complex statistical designs (comparison of regressions between regions). Unless noted otherwise, the sample size for the total data set is 33 sites, for south Asia is 12 sites, and for south-east Asia is 21 sites.

RESULTS AND DISCUSSION

Evaluation of various factors as determinants of the biomass

Biogeography

A hierarchical cluster analysis was run on the 33 study sites based on similarity in primate biomass, species richness, and several habitat factors (rainfall, leaf quality, number of dry months, and tree density). Using a variety of joining criteria (average distance, centroid distance, and Ward's method) always produced a single cluster of south Asian sites (except Kakachi in southern India) distinct from the south-east Asian sites (Fig. 2.3). Within south-east Asia, there was little apparent substructure using any of the joining criteria. The existence of the south Asian cluster is due largely to the far larger number of dry months there than in south-east Asia, but, even when the number of dry months is excluded, a distinct cluster of sites emerges consisting of the south Asian sites except for those in northeast India. These share a low primate diversity (1–2 species) and roughly similar leaf quality. Whether or not these associated variables are linked via ecological causality or shared biogeographic history, it is clear that study sites are not geographically independent of each other at least at the level of south vs. south-east Asia. Thus, for the rest of this chapter, we analyze relationships both for the entire data set and separately within these two major subregions. We generally do not analyze smaller regions because the limited number of sites in each country or part thereof make correlations unreliable.

Habitat types

Within a given geographical area, primate community biomass may vary with habitat type of a site. Even if they fall more or less within the same rainfall and seasonality regime, different habitat types may differ in plant species diversity, productivity, and forage quality, which in turn may control the biomass of the primate communities. As a

(a)

(b)

(c)

(d)

(e)

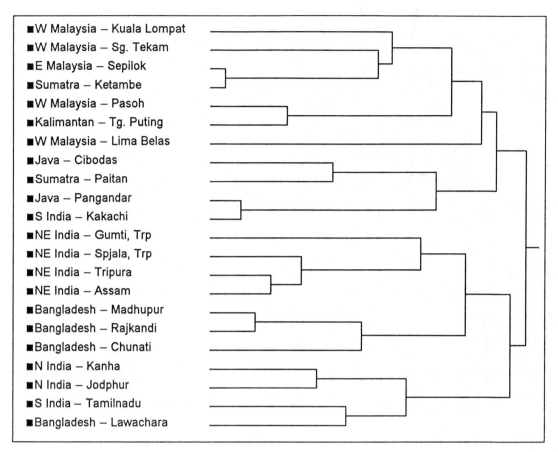

Fig. 2.3. Hierarchical cluster diagram of similarity among South Asian (India and Bangladesh) and South-east Asian (Indonesia and Malaysian) sites.

case study, we present primate communities in two different habitat types in Tripura, north-east India.

The first habitat type is Gumti Wild Life Sanctuary (GWLS), located in the south district of Tripura, north-east India. Most of this habitat has young secondary forests, developed when the area was left undisturbed after shifting cultivation. The average annual precipitation in this area is about 1500 mm. The area has three definite dry months, but, because of the semi-evergreen vegetation type, the area is never completely devoid of foliage at any time during the annual production cycle. Because of excessive leaching of the soil through repeated cultivation and washing out of exposed soil from the steep hilly terrain during rains, as well as wide canopy openings created during the cultivation, the original primary forest of evergreen vegetation is replaced at many sites by more light-demanding secondary forest species. Very few of such species account for a large number of tree counts (species density is 22/ha, and tree density is 280 trees/ha). Small-scale human disturbance, in the form of collection of forest products and

Fig. 2.2. Forest and some primates of Indonesia (a) orangutan, (b) siamang, (c) silvered langur, (d) long-tailed macaque, (e) forest (courtesy of D. J. Chivers).

livestock grazing, ensures that some young regeneration crops are always available (mainly through coppicing or natural seed germination). The presence of dipterocarps may reduce the primate species diversity, but might not affect food availability to the primate community, as most of the dipterocarps are not favored food species.

Gumti is rich in primate species. Seven primate species, out of a total 15 in India, were located in an area of about 105 km^2 surveyed. The population estimates of five species (*Trachypithecus phayrei*, Phayre's langur; *T. pileatus*, capped langur; *Hylobates hoolock*, hoolock gibbon, *Macaca mulatta*, rhesus macaque, and *Nycticebus coucang*, slow loris) were determined through surveys (Gupta, unpublished data), but the populations of *Macaca nemestrina*, and *M. arctoides* (pig- and stump-tailed macaques, respectively) could not be estimated. The biomass of five species worked out at about 60 kg/km^2.

The second habitat type is Sepahijala Wildlife Sanctuary (SWLS), located in the west district of Tripura, northeast India. It covers an area of about 18.5 km^2, and is about 100 km west of GWLS. SWLS is surrounded by human habitation or agriculture fields and has lost connections with the nearby forest tracts. Average rainfall is 1800 mm, and the number of dry months is the same (3 months) as GWLS. The habitat is a semi-evergreen primary forest, which has been degraded over the years, because of human disturbance (mainly due to collection of minor forest products, livestock grazing, illicit cutting and felling of trees). Most of the degraded tracts have been afforested using native and exotic forestry plantation species. The presence of some deciduous forestry plantation species has changed the semi-evergreen nature of the habitat to moist-deciduous patches which remain leafless during the dry months. The oldest forestry plantations, mostly of timber species, are only about 30 years old, and are under tremendous pressure from illicit felling. Gaps created inside these plantations due to felling are being afforested using non-timber fast-growing species, mostly of value for firewood and fodder. The habitat is thus a mix of mature primary forest species, and young regenerating secondary forest species (mostly of forestry plantation origin). Plant species diversity is almost double (41.5 species/ha) that at GWLS, but tree density is marginally lower at about 250 trees/ha.

Primate species diversity at SWLS is lower than at GWLS, as only five primate species have been identified in this habitat. Of the seven species found in GWLS, hoolock gibbons and stump-tailed macaques are absent from SWLS, but the biomass of the five species is almost five times greater (321 kg/km^2) at SWLS than GWLS. In SWLS, most of the primate groups are concentrated at the meeting point of primary forests and plantation forests. No group was seen having its home range entirely within the primary forest or plantation forest patches.

Under normal habitat conditions, one may associate higher primate biomass values with correspondingly higher primate diversity and thus may expect a higher primate diversity in SWLS than GWLS, but the reverse is true. In this example, the differences in seasonality and rainfall in both areas is not sufficiently large to account for the marked differences in primate biomass. This variation may be explained more by differences in habitat quality (historical factors, availability of food and cover resources) and habitat size. SWLS, although smaller in size than GWLS, has a wider spectrum of vegetation types (remnant patches of primary forests, old forestry plantations of timber species with large trees useful as cover for primates, and young forestry plantations of fast-growing native and exotic species). Thus, SWLS could support more groups of those primates which were able to adapt to different feeding regimes. GWLS, being much bigger in area, could accommodate more species, especially those needing a large area to range for food (hoolock gibbon and stump-tailed macaque), whereas the island-type habitat at SWLS perhaps proved too small for these frugivore species. Even at GWLS, the presence of wide gaps due to shifting agriculture favored generalist rhesus macaques and folivorous colobines more than the more frugivorous gibbons and other macaque species. White (1994), studying the biomass of rainforest mammals at five different sites in Gabon, also concluded that differences in biomass of various species at different sites was related to differences in forest structure and composition.

Rainfall

Rainfall patterns may decide the type of vegetation formations (species diversity and tree density) at a given site, although soil, altitude and human interference all have additional influences (Bennett & Davies, 1994). The variation in rainfall patterns in areas of south and south-east Asia is enormous, ranging from as low as 450 mm in western India to above 4000 mm in some areas of Kalimantan and

Table 2.2. *Primate biomass, weather and tree density, diversity and productivity at 33 locations in south and south-east Asia*

Region	Site	No. species	Biomass	Rainfall	Leaf quality	No. dry months	Tree density	Tree diversity	Leaf production	Flower production	Fruit production
W Malaysia	Kuala Lompat	7	1234	2120	0.04	1	371	100	10	5	5
W Malaysia	Sg. Tekam	6	411	2207	0.21	1	547	–	4	2	4
W Malaysia	Lima Belas	5	2342	2558	0.04	1	514	–	4	3	6
W Malaysia	Pasoh	6	670	1878	0.04	1	800	–	4	–	4
W Malaysia	Lesong	6	590	2500	0.03	–	485	–	–	4	–
W Malaysia	Sg. Lelang	5	876	2554	0.21	–	538	–	4	2	3
W Malaysia	Kemasul	7	1281	1546	0.04	–	879	–	4	–	3
W Malaysia	Jengka	5	994	1500	0.03	–	550	–	–	–	–
W Malaysia	Paya Pasir	4	1219	1800	0.04	–	426	–	–	–	–
W Malaysia	Bt. Tarek	6	1953	1800	0.05	–	472	–	–	–	–
W Malaysia	Terengun	5	1236	2000	0.03	–	449	–	–	–	–
E Malaysia	Sepilok	5	251	2977	0.03	1	442	140	6	3	4
E Malaysia	Samunsan	7	605	2977	0.03	1	–	–	.	.	–
Kalimantan	Barito Ulu	8	120	4386	0.04	1	–	–	7	4	7
Kalimantan	Kutai	8	312	2177	0.14	1	–	–	1	3	4
Kalimantan	G. Palung	8	655	–	0.15	–	–	–	–	–	–
Kalimantan	Tg. Puting	7	1005	2400	0.11	0	857	–	6	3	5
Java	Cibodas	4	480	3215	0.08	1	218	–	6	2	3
Java	Pangandaran	3	350	3215	0.04	1	620	–	10	3	4
Sumatra	Paitan	2	370	4217	0.04	0	219	67	–	.	4
Sumatra	Ketambe	6	727	3229	0.05	1	475	–	7	–	5
NE India	Gumti, Trp	5	60	1500	0.59	3	280	22	7	2	5
NE India	Spjala, Trp	5	321	1800	0.31	3	250	41	7	2	7
NE India	Tripura	7	32	2200	0.18	3	260	145	7	2	6
NE India	Assam	5	135	2800	0.25	3	330	–	6	2	5
N India	Kanha	1	370	600	0.3	4	410	–	6	3	4
N India	Jodphur	1	162	450	0.21	6	440	–	3	4	4
S India	Tamilnadu	1	40	1400	0.06	4	560	–	6	3	4
S India	Kakachi	1	613	3084	0.04	2	640	115	7	4	4
Bangladesh	Madhupur	1	461	2280	0.25	3	310	–	3	2	5
Bangladesh	Rajkandi	2	115	2500	0.355	3	220	149	5	3	3
Bangladesh	Lawachara	1	9	2000	0.06	3	290	–	5	3	3
Bangladesh	Chunati	1	13	2200	0.24	5	152	–	3	5	4

Sumatra (Table 2.2). This variation in rainfall may account for variations in vegetation structure and composition, for example, low-height dry deciduous, low tree-species diversity, and low tree density scrub forests in drier regions of western India to tall, emergent evergreen, high tree-species diversity, and high tree density dipterocarp forests in high rainfall areas of south and south-east Asia. Primate community structure appears to vary accordingly, for example, a simple association between one macaque and one colobine species in western India; two macaques and one colobine in southern India; 2–3 macaques, 2–3 colobines and one ape in eastern India and Bangladesh; and two macaques, 4–5 colobines, two gibbons and the orangutans in areas of south-east Asia. Nevertheless, there is no clear relationship between rainfall and primate species diversity across sites

in this area (for all sites, rho = 0.15, P = 0.42; for south Asia alone, rho = 0.20, P = 0.54; for south-east Asia, rho = −0.15, P = 0.51; see also Reed & Fleagle, 1995).

All other things being equal, one might expect a positive linear relationship between rainfall and primate biomass at a given site. Indeed, primate biomass increases weakly with rainfall within the relatively dry region of south Asia, although not significantly so (rho = 0.15, P = 0.64, Fig. 2.4*a*). However, within the high rainfall region of south-east Asia (no site with less than 1500 mm of rain), biomass is significantly negatively related to rainfall (rho = −0.62, P = 0.004, Fig. 2.4*b*). For all sites together, rainfall and primate biomass are not correlated (rho = −0.003, P = 0.99). The negative correlation between high rainfall and high primate biomass at different sites in West Malaysia

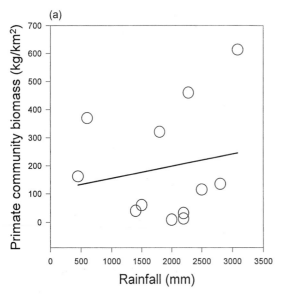

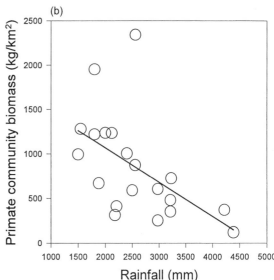

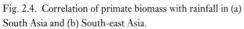

Fig. 2.4. Correlation of primate biomass with rainfall in (a) South Asia and (b) South-east Asia.

and Indonesia is contrary to our expectations and might possibly be due to human-induced changes in vegetation structure and composition (but see Janson & Chapman, chapter 14, this volume, for an alternative view).

Across a wide range of ecological conditions, there should be a significantly positive correlation between rainfall, tree species diversity, and tree density, but in the tropical rainforest sample used here, the relationship across all sites between tree density and rainfall was slightly negative (rho = −0.07, n = 29, P = 0.70). Negative correlations held also for south-east Asian sites (rho = −0.35, P = 0.16, n = 17) and south Asian sites (rho = −0.16, P = 0.62). An ANCOVA of tree density with rainfall as the covariate and region as the factor showed that the negative effect of rainfall on tree density approaches statistical significance (Fig. 2.5, $F[1,26]$ = 3.88, P = 0.06), and there is a highly significant difference in mean tree density between south and south-east Asia at comparable rainfall levels ($F[1,26]$ = 11.46, P = 0.002). Thus, it appears that in the moist tropics, greater rainfall may not lead to higher tree densities but to lower ones, perhaps because trees grow to a larger size and occupy more space individually in moister areas. Data on tree species diversity were not available from most of the sites, thus tree species diversity could not be tested for its association with primate biomass or rainfall in these sites.

Absence of a strong positive correlation between rainfall and primate biomass at almost all the sites could be indirectly related to various types of human disturbances (logging, illicit fellings, shifting cultivation, and so on), that have resulted in alterations in the positive balance between rainfall and tree density, affecting adversely the primate community structure, and hence the biomass. This is implied by a partial correlation run between biomass and rainfall for all sites, while controlling for tree density, which was positive (although weak, P = 0.79).

Tree density

In a typical tropical forest environment, a high value of tree density can be associated with high available food and cover to the primate species inhabiting those areas. It would mean more diverse primate communities in areas of high tree density and hence higher biomass, and vice versa. In many studies it has been found, however, that loss of forest cover may not influence primate densities (and hence biomass) appreciably unless favored food plants are also lost in the process (e.g., study on Tana red colobus by Marsh, 1986; effects of light and heavy logging on red colobus in Kibale, Skorupa, 1986).

We tested whether tree density at a site could be a good

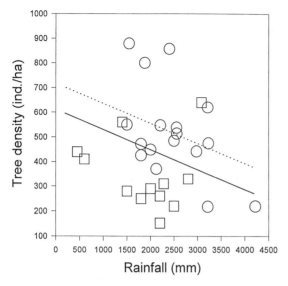

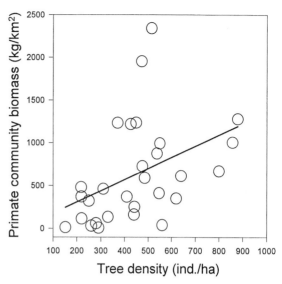

Fig. 2.5. Correlation of tree density with rainfall in south Asia (squares) and south-east Asia (circles).

Fig. 2.6. Correlation of primate biomass with tree density in south and south-east Asia.

predictor of the primate community biomass. Across all sites, primate biomass was positively correlated with tree density (Fig 2.6, rho = 0.52, n = 28, P = 0.004), but this relationship was not consistent within regions. A non-significant negative correlation was found in 11 sites from West Malaysia (rho = −0.25, n = 11, P = 0.47), but in the rest of south-east Asia and in south Asia, the relationships were positive, albeit not significantly so. When analyzing primate biomass using an ANCOVA with tree density as the covariate and regions as the factor, the effect of tree densities was positive, but not significant (P = 0.35).

The negative correlation in West Malaysia could possibly be related to the high representation of the family Dipterocarpaceae at most sites, and conversion of forest into monoculture plantations of commercial plant species. Both of these may contribute positively to high tree density, but they do not necessarily add to the available food resources. In these sites of West Malaysia, a higher tree density did not support high primate density, and hence the lower primate biomass. Marsh & Wilson (1981) suggested that low primate biomass in south-east Asia is due to the predominance of the family Dipterocarpaceae. Laidlaw (1994) noted lowest biomass in sites converted into *Acacia mangium* plantations (e.g., Bt. Tarek, West Malaysia).

The non-significant positive correlations between primate biomass and tree density found in other regions could also be influenced by poor plant species diversity. A habitat with low tree density, but high tree species diversity (e.g., forest edges), can support more diverse primate communities resulting in higher primate biomass, compared to the converse situation with high tree density of few dominant plant species (e.g., GWLS vs. SWLS, Tripura).

Forage quality

Forage quality refers to the nutrient richness of the food consumed. The forage quality of the food eaten by any group of primates may be determined by various chemical tests performed on different plant parts (young leaves, mature leaves, flower buds and flowers, ripe and unripe fruit, etc.). A few studies have determined that the choice of different plant species (and their different parts) is governed by chemical composition (Oates *et al.*, 1980; Davies *et al.*, 1988; Waterman *et al.*, 1988). These studies have shown that the main chemical constituents determining their selection are C (carbon), N (nitrogen), P (phosphorus), TP (total phenolics), TNC (total non-structural carbohydrates), CT (condensed tannin), and ADF (acid detergent fibre). The great variation in the type and concentration of secondary compounds among different species of plants and their parts provides primates the opportunity to avoid potentially toxic effects by eating plants or plant parts that do not contain large amounts of these chemicals (Waterman *et al.,* 1988).

Plants with maximum protein and minimum ADF contents are generally preferred. As a general rule young leaves have high ratios, whereas mature leaves have low ones. Seeds are rich in protein concentration and are, thus, a preferred diet of many primate groups. Most of the secondary forest and light-demanding colonizer species (e.g. Leguminosae) are highly preferred because they have high N, P and ash contents, along with lower contents of digestion inhibitors. Conversely, some plant species have large concentrations of toxic compounds, such as alkaloids, volatile oils, and resins, which prevent animals from consuming those species. Many evergreen species (e.g. Dipterocarpaceae and Rutaceae) invest much in the chemical defenses of their leaves and other plant parts, and thus present the primates with the problem of not choosing them, although most of these plant species might be present in high densities. By and large, different plant species of the same family present in different geographic areas will have the same chemical composition, unless the variations in the climatic, topographic and human disturbance factors on those sites are too severe. Thus, even in the absence of chemical analysis available for all the food plants, a thorough knowledge of the vegetation composition at a given site may be a good determinant of the forage quality for a given primate species.

A positive linear relationship is expected to exist between forage quality, primate diversity, and primate biomass. Forage quality was assessed as high ratios of protein (P) to acid detergent fibre (ADF). The values for P and ADF were calculated as an average of the individual values of young leaves of different tree species from three dominant tree families in each of the study sites. Contrary to our prediction, a statistically significant negative correlation between P/ADF ratio and primate biomass was found across all 33 sites (Fig. 2.7, rho = −0.48, $P = 0.005$), but within south Asia, the relationship was positive albeit not significant (rho = 0.16), whereas in south-east Asia it was weakly negative overall (rho = −0.04). Even within south-east Asia, different regions show different trends, with peninsular Malaysian sites showing a weak negative relationship (rho = −0.07, $P = 0.83$), while the insular south-east Asia had a positive one (rho = 0.42, $P = 0.23$).

Presence of both negative and positive correlations between forage quality and biomass at different sites in this

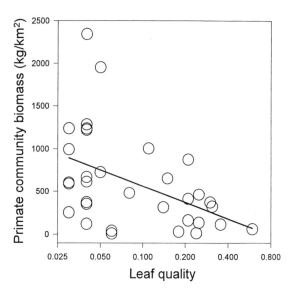

Fig. 2.7. Correlation of primate community biomass and leaf quality.

study suggests that forage quality is not always a predictor of primate biomass at a given site. The statistically significant negative correlation between these two parameters across all 33 sites may be accounted for mainly by vast differences in forage quality and biomass values between south-east Asia and south Asia. Primate community biomass is higher in south-east Asia compared with south Asia, but the reverse is true for forage quality. A part of the explanation for this counter-intuitive result is human disturbance. Most of the sites in south Asia are small compressed areas due to surrounding matrix of human habitation, in comparison with most of the sites in southeast Asia. These compressed sites have lower primate species diversity and smaller group size of existing primate populations, and hence lower primate biomass. Similarly, the odd negative correlation between forage quality and primate biomass in south-east Asia is in part caused by including the sites Sg. Tekam and Sg. Lalang, with very low primate biomass and relatively high forage quality (due to presence of large numbers of Leguminosae trees) relative to other sites in West Malaysia. The lower primate biomass in the latter two sites could be due to a high amount of human-induced disturbances (for example, commercial logging in Sg. Lalang, Laidlaw, 1994).

The weakness of the positive correlations between forage quality and biomass in different sites within regions could be accounted for by low variation in vegetation qual-ity within regions. For instance, in insular south-east Asia, there is a general paucity of high-quality plant families among the three dominant plant families at each site – a total absence of Leguminosae, and very poor representa-tion of species of Verbenaceae (only in two sites) and Moraceae (only one site). Conversely, in south Asia, most sites contained highly-favored food families (Legumi-nosae, Verbenaceae, Anacardiaceac, Moracea, Myrtaceae) among the top three plant families. Representation of these families in dominant form is mainly due to forestry plan-tations of a few native and exotic fast-growing plant species belonging to these families (e.g., in SWLS, GWLS in Tripura, Gupta, in press; in Lawachara, Ahsan, 1994), and large-scale human disturbance leading to the creation of secondary forests with light-demanding tree species, mainly from Leguminosae, Moraceae, and Myrtaceae (e.g., in Chunati, Ahsan, 1994).

Productivity

Productivity is represented here as the number of months of peak availability of three main plant parts constituting the primates' diet, namely, young leaves, fruit and flowers, in each study site. Data on plant productivity and primate biomass were analyzed from all available sites (only 25 out of the 33 sites included in this study), and also separately for south and south-east Asia.

Across all sites, we found a very weak negative correla-tion between biomass and number of months of young leaves (rho = –0.02), while very weak positive correlations occurred between biomass and months of flowers (rho = 0.03) and months of fruit (rho = 0.01). The same general results held within south Asia and south-east Asia sepa-rately, except that the correlation between biomass and months of flowers in south Asia was weakly negative instead of positive.

The total absence of any significant correlation between biomass and duration of peak production of three key plant parts within regions or across all sites could be due to considerable variations in their absolute availability to the primate communities, and also the capabilities of many primate groups to switch to different diets in the absence of any preferred plant part. Information on absolute vege-tation productivity is available only for those few cases where individual primate species have been studied in

sufficient detail. In nearly all those cases the productivity of different consumable food plant parts has been calcu-lated in terms of the percentage of individual trees with the given food plant part in each month. The quantity of pro-duce is generally determined by the percentage of the total individual tree crown area covered by the given plant part. This productivity could vary, however, due to local site factors (soil, rainfall, hill slope, and human disturbance may vary even sometimes within the same study area), so much so, that it may be erroneous to apply the data collected at one site to another.

Even if the productivity is the same in several sites, there may be differences in the synchrony of production across the annual cycle (see Seasonality below). In some areas, one particular plant part may be available on all the trees at the same time (uneven production on an annual cycle), while in another area its production may be spread evenly across the whole year.

Explicit knowledge of the soil types may also help relate the diet of individual primate species with specific plant parts. McKey (1978), studying *Colobus satanas* at two dif-ferent sites, hypothesized that there was a negative rela-tionship between the population density of monkeys and seed eating, and that both were dependent on the soil characteristics at two sites. Maisels *et al.* (1994), studying two sympatric colobines in Zaire, found that seed-eating is more common among colobines living in areas where soils are poor, and suggested that this is mediated through for-est composition, especially the abundance of legumes, and that the development of seed-eating results both from the high availability of nutrient-rich seeds and from the poor quality of mature tree foliage. Oates *et al.* (1990) acknowl-edged that soil conditions have a significant impact on tropical forest vegetation, but argued that it is unlikely that the soil fertility will have any simple, direct, causal link with the density and biomass of primate populations.

The preceding arguments emphasize the importance of gathering more accurate information on the soil types, geology, slopes (in hilly regions), precipitation, association of plant communities, and the nature and intensity of human disturbance, to understand resource availability for a given primate community. Unfortunately, even in the individual species studies available for different sites and species, these data are generally presented only as passing reference. If we had this information available from indi-vidual species' studies for many sites in sufficient detail, it might be possible to generalize resource productivity values to other primate communities with more or less similar climatic, edaphic, topographic and biotic factors.

Seasonality

Seasonality refers to the annual cycle of production in the different tree species making up the given primate community habitat. A less seasonal habitat may have productivity spread across the annual cycle, while in a more seasonal habitat the productivity is restricted to some parts of the annual cycle. Thus, a less seasonal habitat may have more availability of different food types, which may be able to support diverse primate communities, and hence higher biomass, as compared with a seasonal habitat. Many factors (climatic, edaphic, biotic, and so on) may affect this seasonality of production. The one most important factor governing productivity in the tropical vegetation belts harboring most primate communities could be the number of dry and wet months.

To test if variation in the number of dry months across sites accounts for the variation in primate community biomass, in the absence of data on actual forage quality and productivity, the number of dry months (rainfall <100 mm) was related to primate biomass of different primate communities. Across all sites ($n = 25$ with available data), a statistically-significant strong negative correlation (rho = -0.63, $P = 0.0007$) exists between the number of dry months and primate community biomass (Fig. 2.8), consistent with the view that habitats with less seasonal productivity support higher biomass. Similar, but much weaker, negative correlations were found across sites within each region, possibly because the values for number of dry months were very similar across sites within each region. Thus, seasonality may not be a very strong predictor of varying primate community biomass across sites within a climatic region. Clearly, the number of dry months is only a very rough estimate of seasonal food production, which should be perfected by more data on habitat structure.

CONCLUSION

Site-to-site variation in the expected relationships between biomass of primate communities and various habitat parameters at various selected sites in the south and southeast Asian region, may be accounted for by the existence of

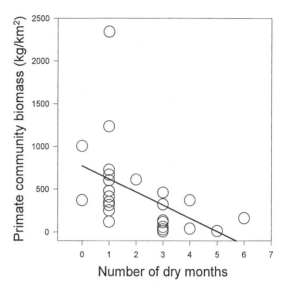

Fig. 2.8. Correlation of the number of dry months with primate community biomass in south and south-east Asia.

human-induced microhabitats for each of those individual primate communities. Depending upon the nature and intensity of disturbances, these microhabitats, regardless of their close geographic locations, vary greatly in various habitat parameters. These local variations may also help to account for variations in primate community structure at broader scales, as well as variation in the use of resources by different primate species at a given site. Thus, management strategies drawn for different primate communities, either based on Bourlière's (1985) 'ecological guilds' classification, or using a more traditional regional approach (Neotropical, African, south Asian, south-east Asian, and so on), or using dietary preferences as a basis (frugivores, folivores, faunivores, and so on), may prove erroneous if they do not account for human-induced disturbance.

A microhabitat approach to the study of primate communities is needed to be able to account for various different local variations, all of which have pronounced effects on the structure and function of communities inhabiting those sites. One main requirement of this model would be to conduct detailed long-term studies of each of the different components of the communities inhabiting those microhabitats. This approach will also help fill in the gaps in existing data even for those sites which have already been covered, but require additional information on a long-term basis (for example, all sites examined in this paper). The second most important aspect of this approach would be to standardize techniques for recording measurements of certain key habitat parameters (described elsewhere in this paper) in relation to primate biomass. A detailed vegetation classification for each different microhabitat is also needed. Focusing on microhabitats will also help in refining the vegetation classification for different sites beyond the general categories now available for most sites. This detailed vegetation classification may then be comparable in its various features (tree species diversity, tree density, productivity, seasonality, and forage quality) with other vegetation types in similar micro-habitats, thus providing much more accurate information on primate communities and their use of resources – a key issue in primate conservation.

ACKNOWLEDGMENTS

We thank John Fleagle, Kaye Reed and Charlie Janson for the invitation to participate in the Primate Communities Workshop in Madison, Wisconsin, and in this resulting volume. DJC's attendance was made possible by the generous grant from the Wenner-Gren Foundation for Anthropological Research. We are indebted to the editors, especially Charlie Janson, for their thorough help with completing our manuscript, and for their tolerance in coping with the postal problems between UK and eastern India. C. P. Heesy assisted with the tables.

REFERENCES

Ahsan, M. F. (1994). *Behavioural ecology of the hoolock gibbon* (Hylobates hoolock) *in Bangladesh*. Ph.D. thesis, University of Cambridge, UK.

Bennett, E. L. & Davies, A. G. (1994). The ecology of Asian colobines. In *Colobine Monkeys: their Ecology, Behaviour and Evolution*, ed. A. G. Davies & J. F. Oates, pp. 129–72. Cambridge: Cambridge University Press.

Bourlière, F. (1985). Primate communities; their structure and role in tropical ecosystems. *International Journal of Primatology*, **6**, 1–26.

Caldecott, J. O. (1986). An ecological and behavioural study of the pig-tailed macaque. *Contributions to Primatology*, **21**, 1–259.

Chivers, D. J. (1974). The siamang in Malaya: a field study

of a primate in tropical rain forest. *Contributions to Primatology*, **4**, 1–335.

Chivers, D. J., ed. (1980). *Malayan Forest Primates: Ten Years' Study in Tropical Rain Forest*. New York: Plenum Press.

Chivers, D. J., ed. (1997). *Project Barito Ulu: the Role of Animals in Forest Regeneration*. Jakarta: Ministry of Forestry.

Dasilva, G. L. (1994). Diet of *Colubus polykomos* on Tiwai Island: selection of food in relation to its seasonal abundance and nutritional qualtity. *International Journal of Primatology*, **15**, 1–26.

Davies, A. G. (1984). *An ecological study of the red leaf monkey* (Presbytis rubicunda) *in the dipterocarp forests of northern Borneo*. Ph.D. thesis, University of Cambridge, UK.

Davies, A. G., Bennett, E. L. & Waterman, P. G. (1988) Food selection by two South-east Asian colobine monkeys (*Presbytis rubicunda* and *Presbytis melalophos*) in relation to plant chemistry. *Biological Journal of the Linnean Society*, **34**, 33–56.

Emmons, L. H. (1984). Geographic variation in densities and diversities of non-flying mammals in Amazonia. *Biotropica*, **16**, 210–22.

Gupta, A. K. (1997). Importance of forestry plantations for conservation of Phayre's langur (*Trachypithecus phayrei*) in north-east India. *Tropical Biodiveristy*, **4**, 187–95.

Gupta, A. K. & Kumar, A. (1994). Feeding ecology and conservation of the Phayre's leaf monkey (*Presbytis phayrei*) in northeast India. *Biological Conservation*, **69**, 301–6.

Janzen, D. H. (1974). Tropical blackwater rivers, animals, and mast fruiting by the Dipterocarpaceae. *Biotropica*, **6**, 69–103.

Laidlaw, R. K. (1994). *The Virgin Jungle Reserves of Peninsular Malaysia: the ecology and dynamics of small protected areas in managed forest*. Ph.D. thesis, University of Cambridge, UK.

Maisels, F., Gautier-Hion, A. & Gautier, J. P. (1994). Diets of two sympatric colobines in Zaire: more evidence on seed eating in forest on poor soils. *International Journal of Primatology*, **15**, 681–701.

Marsh, C. W. (1986). A resurvey of Tana primates and their forest habitat. *Folia primatologica*, **7**, 72–81.

Marsh, C. W. & Wilson, W. L. (1981). *A Survey of Primates in Peninsular Malaysia*. Kuala Lumpur: University Kebangsaan Malaysia.

Mather, R. J. (1992). *A field study of hybrid gibbons in Central Kalimantan, Indonesia*. Ph.D. thesis, University of Cambridge, UK.

McKey, D. B. (1978). Soils, vegetation and seed eating by black colobus monkeys. In *The Ecology of Arboreal Folivores*, ed. G. G. Montgomery, pp 423–37. Washington, DC: Smithsonian Institute Press.

Medway, Lord (1972). Phenology of a tropical rain forest in Malaya. *Biological Journal of the Linnean Society*, **4**, 117–46.

Oates, J. F., Waterman, P. G. & Choo, G. M. (1980). Food selection by the South Indian leaf monkey. *Oecologica*, **45**, 45–56.

Oates, J. F., Whitesides, G. H., Davies, A. G., Waterman, P. G., Green, S. M., Dasilva, G. & Mole, S. (1990). Determinants of variation in tropical forest primate biomass: new evidence from W. Africa. *Ecology*, **71**, 328–43.

Reed, K E. & Fleagle, J. G. (1995). Geographic and climatic control of primate diversity. *Proceedings of the National Academy of Sciences USA*, **92**, 7874–6.

Primack, R. B. (1985). Comparative studies of fruits in wild and cultivated tress of chempedak (*Artocarpus integar*) and terap (*Artocarpus odoratissimus*) in Sarawak, East Malaysia, with additional information on the reproductive biology of the Moraceae in southeast Asia. *Malayan Nature Journal*, **39**, 1–39.

Skorupa, J. P. (1986). Responses of rainforest primates to selective logging in Kibale Forest, Uganda: a summary report. In *Primates: the Road to Self-sustaining Populations*, ed. K. Benirschke, pp. 57–70. New York: Springer-Verlag.

Waterman, P. G., Ross, J A. M., Bennett, E. L. & Davies, A. G. (1988). A comparison of the floristics and leaf chemistry of the tree flora in two Malaysian rain forests and the influence of leaf chemistry on populations of colobine monkeys in the Old World. *Biological Journal of the Linnean Society*, **34**, 1–32.

White, L. J. T. (1994). Biomass of rain-forest mammals in the Lope Reserve, Gabon. *Journal of Animal Ecology*, **63**, 499–512.

Whitesides, G. H., Oates, J. F., Green, S. M. & Kluberdanz, R. P. (1988). Estimating primate densities from transects in a West African rain-forest – a comparison of techniques. *Journal of Animal Ecology*, **57**, 345–67.

Whitmore, T. C. (1975). *Tropical Rain Forests of the Far East*. Oxford: Oxford University Press.

Wilson, C. C. & Wilson W. L. (1975). The influence of selective logging on primates and some other animals in East Kalimantan. *Folia primatologica*, **23**, 245–74.

Wyatt-Smith, J. (1953). Malayan forest types. *Malayan Nature Journal*, **7**, 45–55.

3 · Species coexistence, distribution, and environmental determinants of neotropical primate richness: A community-level zoogeographic analysis

CARLOS A. PERES AND CHARLES H. JANSON

INTRODUCTION

Primates comprise the most intensively investigated mammalian order, with the vast majority of genera having been subjected to at least one single-species ecological study. It is thus paradoxical that primates still remain so poorly studied at the level of species assemblages, from the perspective of either synecological studies restricted to a single site (Terborgh, 1983; Peres, 1993a) or comparisons of multiple sites within or between continents (Bourlière, 1985; Waser, 1986; Peres, 1997a). As a consequence, many questions related to the determinants of local species richness and community assembly rules are yet to be addressed to primates (but see Cowlishaw & Hacker, 1997; Ganzhorn, 1997). This volume, therefore, goes some way towards a timely paradigm shift from the traditional emphasis on behavioral studies of primate groups and individuals to macroecological studies of populations and communities.

In this chapter we present a large-scale analysis of the structure of New World primate communities from the northern tropical forest frontier of southern Mexico to the subtropics of northern Argentina and southernmost Brazil (Appendix 3.1). Our primary goal is to examine the biogeographic determinants of species assemblages at a spatial scale covering the entire platyrrhine primate radiation. We investigate the effects of latitudinal gradients, forest types, extent of forest cover, and amount of rainfall on the central question in mainstream community ecology of how many species can coexist within a given area. Despite the nearly universal pattern of higher species diversity at lower latitudinal regions (Schall & Pianka, 1978; Brown & Gibson, 1983; Stevens, 1989; Pagel et al., 1991; Ruggerio, 1994; Eeley & Lawes, this volume), this relationship has been largely tested using widely distributed taxa occurring

in both tropical and temperate habitats. In New World mammals from Alaska to Patagonia, for instance, 30 years of research have clearly demonstrated and attempted to explain the latitudinal gradient in species richness, which holds consistent even when the effects of continental area are removed (Kaufman, 1995). It remains unclear whether this pattern holds in a group of entirely arboreal tropical vertebrates which by definition are essentially confined to tropical habitats, such as New World primates. Yet neotropical forests, which span nearly 60 degrees latitude, are widely variable in terms of canopy structure, floristic diversity, degree of deciduousness, fruiting seasonality, soil fertility, hydrological cycles, and incidence of extreme climatic events. All of these factors may affect the number of sympatric primate species a forest can accommodate over both evolutionary and ecological time scales.

METHODS

This analysis is based on a comprehensive compilation of 185 forest sites of South and Central America at which the full composition of the local primate assemblage is known (Fig. 3.1; Appendix 3.2). We compiled data on the species composition, latitude, longitude, elevation above sea level (a.s.l.), main forest type, and total annual rainfall for each sampling site. These include a large number of community-wide censuses and single-species studies reporting the occurrence of all sympatric primate species, as well as a standardized series of community-wide surveys covering an entire geographic region (Struhsaker et al., 1975; Freese et al., 1982; Peres, 1997a). However, we excluded several short-term surveys involving low sampling effort, which may have failed to record rare species. We also excluded all hypothetical primate assemblages expected on the basis of

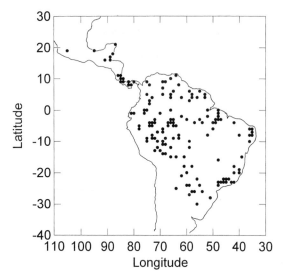

Fig. 3.1. Geographic locations of 185 forest sites of South and Central America at which the full primate species compostion is known. Solid dots may indicate more than one site if they are located less than 54 km apart.

presumed overlaps between species geographic distributions (Rylands, 1987; Ayres & Clutton-Brock, 1992) which largely fail to take into account small-scale range discontinuities due to habitat specificity and other ecological factors. Sampling sites included in this analysis were at least 10 km apart or separated by a major river, and are therefore considered to be demographically independent of one another.

We do not consider patterns of species abundance because population density estimates for all primate species co-occurring in a given area are limited to only two fifths of all neotropical sites at which the complete primate composition is known (see Peres, chapter 15, this volume). Species occurrences were simply described in terms of a 185×21 binary site-by-species matrix consisting of presences (1) or absences (0). Primate species richness is simply defined as the total sum of occurrences reported for a given site. In a few cases this may exclude species which may once have occurred but are no longer present at a given site. This may be the case for large-bodied ateline populations which can be driven to local extinction by overhunting in many parts of their former geographic range (Peres, 1990; in press a; chapter 15, this volume).

Because larger areas will contain a greater number of species, differences in the size of the sampling area used in different studies could present a problem for comparative studies of species richness. Although the spatial scale covered by the strip censuses, broad surveys, and focal-species studies incorporated here is not held constant, primate composition data are reported for areas of between 100–1000 ha. Primate species reported to occur within these areas are thus assumed to be sympatric despite species differences in habitat selection and home range size.

The geographic distribution of site localities incorporated in this analysis is non-random across the wide range of habitat types available for primates in the American tropics. Long-term study sites targeted by field primatologists often reflect a desirable primate community (usually in terms of high species richness and abundance), feasibility of physical access, and occurrence of poorly studied, rare, or endemic species, all of which may introduce biases to the overall spatial distribution of sites at the scale of entire continents. Indeed previous cross-site comparisons of primate communities (Bourlière, 1985; Waser, 1986; Terborgh & van Schaik, 1987; Oates et al., 1990; Fleagle & Reed, 1996) have largely focused on a handful of species-rich and intensively studied, long-term sites, and may suffer from sampling biases and statistical problems associated with small sample sizes. In this analysis, data from 48 of a total of 185 sites were obtained from the three largest programs of standardized primate censuses targeted to site localities in Amazonia and the Guianan Shields ranging from relatively accessible to extremely remote (Muckenhirn et al., 1975; Freese et al., 1982; Peres, 1997a; chapter 15, this volume, unpubl. data). Moreover, any site-selection biases which may affect the location of forest sites considered here are largely consistent throughout the neotropics, rendering the overall compilation reasonably comparable.

This community-level analysis of platyrrhine primates is based on a functional classification of 21 groups of species (hereafter, ecospecies), based on body size, diet, group size, foraging behavior, prey capture techniques, and vertical use of the forest, rather than on species as defined taxonomically (Table 3.1, Appendix 3.1 and Peres, 1997a). Ecospecies corresponded to a single species or a few ecologically equivalent (and mutually exclusive) congeners, usually representing parapatric replacements across sharp biogeographical boundaries. Because of substantial within-genus differences in species ecology, tamarins (Saguinus spp.), titi monkeys (Callicebus spp.), and capuchin

Table 3.1 *Primate ecospecies classification used in this analysis*

Ecospecies (and codes)	Taxonomic species included
1. Pygmy marmosets (Cb)	*Cebuella pygmaea*
2. Marmosets (Cx)	*Callithrix* spp.
3. Saddle-back tamarins (Sf)	*Saguinus fusciollis, S. nigricollis, S. tripartitus*
4. Moustached tamarins (Sx)	*Saguinus mystax, S. labiatus, S. imperator*
5. Midas and bare-faced tamarins (Sm)	*Saguinus midas, S. bicolor, S. inustus, S. geoffroyi, S. oedipus, S. leucopus*
6. Lion tamarins (Le)	*Leontopithecus* spp.
7. Goeldi's monkeys (Cg)	*Callimico goeldii*
8. Squirrel monkeys (Sa)	*Saimiri* spp.
9. Owl monkeys (Ao)	*Aotus* spp.
10. Dusky titi monkeys (Cm)	*Callicebus spp.* (except *C. torquatus*)
11. Collared titi monkeys (Ct)	*Callicebus torquatus*
12. Saki monkeys (Pi)	*Pithecia* spp.
13. Bearded saki monkeys (Ch)	*Chiropotes* spp.
14. Brown capuchins (Ca)	*Cebus apella, C. xanthosternos*
15. White-fronted capuchins (Cf)	*Cebus albifrons*
16. Wedge-capped capuchins (Co)	*Cebus olivaceous, C. capucinus, C. kaapori*
17. Uakaries (Cj)	*Cacajao* spp.
18. Howler monkeys (Al)	*Alouatta* spp.
19. Woolly monkeys (La)	*Lagothrix* spp.
20. Spider monkeys (At)	*Ateles* spp.
21. Woolly spider monkeys (Br)	*Brachyteles* spp.

monkeys (*Cebus* spp.) included more than one, and at most three, ecospecies. In extant neotropical primate assemblages, these are the only polytypic genera which can be represented by more than one sympatric congener coexisting in the same area. All other genera are invariably represented at a given site by only one species regardless of their phenotypic diversification. Species belonging to these genera were thus considered as part of the same ecospecies despite some ecological differences between congeners (Rylands, 1993; Ferrari *et al.*, 1996; Peres, 1993*b*).

The geographic range (in 1000 km^2) of all ecospecies was measured with a HAFF® digital planimeter (accurate to 0.01 cm^2) from distribution maps of individual species and/or genera (such as in Wolfheim, 1983; or the many taxonomic treatises published by Hershkovitz between 1949 and 1988), with additions from more recent distributional data (Ayres & Clutton-Brock, 1992; Pinto & Rylands, 1997; Peres, 1993*c*; Peres, unpubl. data), and collections at the Museu Goeldi, Belém. We also measured the extent of neotropical forest habitats presumed to be available to primates within 5-degree latitudinal bands covering the entire geographic distribution of New World primates. This was done with a digital planimeter using vegetation maps (1:10^6) showing the approximate sixteenth-century distribution of forest cover in South and Central America.

Although this approach disregards contemporary differences in forest habitat conversion across the American tropics, the underlying assumption is that the evolution of neotropical forest faunas is more closely tied to pre-Colombian rather than present-day habitat availability.

Sampling sites and patterns of species distribution

Neotropical primate communities contain an average of 6.0 ± 3.6 species (range = 1–14, $n = 185$), but depauperate communities consisting of only 1–4 species are far more common (42% of the sites) than species-rich communities containing 10–14 species (13% of the sites). Highly diverse communities are thus relatively rare throughout South and Central America, with species richness rapidly tapering off beyond a total of 12 sympatric species (Fig. 3.2). This is

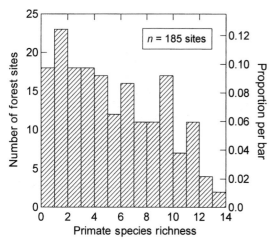

Fig. 3.2. Frequency distribution of neotropical forest sites in terms of their primate species richness.

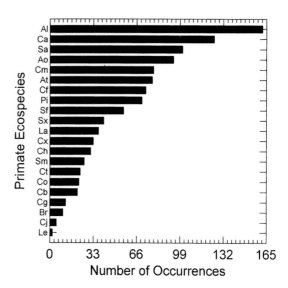

Fig. 3.3. Differential occurrence of neotropical primate ecospecies at 185 forest sites from southern Mexico to northern Argentina. For definition of ecospecies classification and codes see text and Table 3.1.

the case despite the large number of species-rich forest localities targeted by primate censuses in recent years, particularly in western Amazonia (see Appendix 3.2). Indeed the expected sampling bias towards more accessible neotropical forest sites is no longer the case: data on the primate composition are now available for 110 forest sites in Amazonia and the Guianan Shield (59% of all sites, see Appendix 3.2), clearly the most remote forest regions of South America (Fig. 3.1).

Neotropical primate species diverge widely in their geographic range size, habitat specificity, and local population size (Eisenberg, 1979; Ayres & Clutton-Brock, 1992; Peres, 1993a, 1997a), the three main spatial scales defining the ecological distribution of a species (Gaston, 1994; Brown, 1995). It is not surprising, therefore, that primate ecospecies vary nearly 100-fold in their rate of occurrence at forest sites sampled to date (Fig. 3.3). These range from the nearly omnipresent howler monkey, which occurs in dry, moist, and wet tropical forests from southern Mexico to northern Argentina, and at elevations as high as 3200 m a.s.l. in the central Andes (Peres, 1997b), to the highly endemic lion tamarins, which are presently restricted to a small fraction of their former geographic range (Pinto & Rylands, 1997). Along the continuum from the most restricted to the most ubiquitous species, only four ecospecies could be considered as truly widespread and occurring in at least 50% of the communities: *Alouatta*, *Cebus apella*, *Saimiri*, and *Aotus*. With the exception of

Saimiri, these ecospecies occur from Mesoamerica or northern South America to the southern extremes of primate distribution in northern Argentina. Surprisingly, other widely-distributed taxa occurring both within and outside the Amazon basin, such as *Callicebus moloch/ personatus* and *Ateles*, failed to attain this cut-off point, and were only marginally more ubiquitous than several other increasingly rarer taxa.

As expected, therefore, geographic range size was a major determinant of the incidence probability of different ecospecies explaining 87% of the variation in the number of sites at which they occurred ($r = 0.93$, $P < 0.001$, $n = 21$; Fig. 3.4). Contemporary geographic range sizes of neotropical primate ecospecies are widely variable, ranging from 0.02×10^6 km^2 in the case of *Leontopithecus* spp., to 11.07×10^6 km^2 in the case of *Cebus apella* ssp. Wide-ranging ecospecies were thus obviously more likely to be present in a larger number of sites. Although some large-bodied ecospecies can be remarkably widespread (e.g., *Alouatta* spp.), and some small-bodied ones can be highly restricted (e.g., *Callimico goeldii*), body mass is a very poor predictor of the geographic range size of an ecospecies (on log$_{10}$-transformed values: $r^2 = 0.005$, $P = 0.754$, $n = 21$). Ecospecies range size is thus roughly held constant with increasing body size,

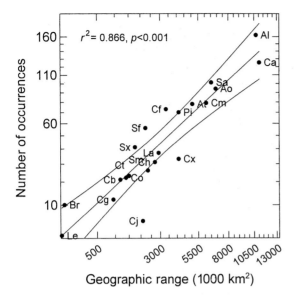

Fig 3.4. Relationship between the geographic range size of neotropical primate ecospecies and the number of sampling sites at which they occurred. For definition of ecospecies codes see Table 3.1.

and both some of the smallest (*Cebuella*, *Callimico*, and *Leontopithecus*) and the largest platyrrhine primate genera (*Brachyteles*) retain the smallest geographic ranges.

The geography of species diversity

Although primate communities become more impoverished with increasing distances from the equator, many low-latitude sites contain only a small fraction of the regional pool of species available, and their local primate diversity can be rivalled or surpassed by those of some high-latitude sites (Figs. 3.5 and 3.6). Indeed the local species packing at low-latitude forests is highly variable, ranging from as few as three to as many as 14 sympatric species. To a large extent this can be attributed to differences in forest types and the habitat diversity of the census area. Western Amazonian seasonally flooded (várzea) forests, for example, contain between only one third and one half of the species co-occurring in adjacent upland (terra firme) forests (Peres, 1993a, 1997a). This is also the case of eastern Amazonian tidal gallery forests which may contain only 2–3 of a regional pool of 6–7 species (Peres, 1989; Peres & Johns, 1992).

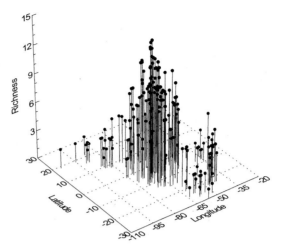

Fig. 3.5. Geographic variation in the local richness of primate species occurring at 185 neotropical forest sites.

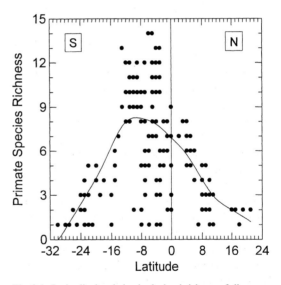

Fig. 3.6. Latitudinal variation in the local richness of all neotropical primate communities documented to date. The line of best-fit indicates the distance weighed least square (WLS) when 100% of the data are used to smooth each value of the curve.

The variation in local primate richness within narrow latitudinal zones thus appears to be largely a function of site differences within and across different habitats (local or α-diversity, and between-habitat or β-diversity, respectively), and to a lesser extent changes in the source biota (regional or γ-diversity). In low-latitude neotropical forests,

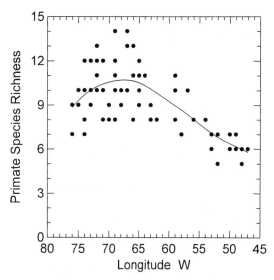

Fig. 3.7. Longtitudinal variation in species richness of Amazonian and Guianan primate communities. The line of best-fit indicates the WLS when 100% of the data are used to smooth each value of the curve.

for example, there is a marked east-west increment in overall mammalian γ-diversity (Voss & Emmons, 1996), which is correlated with primate diversity on large spatial scales (Ayres, 1986; Peres, in press, *b*; Emmons, chapter 10, this volume). Interfluvial regions of western Amazonia, particularly west of the Rio Madeira and Rio Negro, are substantially richer than those of eastern Amazonia. This can be seen in the overall longitudinal gradient of primate richness in Amazonia and the Guianan Shield, which tends to be far richer in western lowland Amazonia, peaking at about 65–75° W (Fig. 3.7).

It is thus undeniable that western Amazonia, particularly in western Brazil and south-eastern Peru, is home to some of the richest primate communities anywhere. As many as 14 sympatric species can be found in terra firme forests of the west bank of the Juruá river, Brazil (Peres, 1988; 1990; 1997*a*). Several other sites containing 12–13 species have also been documented in many parts of western Brazilian Amazonia and the eastern foothills of the Peruvian Andes (Freese *et al.*, 1982; Janson & Emmons, 1990; Puertas & Bodmer, 1993; C. Peres, unpubl. data). From the western Amazonian lowland forests, an increasingly greater number of genera and species appear to drop out towards more seasonal forests either south of this

region (e.g., Bolivian semi-deciduous forests, Paraguayan chaco, the semi-deciduous "cerradão" forests in the southern fringes of Amazonia, and the cerrado scrublands/gallery forest matrix of central Brazil), or north of the region (e.g., Grand Savanna forest enclaves of northern Brazil and southern Venezuela, the Venezuelan and the Colombian llanos, the Orinoco Delta swamp forests, the northern Andean Chocó/Darién moist forests, and the central American tropical dry and moist forests).

Primate communities in the other major closed-canopy tropical forest biome of South America – the Brazilian Atlantic forest – can be surprisingly rich, despite the limited number of genera occurring in the region. Atlantic forest sites contain as few as one and as many as six sympatric species, including representatives of either one of the two genera endemic to this region, *Leontopithecus* and *Brachyteles* (Rylands *et al.*, 1996). Several Amazonian species are represented in the Atlantic forest by ecological counterparts either as congeners (*Callithrix, Callicebus, Cebus, Alouatta*), or as members of different genera serving similar ecological roles (*Saguinus* and *Leontopithecus*, *Lagothrix* and *Brachyteles*). Several Amazonian ecospecies are, however, conspicuously absent in the Atlantic forest, including pygmy marmosets, squirrel monkeys, owl monkeys, the non-*apella* capuchin monkeys, and all pitheciines (sakis, bearded sakis, and uakaries). These functional groups are not obviously replaced in eastern and south-eastern Brazil by any ecological counterparts, which suggest any of several possibilities including lack of colonization opportunities; preempted niche spaces filled by ecologically similar species; or a narrower spectrum of available resources. In addition, whilst no sympatric congeners have ever been reported for any Atlantic forest site, more than one species of *Callicebus, Saguinus*, and *Cebus* often co-occur in many Amazonian and Guianan forests, particularly those in the western part of the basin. Other accounts of the distribution and zoogeography of primates within smaller regions of South and Central America can be found elsewhere (Kinzey, 1982; Ayres & Clutton-Brock, 1992; Ferrari & Lopes, 1996; Norconk *et al.*, 1996; Rylands *et al.*, 1996; Peres, 1997*a*).

Environmental determinants of primate richness

Annual rainfall throughout the neotropics is (i) positively correlated with floristic diversity (Gentry, 1988), (ii) almost

Table 3.2. *Pearson correlation matrix between environmental variables and local primate richness at 185 neotropical forest sites*

	Primate richness	Rainfall	Forest cover	Elevation	Latitude	Longitude
Rainfall	0.391[a]	1.000				
Forest cover	0.482[a]	0.422[a]	1.000			
Elevation	−0.111	0.005	−0.319[a]	1.000		
Latitude	−0.443[a]	−0.446[a]	−0.952[a]	0.302[a]	1.000	
Longitude	−0.205	−0.368[a]	−0.084	0.090	0.183	1.000

Note: [a]Significant Bonferroni probabilities ($P > 0.05$); all correlations are based on the raw data.

certainly negatively correlated with resource seasonality, and (iii) could provide a good univariate predictor of primate richness on a large scale. Indeed rainfall can be a good predictor of primate diversity in relatively dry land masses of the paleotropics and more seasonal regions of the neotropics (e.g., Madagascar: Ganzhorn *et al.*, 1997; Venezuela and the Guianas: Eisenberg, 1979). However, rainfall is not necessarily limiting in much of the primate habitat in South and Central America, which is dominated by evergreen (moist or wet) forests, rather than deciduous (dry) forests. It is not surprising, therefore, that rainfall is only weakly correlated with primate richness when all forest sites are considered ($r = 0.391$, $n = 185$, $P < 0.001$; see Table 3.2). As expected, this correlation becomes gradually stronger as rainfall becomes more limiting within drier neotropical ecoregions as shown by the exclusion of all relatively wet forests receiving >2500 mm/year ($r = 0.443$, $n = 141$, $P < 0.001$), or moist forests receiving >2000 mm/year ($r = 0.573$, $n = 76$, $P < 0.001$). This result is far less than that given by Reed & Fleagle (1995), who provide a r^2-value of 0.67 for a linear relationship between rainfall and primate richness on the basis of only 14 South American sites, only four of which received >2000 mm/year. However, their linear fit to this relationship appears to be best described by a sigmoidal or exponential curve (see Fig. 2. in Reed & Fleagle, 1995), again suggesting that primate richness does not appreciably increase beyond a certain level of rainfall. A similar conclusion can be drawn from a recent analysis by Kay *et al.* (1997) in which a clear positive relationship between local primate richness and rainfall holds true only across South American sites receiving less than 2500 mm/year (64 of 96 sites fitted to a regression line, see Fig. 2a in Kay *et al.*, 1997). Relatively few neotropical sites receiving >3000 mm/year have been

sampled to date, so we cannot confirm whether primate richness appears to decline again in very rainy sites because of a reduction in plant productivity associated with lower soil nutrients (Kay *et al.*, 1997) or other reasons. Several extremely wet high-elevation sites including cloud forests both along the Andes and the Atlantic forest, for example, might sustain a low primate richness because of their low tree diversity, unfavorable floristic composition, and occurrence of extreme weather events at supra-annual cycles, rather than high annual rainfall *per se*. Nevertheless beyond the 2500 mm/year cut off value, other large-scale environmental variables could easily override the importance of rainfall on primate community structure, so we now turn to these variables.

Because of the gradual increase in species richness towards the equator in a number of phylogenetically independent taxa (Schall & Pianka, 1978; Stevens, 1989), we expected latitude to be one of the most important predictors of neotropical primate diversity. This would be consistent with the geographical pattern of species packing found in mammalian faunas right across several continents both in the New (Ruggerio, 1994; Kaufman, 1995) and Old World (Pagel *et al.*, 1991; Cowlishaw & Hacker, 1997; Eeley & Lawes, chapter 12, this volume; but see Smith *et al.*, 1994).

Although the number of primate species found at any one site was clearly negatively correlated with latitude ($r = -0.443$, $n = 185$, $P < 0.001$), distance from the equator *per se* was not necessarily a good predictor of neotropical primate richness. Extent of forest cover within 5-degree latitudinal bands, for example, provided an even better correlate of species richness. In a stepwise multiple regression, we thus examined the combined effects of several environmental variables on primate richness including total rainfall, extent of forest cover, forest elevation, latitude,

Table 3.3. *Stepwise multiple regression considering the effects of total rainfall, extent of forest cover within 5-degree latitudinal bands, elevation, latitude, and longitude on neotropical primate species richness*[a]

Effect	Coefficient	SE	Standard coefficient	d.f.	*F*-value	*P*
Terms retained						
Total fainfall	0.799	0.124	0.391	1	41.3	0.000
Forest cover	0.483	0.077	0.557	1	39.6	0.000
Elevation	0.122	0.035	0.198	1	12.1	0.001
Latitude	0.172	0.083	0.188	1	4.3	0.04
Terms removed[b]	Part. Corr.					
Longitude	0.116	–	–	1	2.5	0.118

Note: [a]All variables were $\log_{10}$-transformed; r = 0.671; r^2 = 0.451; n = 185 sites.
[b]Backward elimination procedure at a *p*-to-remove value = 0.05.

and longitude, even though it is not obvious how the latter variable could affect species richness. Continental longitude was the only one of these variables that could be dropped from the regression equation without appreciably decreasing the coefficient of determination of primate species richness (Table 3.3). The number of primate species found in the neotropics was in fact best predicted by extent of forest cover and total rainfall, and to a considerably lesser extent by elevation and latitude ($r = 0.671$). This relationship is described by log(richness) = −3.309 + 0.483 × log(forest area) + 0.799 × log(rainfall) + 0.122 × log(forest elevation) + 0.172 × log(latitude).

The importance of forest area was confirmed when we compared the mean and maximum primate richness within 5-degree latitudinal zones with that expected from the extent of forest cover in the same zones, on the assumption that the local richness of hypothetically constructed primate communities is restricted by the same ecological parameters as those which must operate in "real-world" communities. Both mean and maximum primate richness observed across different latitudinal zones was closely matched by those predicted by the extent of forest cover in the same zones (mean richness: paired *t*-test, $t = 0.001$, Bonferroni adjusted $P = 1.0$; maximum richness: $t = 0.426$, $P = 0.68$; Fig. 3.8). In other words, neotropical primate communities are most diverse where primate forest habitat is most extensive, and presumably most stable in evolutionary time.

To a large degree this may explain why primate richness peaks at around 6° S, and decidedly fails to decline

symmetrically on either side of the equator (Fig. 3.6). The richest New World primate communities are thus found in a region south of the Amazon between 3 and 13° S, where the longitudinal extension of neotropical forest cover is greatest. The peak in neotropical primate diversity may ultimately reflect the very shape of South America, which is widest at low latitudes south of the equator. However, a large proportion of the variation in the availability of primate habitat across different latitudinal zones can ultimately be attributed to macroclimatic consequences of distance from the equator (e.g., strength of the dry season; extremes in cold weather; intensity and periodicity of forests), and to a large degree these variables should be intercorrelated.

Forest types and primate community similarity

The profound influence of habitat type on primate community structure can be seen in the gradual decline in species richness from the relatively aseasonal terra firme and alluvial forests of Amazonia to more seasonal forests towards either the northern or southern frontier of the neotropical primate distribution (Table 3.4). Forest types, however, essentially incorporate several environmental variables operating on different spatial scales, including rainfall, hydrological regimes, soil types, elevation, latitude, and distance from the coast. Primate richness, however, appears to be ultimately determined by both ecological and historical factors associated with regional geomorphology and climate. It is therefore not surprising that similar forest types within major geographic regions tend to share a

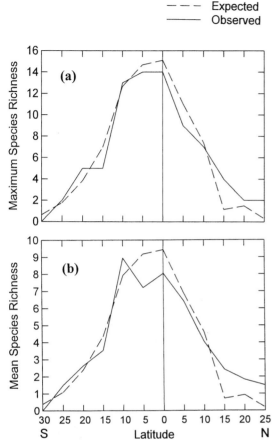

- - - Expected
——— Observed

Fig. 3.8. Observed patterns of (a) maximum and (b) mean primate species richness in the neotropics compared to those expected on the basis of the extent of forest habitat available for primates across different 5-degree latitudinal zones.

similar primate species composition. A cluster analysis resolving the degree of community similarity among all forest sites[1] shows that they are largely clustered according to forest type and geographic regions, despite the fact that patterns of species abundance were not considered[2] (Fig. 3.9). Given the vagaries of species colonization and extinction at different sites, there were relatively few

[1] Forest sites were grouped or separated according to the average linkage method of hierarchical clustering based on the pairwise Euclidean distances derived from the binary matrix describing the primate species composition at all sites.
[2] For an analysis of how forest types affect primate community biomass within a given geographic region see Peres, chapter 15, this volume.

exceptions to the expected biogeographic groupings of sites (approximately a 10% mismatch from that expected on the basis of ecoregions alone). Moreover, forest types in disparately different geographic regions can share a similar, if not the same, species composition quite irrespective of their history.

In general, there is a tendency for gradual community impoverishment associated with the deletion of small-bodied taxa (e.g., callitrichines) and the persistence of at least one large-bodied (ateline) species, at extreme latitudinal or elevational zones associated with colder and/or drier climates. Such predictable patterns of species deletion is illustrated, for example, by the convergence in ecospecies composition occurring in montane forests of Central America and the Brazilian Atlantic forests. This can also be seen in the species composition in environments characterized by highly seasonal rainfall such as the deciduous forests of the Paraguayan chaco and the Venezuelan Llanos, which also share similar sets of species. The ubiquitous *Cebus–Alouatta* association, often to the exclusion of all other species, is typical of high-latitude and extremely seasonal forests subjected to similar ecological scenarios but entirely different histories. Within the distribution of squirrel monkeys, the *Saimiri–Alouatta* association also provides a stable community signature for the relatively impoverished gallery, riparian, tidal, and flooded forests wherever these habitats might occur.

Indeed forest type appears to be the most decisive factor in explaining the variation of primate species richness throughout the neotropics. This was particularly the case when all key environmental variables retained in a stepwise multiple regression (see Table 3.3) were entered as covariates of forest type in an analysis of variance. The combination of these variables and forest types thus accounted for 68% of the overall variation in the number of primate species occurring throughout the neotropics (Table 3.5).

DISCUSSION

Compared to Old World primate assemblages, those in the New World are noticeably distinctive in their (i) bias towards small-bodied insectivores and frugivores, (ii) conspicuous absence of terrestrial species and more than one nocturnal species, and (iii) paucity of large-bodied folivores (Bourlière, 1985; Terborgh & van Schaik, 1987;

Table 3.4. *Number of sympatric primate species co-occurring within different major forest types found in the neotropics*

Forest type	Mean	SE	N
Amazonian terra firme forest	8.8	0.3	67
Amazonian alluvial forest	10.1	0.7	11
Amazonian flooded forest	4.8	0.8	9
Evergreen lowland Atlantic forest	3.3	0.7	11
Evergreen lowland Mesoamerican forest	2.3	0.8	9
Montane/semi-montane evergreen forest	4.6	0.7	13
Semi-deciduous seasonal forest	5.0	0.4	37
Deciduous seasonal forest	2.2	0.5	22
Gallery forest (in a savanna matrix)	2.2	1.1	5
Araucaria pine forest	1.0	2.4	1

Fleagle & Reed, 1996). Unlike the paleotropics, members of polytypic genera coexisting at the same site are also relatively rare in the neotropics. At most only two sympatric congeners can be found anywhere (in one callitrichine and two cebid genera), and even these only in Amazonia and the Guianan Shield. Moreover, even the most diversified platyrrhine genera are always represented by only one species in the case of *Callithrix*, and at most two species in the case of *Saguinus*. In contrast, as many as six species of guenons (*Cercopithecus* spp.) can be found coexisting at a single site in mainland Africa (Gautier-Hion, 1988), and to a lesser degree this is the case for *Hapalemur* in Madagascar (P. Wright, pers. comm.). This would seem to indicate that different neotropical primate genera represent the best taxonomic denominator of ecological resolution. Indeed the fact that ecological space in this adaptive radiation is best resolved at the level of genus supports recent molecular clock estimates generally suggesting considerably longer divergence times between platyrrhine genera, compared to primate radiations in the Old World (Porter *et al.*, 1997; see Fleagle & Reed, chapter 6, this volume).

Latitudinal gradients revisited

The latitudinal variation in species richness in New World primates essentially confirms the nearly universal pattern already shown for other mammals (Brown & Gibson, 1983; Stevens, 1989; Pagel *et al.*, 1991; Kaufman, 1995). Several factors have been proposed to explain latitudinal gradients in species richness, including the latitudinal variation in habitat seasonality, extremes of temperatures, net primary productivity (energy levels and evapotranspiration), photoperiodicity, habitat and floristic diversity, evolutionary lag since the end of the last Ice age, competition, and levels of parasite diversity. In neotropical primates, the reduction in species richness at high latitudinal and elevational zones could be explained on the basis of greater resource seasonality, lower floristic diversity, and the substantial decrease in available area of tropical forest at increasing distances from the equator. Other large-scale geographic patterns emerging from this analysis of neotropical primate communities, however, cannot be easily explained by contemporary ecological factors. For example, neither the data considered here nor the recent fossil record show that primate assemblages in the more aseasonal northern regions of the Atlantic forest were ever far richer than those in south-eastern Brazil. Pre-Colombian extinctions of at least two Atlantic forest platyrrhine species weighing as much as 20–23 kg (Cartelle, 1993; Hartwig, 1995; Hartwig & Cartelle, 1996) were most likely caused by natural habitat changes and selective overkill by paleoindians, but there is no reason to assume that these processes were restricted to the "northern" Atlantic forest. Indeed, the north–south Atlantic forest corridor spanning over 20 degrees latitude essentially shows a relatively homogeneous primate fauna (Kinzey, 1982; Rylands *et al.*, 1996). Those genera now endemic to small areas in south-eastern Brazil (i.e., *Leontopithecus* and *Brachyteles*) were once far more widespread, occurring well beyond the northern limits of their present distribution (Cartelle, 1993; Rylands *et al.*, 1996). Diversity gradients in the Atlantic forest primate fauna, therefore, do not necessarily reflect those of the overall plant community, which tends to be far richer at lower latitudes (M. Tabarelli, unpubl. data.).

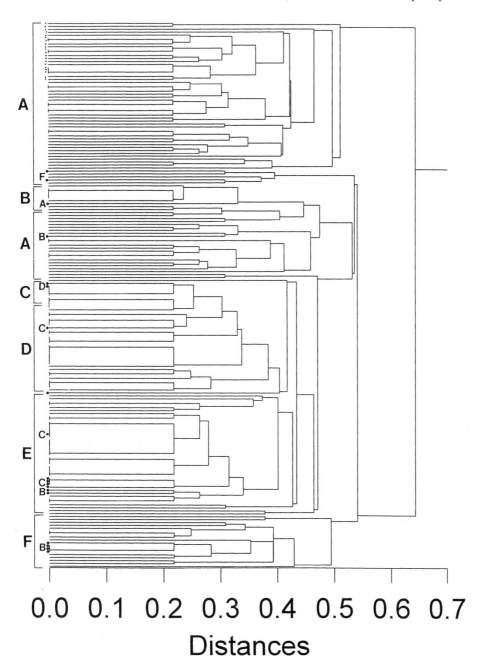

Fig. 3.9. Cluster analysis of ecospecies similarity between 185 neotropical primate communities. Community similarity decreases with increasing Euclidean distances as resolved on the basis of the average linkage method. A cluster analysis based on Jaccard coefficients of similarity produced similar results. Forest sites were grouped according to biogeographic regions: (A) western and central Amazonia; (B) eastern Amazonia; (C) Paraguayan chaco, southern Bolivia, northern Argentina, and southern Brazil; (D) Atlantic forest and central Brazil; (E) Central America, north-western Columbia, and western Ecuador; and (F) the Guianan Shields. Solid dots labeled by small letters represent exceptions (mismatches) to major site groupings (large letters).

Table 3.5. *Analysis of variance considering the effects of forest types and other key environmental covariates on neotropical primate richness[a]*

Source	Sum-of-squares	d.f.	Mean-square	F-ratio	P
Forest type	4.536	9	0.504	13.226	0.000
Rainfall	0.559	1	0.559	14.707	0.016
Elevation	0.502	1	0.502	13.217	0.000
Extent of forest cover	0.993	1	0.993	26.136	0.000
Latitude	0.546	1	0.546	14.361	0.000
Error	6.534	172	0.038		

Note: [a] On $\log_{10}$-transformed values; $r = 0.822$, $r^2 = 0.676$, $n = 185$.

Other factors partly correlated with latitude, including forest types, extent of forest cover, and rainfall within latitudinal zones, were important predictors of local primate diversity. However, beyond 2000 mm of rain per year, primate richness was at best weakly correlated with rainfall in our data. This low correlation is consistent with the curvilinear effect of rainfall on primate richness found in a previous study by Kay *et al.* (1997), which they suggest is related to reduced productivity in very wet sites. However, the causal mechanisms for this relationship remain unclear since relatively few forest sites in our data set were clearly very wet (only 15 of 185 sites with >3000 mm/year), and those are confounded by other factors such as forest elevation. In any case, neotropical primates are not the best taxonomic group to examine this relationship because they are by definition restricted to a rather narrow range of terrestrial primary productivity.

The conclusion that species assemblages were found to be richer in regions containing more forest habitat supports patterns uncovered for other taxonomic groups (e.g., birds: Rahbek, 1997; Blackburn & Gaston, 1997; bats: Findley, 1993) as well as entire primate radiations on different continents (Reed & Fleagle, 1995). It remains unclear, however, whether this reflects classic theory on species–area relationships (MacArthur & Wilson, 1967; MacArthur, 1972; Cody, 1975), or other environmental gradients correlated with extent and stability of tropical forest cover over evolutionary time. In essence, this highlights the age-old question of whether low-latitude regions are more species-rich because tropical (or equatorial) stability favors higher rates of speciation or lower rates of extinction (Jablonski, 1993; Roy *et al.*, 1998). Evolutionary opportunities for primates occurring in low-latitude neotropical regions could result from key allopatric mechanisms of speciation such as riverine barriers (e.g., Ayres & Clutton-Brock, 1992; Peres *et al.*, 1996) which may not be available at higher latitudes (Kinzey, 1982). However, such classic trajectories of allopatric speciation rarely break into novel ecological opportunities, and need not assure a larger primate niche space in existing assemblages of potential competitors. In contrast, Amazonian primates have had numerous dispersal opportunities to colonize other forest biomes during the last 2 million years. Such migration must have occurred during more favorable climatic periods, with at least some species filtering through the northeastern Brazilian connection between eastern Amazonia and the northern Atlantic forest, or via the central Brazilian habitat corridors such as gallery forests and patches of mesic vegetation within the matrix of *cerrado* scrublands. It is clear that not all of those dispersal events were successful as can be seen in several Amazonian ecospecies conspicuously missing from forests of Central America and eastern, south-eastern and southern Brazil. The lower local species richness in those regions could result from greater probability of local extinctions due to limited habitat availability for an arboreal group of tropical organisms and perhaps a less stable forest cover in geological time. In high-latitude regions, such evolutionary factors interact with lower floristic diversity, greater seasonality in resource availability, and relatively frequent incidence of severe climatic events (e.g., frosts) which can limit the distribution of small-bodied species.

ACKNOWLEDGMENTS

Primate surveys carried out by C. Peres and collaborators have been funded by the World Wildlife Fund (1987–1988), Wildlife Conservation Society (1991–1995), and the Josephine Bay and Michael Paul Foundations (1996–1998). We thank the Brazilian Oil Company (Petrobrás, S.A.) for providing helicopter transportation to several survey sites which were otherwise inaccessible. We are particularly grateful to L. Brandão, P. Develey, A. Calouro, L. M. Petroni, S. Knapp, A. P. Carmignoto, N. Rocha and L. Sainz who kindly made unpublished primate checklists available to us. We thank C. Chapman, J. Ganzhorn, and A. Carkeek for making constructive comments on a preliminary version of the manuscript.

REFERENCES

Ayres, J. M. (1986). *Uakaris and Amazonian flooded forest.* PhD thesis, University of Cambridge, UK.

Ayres, J. M. & Clutton-Brock, T. H. (1992). River boundaries and species range size in Amazonian primates. *American Naturalist,* **140**, 531–7.

Blackburn, T. M. & Gaston, K. J. (1997). The relationship between geographic area and the latitudinal gradient in species richness in New World birds. *Evolutionary Ecology,* **11**, 195–204.

Bourlière, F. (1985). Primate communities: their structure and role in tropical ecosystems. *International Journal of Primatology,* **6**, 1–26.

Brown, J. H. (1995). *Macroecology.* Chicago: University of Chicago Press.

Brown, J. H. & Gibson, A. C. (1983). *Biogeography.* St Louis, MO: Mosby.

Cartelle, C. (1993). Achado de *Brachyteles* do Pleistoceno final. *Neotropical Primates,* **1**, 8.

Cody, M. L. (1975). Towards a theory of continental species diversities: bird distributions over mediterranean habitat gradients. In *Ecology and Evolution of Communities,* Ed. M. L. Cody & J. M. Diamond, pp. 214–57, Cambridge, Mass: Harvard University Press.

Cowlishaw, G. & Hacker, J. E. (1997). Distribution, diversity, and latitude in African primates. *American Naturalist,* **150**, 505–12.

Eisenberg, J. F. (1979). Habitat, economy and society: some correlations and hypotheses for the neotropical primates. In *Primate Ecology and Human Origins: Ecological Influences on Social Organization,* Ed. I. S. Bernstein & E. O. Smith, pp. 215–62. New York: Garland Press.

Ferrari, S. F. & Lopes, M. A. (1996). Primate populations in eastern Amazonia. In *Adaptive Radiations in Neotropical Primates,* Ed. M. Norconk, A. Rosenberger & P. Garber, pp. 53–68. New York: Plenum Press.

Ferrari, S. F., Corrêa, H. K. M. & Coutinho, P. E. G. (1996). Ecology of the "southern" marmosets (*Callithrix aurita* and *Callithrix flaviceps*). In *Adaptive Radiations in Neotropical Primates,* eds. M. Norconk, A. Rosenberger & P. Garber, pp. 157–71. New York: Plenum Press.

Findley, J. S. (1993). *Bats: A Community Perspective.* Cambridge: Cambridge University Press.

Fleagle, J. G. & Reed, K. E. (1996). Comparing primate communities: a multivariate approach. *Journal of Human Evolution,* **30**, 489–510.

Freese, C. H., Heltne, P. G., Castro, N. R. & Whitesides, G. (1982). Patterns and determinants of monkey densities in Peru and Bolivia, with notes on distributions. *International Journal of Primatology,* **3**, 53–90.

Ganzhorn, J. U. (1997). Test of Fox's assembly rule for functional groups in lemur communities in Madagascar. *Journal of Zoology,* **241**, 533–42.

Ganzhorn, J. U., Malcomber, S., Andrianantoaina, O. & Goodman, S. M. (1997). Habitat characteristics and lemur species richness in Madagascar. *Biotropica,* **29**,331–43.

Gaston, K. J. (1994). *Rarity.* London: Chapman & Hall.

Gautier-Hion, A. (1988). Polyspecific associations among forest guenons: ecological, behavioural and evolutionary aspects. In *A Primate Radiation: Evolutionary Biology of the African Guenons,* Ed. A. Gautier-Hion, F. Bourliere, J.-P. Gautier & J. Kingdon, pp. 452–76. Cambridge: Cambridge University Press.

Gentry, A. H. (1988). Tree species richness of upper Amazonian forests. *Proceedings of the National Academy of Sciences, USA,* **85**, 156–9.

Hartwig, W. G. (1995). A giant New World monkey from the Pleistocene of Brazil. *Journal of Human Evolution,* **28**, 189–95.

Hartwig, W. C. & Cartelle, C. (1996). A complete skeleton of the giant South American primate *Protopithecus. Nature,* B381, 307–11.

Hershkovitz, P. (1977). *Living New World Monkeys* (Platyrrhini), Vol. 1. University of Chicago Press, Chicago.

Hershkovitz, P. (1983). Two new species of night monkeys, genus *Aotus* (Cebidae, Platyrrhini): a preliminary report in *Aotus* taxonomy. *American Journal of Primatology,* **4**, 209–43.

Hershkovitz, P. (1987). The taxonomy of South American sakis, genus *Pithecia* (Cebidae, Platyrrhini): a preliminary report and critical review with the description of a new species and a new subspecies. *American Journal of Primatology*, **12**, 387–468.

Hershkovitz, P. (1990). Titis, New World monkeys of the genus *Callicebus* (Cebidae, Platyrrhini): a preliminary taxonomic review. *Fieldiana (Zoology)*, **55**, 1–109.

Jablonski, D. (1993). The tropics as a source of evolutionary novelty through geological time. *Nature*, **364**, 142–4.

Janson, C. H. & Emmons, L. H. (1990). Ecological structure of the non-flying mammal community at the Cocha Cashu Biological Station, Manu National Park, Peru. In *Four Neotropical Rainforests*, Ed. A. Gentry, pp. 314–38. New Haven: Yale University Press.

Kaufman, D. M. (1995). Diversity of mammals – universality of the latitudinal gradients of species and bauplans. *Journal of Mammalogy*, **76**, 322–34.

Kay, R. F., Madden, R. H., van Schaik, C. & Higdon, D. (1997). Primate species richness is determined by plant productivity: implications for conservation. *Proceedings of the National Academy of Sciences, USA*, **94**, 13023–27.

Kinzey, W. G. (1982). Distribution of primates and forest refuges. In *Biological Diversification in the Tropics*, Ed. G. T. Prance, pp. 445–482. New York: Columbia University Press.

MacArthur, R. H. (1972). *Geographical Ecology: Patterns in the Distribution of Species*. New York: Harper & Row.

MacArthur, R. H. & Wilson, E. O. (1967). *The Theory of Island Biogeography*. Princeton: Princeton University Press.

Muckenhirn, N. A., Mortensen, B. K., Vessey, S., Fraser, C. E. O. & Singh, B. (1975). *Report on a Primate Survey in Guyana*. Washington, DC: Pan American Health Organization.

Norconk, M. A., Sussman, R. W. & Phillips-Conroy, J. (1996). Primates of Guayana Shield forests: Venezuela and the Guianas. In *Adaptive Radiations in Neotropical Primates*, Ed. M. Norconk, A. Rosenberger & P. Garber, pp. 69–83. New York: Plenum Press.

Oates, J. F., Whitesides, G. H., Davies, A. G., Waterman, P. G., Green, S. M., Dasilva, G. L. & Mole, S. (1990). Determinants of variation in tropical forest primate biomass: new evidence from west Africa. *Ecology*, **71**, 328–43.

Pagel, M. D., May, R. M. & Collie, A. R. (1991). Ecological aspects of the geographical distribution and diversity of mammalian species. *American Naturalist*, **137**, 791–815.

Peres, C. A. (1988). Primate community structure in Brazilian Amazonia. *Primate Conservation*, **9**, 83–7.

Peres, C. A. (1989). A survey of a gallery forest primate community, Marajó Island, Pará, *Silvestre Neotropical*, **2**, 32–7.

Peres, C. A. (1990). Effects of hunting on western Amazonian primate communities. *Biological Conservation*, **54**, 47–59.

Peres, C. A. (1993a). Structure and spatial organization of an Amazonian terra firme forest primate community. *Journal of Tropical Ecology*, **9**, 259–76.

Peres, C. A. (1993b). Notes on the ecology of buffy saki monkeys (*Pithecia albicans*, Gray 1860): a canopy seed-predator. *American Journal of Primatology*, **31**, 129–40.

Peres, C. A. (1993c). Notes on the primates of the Juruá River, western Brazilian Amazonia. *Folia Primatologica*, **61**, 97–103.

Peres, C. A. (1997a). Primate community structure at twenty western Amazonian flooded and unflooded forests. *Journal of Tropical Ecology*, **13**, 381–405.

Peres, C. A. (1997b). Effects of habitat quality and hunting pressure on arboreal folivore densities in neotropical forests: a case study of howler monkeys (*Alouatta* spp.). *Folia Primatologica*, **68**, 199–222.

Peres, C. A. (in press a). Impact of subsistence hunting on vertebrate community structure in Amazonian forests: a large-scale cross-site comparison. *Conservation Biology*.

Peres, C. A. (in press b). Nonvolant mammal community structure in different Amazonian forest types. In *Mammals of the Neotropics*, Vol. 3, Ed. J. F. Eisenberg & K. H. Redford. University of Chicago Press, Chicago.

Peres, C. A. & Johns, A. D. (1992). Patterns of primate mortality in a drowning forest: lessons from the Tucuruí Dam, eastern Amazonia. *Primate Conservation*, **13**, 7–10.

Peres, C. A., Patton, J. L. & da Silva, M. N. F. (1996). Riverine barriers and gene flow in Amazonian saddle-back tamarins. *Folia Primatologica*, **67**, 113–24.

Pinto, L. P. S. & Rylands, A. B. (1997). Geographic distribution of the golden-headed lion tamarin, *Leontopithecus chrysomelas*: implications for its management and conservation. *Folia Primatologica*, **68**, 161–80.

Porter, C. A., Page, S. L., Czelusniak, J., Schneider, H., Schneider, M. P. C., Sampaio, I. & Goodman, M. (1997). Phylogeny and evolution of selected primates as determined by sequences of the epsilon-globin locus and 5′ flanking regions. *International Journal of Primatology*, **18**, 261–95.

Puertas, P. & Bodmer, R. E. (1993). Conservation of a high diversity primate assemblage. *Biodiversity and Conservation*, **2**, 586–93.

Rahbek, C. (1997). The relationship among area, elevation, and regional species richness in neotropical birds. *American Naturalist*, **149**, 875–902.

Reed, K. E. & Fleagle, J. G. (1995). Geographic and climatic control of primate diversity. *Proceedings of the National Academy of Sciences*, **92**, 7874–6.

Roy, K., Jablonski, D., Valentine, J. W. & Rosenberg, G. (1998). Marine latitudinal diversity gradients: tests of causal hypotheses. *Proceedings of the National Academy of Sciences*, **95**, 3699–702.

Ruggerio, A. (1994). Latitudinal correlates of the sizes of mammalian geographic ranges in South America. *Journal of Biogeography*, **21**, 545–59.

Rylands, A. B. (1987). Primate communities in Amazonian forests: their habitats and food sources. *Experienta*, **43**, 265–79.

Rylands, A. B. (1993). The ecology of the lion tamarins, *Leontopithecus*: some intrageneric differences and comparisons with other callitrichids. In *Marmosets and Tamarins: Systematics, Behaviour, and Ecology*, Ed. A. B. Rylands, pp. 296–313. Oxford: Oxford University Press.

Rylands, A. B., da Fonseca, G. A. B., Leite, Y. L. R. & Mittermeier, R. A. (1996). Primates of the Atlantic forest. In *Adaptive Radiations in Neotropical Primates*, Ed. M. Norconk, A. Rosenberger & P. Garber, pp. 21–51. New York: Plenum Press.

Schall, J. J. & Pianka, E. R. (1978). Geographic trends in numbers of species. *Science*, **201**, 679–86.

Smith, F. D. M., May, R. & Harvey, P. H. (1994). Geographical ranges of Australian mammals. *Journal of Animal Ecology*, **63**, 441–50.

Stevens, G. C. (1989). The latitudinal gradient in geographic range: how so many species coexist in the tropics. *American Naturalist*, **133**, 240–6.

Struhsaker, T. T., Glander, K., Chirivi, H. & Scott, N. J. (1975). A survey of primates and their habitats in northern Colombia. In *Primate Censusing Studies in Peru and Colombia. PAHP Report to the National Academy of Sciences on the Activities of Project AMRO-0719*. Washington, DC: WHO.

Terborgh, J. (1983). *Five New World Primates: a Study in Comparative Ecology*. Princeton: University Press Princeton.

Terborgh, J. & van Schaik, C. P. (1987). Convergence vs. nonconvergence in primate communities. In *Organization of Communities, Past and Present*, Ed. J. H. R. Gee & P. S. Giller, pp. 205–26. Oxford: Blackwell Scientific.

Voss, R. S. & Emmons, L. H. (1996). Mammalian diversity in neotropical lowland rainforests: a preliminary assessment. *Bulletin of the American Natural History Museum*, **230**, 1–115.

Waser, P. M. (1986). Interactions among primate species. In *Primate Societies*, Ed. B. B. Smuts, D. L. Cheney, R. M. Seyfarth, R. W. Wrangham & T. T. Struhsaker, pp. 210–26. Chicago: University of Chicago Press.

Wolfheim, J. H. (1983). *Primates of the World: Distribution, Abundance and Conservation*. Seattle: University of Washington Press.

APPENDIX 3.1. LIST OF SOUTH AMERICAN PRIMATE ECOLOGICAL SPECIES GROUPS (SEE TABLE 3.1) AND THEIR CHARACTERISTICS

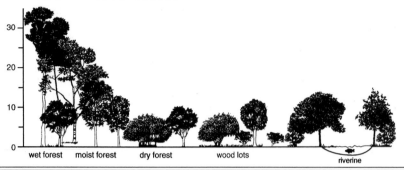

	Endangered status	Forest type					Major foods	Major regions
		Wet	Moist	Dry	Woodlots	Riverine		
Family Cebidae								
Cebuella pygmaea	L	X	X			X	G,F,N,I,	AM
Callithrix spp.	L, V, E		X	X	X		G,F,I,	AM
Saguinus fuscicollis, *S. nigricollis, S. tripartitus*	L	X	X				F,I,N,G	AM
Saguinus mystax, S. labiatus, *S. imperator*	L, V	X	X				F,I,N,G	AM
Saguinus midas, S. bicolor, *S. inustus, S. geoffroyi,* *S. oedipus, S. leucopus*	L, V, T, E	X	X	X	X		F,I,G,	CA,AM,SB
Leontopithecus spp.	E, C	X	X				F,I,	SB
Callimico goeldii	E	X	X				I,V,F	AM
Saimiri spp.	L, E	X	X			X	I,F,N	AM,CA
Aotus spp.	L	X	X	X	X	X	F,I,N	AM,CA
Cebus apella, C. xanthosternos	L	X	X				F,I,S,N,P	AM
Cebus albifrons	L	X	X			X	F,I,S,N	AM
Cebus olivaceous, C. capucinus, *C. kaapori*	L	X	X	X	X		F,I,S,N,	GY,CA
Family Atelidae								
Callicebus spp. (except *C. torquatus*)	L, V, E	X	X			X	F,L,I	AM
Callicebus torquatus	L	X	X			X	F,IL	AM
Pithecia spp.	L	X	X				S,F,L	AM
Chiropotes spp.	E	X	X				S,F,L	GY
Cacajao spp.	E					X	S,F,L	AM
Alouatta spp.	L, V, C	X	X	X	X	X	L,F	ALL
Lagothrix spp.	V, E	X	X				F,L,S	AM
Ateles spp.	L, V, E	X	X				F,L	CA,AM,GY
Brachyteles spp.	E	X	X				F,L	SB

Note: G, gum; N, nectar; F, fruit; I, insects; V, small vertebrates; L, leaves; S, seeds, P, pith and meristems; CA, Central America; AM, Amazonia; GY, Guyanan shield; SB, Southeastern Brazil. Endangered status: L, lower risk; V, vulnerable; T, threatened; E, endangered; C, critically endangered.

APPENDIX 3.2. SOURCES OF DATA USED IN THIS ANALYSIS. A FULL REFERENCE LIST CAN BE OBTAINED FROM THE FIRST AUTHOR UPON REQUEST

Country and region	Forest site	Sources
Eastern Amazonian Brazil	Eastern Marajó Is., Pará	Peres 1989
Eastern Amazonian Brazil	Northeastern Marajó Is., Pará	Ayres *et al.* 1989
Eastern Amazonian Brazil	Rio Cajarí, Amapá	C. Peres & H. Nascimento, unpubl. data
Eastern Amazonian Brazil	Kayapó Reserve, Pará	C. Peres, unpubl. data
Eastern Amazonian Brazil	Pucuruí, Pará	Johns 1986
Eastern Amazonian Brazil	Remansinho, Pará	Johns 1986
Eastern Amazonian Brazil	Vila Braba, Pará	Johns 1986
Eastern Amazonian Brazil	Base 4, Pará	Ghilardi & Alho 1990
Eastern Amazonian Brazil	Serra de Carajás, Pará	Johns 1986
Eastern Amazonian Brazil	Tailândia, Pará	Ferrari & Lopes 1997
Eastern Amazonian Brazil	Rio Capim, Pará	Ferrari & Lopes 1997
Eastern Amazonian Brazil	Irituia, Pará	Ferrari & Lopes 1997
Eastern Amazonian Brazil	Gurupí, Pará	Ferrari & Lopes 1997
Eastern Amazonian Brazil	Caxiuanã, Pará	Ferrari & Lopes 1997
Eastern Amazonian Brazil	Rio Iriri, Pará	Martins *et al.* 1998
Eastern Amazonian Brazil	Belém, Pará	Pine 1973
Western Amazonian Brazil	Urucu, Amazonas	Peres 1993
Western Amazonian Brazil	Igarapé Açu, Amazonas	Peres 1988
Western Amazonian Brazil	SUC-I, Amazonas	Peres 1988
Western Amazonian Brazil	Oleoduto, Amazonas	C. Peres, unpubl. data
Western Amazonian Brazil	Igarapé Curimatá, Amazonas	C. Peres & H. Nascimento, unpubl. data
Western Amazonian Brazil	São Domingos, Acre	Peres 1988
Western Amazonian Brazil	Kaxinawá Reserve, Acre	Peres 1997
Western Amazonian Brazil	Riozinho, Amazonas	Peres 1997
Western Amazonian Brazil	Porongaba, Acre	Peres 1997
Western Amazonian Brazil	Sobral, Acre	Peres 1997
Western Amazonian Brazil	Condor, Amazonas	Peres 1997
Western Amazonian Brazil	Penedo, Amazonas	Peres 1997
Western Amazonian Brazil	Altamira, Amazonas	Peres 1997
Western Amazonian Brazil	Barro Vermelho, Amazonas	Peres 1997
Western Amazonian Brazil	Fortuna, Amazonas	Peres 1997
Western Amazonian Brazil	Igarapé Jaraqui, Amazonas	Peres 1997
Western Amazonian Brazil	Vira Volta, Amazonas	Peres 1997
Western Amazonian Brazil	Vai Quem Quer, Amazonas	Peres 1997
Western Amazonian Brazil	Sacado, Amazonas	Peres 1997
Western Amazonian Brazil	Nova Empresa, Amazonas	Peres 1997
Western Amazonian Brazil	Boa Esperança, Amazonas	Peres 1997
Western Amazonian Brazil	Barro Vermelho II, Amazonas	Peres 1997
Western Amazonian Brazil	Lago da Fortuna, Amazonas	Peres 1997
Western Amazonian Brazil	Lago Teiú, Amazonas	Ayres 1986
Western Amazonian Brazil	Lago Mamirauá, Amazonas	Ayres 1986
Western Amazonian Brazil	Ponta da Castanha, Amazonas	Johns 1986
Western Amazonian Brazil	Açaituba, Amazonas	Johns 1986
Western Amazonian Brazil	Rio Iaco, Acre	Martins 1992
Western Amazonian Brazil	Fazenda União, Acre	Martins 1992

APPENDIX 3.2. (*cont.*)

Country and region	Forest site	Sources
Western Amazonian Brazil	Antimarí N.F., Acre	Calouro 1996
Western Amazonian Brazil	Samuel, Rondônia	Lemos de Sá, 1996
Western Amazonian Brazil	Rio Jamarí, Rondônia	da Rocha *et al.* 1992
Western Amazonian Brazil	Tres Irmãos, Rondônia	Ferrari *et al.* 1996
Western Amazonian Brazil	Lago dos Reis, Rondônia	Ferrari & Lopes 1992
Western Amazonian Brazil	Aripuanã, Mato Grosso	Rylands 1982
Western Amazonian Brazil	Iquê-Juruena E.S., Mato Grosso	Setz & Milton 1985
Central Amazonian Brazil	Tapajós N.P., Pará	Branch 1983, Ayres & Milton 1981
Central Amazonian Brazil	Manaus, Amazonas	Rylands & Keuroghlian
Central Amazonian Brazil	Jaú N.P., Amazonas	Neo Prim 3(2) 1995
Central Amazonian Brazil	Alter do Chão, Pará	Albernaz 1993
Northern Amazonian Brazil	Pico da Neblina, Amazonas	Boubli 1994, P. Boubli, pers. comm.
Northern Amazonian Brazil	Maracá Island, Roraima	Mendes-Pontes 1996
Central Brazil	Acurizal, Mato Grosso	Schaller 1983
Central Brazil	P.N. Brasília, D.F.	C. Peres, unpubl. data
Central Brazil	IBGE Reserve, D.F.	Fonseca & Lacher 1984
Northeastern Brazil	Una, southern Bahia	Rylands 1982
Northeastern Brazil	Extremos, Rio Grande do Norte	Meireles *et al.* 1992
Northeastern Brazil	Nisia Floresta, Rio Grande do Norte	Digby 1995
Northeastern Brazil	Usina Sinimbu	Langguth *et al.* 1985
Northeastern Brazil	Dois Irmãos, Pernambuco	Mendes-Pondes & Cruz 1995
Northeastern Brazil	Tapacurá, Pernambuco	Scanlon *et al.* 1989
Northeastern Brazil	Ouricuri, Pernambuco	Mares *et al.* 1985
Northeastern Brazil	Pacatuba, Paraíba	Bonvicino 1989
Northeastern Brazil	Sete Cidades N.P., Piauí	Digby *et al.* 1996
Southeastern Brazil	CVRD, Espírito Santo	Chiarello 1995
Southeastern Brazil	RBAR, Espirito Santo	Pinto *et al.* 1993
Southeastern Brazil	Sooretama, Espírito Santo	Kinzey & Becker 1983
Southeastern Brazil	Caratinga, Minas Gerais	K. Strier and S. Ferrari, pers. comm.
Southeastern Brazil	São Francisco Xavier, São Paulo	Antonietto *et al.* 1994
Southeastern Brazil	Intervales, São Paulo	L.M. Petroni, pers. comm.
Southeastern Brazil	Rio Doce Park, Minas Gerais	Neot. Prim. 2(4) 1994
Southeastern Brazil	Rio Jequitinhonha, Minas Gerais	Rylands *et al.* 1988
Southeastern Brazil	Faz. Monte Alegre, Minas Gerais	Muskin 1984
Southeastern Brazil	Ibitipoca S.P., Minas Gerais	Fontes *et al.* 1996
Southeastern Brazil	Barreiro Rico, São Paulo	Torres de Assumpção 1983
Southeastern Brazil	Santa Genebra, São Paulo	Chiarello 1994
Southeastern Brazil	Bananal E.E., São Paulo	L. Brandão, pers. comm.
Southeastern Brazil	São Sebastião, São Paulo	Oliveira & Manzatti 1996
Southeastern Brazil	Ilhabela Island, São Paulo	Olmos 1994
Southeastern Brazil	Poço das Antas, Rio de Janeiro	C. Peres, unpubl. data
Southeastern Brazil	Cach. de Macacu, Rio de Janeiro	S. Knapp, pers. comm.
Southern Brazil	Aracurí, Rio Grande do Sul	Marques 1996
Southern Brazil	Estância Casa Blanca, Rio Grande do Sul	Bicca-Marques 1990
Southeastern Peru	Cocha Cashu	Terborgh 1983, Janson & Emmons 1990
Southeastern Peru	Cuzco Amazonico	Woodman *et al.* 1991
Southeastern Peru	Tambopata-Tavara	Conservation International 1994

APPENDIX 3.2. (*cont.*)

Country and region	Forest site	Sources
Southeastern Peru	Yomiwato	Mitchell & Raez–Luna 1991
Southeastern Peru	Diamante	Mitchell & Raez–Luna 1991
Eastern Peru	Cahuana Island	Sioni 1986
Eastern Peru	Tahuayo–Blanco	Puertas & Bodmer 1993
Eastern Peru	Yavarí Miri	Puertas & Bodmer 1993
Eastern Peru	Ampiyacu	Freese *et al.* 1982
Eastern Peru	Orosa	Freese *et al.* 1982
Eastern Peru	Samiria	Freese *et al.* 1982
Eastern Peru	Isla de Iquitos	Moya 1989
Eastern Peru	Upper Nanay	Freese *et al.* 1982
Eastern Peru	Yarapa River	Ramirez 1984
Northern Peru	Moyobamba	Freese *et al.* 1982
Northern Peru	Cordillera Colán	Butchard *et al.* 1995
Eastern Peru	Itaya	Freese *et al.* 1982
Central Peru	von Humboldt	Freese *et al.* 1982
Central Peru	Panguana	Freese *et al.* 1982
Eastern Peru	Balta	Patton *et al.* 1982
Northern Bolivia	Ixiamas	Freese *et al.* 1982
Northern Bolivia	El Triunfo	Freese *et al.* 1982
Northern Bolivia	Riberalta	Freese *et al.* 1982
Northern Bolivia	Puerto Oro	Cameron *et al.* 1989
Northern Bolivia	Cobija	Freese *et al.* 1982
Northern Bolivia	El Beni Reserve, Beni	Garcia & Braza 1987
Northern Bolivia	Alto Madidi	Conservation International 1991
Northern Bolivia	Bella Flor, Pando	Christen & Geissmann 1994
Northern Bolivia	Puerto Rico, Panda	Pook & Pook 1982
Northern Bolivia	Rio Acre, Pando	Christen & Geissmann 1994
Northern Bolivia	Tres Corazones, Pando	Christen & Geissmann 1994
Northern Bolivia	Tres Estrellas, Pando	Christen & Geissmann 1994
Southeastern Bolivia	San José	Freese *et al.* 1982
Southern Bolivia	Santa Cruz	Norca Rocha & Lila Sainz, pers. comm.
Northern Paraguay	Defensores del Chaco N.P.	Stallings 1989
Western Paraguay	Tinfunque N.P.	Stallings 1989
Western Paraguay	Tenente Enciso N.P.	Stallings 1989
Eastern Paraguay	Cerro Corá N.P.	Stallings 1989
Southern Paraguay	Ybycui N.P.	Stallings 1989
Eastern Paraguay	Golondrina Ranch	Wright 1996
Eastern Paraguay	Mbaracayú	Hill 1996
Central Suriname	Raleigh–Voltzberg	Mittermeier & van Roosmalen 1981
Northern Suriname	Brakopondo	Eisenberg & Thorington 1973
Central Guyana	Berbice River	Muckenhirn *et al.* 1975
Central Guyana	Apoteri Area	Muckenhirn *et al.* 1975
Southern Guyana	East Kanuku Montains	Conservation International 1993
Northern Guyana	Kartabo	Beebe 1925
French Guiana	Saut Petit	Pack 1993
French Guiana	Saut Pararé	Guillotin *et al.* 1994
French Guiana	Les Nouragues	Simmen 1992, Drubbel & Gautier 1993

APPENDIX 3.2. (*cont.*)

Country and region	Forest site	Sources
Colombia	El Tuparro	Defler & Pintor
Colombia	Coraza–Montes	Patiño & Ossa 1994
Southeastern Colombia	Pto Japón	Izawa 1976
Southeastern Colombia	Tinigua N.P.	Stevenson *et al.* 1994
Southeastern Colombia	La Macarena	Nishimura *et al.* 1995
Colombia	Puerto Rico	Green 1978
Colombia	Magdalena Medio	Vargas & Solano 1996
Colombia	Hacienda Barbascal	Mason 1996
Colombia	Ventura	Bernstein *et al.* 1976
Southeastern Colombia	E.B. Caparú	Defler 1996
Eastern Ecuador	Cuyabeno	de la Torre *et al.* 1995
Eastern Ecuador	Tambococha	Albuja 1994
Western Ecuador	Machalilla N.P.	Conservation International 1992
Western Ecuador	Jauneche	Conservation International 1992
Northern Venezuela	Hato Masaguaral	Sekulic 1982
Northern Venezuela	Hato Pinero	Miller 1991
Northern Venezuela	Hato El Frio	Braza *et al.* 1981
Northern Venezuela	Isla Margarita	Sanz & Marquez 1994
Central Venezuela	Guri Lake	Kinzey & Norconk 1993
Northern Venezuela	Guatopo	Eisenberg *et al.* 1979
Southern Venezuela	Rio Manapiare	Teaford & Runestad 1993
Southern Venezuela	Tawadu	Handley 1976, Castellanos & Chaunin 1996
Southern Venezuela	Cunucunuma	Handley 1976
Northern Argentina	Rio Riachuelo	Rumiz 1989
Northern Argentina	El Rey	Brown & Zunino 1990
Northern Argentina	Formosa	Di Bitteto & Arditi 1993
Northern Argentina	Iguazu N.P.	Brown & Zunino 1990
Guatemala	Tikal	Bramblett *et al.* 1980, Bolin 1981
Southern Mexico	Los Tuxtlas	Estrada & Coates-Estrada 1988
Southern Mexico	Lacondona	Medellin 1994
Southern Mexico	Chamela	Ceballos & Miranda 1986
Southern Mexico	Yucatan	Watts *et al.* 1986
Panama	Chiquiri	Baldwin & Baldwin 1977
Panama	Barro Colorado Island	Glanz 1990
Panama	Coiba Island	Baldwin & Baldwin 1977
Panama	Hacienda Barqueta	Baldwin & Baldwin 1972
Panama	Rodman Site	Dawson 1979
Belize	Bermuda Landing	Horwich 1983
Belize	Columbia River	Conservation International 1993
Costa Rica	Santa Rossa	Fedigan *et al.* 1988
Costa Rica	La Pacifica	Glander 1992
Costa Rica	La Selva	Timm 1994
Costa Rica	Palo Verde	Massey 1987
Costa Rica	Corcovado	Pineros 1993
Costa Rica	Cabo Blanco	Lippold 1989
Costa Rica	Refugio Curucuru	Tomblin & Canford 1994

4 • Primate communities: Madagascar

JÖRG U. GANZHORN, PATRICIA C. WRIGHT AND JONAH
RATSIMBAZAFY

INTRODUCTION

Madagascar, "La Grande Ile" off the coast of south-east Africa, is the fourth largest island on earth. Its 587 000 km² are topped only by the islands of Greenland, New Guinea and Borneo. The island broke off from Africa some 150 to 160 million and from India some 88 to 95 million years ago. Although Madagascar is separated from Africa only by the Mozambique Channel, no more than about 300 to 450 km wide, the prevailing winds and ocean currents were and still are unfavorable for repeated colonization of the island (Krause *et al.*, 1997). Due to the long isolation and low rate of colonization events, the flora and fauna of Madagascar underwent impressive adaptive radiations, resulting in one of the world's most diverse arrays of endemic plants and animals (Myers, 1986; Mittermeier, 1988; Table 4.1).

Based on phytogeographic criteria, the evergreen forests of eastern Madagascar are distinguished from the deciduous formations of the west and south (Du Puy & Moat, 1996; Lowry *et al.*, 1997; Fig. 4.1). The evergreen rainforests of the east receive between 1500 and more than 3000 mm of rain per year. The deciduous forests of the west and extreme north of Madagascar are subject to a distinct dry season of four to eight months without rain and annual precipitation of 500 to 2000 mm. In both vegetation types, annual rainfall decreases from the north to the south. Parts of the south and south-west of the island receive less than 500 mm of rain per year at irregular intervals with an extended dry season for more than eight months. Here, habitats are represented by dry deciduous, riverine and spiny forest characterized by Didieraceae and other succulent plants.

Among the many Malagasy taxa consisting almost exclusively of endemic species (see Table 4.1), lemurs stick out as their group consists only of endemic species. Two

lemur species have been introduced to the neighboring islands of the Comores, where they face immediate threat of extinction (Tattersall, 1998). The lemurs make Madagascar the country with the third highest primate diversity on earth. This tremendous radiation might have been possible due to the poor representation of other mammalian competitors. Apart from 14 genera of lemurs, only 25 other genera of non-flying mammals are described for the island. This is in sharp contrast to the situation of Borneo, an island of comparable size, where eight primate genera share their habitat with some 44 other mammalian genera (see Ganzhorn, chapter 8, this volume for more details). Today, lemurs are represented by five endemic primate families with 14 genera and about 33 species. They range from 30 g to about 6–7 kg, encompass all dietary categories from omnivorous to exclusively folivorous species, and include some of the world's most specialized primate species, such as the gentle lemur *(Hapalemur griseus alaotrensis)* living exclusively in the reed beds of Lac Alaotra or the aye-aye *(Daubentonia madagascariensis)* (Harcourt & Thornback, 1990; Mittermeier *et al.,* 1994; Table 4.2).

The primate fauna of Madagascar and their habitats have undergone severe changes for the last few thousand years. Several thousand years prior to the arrival of humans some 2000 years ago, some parts of Madagascar became drier, leading eventually to the extinction of birds and mammals that require more mesic conditions (Dewar, 1984; Straka, 1996; Burney, 1997; Goodman & Rakotondravony, 1996; Goodman & Rakotozafy, 1997). The arrival of humans accentuated this process and eliminated at least one-third of the lemur species of the island in recent times. All lemur species known to have gone extinct during the last few hundred or few thousand years were larger than any of the still existing species (Richard & Dewar, 1991;

Table 4.1. *Estimated number of species and degree of endemism among selected plant and animal groups from Madagascar*

Group	Number of species	% Endemism
Native vascular plants	> 9345	81 %
Mammals: Lemurs	> 33	100 %
Birds	> 256	53 %
Reptiles	> 300	95 %
Amphibians	> 170	99 %

Source: Data from Phillipson (1994, 1996), Mittermeier *et al.* (1994) and Langrand & Wilmé (1997).

Godfrey *et al.*, 1997). The survival of the persisting species is still threatened (Mittermeier *et al.*, 1992, 1994; MacPhee & Marx, 1997; Wright, 1997). It is therefore not only of academic interest, but also of utmost importance for conservation to come to a better understanding of the factors which contribute to the abundance of lemurs and the number of potentially coexisting species. The goals of this chapter are to describe variations in the abundance of species and their community characteristics on a national level and link some of this variation to environmental factors.

We use two sites (one wet- and one dry-forest site; Fig. 4.2) for more detailed comparisons and analyse potential causal relationships between the environmental factors identified above and lemur community characteristics.

Fig. 4.1. Vegetation types of Madagascar and similarities in lemur species composition (from Ganzhorn, in press). Wet forests are mainly in the east and dry forests are found in the west and south.

Table 4.2 *The living lemurs of Madagascar and some of their life history characteristics*

Species	Body mass [kg]	Activity	Diet	Evergreen wet	Marsh	Deciduous dry	Riverine	Spiny forest
Microcebus murinus	0.06	N	OM	+		+++	+++	+++
M. cf. myoxinus	0.03	N	OM			+++		
M. ravelobensis	0.06	N	OM			+++		
M. rufus	0.04	N	OM	+++				
Allocebus trichotis	0.09	N	OM	+++				
Cheirogaleus medius	0.28	N	OM	+		+++	+++	
C. major	0.4	N	OM	+++				
Mirza coquereli	0.3	N	OM	+		+++	+++	
Phaner furcifer	0.46	N	OM	+		+++	+++	
Lepilemur dorsalis	0.5	N	FO	+++				
L. edwardsi	0.91	N	FO			+++	+++	
L. leucopus	0.61	N	FO	+			+++	+++
L. microdon	0.97	N	FO	+++				
L. mustelinus	0.78	N	FO	+++				
L. ruficaudatus	0.77	N	FO			+++	+++	
L. septentrionalis	0.75	N	FO	+++		+++	+++	
Eulemur coronatus	1.2	C	FR	+++		+++	+++	
E. fulvus	2.0–2.3	C	FR	+++		+++	+++	
E. macaco	1.8	C	FR	+++		+++	+++	
E. mongoz	1.5	C	FR	+		+++	+++	
E. rubriventer	2.0	C	FR	+++				
Hapalemur aureus	1.5	D	FO	+++				
H. griseus	0.7–1.4	D	FO	+++	+		+++	
H. simus	2.6	C	FO	+++				
Lemur catta	2.2	D	FR			+++	+++	+++
Varecia variegata	3.5	D	FR	+++				
Avahi laniger	1.2	N	FO	+++				
Avahi occidentalis	0.79	N	FO			+++	+++	
Indri indri	6.3	D	FO	+++				
Propithecus diadema	6.1	D	FR	+++				
P. tattersalli	3.5	D	FR	+++		+++		
P. verreauxi	3.0–4.3	D	FR	+		+++	+++	+++
Daubentonia madagascariensis	2.6–3.5	N	OM	+++		+++	+++	

Note: N, nocturnal; C, cathemeral; D, diurnal; OM, omnivorous; FO, folivorous; FR, frugivorous.
Source: Data from Harcourt & Thornback (1990), Ganzhorn & Kappeler (1993), Mittermeier *et al.* (1994), van Schaik & Kappeler (1996), Smith & Jungers (1997), Ganzhorn (1998), Wright *et al.* (1997), Zimmerman *et al.* (1998).

Fig. 4.2. Comparison of the eastern rainforest site of
Ranomafana and a characteristic species, the golden bamboo
lemur (left), with the western dry forest site of Kirindy and a
characteristic species, Verreaux's sifaka (right) (Ranomafana
photo courtesy of F. White; Kirindy photos courtesy of
S. Sommer).

VARIATIONS IN THE ABUNDANCE OF SPECIES AND THEIR COMMUNITY CHARACTERISTICS

In broad terms, the distribution of lemur species follows the distribution of vegetation types. Thus, lemur communities of the dry forests are clearly distinct from those of the wet forests (Richard & Dewar, 1991; Fig. 4.1).

However, the link between lemur species and specific floristics does not apply to all lemur species. The fork-marked lemur (*Phaner furcifer*) and the woolly lemur (*Avahi laniger*) are found in both eastern wet forests and in the western dry deciduous forest. This might indicate, that, at least for some species, the present distribution does not reflect adaptations to specific vegetation formations exclusively, but rather are the consequence of historical colonization and extinction processes.

Historical biogeographic processes are difficult to reconstruct and at present still open to discussion (Martin, 1972; Tattersall, 1982; Thalmann & Rakotoarison, 1994; Ganzhorn, 1998). This is in part due to the high degree of forest fragmentation, outstanding anthropogenic changes in the recent past, poor knowledge of precise lemur distributions and conflicting interpretation of Pleistocene changes in forest cover (Burney, 1997; Wells, in press).

Leaving the historical processes aside, the present lemur communities of the eastern wet and western dry forest show some distinct characteristics, which ask for explanations:

1. The communities of the eastern wet forest have more species than communities of the western dry deciduous forest.
2. Wet forests have more diurnal and cathemeral species than dry forests.
3. The only mainly frugivorous species (*Varecia variegata*) and the bamboo eating lemurs (*Hapalemur* spp.) occur only in the east (with an isolated population of *Hapalemur griseus* in Bemaraha). In contrast, the mainly gum eating *Phaner furcifer* is common in the west, but occurs only at few sites in evergreen wet forest. *Mirza coquereli*, an omnivorous/predatory species is widespread in the dry and evergreen forests of the west, but does not occur in the east.
4. Densities of lemurs are higher in the dry than in the wet forests.

5. Despite higher species number, biomass of all lemurs are lower in wet than in dry forests.

THE NUMBER OF LEMUR SPECIES PER COMMUNITY

Sites in the east contain between 2 and 13 lemur species (median = 9.5; quartiles: 6 to 11 species; $n = 16$). In contrast, eight communities of the deciduous forests contain between 4 and 8 species with a median of 5.5 species (quartiles: 5 to 7; see also Fig. 4.3). Apart from the taxonomic pattern, there are several regularities in the number of lemur species found in any given community and their population densities. First, the number of lemur species increases with increasing floristic richness of tree species, which in turn increases with annual precipitation (Fig. 4.3; Reed & Fleagle, 1995; Ganzhorn et al., 1997). This correlation is consistent with the idea that increased habitat heterogeneity provides more different options to specialize and thus to avoid interspecific competition. Potential resource axes for differential resource utilizations are food, habitat and temporal niche diversification (Schoener, 1974; exemplified for lemur communities by Charles-Dominique et al., 1980; Ganzhorn, 1989). At even higher habitat diversity, essential resources might be diluted by components that can not be used. This is suggested by the finding that within a single contiguous dry deciduous forest, lemurs were found more frequently in microhabitats of relatively high tree species richness, if the surrounding forest had few tree species; but the animals preferred microhabitats with relatively few tree species in a forest block of high tree species richness (Ganzhorn et al., 1997). This humped relationship between consumer species richness and resource diversity resembles the relationship shown also for rodents in the desert of Israel (Abramsky & Rosenzweig, 1984). Thus, in floristically very diverse areas the density of consumer species could decline (though not necessarily). At least on a small scale, the reduced density of primates might then also result in reduced number of species found per unit area and per unit survey time (Ganzhorn et al., 1997; Kay et al., 1997).

The representation of diurnal, cathemeral and nocturnal species differs between the east and the west. Nocturnal species live in similar species numbers in both forest types (Table 4.3). The assumed increase in habitat diversity is not reflected in increased numbers of nocturnal

Table 4.3. *Genera and number of species (in brackets) with different activity patterns in wet and dry forests*

	Wet forest	Dry forest
Diurnal		*Lemur* (1)
	Propithecus (3)	*Propithecus* (2)
	Hapalemur (2)	*Hapalemur* (1)
	Varecia (1)	
	Indri (1)	
Cathemeral	*Eulemur* (4)	*Eulemur* (4)
	Hapalemur (1)	
Nocturnal		*Mirza* (1)
	Avahi (1)	*Avahi* (1)
	Lepilemur (3)	*Lepilemur* (5)
	Microcebus (2)	*Microcebus* (> 3)
	Cheirogaleus (2)	*Cheirogaleus* (1)
	Phaner (1)	*Phaner* (1)
	Daubentonia (1)	*Daubentonia* (1)
	Allocebus (1)	

Source: Data on activity patterns from van Schaik & Kappeler (1996) except for *Hapalemur* (Wright, 1986; Overdorff, 1993, 1996; Overdorff *et al.*, 1997).

species in the wet forest. Rather, the diurnal and cathemeral species (which are also the larger species) diversified in the wet forest to a greater extent than in the dry forest. This indicates that the niches available for the small nocturnal species (up to about 1 kg) are very similar in the wet and in the dry forest. But the wet forest offers additional options for species diversification which are not available in the dry forest. This is exemplified by two of the best studied lemur communities (Table 4.4): Ranomafana in the eastern rain forest (Wright *et al.*, 1997) and Kirindy/CFPF in the dry deciduous forest (Ganzhorn & Sorg, 1996). There are about equal numbers of omnivorous nocturnal lemur species in both types of forest. The rain forest site houses more folivorous and frugivorous species. Very few lemur communities of the dry forest contain congeneric species, but almost every rain forest site has two species of *Eulemur* and up to three species of *Hapalemur* coexist at Ranomafana and also at Andringitra.

Apart from the coexistence of congeneric species, the larger number of diurnal/cathemeral species in the wet forest is due to the addition of new genera, such as *Varecia variegata* and *Indri indri* (Table 4.3). Together with the genus *Hapalemur*, these genera are considered to belong to some of the most specialized extant lemur species. *Hapalemur*

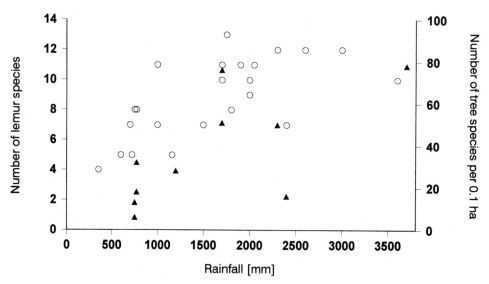

Fig. 4.3. Number of lemur species in relation to annual rain fall and tree species richness in 0.1 ha plots (extended from Ganzhorn *et al.*, (1997) with data from Hawkins (1994), contr. to Goodman (1996*a*), Ratsimbazafy & Wright (unpubl.)). Open circles are values for tree species; black triangles are values for lemur species.

Table 4.4. *The lemur communities of Ranomafana and Kirindy/CFPF as examples of communities in the eastern rain and western dry deciduous forest*

Site	Folivores	Frugivores	Omnivores
Ranomafana	*Lepilemur* sp.	*Eulemur rubriventer*	*Microcebus rufus*
	Avahi laniger	*Eulemur fulvus*	*Cheirogaleus major*
	Hapalemur griseus	*Varecia variegata*	*Daubentonia madagascariensis*
	Hapalemur aureus	*Propithecus diadema*	
	Hapalemur simus		
Kirindy/CFPF	*Lepilemur ruficaudatus*	*Eulemur fulvus*	*Microcebus cf. myoxinus*
		Propithecus verreauxi	*Microcebus murinus*
			Cheirogaleus medius
			Mirza coquereli
			Phaner furcifer

spp. specialize in various ways of feeding on bamboo (Glander *et al.*, 1989) which seems to occur with higher abundance and diversity in wet than in dry forests. Similarly, in the wet forest, congeneric species of *Eulemur* show different utilization of fruit and flower resources in particular (Overdorff, 1993) and the presence of the frugivorous *Varecia variegata* in the east may only be possible due to the year round supply of fleshy fruit (White *et al.*, 1995; Balko, 1997). In the west, fruit is available year round, even with a slight peak during the dry season (Sorg & Rohner, 1996). But in the west, fruits available during that season are dry and fiberous with no or only very little flesh. This may not provide enough usable energy for a specialized frugivore. Thus, *Hapalemur* spp. are bamboo specialists, *Varecia variegata* is the most frugivorous of all extant lemur species and *Indri indri* is one of the two largest folivores of the island. According to our previous line of arguments, this addition of new specializations should be facilitated in more diverse habitats, such as in wet forests (Fig. 4.3). While fleshy fruits seem to be more abundant in the east, gum might be more readily available in the dry forests, as indicated by the distribution and abundance of gum eating *Phaner furcifer* in the west. In Australia, gum production and its actual availability as a food item seems to be related in various ways to temperature, rainfall and whether or not the gums and saps are water soluble and thus washed off the trees by heavy rains (Smith, 1984). Similar data are lacking for Madagascar. Thus, even though the abundance of *Phaner furcifer* correlates positively with the number of gum producing trees in western Madagascar

(Charles-Dominique & Petter, 1980; Ganzhorn & Kappeler, 1996), it is questionable whether gum availability can provide a causal explanation for the limited representation of *Phaner furcifer* in the wet forests. After all, the species occurs in high densities in the wet forests of Montagne d'Ambre and is also known from Zahamena, another wet forest with substantial annual precipitation.

The lack of *Mirza coquereli* in the east is even more startling. In the west, this species is omnivorous with a strong component of predation on vertebrates (Petter *et al.*, 1977; Hladik *et al.*, 1980; Kappeler, pers. comm.). There is no ecologically equivalent lemur species in the east which might fill this niche and thus possibly competitively exclude *Mirza* from eastern wet forest sites. However, *Galidia elegans*, a 700 g carnivore of the eastern wet forest could be a competitor for meat and *Cheirogaleus major* with about the same body mass as *Mirza*, could outcompete *Mirza* for fruit, as does its smaller sister species, *C. medius*, in the dry forest (Wright & Martin, 1995; Ganzhorn & Kappeler, 1996; Colquhoun, 1998).

However, there are certainly other factors in addition to floristic diversity that are likely to contribute to species diversity in Madagascar and elsewhere. Ganzhorn (chapter 8, this volume) linked the large range of body mass in the lemur radiation to reduced competition from other mammals. This may only be part of the truth and competition may actually be not all that important in limiting the number of sympatric primate species in any given forest. In Madagascar, there are no competing vertebrate herbivores, the non-volant mammalian insectivores, the tenrecs, are mainly terrestrial and found rarely above one meter in the vegetation, and the guild of frugivorous birds and bats is depauperate (Goodman & Ganzhorn, 1997; Wright, 1997). Ranomafana and Andringitra have four to six species of bats (Goodman, 1996b; Wright unpubl.). In contrast, the Amazon bat faunas number about 100 species (Findley, 1993) and Barro Colorado Island has 70 (Kalko et al., 1996). Also, 106 bird species occur in Ranomafana, compared to about 1000 in Manu National Park (Peru; Thiollay, 1994). Yet, in present day primate communities, the upper limit of sympatric species is 12 to 17, no matter what happens with birds, bats and other mammals. The largest number of sympatric primate species in the world are or have been in Central Zaire and Madagascar (Ampazambazimba; MacPhee et al., 1985; Wright, 1997). The biological bases for this ceiling in the number of sympatric primate species are unknown.

POPULATION DENSITIES

Lemur populations show a close match between species body mass and their mean population densities (Fig. 4.4a). This allometric relation holds across the different types of vegetation. Figure 4.4 is based on mean densities of species. Thus, each species entered the analysis only once. This allometric relationship allows us to examine whether or not population densities of lemur populations are different in wet and dry forest once the effect of body mass has been accounted for. For this, we calculated the residuals of single lemur populations between the density predicted by

the regression and the actual population density of any given lemur species. Figure 4.4b shows, that, on average, lemurs of the same body mass reach higher population densities in the dry than in the wet forest.

In order to come to a better understanding of the environmental factors possibly responsible for these differences in Madagascar, we once again use data from Ranomafana (evergreen wet forest) and Kirindy/CFPF (dry deciduous forest), two of the better known forest ecosystems in Madagascar for more detailed comparisons.

COMPARISON OF RAIN AND DRY FOREST HABITATS: FOOD QUANTITY

Surprisingly the number of trees ≥ 10 cm diameter at breast height (DBH) in standard plots of Missouri Botanical Garden (20×500 m) does not differ substantially between Ranomafana and Kirindy (Fig. 4.5). In Figure 4.5 the numbers for Kirindy differ from the number of trees per "1 ha plot" given previously (Abraham et al., 1996) and shown in Table 4.5, because, first, the data here represent two other forest plots of the same area, and second, here, we consider only trees ≥ 10 cm DBH (equivalent to 31.4 cm girth), while Abraham et al. considered trees ≥ 30 cm girth. In N5 and CN8 (two plots in the Kirindy Forest/CFPF) another 68 and 60 trees were between 30 and 31.4 cm girth on the one hectare plots. This reflects the high proportion of small trees in the dry forest and thus one of the main structural differences between wet and dry forests, which is not represented appropriately if only trees ≥ 10 cm DBH

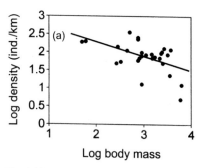

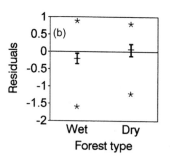

Fig. 4.4.(a). Least square regression between body mass (g) and mean population density (number per km²) of lemurs based on mean density per species ($y = -0.34x + 2.78$; $R^2 = 0.26$; $n = 26$; $P < 0.01$). Values are means per species, averaged across all population densities known from Madagascar.

(b) Difference between the population density predicted by the regression from Fig. 4.4(a) and the population density of distinct populations in wet ($n = 54$) and in dry forests ($n = 37$); t-test: $t = 2.56$; d.f. = 89, $P = 0.01$. Values are means, 95% confidence limits, minima and maxima.

Table 4.5. *Forest characteristics of rain and dry forest sites*

	Ranomafana (Vatoharanana)		Kirindy/CFPF (N5)	
Mean growth rate (DBH/year)	4.71 ± 3.73 mm/year (67 trees, 6 years)		0.84 ± 0.69 mm/year (151 trees, 3 years)	
Probability of fruiting for trees used by lemurs for fruit eating	39.6% ± 16.3% (*n* = 25 species with 4 ind./spec.; 5 years of records)		71.7% ± 19.1% (*n* = 11 species with 1 or 2 ind./spec.; 4 to 6 years of records)	
Number of tree families/species/stems considered	37+/112+/660		23+/45+/778	
	Species	*Stems*	*Species*	*Stems*
Trees within the 1 ha plot used by diurnal lemurs for fruit eating (660 and 778 trees considered for Ranomafana and Kirindy/CFPF, respectively)	70 (62%)	480 (73%)	14 (61%)	140 (18%)
Trees within the 1 ha plot used by diurnal lemurs for flower and leaf eating (660 and 778 trees considered for Ranomafana and Kirindy, respectively)	31 (28%)	156 (24%)	13 (57%)	446 (57%)
Crown diameter of fruit trees used by *Eulemur* spp. and *Propithecus* spp. (mean ± s.d.)	4.09 m ± 2.43 m *n* = 109		6.75 m ± 3.57 m *n* = 48	

Source: Data for Ranomafana from Wright (unpubl.); data for Kirindy/CFPF from Rohner & Sorg (1986/1987), Ganzhorn (1995*b*, unpublished), Sorg & Rohner (1996), Abraham *et al.* (1996) and Carrai (1996); note that N5 here represents the plot by Abraham *et al.* (1996), measuring only 0.93 ha with trees ≥ 30 cm girth; this is a different plot than shown in Fig. 4.4.

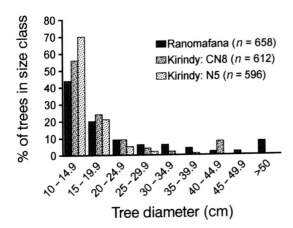

Fig. 4.5. Size distribution of trees ≥ 10 cm DBH on 1 ha plots (MBG-standards) in the wet forest (Ranomafana: Vatoharana; Wright (unpubl.) and two plots in the dry deciduous forest (Kirindy/CFPF; CN8 is a riverine forest; N5 is a plot on an edaphically dry site; Bienert & Ganzhorn (unpubl.).

are considered. Nevertheless, the size distributions differ significantly between all three forest plots (Kolmogorov–Smirnov test: $P < 0.001$ for all comparisons). The size diversity of tree diameters is highest in the wet forest, followed by the riverine forest (CN8) and by the dry forest growing away from the river (N5).

Within the 1 ha plot of Ranomafana, 70 tree species with 480 stems provide fruit for *Propithecus diadema* and *Eulemur* spp. and 31 tree species with 156 stems can be used for leaf eating.

Of the 45 tree species with 778 stems found in the 0.93 ha plot in Kirindy (N5), 13 species (140 stems) provide fruit and 13 species provide leaves and flowers (446 stems) as food for *Propithecus verreauxi* and *Eulemur fulvus* (Table 4.5). Thus, fruit availability based on the number of trees is higher in the wet forest than in the dry forest. In the dry forest more trees are used for leaf eating. But congeneric lemur species of similar body size feed on smaller sized tree crowns in the wet than in the dry forest (Table 4.5; Fig. 4.2). Thus, even though the density of trees with lemur fruit is about 3.4 times higher in the wet forest, the total amount of fruit available may not be all that much different between forest types. These small crown diameters as well as the higher investment in wood growth compared to the dry forest, may reflect smaller overall crown sizes in wet forests, possibly due to impacts of cyclones, which may be one of the most important agents affecting Malagasy forest ecosystems in general (Ganzhorn, 1995a).

In addition, fruiting is highly unpredictable in the wet forest. In Ranomafana, fruit trees used by lemurs produce fruit only in about 40 % of all years, alternating with years with almost complete failures of fruit production. The low probability of fruiting applies to other tree species of the wet forest as well (Gachet, 1969). This means, that in the wet forest, frugivorous lemurs have to go through bottlenecks of food availability at least every other year. In the dry forest, trees fruit much more regularly than in the wet forest. We postulate, that this difference in fruiting probability has a major impact on the biomass of frugivorous lemurs in different areas.

SOIL QUALITY AS A POSSIBLE REASON FOR FRUITING PATTERNS

Soils in Ranomafana have higher concentrations of carbon and nitrogen than soils of the Kirindy forest, indicating substrate for intensive microbial activity in Ranomafana but not in Kirindy. But the exchange capacity for positive ions is very low in the wet forest (Table 4.6). In concert with the low pH and low concentrations of potassium, trees in Ranomafana may not have sufficient supplies of minerals to allow regular fruiting. In contrast, tree growth in Kirindy is limited by water availability for about six to eight months per year, when it does not rain. Water retention time in the Kirindy Forest is five to seven days at the most (Sorg, 1996). This means that trees have to stop growing after about one week without rain. The same might apply to microbial activity. In the Kirindy Forest, the low speed of decomposition during the dry season, probably little to no uptake of minerals during that time, the narrow time frame for tree growth during the short rainy season and the flat terrain, which reduces leaching of minerals, may result in sufficient mineral reservoirs for fruit production, once there is enough water.

This argument may also be relevant for folivorous primates. In this group, the biomass of folivorous primates is correlated positively with leaf quality and pH of the soil, and negatively with annual rainfall (Oates et al., 1990; Ganzhorn, 1992). Thus, low mineral availability in acid soils together with leaching due to high rainfall may simply reduce the growth conditions for trees, resulting in poor foliage for folivores and unpredictable fruit production for frugivores in the wet forest.

The comparison of growth rates of the trees in the wet and in the dry forest as shown in Table 4.5 is hampered by uneven size distribution of the sample trees. In Ranomafana all trees sampled were larger than 10 cm DBH. In Kirindy, only 11 of the 151 trees considered were larger than 10 cm DBH (in Kirindy, there was no difference in mean annual growth rates between trees smaller or larger than 10 cm DBH). It is clear that growth rate is related to DBH (Clark, 1994), but it also depends on sun exposure (Ganzhorn, 1995b) and the surrounding vegetation. None of these has been measured in a way that would allow inter-site comparisons. Nevertheless we believe that our comparisons reflect reality. In the dry forest, even the larger and fastest growing tree species reach annual increments in DBH of only about 1 mm, with maxima of 2 to 3 mm (Deleporte et al., 1996).

The higher growth rates of trees found in the wet forest is not incompatible with the idea of lower soil fertility and smaller crowns in the wet compared to the dry forest. Wood can be produced mainly with water and CO_2, while the production of fruit and high quality leaves requires more minerals. The high investment of rainforest trees into stem wood and the small sizes of crowns may reflect constraints imposed upon the trees by cyclones at Madagascar's east coast.

SUMMARY

This review revealed some unexpected similarities and differences between wet and dry forests of Madagascar. To avoid misunderstandings, it should be kept in mind that "wet forests" here are represented almost exclusively by mid-altitude forests while data for "dry forests" are exclusively based on low altitude forests. Thus, the comparisons may not be representative for all of Madagascar's forests.

Lemurs reach higher population densities and overall biomass in dry than in wet forests, but the number of species is lower at the dry sites. A similar inverse relationship between species number and biomass has also been

Table 4.6. *Soil characteristics of Ranomafana[a] and Kirindy/CFPF[b]*

	Depth (cm)	pH	% C	% N	C/N	Exchange capacity for positive ions (m equiv/kg)	P (water soluble) (mg/kg soil)	K (water soluble) (mg/kg soil)
Ranomafana	0 to 6	3.6	13.74	0.77	17.8	0.08 to 7.59	1.0 to 5.6	0.07 to 0.16
Ranomafana	6 to 18	4.1	4.94	0.33	15.0			
Wet forest comparisons (Oates *et al.*, 1990)	0–20	3.4 to 6.0					4.5 to 22.4	0.004 to 0.08
Kirindy/CFPF	10	6.0 to 6.5	0.53 to 2.15	0.03 to 0.15	13 to 21	43.7 to 157.2	0.2–0.9	11.7 to 35.0
Kirindy/CFPF	30	4.9 to 6.9				26.6 to 66.6	0.2 to 0.6	4.2 to 15.8
Dry forest comparison (Maass, 1995)		4.3 to 6.3	2.07 to 2.32	0.09 to 0.21		59.1	9.4 to 22.5 (available P)	35.2 to 162.8 (exchangeable K)

Note: [a]From Johnson & Buol, in press.

[b]Range of four sites; from Felber, 1984.

described by Peres (1997) for the Amazon (see also Gupta & Chivers, chapter 2 and Janson & Chapman, chapter 14, this volume).

These relationships between primary productivity, richness of consumer species and their biomass probably reflect complex and regionally varying causalities which are poorly understood for primates. For other organisms, nutrient-rich soils have lower plant species richness but higher biomass than poorer soils, and high species richness in the open ocean seems to be associated with low productivity (reviewed by Begon *et al.*, 1996). Conceptually, it makes sense to link the ascending part of the humped relation between productivity and consumer diversity to increased resource availability. But it is not clear why consumer species tend to decline in richer habitats (reviewed by Abramsky & Rosenzweig, 1984). In Madagascar, the higher number of consumer species in the wet forest may in part be linked to increased number of tree species with increasing rainfall. The resulting increase in microhabitat diversity might allow increased lemur species packing in the wet forest.

The densities of trees with DBH greater than 10 cm are very similar in dry and in wet forests. But the average crown size of food trees is smaller in the wet than in the dry forest and fruiting is much more irregular. Thus, overall plant productivity of everything that costs more than just water and CO_2 may be lower in the wet than in the dry forest. In part, this may contribute to the lower biomass of lemurs in wet compared to dry forests. However, this difference, in particular, may be a consequence of comparing wet and dry forest from different altitudes.

The most significant difference between wet and dry forests is a much lower fruiting probability of trees from the wet forest. On average they bear fruit only once every three years, while dry forest trees fruit much more predictably in most years. This difference may result from soil characteristics. Soil from a wet forest site is extremely poor in exchangeable minerals while soil from a dry forest site is not. Under these conditions, one load of fruits might exhaust the tree for the next few years to come, such as described for many mast fruiting trees (reviewed by Kelly, 1994). The resulting bottlenecks in food availability together with poor food quality, again due to low soil fertility, might then have a major impact on the carrying capacity of Madagascar's forests for lemurs.

ACKNOWLEDGMENTS

We thank J. Fleagle, K. Reed and C. Janson for their invitation to participate in the workshop on primate communities. Financial support of the Wenner-Gren-Foundation and the Deutsche Forschungsgemeinschaft is gratefully acknowledged. We thank E.L. Simons for sharing unpublished data on the distribution of *Daubentonia*. J. Fleagle and C. Janson provided very helpful comments on the manuscript. S. Hänel drew the artwork.

REFERENCES

Abraham, J.-P., Rakotonirina, B., Randrianasolo, M., Ganzhorn, J. U., Jeannoda, V. & Leigh, E. G. (1996). Tree diversity on small plots in Madagascar: a preliminary review. *Revue d'Ecologie,* **51**, 93–116.

Abramsky, Z. & Rosenzweig M. L. (1984). Tilman's predicted productivity-diversity relationship shown by desert rodents. *Nature,* **309**, 150–1.

Balko, E. A. (1997). *A behaviorally plastic response to forest composition and habitat disturbance by Varecia variegata in Ranomafana National Park, Madagascar.* Ph.D. Dissertation. SUNY at Syracuse.

Begon, M., Harper, J. L. & Townsend, C. R. (1996). *Ecology.* Oxford: Blackwell.

Burney, D. A. (1997). Theories and facts regarding Holocene environmental change before and after human colonization. In *Natural Change and Human Impact in Madagascar,* ed. S. M. Goodman & B. D. Patterson, pp.75–89. Washington, DC: Smithsonian Institution Press.

Carrai, V. (1996). *Eco-etologia di* Propithecus verreauxi verreauxi *nella foresta di Kirindy (Morondava, Madagascar).* Masters Thesis. University of Pisa, Italy.

Charles-Dominique, P. & Petter, J. J. (1980). Ecology and social life of *Phaner furcifer.* In *Nocturnal Malagasy Primates: Ecology, Physiology and Behavior,* ed. P. Charles-Dominique, H. M. Cooper, A. Hladik, C. M. Hladik, E. Pages, G. F. Pariente, A. Petter-Rousseaux, J. J. Petter & A. Schilling, pp. 75–96. New York: Academic Press.

Charles-Dominique, P., Cooper, H. M., Hladik, A., Hladik, C. M., Pages, E., Pariente, G. F., Petter-Rousseaux, A., Petter, J. J., & Schilling, A. (ed.) (1980). *Nocturnal Malagasy Primates: Ecology, Physiology, and Behavior.* New York: Academic Press.

Clark, D. A. (1994). Plant demography. In *La Selva: Ecology and Natural History of a Neotropical Rainforest,* ed. L. A. McDade, K. S. Bawa, H. A. Hespenheide & G. S. Hartshorn, pp. 90–105. Chicago: University of Chicago Press.

Colquhoun, I. C. (1998). The lemur community of Ambato Massif: an example of the species richness of Madagascar's classified forests. *Lemur News,* **3**, 11–14.

Deleporte, P., Randrianasolo, J. & Rakotonirina (1996). Sylviculture in the dry dense forest of western Madagascar. In *Primate Report,* **46-1**: *Ecology and Economy of a Tropical Dry Forest in Madagascar,* ed. J. U. Ganzhorn & J.-P. Sorg, pp. 89–116. Göttingen.

Dewar, R. E. (1984). Extinctions in Madagascar. In *Quaternary Extinctions,* ed. P. S. Martin & R. G. Klein, pp. 574–593. Tuscon: University of Arizona Press.

Du Puy, D. J. & Moat, J. (1996). A refined classification of the primary vegetation of Madagascar based on the underlying geology: using GIS to map its distribution and to assess its conservation status. In *Biogeography of Madagascar,* ed. W. R. Lourenco, pp. 205–18. Paris: ORSTOM.

Felber, H. R. (1984). *Influence des principales propriétés physico-chimiques des sols et de la structure des peuplements sur le succès de la régénération naturelle d'essences représentatives sur des layons de débardage dans une forêt sèche de la côte occidentale de Madagascar.* Master thesis. Fachbereich Waldbau, ETH Zurich.

Findley, J. S. (1993). *Bats: A Community Perspective.* Cambridge: Cambridge University Press.

Gachet, C. (1969). Résultats de deux périodes d'observations phénologiques (Périnet et Betsipotika). Centre Technique Forestier et Tropical (Madagascar). *Etude Sol et Forêt N° 47,* Antananarivo.

Ganzhorn, J. U. (1989). Niche separation of the seven lemur species in the eastern rainforest of Madagascar. *Oecologia (Berlin),* **79**, 279–86.

Ganzhorn, J. U. (1992). Leaf chemistry and the biomass of folivorous primates in tropical forests. *Oecologia (Berlin),* **91**, 540–7.

Ganzhorn, J. U. (1995a). Cyclones over Madagascar: Fate or fortune? *Ambio,* **24**, 124–5.

Ganzhorn, J. U. (1995b). Low level forest disturbance effects on primary production, leaf chemistry, and lemur populations. *Ecology,* **76**, 2084–96.

Ganzhorn, J. U. (1998). Nested patterns of species composition and its implications for lemur biogeography in Madagascar. *Folia Primatologica,* **69**, 332–41.

Ganzhorn, J. U. (in press). Lemurs as indicators for assessing biodiversity in ecosystems of Madagascar: why it does not work. In *Aspects of Biodiversity on Ecosystems to Analysis of Different Complexity Levels in Theory and Praxis, Tasks for Vegetation Science Edition,* ed. A. Kratochwil. Dordrecht: Kluwer Academic Publishers.

Ganzhorn, J. U. & Kappeler, P. M. (1993). Lemuren Madagaskars: Tests zur Evolution von Primatengemeinschaften. *Naturwissenschaften,* **80**, 195–208.

Ganzhorn, J. U. & Kappeler, P. M. (1996). Lemurs of the Kirindy Forest. In *Primate Report,* **46-1**: *Ecology and Economy of a Tropical Dry Forest in Madagascar,* ed. J. U. Ganzhorn & J.-P. Sorg, pp. 257–274. Göttingen.

Ganzhorn, J. U. & Sorg, J.-P. (ed.) (1996). *Primate Report*, **46-1**: *Ecology and Economy of a Tropical Dry Forest*, Göttingen.

Ganzhorn, J. U., Malcomber, S., Andrianantoanina, O. & Goodman, S. M. (1997). Habitat characteristics and lemur species richness in Madagascar. *Biotropica*, **29**, 331–43.

Glander, K. E., Wright, P. C., Seigler, D. S., Randrianasolo, V. & Randrianasolo, B. (1989). Consumption of cyanogenic bamboo by a newly discovered species of bamboo lemur. *American Journal of Primatology*, **19**, 119–24.

Godfrey, L. R., Jungers, W. L., Reed, K. E., Simons, E. L. & Chatrath, P. S. (1997). Inferences about past and present primate communities in Madagascar. In *Natural Change and Human Impact in Madagascar*, ed. S. M. Goodman & B. D. Patterson, pp. 218–56. Washington, DC: Smithsonian Institution Press.

Goodman, S. M., (ed.) (1996a). A floral and faunal inventory of the eastern slopes of the Réserve Naturelle Intégrale d'Andringitra, Madagascar: with reference to elevational variation. *Fieldiana: Zoology, New Series Edition. Volume 85*. Chicago: Field Museum of Natural History.

Goodman, S. M. (1996b). Results of a bat survey of the eastern slopes of the Réserve Naturelle Intégrale d'Andringitra, Madagascar. In *Fieldiana: Zoology Edition,* **85***: A Floral and Faunal Inventory of the Eastern Side of the Reserve Naturelle Intégrale d'Andringitra, Madagascar: With Reference to Elevational Variation*, ed. S. M. Goodman, pp. 284–8. Chicago: Field Museum Natural History.

Goodman, S. M. & Ganzhorn, J. U. (1997). Rarity of figs (*Ficus*) on Madagascar and its relationship to a depauperate frugivore community. *Revue d'Ecologie*, **52**, 321–9.

Goodman, S. M. & Rakotondravony, D. (1996). The Holocene distribution of *Hypogeomys* (Rodentia: Muridae: Nesomyinae) on Madagascar. In *Biogéographie de Madagascar*, ed. W. R. Lourenco, pp. 283–93. Paris: ORSTOM.

Goodman, S. M. & Rakotozafy, L. M. A. (1997). Subfossil birds from coastal sites in western and southwestern Madagascar: a paleoenvironmental reconstruction. In *Natural Change and Human Impact in Madagascar*, ed. S. M. Goodman & B. D. Patterson, pp. 257–79. Washington, DC: Smithsonian Institution Press.

Harcourt, C. & Thornback, J. (1990). *Lemurs of Madagascar and the Comoros*. Gland: IUCN.

Hawkins, A. F. W. (1994). *Isalo Faunal Inventory*. Final Report to ANGAP. Landell Mills, Ltd.

Hladik, C. M., Charles-Dominique, P. & Petter, J. J. (1980). Feeding strategies of five nocturnal prosimians in the dry forest of the west coast of Madagascar. In *Nocturnal Malagasy Primates*, ed. P. Charles-Dominique, H. M. Cooper, A. Hladik, C. M. Hladik, E. Pages, G. F. Pariente, A. Petter-Rousseaux, J. J. Petter & A. Schilling, pp. 41–73. New York: Academic Press.

Johnson, B. K. & Buol, S. W. (in press). Soil survey and characterization of the Ranomafana National Park Region. In *Biodiversity Research in the Rain Forest of Madagascar Ranomafana National Park*, ed. P. C. Wright. Covelo, CA: Island Press.

Kalko, E. K. V., Handley, C. O. J. & Handley, D. (1996). Organization, diversity and long-term dynamics of a neotropical bat community. In *Long-Term Studies of Vertebrate Communities*, ed. M. L. Cody & J. A. Smallwood, pp. 503–54. London: Academic Press.

Kay, R. F., Madden, R. H., van Schaik, C. & Higdon, D. (1997). Primate species richness is determined by plant productivity: implications for conservation. *Proceedings of the National Academy of Sciences USA*, **94**, 13023–7.

Kelly, D. (1994). The evolutionary ecology of mast seeding. *Trends in Ecology and Evolution*, **9**, 465–70.

Krause, D. W., Hartmann, J. H. & Wells, N. A. (1997). Late Cretaceous vertebrates from Madagascar: implications for biotic change in deep time. In *Natural Change and Human Impact in Madagascar*, ed. S. M. Goodman & B. D. Patterson, pp. 3–43. Washington, DC: Smithsonian Institution Press.

Langrand, O. & Wilmé, L. (1997). Effects of forest fragmentation on extinction patterns of the endemic avifauna on the central high plateau of Madagascar. In *Natural Change and Human Impact in Madagascar*, ed. S. M. Goodman & B. D. Patterson, pp. 280–305. Washington, DC: Smithsonian Institution Press.

Lowry, P. P. II, Schatz, G. E. & Phillipson, P. B. (1997). The classification of natural and anthropogenic vegetation in Madagascar. In *Natural Change and Human Impact in Madagascar*, ed. S. M. Goodman & B. D. Patterson, pp. 93–123. Washington, DC: Smithsonian Institution Press.

Maass, J. M. (1995). Conversion of tropical dry forest to pasture and agriculture. In *Seasonally Dry Tropical Forests*, ed. S. H. Bullock, H. A. Mooney & E. Medina, pp. 399–422. Cambridge: Cambridge University Press.

MacPhee, R. D. E. & Marx, P. A. (1997). The 40,000-year plague: humans, hyperdisease, and first-contact extinctions. In *Natural Change and Human Impact in Madagascar*, ed. S. M. Goodman & B. D. Patterson, pp. 169–217. Washington, DC: Smithsonian Institution Press.

MacPhee, R. D. E., Burney, D. A. & Wells, N. A. (1985). Early Holocene chronology and environment of Ampasambazimba, a Malagasy subfossil lemur site. *International Journal of Primatology*, **6**, 463–89.

Martin, R. D. (1972). Adaptive radiation and behaviour of the Malagasy lemurs. *Philosophical Transaction of the Royal Society London B*, **264**, 295–352.

Mittermeier, R. A. (1988). Primate diversity and the tropical forest. In *Biodiversity*, ed. E. O. Wilson, pp.145–54. Washington, DC: National Academy Press.

Mittermeier, R. A., Konstant, W. R,. Nicoll, M. E. & Langrand, O. (1992*). Lemurs of Madagascar: An Action Plan for their Conservation 1993–1999*. Gland, Switzerland: IUCN.

Mittermeier, R. A., Tattersall, I., Konstant, W. R., Meyers, D. M & Mast, R. B. (1994). *Lemurs of Madagascar*. Washington, DC: Conservation International.

Myers, N. (1986). Tropical deforestation and a mega-extinction spasm. In *Conservation Biology: The Science of Scarcity and Diversity*, ed. M. E. Soulé, pp. 394–409. Sunderland, MA: Sinauer Associates.

Oates, J. F., Whitesides, G. H., Davies, A. G., Waterman, P. G., Green, S. M., Dasilva, G. L. & Mole, S. (1990). Determinants of variation in tropical forest primate biomass: new evidence from West Africa. *Ecology*, **71**, 328–43.

Overdorff, D. J. (1993). Similarities, differences, and seasonal patterns in the diets of *Eulemur rubriventer* and *Eulemur fulvus rufus* in the Ranomafana National Park, Madagascar. *International Journal of Primatology*, **14**, 721–53.

Overdorff, D. J. (1996). Ecological correlates to social structure in two lemur species in Madagascar. *American Journal of Physical Anthropology*, **100**, 487–506.

Overdorff, D. J., Strait, S. G. & Telo, A. (1997). Seasonal variation in activity and diet in a small bodied folivorous primate, *Hapalemur griseus*, in southeastern Madagascar. *American Journal of Primatology*, **43**, 211–23.

Peres, C. A. (1997). Primate community structure at twenty western Amazonian flooded and unflooded forests. *Journal of Tropical Ecology*, **13**, 381–405.

Petter, J. J., Albignac, R. & Rumpler, Y. (eds.) (1977*). Faune de Madagascar: Mammifères Lémuriens. Volume 44*. Paris: OSTOM CNRS.

Phillipson, P. B. (1994). Madagascar. In *Centres of Plant Diversity. Volume 1S*, ed. D. Davis, V. H. Heywood & A. C. Hamilton, pp. 271–81. Gland: WWF and IUCN.

Phillipson, P. B. (1996). Endemism and non-endemism in the flora of south-west Madagascar. In *Biogeography of Madagascar*, ed. W. R. Lourenco, pp. 125–36. Paris: ORSTOM.

Reed, K. E. & Fleagle, J. G. (1995). Geographic and climatic control of primate diversity. *Proceedings of the National Academy of Sciences USA*, **92**, 7874–6.

Richard, A. F. & Dewar, R. E. (1991). Lemur ecology. *Annual Review Ecology and Systematics*, **22**, 145–75.

Rohner, U. & Sorg, J.-P. (1986/1987). Observations phénologiques en forêt dense sèche. T.1/T.2 Réimpression 1989. CFPF Morondava, Fiches techn. 12/13.

Schoener, T. (1974). Resource partitioning in ecological communities. *Science*, **185**, 27–39.

Smith, A. (1984). Diet of Leadbeater's Possum, *Gymnobelideus leadbeateri* (Marsupialia). *Australian Wildlife Research*, **11**, 265–73.

Smith, R. J. & Jungers, W. L. (1997). Body mass in comparative primatology. *Journal of Human Evolution*, **32**, 523–59.

Sorg, J.-P. (1996). L'étude de la végétation, un outil au service de l'amenagement et de la gestion des ressources forestières à Madagascar. *Akon'ny Ala*, **18**, 26–36.

Sorg, J.-P. & Rohner, U. (1996). Climate and phenology of the dry deciduous forest at Kirindy. In *Ecology and Economy of a Tropical Dry Forest in Madagascar*, *Primate Report*, **46-1**, ed. J. U. Ganzhorn & J.-P. Sorg, pp. 57–80. Göttingen.

Straka, H. (1996). Histoire de la végétation de Madagascar oriental dans les derniers 100 millénaires. In *Biogéographie de Madagascar*, ed. W. R. Lourenco, pp. 37–47. Paris: ORSTOM.

Tattersall, I. (1982). *The Primates of Madagascar*. New York: Columbia University Press.

Tattersall, I. (1998). Lemurs of the Comoro Archipelago: status of *Eulemur mongoz* on Mohéli and Anjouan, and of *Eulemur fulvus* on Mayotte. *Lemur News*, **3**, 15–17.

Thalmann, U. & Rakotoarison, N. (1994). Distribution of lemurs in central western Madagascar, with a regional distribution hypothesis. *Folia Primatologica*, **63**, 156–61.

Thiollay, J. M. (1994). Structure, density and rarity in an Amazonian rainforest bird community. *Journal of Tropical Ecology*, **10**, 449–81.

van Schaik, C. P. & Kappeler, P. M. (1996). The social system of gregarious lemurs: lack of convergence with anthropoids due to evolutionary disequilibrium? *Ethology*, **102**, 915–41.

Wells, N. (in press). Geology and soils: rain forest stability from the Eocene to the Holocene in eastern Madagascar. In *Biodiversity Science and Village Economics: Madagascar's Ranomafana National Park*, ed. P. C. Wright. Covelo, CA: Island Press.

White, F. J., Overdorff, D. J., Balko, E. A. & Wright, P. C. (1995). Distribution of ruffed lemurs (*Varecia variegata*) in Ranomafana National Park. *Folia Primatologica*, **64**, 124–31.

Wright, P. C. (1986). Diet, ranging behavior and activity pattern of the gentle lemur (*Hapalemur griseus*). *American Journal of Physical Anthropology*, **69**, 283–4.

Wright, P. C. (1997). The future of biodiversity in Madagascar: a view from Ranomafana National Park. In *Natural Change and Human Impact in Madagascar*, ed. S. M. Goodman and B. D. Patterson, pp. 381–405. Washington, DC: Smithsonian Institution Press.

Wright, P. C. & Martin, L. M. (1995). Predation, pollination and torpor in two nocturnal prosimians: *Cheirogaleus major* and *Microcebus rufus* in the rainforest of Madagascar. In *Creatures of the Dark*, ed. L. Alterman, G. A. Doyle & M. K. Izard, pp. 45–60. New York: Plenum Press.

Wright, P. C., Heckscher, S. K. & Dunham, A. F. (1997). Predation on Milne-Edward's Sifaka (*Propithecus diadema edwardsi*) by the Fossa (*Cryptoprocta ferox*) in the rain forest of southeastern Madagascar. *Folia Primatologica*, **68**, 34–43.

Zimmermann, E., Cepok, S., Rakotoarison, N., Zietemann, V. & Radespiel, U. (1998). Sympatric mouse lemurs in north-west Madagascar: a new rufous mouse lemur species (*Microcebus ravelobensis*). *Folia Primatologica*, **69**, 106–14.

5 • Primate diversity

JOHN G. FLEAGLE, CHARLES H. JANSON AND KAYE E. REED

The chapters in this section have provided a series of geographically focused reviews of the diversity of ecological communities in the four major biogeographical regions currently inhabited by non-human primates – Africa, Asia, Madagascar, and the neotropics of South and Central America. The comparative data that are available for these different regions are by no means uniform. As a result, these chapters not only offer an opportunity to compare the similarities and differences in the primate communities of these different regions, but also highlight the gaps in our current knowledge. Despite differences in the scope and focus of the different chapters, several general themes emerge in the intraregional studies found in this section.

In all of the chapters, the authors emphasized that there are very few primate communities in the world today that have not been affected in some way by human activity (see also Tutin & White, chapter 13, Peres, chapter 15 and Struhsaker, chapter 17, this volume). In addition, it is important to keep in mind that many of the sites where primates have been most thoroughly studied have often been chosen precisely because they have extremely high numbers of species and often easy access from roads. As a result all comparisons either within regions or between them need to make some attempt to take these factors into consideration.

One aspect that is most readily comparable among primate communities is α diversity (the number of species in a community). Of the four major biogeographic regions, Africa seems to have the highest levels of α diversity found anywhere in the world with up to 17 species in some communities. At the opposite extreme, Asian communities generally have the lowest maximum diversity with few communities having more than eight sympatric species. The most diverse communities in Madagascar and South America fall between these extremes. The reasons for such high numbers in Africa are not immediately obvious; however, many of the richest African "sites", seem to be composed of a complex mosaic of different habitats (Bourliere, 1985). Indeed, this raises the broader issue of how one identifies a single community as the diversity of virtually any "community" will increase as one expands the radius of included habitats (see especially Chapman *et al.*, chapter 1, this volume).

Most of the chapters in this section found that α diversity is positively associated with rainfall (see also Reed & Fleagle, 1995; Emmons, chapter 10, this volume). However, the strength of this relationship was quite variable. The relationship between α diversity and rainfall is certainly not a linear one (Kay *et al.*, 1997) and at different rainfall levels the relationship may be positive, negative or insignificant. The complexity of this relationship only emphasizes the need to understand the proximate factors linking rainfall to primate diversity, including the relationship between rainfall and plant productivity and plant species diversity as well as the role of soil type and seasonality of rainfall. In general, α diversity seems to be driven largely by the numbers of frugivorous taxa.

In contrast, α diversity at sites within geographic regions is generally not positively correlated with biomass. Rather they tend to be negatively correlated. Several studies in this section found that community biomass was greater and species diversity lower in dryer forests (see also Janson & Chapman, chapter 14, this volume). Biomass seems to be largely driven by the abundance of folivores - howling monkeys in South America and colobines in Africa and Asia (see also Reed, chapter, 7, this volume).

In examining the distribution of individual species within each of the geographical regions, most authors found no general pattern from either body size or diet to predict the species with the broadest geographical range.

In reviewing the diversity of primate communities within different biogeographical regions and reviewing the likely factors that have led to this diversity, all of the authors identify the same suite of factors – geological and biogeographic history, availability of resources, competition with other vertebrates, and local extinctions due to human activities. These factors are all addressed in more detail in later chapters of this volume.

REFERENCES

Bourlière, F. (1985). Primate communities: their structure and role in tropical ecosystems. *International Journal of Primatology*, **6**, 1–26.

Kay, R. F., Madden, R. H., van Schaik, C. & Higdon, D. (1997). Primate species richness is determined by plant productivity: implications for conservation. *Proceedings of the National Academy of Sciences USA*, **94**, 13023–7.

Reed, K. E. & Fleagle, J. G. (1995). Geograpic and climatic control of primate diversity. *Proceedings of the National Academy of Sciences USA*, **92**, 7874–7876.

6 • Phylogenetic and temporal perspectives on primate ecology

JOHN G. FLEAGLE AND KAYE E. REED

INTRODUCTION

As many studies in this volume and elsewhere have noted, there are persistent differences in the ecological characteristics of the individual species assemblages found on different continents (Terborgh & van Schaik, 1987; Fleagle & Reed, 1996; Kappeler & Heymann, 1996). For example, the primates of the Neotropics tend to be smaller and less ecologically diverse than those of other continents, while Madagascar has a larger number of folivores than other biogeographical areas. It seems almost certain that the differences between the primate assemblages of different biogeographical regions are the result of many causal factors and their interactions, including differences in productivity, in the composition of the plant communities, in climate and soil, and in the potential for competition with other groups of vertebrates. These factors are examined in other chapters of this volume.

However, the primate assemblages we see today are not simply epiphenomena of present ecological conditions. Rather, they are also the product of evolutionary and ecological processes that have been ongoing for millions of years. Environments are not stable today, and they never have been; they are constantly changing. The temporal scale of environmental change ranges from decades and centuries for human-induced activities of habitat destruction such as logging, land clearing, hunting, or introduction of exotic species (Struhsaker, chapter 17, this volume) and also for natural epidemics. Global climatic cycles seem to have periodicities ranging from a few years, such as the El Niño phenomena to tens of thousands of years for the glacial cycles that have dominated the past two million years (Tutin & White, chapter 13, this volume; Potts, 1994). On the even larger scale of geological epochs, there have been dramatic changes in continental interconnections and in sea level (Haq *et al.*, 1987) that have altered both global and regional climates and led to major biogeographical changes in the distribution of fauna (Brown & Lomolino, 1998).

Thus primate assemblages have histories, both as individual species and as communities. The assemblage of species (either taxonomic or ecological) that we find in a 100 000 hectare block of land today is almost certainly not the same we would have encountered 5000 years ago, 50 000 years ago, 5 million years ago, or 50 million years ago. Just as local habitats have changed throughout earth history, so have the species occupying those habitats. Although to some degree the species found in any local fauna certainly reflect local environmental conditions, the relationship between climate change and faunal change at a global level is by no means simple – there are many other factors involved. In addition to climate change, faunal change may also result from factors such as the chance timing of adaptive breakthroughs in individual lineages. More dramatically, the global faunal changes that took place at the Cretaceous–Tertiary boundary and the Pleistocene extinctions that took place in many parts of the world with the arrival of humans cannot be simply explained by climate (Martin & Klein, 1984), nor can the faunal changes that took place in North America and South America with the appearance of the Panama Land Bridge in the early Pliocene (Stehli & Webb, 1985).

Just as there is evidence of major faunal change for reasons other than major climate change, there is also evidence of faunal similarity that, at some level, is independent of current environmental differences. Thus, although there are certainly examples of ecological and morphological convergence between rainforest mammals in different parts of the world (Terborgh, 1992) most studies of primate communities have emphasized the differences between primates in different regions (Kappeler & Heymann, 1996), and similarities within regions (Fleagle & Reed, 1996). For example, although there are dramatic ecological differences between the habitats of the eastern rainforest and the western dry forests of Madagascar (Ganzhorn, Wright & Ratsimbazafy, chapter 4, this volume), the primate assemblages of the two regions are still much more similar ecologically to one another (Figs 6.7 and 6.8) than to primate assemblages in other parts of the world (Fleagle & Reed, 1996). While this may partly reflect characteristic features of the Malagasy environment, it also reflects the nature of the primates that inhabit Madagascar due to historical and biogeographical circumstances.

The goal of this chapter is to examine modern primate assemblages in different parts of the world from a temporal and phylogenetic perspective, to examine the relationship between phylogenetic history and the patterns of ecological diversity we find today in the primate assemblages of different continents. Our major hypothesis is that the patterns of ecological diversity we find in different parts of the world today reflect the deep evolutionary history of these different faunas. The patterns of ecological diversity we see today in primate faunas can only be understood through an understanding of their evolutionary history. More specifically, for the living primate assemblages of Africa, Asia, Madagascar and South America we can ask: Are the characteristic features of body size, diet and locomotion among the primate faunas of different biogeographical areas today recent phenomena dating from the last few thousands of years, or do many of these differences represent patterns of species diversity that have been established for tens of millions of years? Did current assemblages appear more or less in present form through a single explosive radiation, or are they the result of gradual accretion of taxa? Are the ecological differences in primate communities that we see today remnants of larger faunas that have been depleted through recent climatic change or human activities? Is there evidence of major differences in the rate of ecological diversification in different regions occupied by primates?

There are two main types of evidence that can be used to reconstruct the history of individual primate faunas. One source is the fossil record. The fossil record can provide positive evidence of the presence of primates found in different parts of world during various times in the past (Fleagle & Kay, 1985; Fleagle et al., 1997; Fleagle, 1999). The fossil record also can provide several types of information about the comparative ecology of extinct species. Since most fossil species are known from teeth, body size (Fleagle, 1978, 1999) and diet (Kay & Covert, 1984; Strait, 1997; Ungar, 1998) can be reconstructed relatively well for most species and provide the best clues to ecology. Parts of the postcranial skeleton are less common, but can provide clues to locomotor behavior for some extinct taxa (Bown et al., 1982).

Nevertheless, the fossil record is woefully incomplete; there are many gaps in the record for many millions of years in most parts of the world and there are many geographical areas for which we have virtually no record. In general, the fossil record can clearly provide positive evidence for the presence of particular groups of primates and species with particular ecological characteristics, but the absence of fossil evidence does not necessarily indicate an absence of particular groups or ecotypes if we have other reasons to suspect their presence. For example, there are fossil tarsiers in Asia from the Eocene and Miocene, in addition to several extant species. Despite the absence of any fossil tarsiers from the Oligocene, Pliocene and Pleistocene, it seems likely that they were present in parts of Asia during these time periods as well.

A second source of evidence about the history of primate faunas comes from the biomolecular phylogenetic studies that enable us to reconstruct the timing of the first appearances of particular groups of living primates (Porter et al., 1997). Using this information we can at least postulate the existence of a particular taxon even if we have no evidence of its presence in the fossil record or direct evidence of its paleoecology. For example, most studies of the evolutionary history of platyrrhine primates indicate that the lineage leading to the genus *Cebus* has been present as a separate lineage from *Saimiri*, its closest living relative, for at least 15 to 20 million years, despite the absence of any capuchin-like monkeys in the fossil record. Thus, there were capuchin monkeys of some sort somewhere in the neotropics for much of the past 15 million years, even if we know nothing about the past distribution of this lineage. Together these two sources of information can provide a broad temporal and biogeographic framework for examining the history of the primates we see in the different parts of the world today.

OVERVIEW OF THE PRIMATE FOSSIL RECORD

From a very broad perspective, the fossil record of primates can be seen as a cascade of adaptive radiations appearing in successive epochs throughout the last 55 million years, with the living faunas representing, to some degree, remnants or expansions of each of the radiations (Fig. 6.1; Fleagle & Reed, 1997; Fleagle, 1999).

The first primates that can be clearly related to living members of our order date from the beginning of the Eocene Epoch, approximately 55 million years ago. In many adaptive features as well as their likely phylogenetic affinities, these earliest primates were most similar to extant prosimians – the tarsiers of south-east Asia and the strepsirhine primates (lemurs and lorises) of Madagascar, Africa and Asia. In the Eocene of North America, Europe and Asia there were numerous tarsier-like omomyoids and

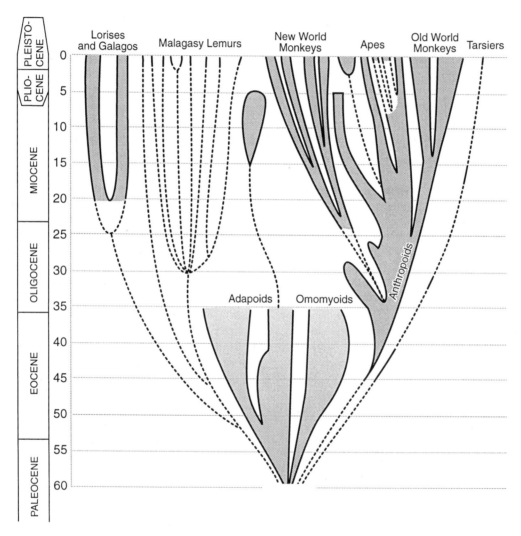

Fig. 6. 1. A geological time chart showing major events in primate evolution and the distribution of fossil primates throughout time and space (from Fleagle, 1989).

lemur-like adapoids. In general the prosimians of the Eocene, like those alive today, were relatively small primates with most species less than 1 kg in estimated mass and only a few as large as 5 kg (Fig. 6.2). Only toward the later part of the Eocene do we find a few larger species. In some areas of western North America and Western Europe there are fossil assemblages with more than a half dozen synchronic species. Although more poorly known, Asian primate faunas seem to be roughly similar to those of North America and Europe but differ in having true tarsiers and a unique radiation of adapoids (Beard 1998a,b). The Eocene primate assemblages include both diurnal and nocturnal species as well as quadrupedal and leaping forms. The fossil prosimians that are abundant throughout much of the Eocene in North America and Europe are absent from Oligocene faunas, and most probably became extinct because of a dramatic cooling of global temperatures that took place at the Eocene–Oligocene boundary.

The end of the Eocene Epoch and the early part of the Oligocene is characterized by the first appearance of anthropoid primates in Northern Africa and probably Asia (Simons & Rasmussen, 1994; Kay et al., 1997). The Eocene higher primates from North Africa and Asia are generally very small, tarsier or marmoset-sized species, but by the early Oligocene there was a range of species in North Africa (and Oman) that were comparable in size and adaptations to modern platyrrhines (Fig. 6.2). The Eocene anthropoids were probably largely insectivorous and frugivorous. The few postcranial elements indicate quadrupedal and leaping habits. The early Oligocene anthropoids show a diversity of dietary and locomotor adaptations comparable to those of modern platyrrhines with mostly frugivorous and seed-eating diets, few if any folivores, and all arboreal quadrupeds and leapers (Fleagle & Kay, 1985).

The end of the Oligocene and beginning of the Miocene Epoch (23 mya) marks the first appearance of fossil platyrrhines in South America and the first appearance of hominoids (apes in the broadest sense) and Old World monkeys in Africa. Early and middle Miocene fossil platyrrhines are difficult to assign conclusively to modern groups of platyrrhines, but by the later Miocene most living families and perhaps even a few modern genera are present in the fossil record (Fleagle et al., 1997).

Although both apes and monkeys first appear in the early Miocene of Africa, species that can be linked with extant families appear much later in the Epoch (Delson,

1994). There are also early lorisoids, probably both galagos and lorises, from the early part of the Miocene of Africa, and modern lorises from the late Miocene of Asia.

The end of the Miocene and beginning of the Pliocene Epoch (5 mya) is marked by the major radiation of Old World monkeys in Africa and Eurasia. The only fossil record for Malagasy prosimians is from the latest Pleistocene and historical times although the radiation is obviously much more ancient.

THE TIMING OF PRIMATE RADIATIONS FROM MOLECULAR EVIDENCE

Attempts to calibrate primate and human evolution using a 'molecular clock' go back over three decades (Sarich & Wilson, 1967). However, the amount of data and the sophistication of the analyses going into such reconstructions have increased vastly in recent years. Nevertheless, there remains some diversity in estimates for some events. The two major difficulties are the fact that different molecules clearly evolve at separate rates giving many different clocks and the basic problem that any molecular clock needs to be calibrated. There have been several recent attempts to provide time scales for primate evolution using molecular phylogeny (Purvis, 1995; Porter et al., 1997; Yoder, 1997; Goodman et al., 1998; Stewart & Disotell, 1998). When calibrated to a platyrrhine–catarrhine divergence of 35 mya or a cercopithecoid–hominoid split of 25 mya, the dates suggested by molecular analyses generally accord reasonably well with the first appearances of most taxa in the fossil record. Nevertheless, there remain some striking discrepancies in divergence dates provided by molecular phylogenies, especially among the strepsirrhines. For example, Yoder (1997) suggests 50 mya as the divergence of galagos and lorises while Porter et al. (1997) give a date of 23 mya, more in accordance with the fossil evidence. Moreover, some authorities argue that the currently estimated divergence dates that accord reasonably well with the fossil record are all much too young (Martin, 1993). For the analyses in this paper we have used divergence times taken primarily from Porter et al. (1997) and Stewart & Disotell (1998). These accord well with the fossil record. The most notable gaps are in estimates for the divergence times of individual species of Malagasy primates and catarrhines; for these we have made estimates based on uncalibrated molecular phylogenies (Yoder, 1997;

Taxonomic Composition

Body Size Distribution

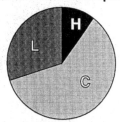

Present Kibale Forest

H = Hominoids
C = Cercopithecoids
L = Lorisoids

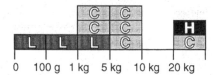

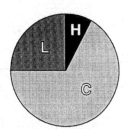

Pliocene 3.0 mya Omo, Shungura B

H = Hominoids
C = Cercopithecoids
L = Lorisoids

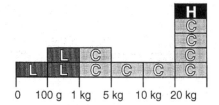

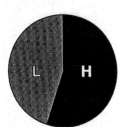

Early Miocene 20 mya Songhor, Kenya

H = Hominoids
L = Lorisoids

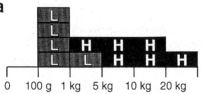

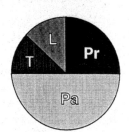

Early Oligocene 33 mya Fayum, Quarries I,M

Pa = Parapithecoids
Pr = Propliopithecoids
L = Lorisoids
T = Tarsioids

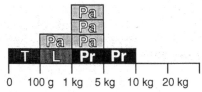

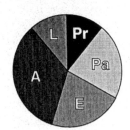

Late Eocene 36 mya Fayum, L-41

Pa = Parapithecoids
Pr = Propliopithecoids
E = Early Anthropoids
L = Lorisoids
A = Adapoids

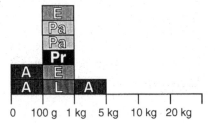

Fig. 6. 2. The size range of primates, based on tooth size,
throughout the past 65 million years (from Fleagle, 1999).

Disotell, 1996; Morales & Melnick, 1998) with the divergence date from their sister genera as an upper limit. As the timing of these divergences are probably all relatively recent, it is unlikely that the estimates are off by more than a few million years.

Using this general summary of primate evolution (see Fleagle, 1999, for detailed references) and the timing of major phylogenetic events in primate history, we can now attempt to reconstruct the history of the primate faunas in each of the four major continental areas – Africa, Asia, Madagascar, and the Neotropics.

Africa

The fossil record of primates is richer from Africa than from any other continent. Nevertheless, our knowledge of fossil primates from the past 60 million years on that continent is still based on limited remains that are widely and unevenly distributed in time and space. Fossil mammals, including primates, are known only from North Africa during the Paleocene, Eocene and early Oligocene. Northern and especially eastern Africa have extensive fossil records through much of the Miocene, Pliocene and Pleistocene. Only a few limited Miocene faunas and abundant Pliocene and Pleistocene faunas are known from southern Africa. Central and western Africa have virtually no fossil record.

There are four families of nonhuman primates found throughout Africa today (Figs. 6.3 and 6.4): galagos (Galagidae), lorises (Lorisidae), Old World monkeys (Cercopithecidae, including colobines and cercopithecines) and great apes (Pongidae). It is reasonable to assume that these are all of African origin as all have close relatives among fossil taxa from Africa, but this is a currently debated issue (see Begun, 1997; Stewart & Disotell, 1998). Fossil galagos attributable to modern genera date from the Pliocene, approximately 2–4 million years ago, an age that is compatible with molecular estimates for the divergence of several modern species (Masters *et al.*, 1994); however, there are several genera and species from the early Miocene 17–19 mya that have been consistently identified as fossil galagos even though they have more primitive limb bones than the modern genera and species. Likewise, there are several early Miocene genera and species that have traditionally been identified as fossil lorises, again without many of the limb specializations characteristic of the living species.

Fig. 6. 3.   Primates of the Taï Forest, Ivory Coast

Many living genera of Old World monkeys (*Cercopithecus, Papio, Cercocebus, Colobus*) have been identified from the Pliocene and Pleistocene (~ 4–1 mya) deposits of eastern and southern Africa. The first appearance of the two living subfamilies, colobines and cercopithecines, date to the late Miocene (10–7 mya), and the earliest Old World monkeys first appear approximately 19–15 million years ago in eastern and northern Africa. The living African apes, chimpanzees and gorillas, have no fossil record and molecular studies indicate they diverged from one another (and from humans) only in the last 10–5 million years. Nevertheless, throughout much of the Miocene (23 to 5 mya) there have been numerous ape-like fossil anthropoids. These fossil apes ranged from species the size of a small Old World monkey to others the size of a female gorilla. Some almost certainly had dietary habits comparable to living great apes although only a few seem to share

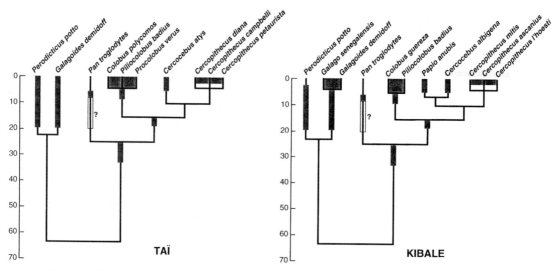

Fig. 6. 4. Molecular phylogeny of the primates of the Taï Forest, Ivory Coast and Kibale Forest, Uganda showing molecular divergence times. Broad bands indicate times that these taxa or relatives are documented in the African fossil record.

locomotor similarities with modern pongids. Several fossils from middle and late Miocene have been frequently identified as the sister taxa of modern African apes (Ishida & Pickford, 1997).

Comparative analyses of fossil African primates indicate that African primate assemblages have changed in both their taxonomic composition and their ecological diversity during the past 35 million years (Figs. 6.1 and 6.2). The late Eocene and Oligocene assemblages were very different from modern African assemblages in body size, dietary habits and locomotor diversity as well as phylogenetic composition. The diverse late Eocene community known from the Fayum in Egypt was composed of numerous tiny (100–300 g) anthropoids and prosimians with frugivorous and insectivorous diets. The largest primate was a folivorous adapoid prosimian with an estimated mass of about 3000 g. The early Oligocene fauna was taxonomically different from living faunas, and ecologically much like a modern platyrrhine fauna with numerous arboreal, largely frugivorous anthropoids ranging size between 300 g and 6000 g as well as a few prosimians, including a fossil tarsier (Fig. 6.2). The early Miocene assemblages of primitive hominoids, lorisoids, and a few monkeys were more similar to modern assemblages in body size and trophic diversity despite lacking modern families of Old World monkeys or

apes (Fig. 6.2; Fleagle & Kay, 1985; Fleagle, 1999). The primate assemblages from the Pliocene and Pleistocene of eastern and southern Africa are dominated by mangabeys, baboons and geladas, and large colobines with limited remains of smaller colobines and a few fossil galagos. The guenons that dominate modern African faunas are known from almost no fossils and there are no fossil chimpanzees or gorillas. Although the modern size range of African primates is not so different from that 20 million years ago or from that of some of the Pliocene and Pleistocene faunas, the composition of most modern primate assemblages is unlike anything known from the fossil record and all fossil assemblages are generally very different from any living assemblage. For example, many have noted that in the early Miocene Old World monkeys were very rare and the monkey-like species were either primitive apes or basal catarrhines whereas today monkeys are much more diverse than apes (Andrews, 1981; Fleagle, 1999). The Pliocene and Pleistocene faunas generally had much larger Old World monkey species than those of today and lack any substantial evidence of diversity among small guenons and prosimians. It seems most probable that these differences reflect not only taphonomic and collection biases, but also the recent radiation of many characteristic groups found among living faunas – cercopithecine monkeys and galagos

– and the fact that most of the fossil record is from northern, eastern, and southern Africa whereas the largest assemblages of living primates are found in the rainforests of central and western Africa. Although most fossil primates are from areas outside of present day tropical forests, paleoecological evidence from soils, botanical remains (including buttressed trees), and faunal structure indicates that Oligocene and Miocene primates were tropical forest species (Bown *et al.*, 1982).

Thus, the fossil record provides no direct evidence that the assemblages of primates found in the tropical forests of Africa today have any great temporal depth. Both paleontological and molecular evidence indicate that the phylogenetically diverse assemblage of primates that characterize most modern African faunas has accumulated over many epochs. Although the living genera and families are likely to have all originated in Africa sometime during the past 50 million years (but see Stewart & Disotell, 1998), the different groups have been separated from one another for varying durations (Fig. 6.4). The lineage leading to galagos and lorises has probably been distinct from anthropoids for 50 million years. The galagos and lorises have been distinct in some form for at least 20 million years, as have the monkeys and apes. However, neither the earliest apes nor the earliest monkeys possessed many of the adaptations that characterize living members of those groups. The two subfamilies of Old World monkeys have probably been distinct for 15 million years or so and guenons have been separated from baboons and mangabeys for approximately 10 million years. Individual species of colobines, guenons and galagos are estimated to have been separated from one another for less than 5 million years.

In summary, although there have been diverse assemblages of tropical forest primates in Africa for at least 35 million years, the modern assemblages are certainly more taxonomically (and perhaps more ecologically) diverse than any of the assemblages known from earlier epochs in that they contain a mixture of lineages that have appeared sequentially over the past 50 million years. Moreover, one could argue that African primate assemblages are perhaps expanding in geological time since many modern assemblages contain recent radiations of several sympatric species in both old (galagos) and relatively recent *(Cercopithecus)* groups (see Eeley & Lawes, chapter 12 and Chapman *et al.*, chapter 1, this volume).

Asia

Asia and Africa have been connected with one another sporadically for much of the last 20 million years and intermittently, perhaps through Europe, since the Eocene. Of the four major continental areas, Africa and Asia have the greatest overlap in primate taxa, sharing three families – Lorisidae, Cercopithecidae (including both cercopithecines and colobines), and Pongidae (great apes). Asia has two families not found in Africa today – the lesser apes (Hylobatidae) and the tarsiers (Tarsiidae) (Figs 6.5 and 6.6). As in Africa (and Europe) there was an extensive Miocene radiation of primitive apes and/or basal catarrhines. In addition, the adapoid prosimians which disappeared in North America, Europe, and Africa by the Oligocene persisted in Asia until the latest Miocene.

Tarsiers are clearly the most ancient and distinctive group of Asian primates and almost certainly originated in the region. Fossil tarsiers, virtually indistinguishable from the modern genus are known from the Eocene (45 mya) in China and the Miocene of Thailand (Beard, 1998). Fossil

Fig. 6. 5.   The primates of the Krau Game Reserve, Malaysia.

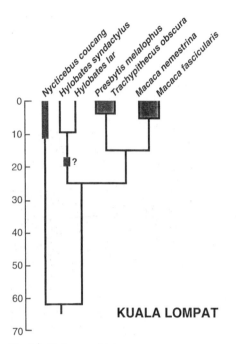

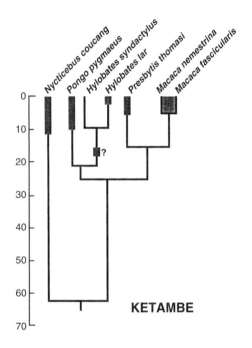

Fig. 6. 6 Phylogeny of Asian primates from the Krau Game
Reserve, Malaysia and Ketambe Reserve, Sumatra showing
molecular divergence times. Broad bands indicate times that
these taxa or relatives are documented in the Asian fossil record.

lorises are known from the later Miocene (12–10 mya of
northern Pakistan; MacPhee & Jacobs, 1986). Fossil apes
probably closely related to the Asian orangutan date from
roughly the same time (Ward, 1997). However in the later
Miocene there were several species of large ape, and there
were apes much larger than the living orangutan (including *Gigantopithecus* and larger orangutans) throughout
much of the Pliocene and Pleistocene in many parts of
southern and eastern Asia. The fossil history of the lesser
apes (Hylobatidae) is a frustrating and difficult problem in
primate evolution (Fleagle, 1984). There are fossil gibbons
from Pleistocene deposits in China and south-east Asia.
Beyond that there is no undisputed evidence for fossil gibbons. Although there have been fossil apes that resemble
living lesser apes in many aspects of their anatomy for nearly
20 million years in Asia and Europe, it is generally accepted
that they are not ancestral to living lesser apes. There are
Pleistocene fossils attributed to several living genera of
colobines and cercopithecines from many parts of Asia.
The earliest fossil colobines and cercopithecines, often allocated to living genera, first appear in Asia approximately

7 to 5 million years ago, soon after their divergence from
their sister taxa in Africa (Barry, 1987; Delson, 1994; 1996;
Morales & Melnick, 1998; Collura & Stewart, 1996).

To date there have been no comparative studies of the
primate assemblages from any Asian localities, probably
because most are so poorly known. Nevertheless, there are
indications that some Eocene faunas had remarkable numbers of extremely tiny (< 20 g) primates unlike anything
known today (Gebo *et al.*, 1998). Otherwise there are no
fossil faunas with large numbers of synchronic primate
taxa comparable to those of Africa or South America, and
most fossil faunas are both taxonomically and ecologically
limited compared with living Asian faunas. For example,
the later Miocene faunas of the Indian and Pakistan
Siwaliks generally have no more than two fossil apes, a
prosimian, an adapoid or two, and no monkeys.

As with Africa, there is no evidence that the modern
faunas of Asia are very old as a unit: rather, individual
parts have accumulated over the epochs (see also Gupta &
Chivers, chapter 2, this volume). Indeed, the major components, macaques and colobine monkeys, are first known

from the latest Miocene or early Pliocene (Fig. 6.6). Earlier faunas were composed of a few species of more primitive and generalized taxa, although some such as pongid great apes were more diverse (taxonomically and ecologically) and widespread than today. Like Africa, Asia contains a phylogenetically diverse assemblage of extant primates some of which have been distinct from one another for 50 million years or more – tarsiers, lorises, and anthropoids (Fig. 6.6). The apes diverged from the monkeys roughly 25 million years ago, the lesser apes diverged from the orangutan slightly more recently (18 mya), and the colobines and cercopithecines have been distinct for approximately 15 million years. Siamang and gibbons have been separate for about 10 million years. The different genera of leaf monkeys have been separate for less time as have the macaques. Finally, modern Asian faunas usually contain several closely related species of either the same (*Macaca*, *Hylobates*) or closely related genera (colobines). As in the African assemblages, Asia's primate fauna have changed continuously throughout the past 50 million years. There is a very hierarchical pattern of divergence times among the extant taxa as the modern faunal elements have slowly accumulated.

Madagascar

The living prosimians of Madagascar are totally unique to that island at the family level and seem to represent a very old endemic radiation (Yoder, 1997; Figs. 6.7, 6.8 and 6.9). However, any attempt to reconstruct the history of living Malagasy primate assemblages must be based totally on molecular estimates of phylogenetic divergences as there is no fossil record older than the last part of the Pleistocene. Nevertheless, the recent fossil record clearly demonstrates that the extant Malagasy fauna is a meager subset of the taxa present as recently as 1000 years ago. The most striking feature of the very recent fossil record is that it documents an extraordinary radiation of large, now-extinct lemurs that greatly expanded the ecological diversity of primates on that island (Godfrey *et al.*, 1997). Moreover, these large extinct species have not obviously been replaced by living species; rather, they were living alongside many of the same taxa alive today. With the exception of cheirogaleids, all living Malagasy genera are known from fossils of the same age as the giant extinct lemurs. In addition, some rare extant taxa are more abundant in the

Fig. 6.7. Diurnal and nocturnal primates of Ranomafana National Park, Madagascar.

fossil record (*Hapalemur simus*); one monotypic genus (*Daubentonia*) had a second, much larger, species in the recent past; and several extant taxa had much larger geographic ranges in the past (*Indri, Hapalemur, Daubentonia*). By any criteria, the living Malagasy primate assemblages are more taxonomically and ecologically limited than those of the very recent past. How this recent extinction has affected the biogeography, ecology, and behavior of the extant species is not clear (see van Schaik & Kappeler, 1996).

Despite the absence of an old fossil record, we can be sure that the living Malagasy primates document an old diversification (Fig. 6.9). Madagascar has been separated from the African mainland for over 100 million years. There is no paleontological indication when primates first reached Madagascar, but they radiated to become the most ecologically diverse group of mammals on the island (see Ganzhorn *et al.*, chapter 4, this volume). The most recent molecular estimates indicate that the Malagasy

Fig. 6. 8.   Diurnal and nocturnal primates of Marazolaza Reserve near Morondava, Madagascar.

strepsirhines have been separated from the African lorises and galagos since the beginning of the Eocene. However, molecular analyses also suggest that most of the living families have been distinct from one another for about 30 million years with only the aye-aye (*Daubentonia*) showing an older divergence (Porter *et al.*, 1997; Yoder, 1997). Whether this 30 million-year-old radiation reflects the first arrival of prosimians on Madagascar, some major climate event, or a methodological artifact is unclear, but has potential implications for the history of the fauna (see below). Divergences among species within families have generally not been calculated. Nevertheless, the phylogenetic composition of Malagasy assemblages is unusual compared with those of Africa and Asia in having relatively more taxa with relatively similar old divergences and fewer clusters of sympatric congeneric or closely related species. Rather, most congeneric species (except *Hapalemur* and *Eulemur*) are allopatric (see Ganzhorn *et al.*, chapter 4, this volume). As a result, the divergence times of the living Malagasy assemblages lack the hierarchical pattern seen in Africa and Asia. Rather, the divergence times among all but congeneric species are all relatively long, reflecting an explosive phylogenetic radiation of families about 30 million years ago.

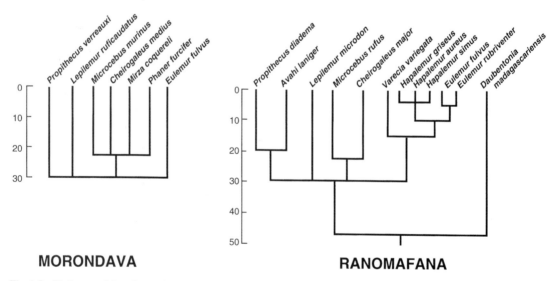

Fig. 6. 9.   Phylogeny of the primates from Ranomafana National Park and Marozalaza Reserve showing molecular divergence times.

South America

Compared with the primates of other major continental areas, the platyrrhine primates of the neotropics (Fig. 6.10)

are a much younger (Fig. 6.11) and more uniform radiation (Fleagle & Reed, 1996). For most of the Cenozoic Era (the last 65 million years), South America, like Madagascar, was an island continent with a mammalian fauna that was distinct from that of other continents (Simpson, 1980). The earliest primates in South America date from approximately 25 million years ago (Fig. 6.1), almost certainly as a waif dispersal from some other continent, probably Africa (Hartwig, 1994). This timing of the first platyrrhines is consistent with the presence of many platyrrhine-like anthropoids in Africa 35–30 million years ago. The divergence dates for modern platyrrhine lineages, based on both fossils and molecular analyses, are all within the last 20 million years or so. The earliest sites containing fossil platyrrhines are from the southernmost part of the continent. Although many have been argued to show close links with living genera, the evidence is usually equivocal and hotly debated. The relatively few early species can not be unequivocally linked to living subfamilies. However, by approximately 13 to 12 million years ago, the La Venta fauna from Colombia has an assemblage of primates that includes presumed members of most modern clades, including howling monkeys, pitheciines, squirrel monkeys, owl monkeys, and possible several lineages of callitrichines. This later Miocene assemblage is taxonomically similar to modern assemblages and also has a similar dietary diversity as estimated from dental morphology of the fossil species (Fleagle et al., 1997). Fossil platyrrhines younger than 12 mya are very rare. The major fossil evidence for any striking differences between fossil platyrrhine

Fig. 6.10. Primates of the Raleighvallen–Voltsberg Nature Reserve, Suriname.

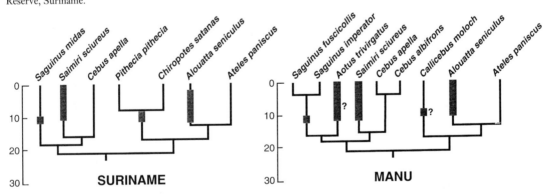

Fig. 6.11. Phylogeny of the primates from the Raleighvallen–Voltsberg Nature Reserve, Suriname and Manu National park, Peru showing molecular divergence times. Broad bands indicate times when these taxa or close relatives are documented in the fossil record.

Table 6. 1. *Different measures of ecological diversity within communities from four biogeographical regions*

	Number of taxa	× Divergence time	Polygon area	× Centroid distance	× Ecol. dist. total	× Ecol. dist. 2 components
Suriname	7	18. 95	149	0. 790	0. 408	0. 830
Manu	9	18. 54	149	0. 720	0. 472	0. 680
Ranomafana	12	27. 82	491	1. 380	0. 612	1. 340
Morondava	7	27. 48	460	4. 350	0. 620	1. 360
Taï	10	31. 78	598	1. 140	0. 613	1. 160
Kibale	11	29. 11	577	1. 160	0. 670	1. 180
Lompat	7	31. 09	432	1. 230	0. 466	1. 250
Ketambe	7	32. 79	629	1. 240	0. 540	1. 290
Correlation			0. 899	0. 658	0. 323	0. 717

assemblages and modern assemblages are in the presence of one or more atelines over twice as large as any living platyrrhine (but definitely arboreal) from latest Pleistocene or Recent caves in Brazil (Hartwig & Cartelle, 1996) and a small fauna of very unusual monkeys from the Caribbean where there are currently no non-human primates (Ford, 1990; Horovitz & MacPhee, 1998).

Compared with those of other continental areas, the primates of the Neotropics are both more ecologically uniform (Fleagle & Reed, 1996) and more recent in their likely divergence dates (Fig. 6.11; Porter *et al.*, 1997). Thus the split between the two major phyletic groups, atelids and cebids (Schneider *et al.*, 1993) dates to just over 20 million years, and major groups within these two families diverged only slightly more recently – atelines and pitheciines (including *Callicebus*) at 17 mya, cebines, *Aotus*, and callitrichines at 17–18 mya, and *Cebus* and *Saimiri* at 16.3 (Fig. 6.9). This pattern of an early–middle Miocene diversification of virtually all modern lineages has long been accepted for platyrrhines based on both the fossil record (Delson & Rosenberger, 1980) and immunological data (Sarich & Cronin, 1980). Compared with the assemblages of other continents, neotropical primates are more similar to Malagasy assemblages in having few sympatric congeneric species (see Peres & Janson, chapter 3, this volume). Like the primates of the other island continent, platyrrhine assemblages lack a hierarchical structure of divergence times. Rather, they show evidence of an explosive radiation between 20 and 16 million years ago such that most taxa are equally distant from other members of the assemblage.

ECOLOGICAL DIVERSITY

In a recent paper Fleagle & Reed (1996) compared the ecological diversity of primate communities from eight localities in four major continental areas using quantitative data of ten ecological variables for 70 species collated from naturalistic field studies. To account for the intercorrelations among the ecological variables, we summarized the ecological diversity among the individual species using a principal coordinates analysis and found that the first two components accounted for roughly 55% of the ecological information contained in the 70 × 10 matrix. When the species making up individual assemblages were plotted on the multivariate ecological axes the assemblages from different geographical regions (Africa, Asia, Madagascar, South America) showed very different distributions in ecological space (Figs. 6.12 and 6.13). However, assemblages from the same biogeographical regions were very similar even though in many cases they had few species in common and different numbers of species. In comparing the overall ecological diversity of the different assemblages using three different measures (Table 6.1), we found that the South American assemblages were much less diverse than those of other regions in all measures, but that diversity among the other assemblages differed according to the measure. For example, the Malagasy assemblages which showed a distinct 'bimodal' plot of the species had higher average distance from the centroid, and the largest average distance between individual species for the first two components. However the African localities and one Asian locality (Ketambe) showed the highest values in the area of the two dimensional

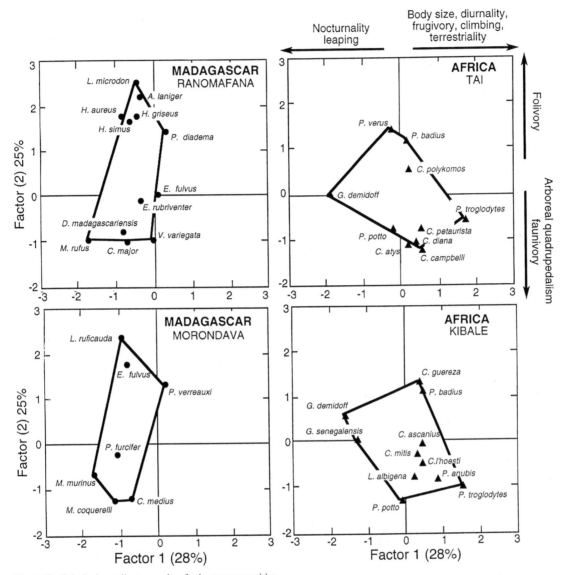

Fig. 6. 12. Principal coordinates results of primate communities of Madagascar and Africa (from Fleagle & Reed, 1996).

polygon encompassed by component species. In this study we have used these same ecological data to examine the relationship between ecological diversity and phylogeny.

PHYLOGENY AND ECOLOGY

One of the most striking results of the earlier principal coordinates analysis of ecological data for the 70 species in

the community analysis (Fleagle & Reed, 1996) was that closely related taxa usually clustered together based on quantitative ecological variables alone (Fig. 6.14). Thus, galagos cluster together on one side of the plot; gibbons and siamang with other apes on another; colobines group together near the center; cheirogaleids clustered in the lower left quadrant; and cercopithecines group together in the lower right quadrant. Clearly the ecological variables

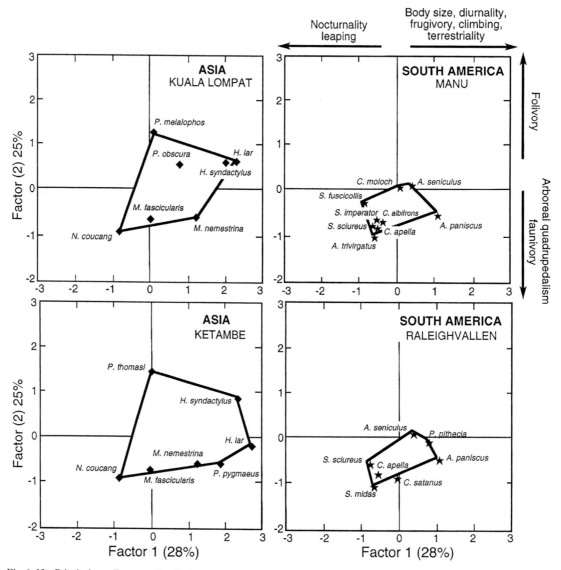

Fig. 6. 13. Principal coordinates results of primate communities from Asia and South America (from Fleagle & Reed, 1996).

are clustering phylogenetic groups. Nevertheless, it is clear from this plot that there is also overlap between very divergent phylogenetic groups in this ecological space, demonstrating that there has been ecological convergence in the variables measured. For example cheirogaleids, lorises and many platyrrhines overlap as small, arboreal frugivorous and insectivorous quadrupeds; colobines and some indriids overlap as medium-sized diurnal arboreal,

leaping folivore-frugivores, and several platyrrhines, cercopithecines, and lemurids overlap as medium-sized, arboreal, diurnal quadrupedal frugivores. Thus the earlier ecological analysis provides evidence for both ecological clustering of phylogenetic groups and also ecological convergence among very distantly related groups. In this study we can use these same ecological data to examine the extent to which ecological distance and diversity are related to the

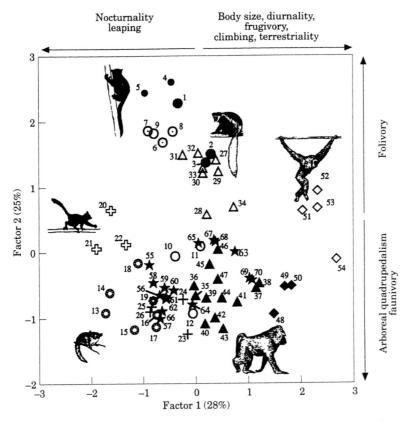

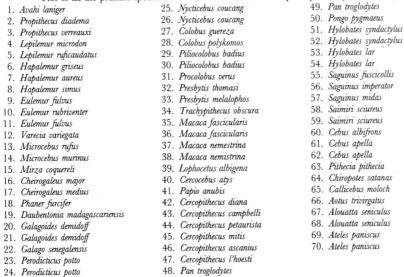

Plot of all the primate species on the first two factors of the principal components analysis.

1. *Avahi laniger*	25. *Nycticebus coucang*	49. *Pan troglodytes*
2. *Propithecus diadema*	26. *Nycticebus coucang*	50. *Pongo pygmaeus*
3. *Propithecus verreauxi*	27. *Colobus guereza*	51. *Hylobates syndactylus*
4. *Lepilemur microdon*	28. *Colobus polykomos*	52. *Hylobates syndactylus*
5. *Lepilemur ruficaudatus*	29. *Piliocolobus badius*	53. *Hylobates lar*
6. *Hapalemur griseus*	30. *Piliocolobus badius*	54. *Hylobates lar*
7. *Hapalemur aureus*	31. *Procolobus verus*	55. *Saguinus fuscicollis*
8. *Hapalemur simus*	32. *Presbytis thomasi*	56. *Saguinus imperator*
9. *Eulemur fulvus*	33. *Presbytis melalophos*	57. *Saguinus midas*
10. *Eulemur rubriventer*	34. *Trachypithecus obscura*	58. *Saimiri sciureus*
11. *Eulemur fulvus*	35. *Macaca fascicularis*	59. *Saimiri sciureus*
12. *Varecia variegata*	36. *Macaca fascicularis*	60. *Cebus albifrons*
13. *Microcebus rufus*	37. *Macaca nemestrina*	61. *Cebus apella*
14. *Microcebus murinus*	38. *Macaca nemistrina*	62. *Cebus apella*
15. *Mirza coquereli*	39. *Lophocetus albigena*	63. *Pithecia pithecia*
16. *Cheirogaleus major*	40. *Cercocebus atys*	64. *Chiropotes satanas*
17. *Cheirogaleus medius*	41. *Papio anubis*	65. *Callicebus moloch*
18. *Phaner furcifer*	42. *Cercopithecus diana*	66. *Aotus trivirgatus*
19. *Daubentonia madagascariensis*	43. *Cercopithecus campbelli*	67. *Alouatta seniculus*
20. *Galagoides demidoff*	44. *Cercopithecus petaurista*	68. *Alouatta seniculus*
21. *Galagoides demidoff*	45. *Cercopithecus mitis*	69. *Ateles paniscus*
22. *Galago senegalensis*	46. *Cercopithecus ascanius*	70. *Ateles paniscus*
23. *Perodicticus potto*	47. *Cercopithecus l'hoesti*	
24. *Perodicticus potto*	48. *Pan troglodytes*	

Fig. 6. 14. Principal coordinates analysis of ecological variables of 70 taxa from eight primate communities from Africa, Asia, Madagascar, and South America showing phylogenetic clustering of taxa (from Fleagle & Reed, 1996).

phylogenetic patterns found in these primate communities from the four different biogeographic regions. Do the faunas characterized by explosive phylogenetic radiations (Madagascar and South America) show a different relationship between ecological distance and phylogenetic distance than the two faunas (Africa and Asia) with more hierarchically arranged assemblages that have slowly accumulated over recent epochs? Has the evolution of ecological diversity taken place at different rates in different biogeographical regions?

Phylogenetic distance and ecological distance

To evaluate the overall relationship between phylogenetic distance and ecological distance, we compared the 70×70 matrix of Euclidean Distances between pairs of taxa using the ecological data with the 70×70 matrix of Divergence Times based on molecular systematics as summarized above. The two matrices have a correlation value of 0.595 using the Mantel statistic which is highly significant (<0.002). Thus, for the sample as a whole, ecological distance between pairs of species is associated with the estimated time of phylogenetic divergence, but there is nevertheless considerable ecological diversity that is not correlated with divergence times (see also Wright & Jernvall, chapter 18, this volume). In order to evaluate the relationship between ecological diversity and phylogeny at the level of individual assemblages and biogeographical regions, we conducted several additional analyses.

Phylogenetic diversity and ecological diversity among communities

As noted above, primate communities in different continents show differences in total ecological diversity. In particular, South American communities show much less diversity than do communities from Madagascar, Africa and Asia (Table 6.1). To what extent is this ecological diversity among different assemblages associated with differences in the phylogenetic diversity of these same assemblages? As discussed above, total ecological diversity within individual primate assemblages has been quantified using several different measures: (1) average total ecological distance between species; (2) average distance for the first two axes of the principal coordinates analysis; (3) average distance from the centroid; and (4) area of the polygon

described by the position of the individual species in the first two axes of the principal coordinates analysis. To examine the relationship between phylogenetic diversity and ecological diversity we have examined the correlation of each of the measures of ecological diversity in each assemblage with average divergence time between individual species for each assemblage (Table 6.1; Fig. 6.15).

Each measure of ecological diversity is positively correlated with average divergence time. However, the strength of the correlation varies considerably among measures. The two South American assemblages show the lowest average divergence times just as they show the lowest values of overall assemblage ecological diversity in virtually all of the measures, while the average divergence times of the other communities are all relatively similar. Of the four measures of total ecological diversity, the area of the polygon in two-dimensional ecological space and the average interspecific differences in two-dimensional space show the highest correlations with average divergence time, the former having an R^2 value of 0.90. Average distance from the centroid is slightly lower and total ecological distance is the lowest.

Thus, for the eight communities sampled here, overall ecological diversity for the different assemblages is positively correlated with average divergence time although some measures of assemblage diversity show much higher correlations than others. However, to address the question of how regional phylogenetic history relates to ecological diversity of individual communities we need to examine the relationship between ecological distance and divergence time in the species making up each assemblage.

Ecological distance and phylogenetic distance within assemblages

To examine the relationship between ecological diversity and phylogenetic diversity within assemblages we computed the least squares regression between the pairwise Euclidean distances based on the ecological variables and the pairwise divergence times for the species making up each of the eight assemblages from four biogeographic areas. Even though the data points are certainly not independent, this regression nevertheless provides a crude estimate of the pattern in the data from different regions. Although for each assemblage there was a positive and significant relationship between ecological distance and

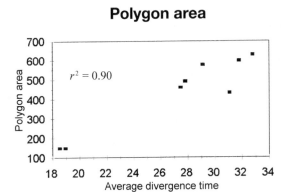

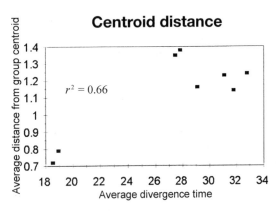

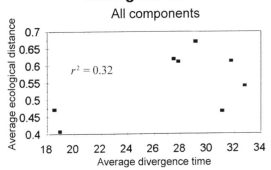

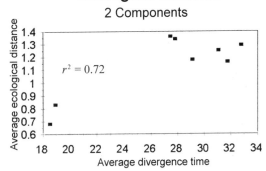

Fig. 6. 15.   Regression of ecological diversity measures and average divergence time between taxa for eight communities from Africa, South America, Madagascar, and Asia. (Data from Table 6. 1, and Fleagle & Reed, 1996.)

divergence times within communities, there are considerable differences in the level of the correlation in the assemblages from different regions.

In Africa and in Asia, the two continental areas in which there is a very hierarchical pattern of divergence times and the sympatric species in the communities represent diverse radiations from many different epochs, there is a very high correlation between ecological distance and divergence times (Fig. 6.16). In both of these continents, the most phylogenetically divergent species (the nocturnal lorises and galagos) are also the most ecologically distinct whereas the more recently related taxa (species of *Cercopithecus*, *Macaca*, *Hylobates*, and colobines) are very similar ecologically.

In contrast, the correlation between ecological distance and divergence times is much lower for the Malagasy and South American assemblages (Fig. 6.17). Although the Malagasy primates probably represent an older radiation than that of the New World platyrrhines, the pattern of phylogenetic divergences is similar in both with many clades of similar age showing very different amounts of ecological similarity and difference. Apart from congeneric pairs (only for *Saguinus* and *Cebus*), most divergence times among platyrrhine taxa are between 16 and 22 mya (Fig. 6.11). Yet this same divergence time has produced pairs of very similar taxa such as *Aotus* and *Callicebus* and very different pairs such as *Saguinus fuscicollis* and *Alouatta*. Likewise, most of the Malagasy families seem to have diverged at roughly the same time, approximately 30 million years ago (Fig. 6.11). Yet a divergence time of 30 million years

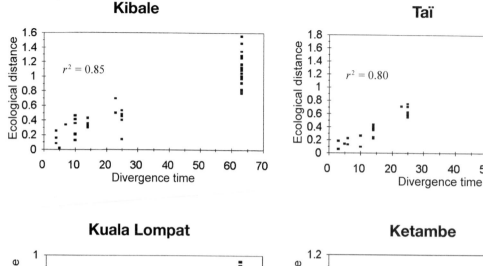

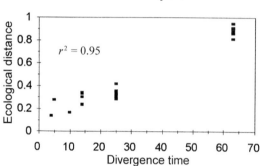

Fig. 6. 16. Regressions of ecological distance and divergence
time for all pairs of taxa in African and Asian communities.

includes pairs such as *Lepilemur* and *Avahi* which are
extremely similar ecologically, and other pairs such as *Pro-pithecus* and *Microcebus* or *Lepilemur* and *Cheirogaleus*
which are very divergent.

Although the phylogenetic relationships and divergence
times of the Malagasy species are less well resolved than
those of the other regions (compare Porter *et al.*, 1997 with
Yoder, 1997) and can not be corroborated with any fossil
record, it seems unlikely that the low correlation reflects
just poor resolution of the molecular data since the two
major studies yield the same pattern. The relationships
and divergence times for the South American platyrrhines
appear to be quite well resolved for virtually all taxa (see
Porter *et al.*, 1997; Schneider *et al.*, 1993; 1996; Horovitz &
Meyer, 1997) and platyrrhines have the same low correlation. Likewise, there is no doubt that our characterizations
of species-specific ecology and hence ecological distances

do not capture all the subtleties of ecological differences.
For example, the odd and morphologically distinctive aye-
aye is characterized as a medium-sized, nocturnal, arboreal, quadrupedal frugivore-insectivore, and pitheciines
are characterized as fruit eaters rather than specialist seed
predators. However, such simplified ecological characterizations should have the same effect on all assemblages.

It seems much more likely that the different levels of
correlation between ecological distance and divergence
times found in Africa and Asia compared with those for the
island faunas of Madagascar and South America reflect the
different types of phylogenetic radiations resulting from
their biogeographical histories. Both Madagascar and South
America were separated from other continents throughout
almost all of the Age of Mammals, and for each the extant
taxa seem to be largely the result of a single major radiation of non-human primates in which different lineages

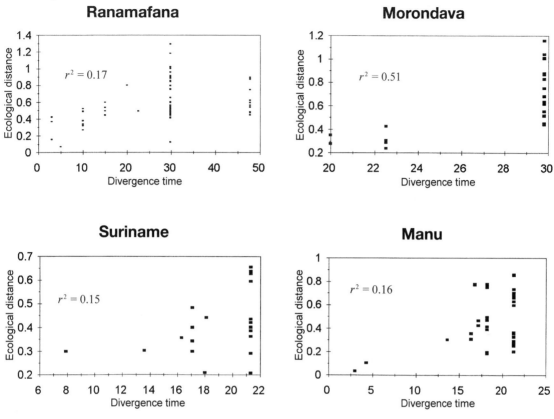

Fig. 6. 17. Regressions of ecological distance and divergence time for all pairs of taxa in Malagasy and South American communities.

radiated into a variety of niches, some of them distinct, some of them overlapping to various degrees. The biogeographical history of Africa and Asia has been quite different during this time as these continents have been repeatedly connected to varying degrees with one another as well as with Europe. Throughout the past 60 million years there have been several periods both of geographical isolation and of faunal interchange among these continents. The fossil record of Old World primates shows a continuous turnover of dominant primate taxa as new groups evolve and replace older, more generalized lineages or occupy distinct ecological niches alongside the older elements of the fauna. Both the fossil record (Delson & Rosenberger, 1984) and the molecular divergence times clearly show this layered or accretionary development of the extant assemblages. For individual communities, the

correlation between ecological distance and divergence times seems to reflect the biogeographical history of the fauna.

Although the correlations of the relationship between pairwise divergence times and ecological distances vary considerably between communities according to the biogeographical history of the region, the slope of the regression line is very similar for all regions. When the species from both sites in each region are combined, the slope of ecological divergence as a function of divergence time in millions of years is 0.015 for Africa, 0.012 for Asia, 0.018 for South America and 0.009 for Madagascar with the aye-aye or 0.016 with the aye-aye excluded from the analysis. If the aye-aye is excluded from the Malagasy data because it appears to be part of a separate radiation, or because its unique features are not adequately captured by our data set (see Ganzhorn, chapter 8, this volume), these slopes are not significantly different in an analysis of covariance. Thus the mean rate of ecological divergence between species is the same in all continents, it is only the pattern that changes according to biogeographic history.

DISCUSSION AND SUMMARY

A major dichotomy in the evolutionary history of New World monkeys and that of Old World primates has been argued for many years based on the fossil record (Delson & Rosenberger, 1984) and has most recently been confirmed by the results of molecular phylogeny (Schneider & Rosenberger, 1996; Stewart & Disotell, 1998). However, the extension of these evolutionary patterns of species radiations to the patterns of ecological diversity and its evolution raises many questions about adaptive radiations, paleoecology, and major geological events that can not be readily addressed from the current data.

First of all, the finding of a similar rate of ecological diversification and strong biogeographic differences in the relationship between phylogenetic distance and ecological distance, in no way detracts from the role of local ecological conditions in selecting which species can survive at a given locality. However, these macroevolutionary patterns linking phylogeny and ecological adaptations do suggest that there may be some broad constraints on the pathways of ecological evolution that primates have taken on different continents. Thus the relatively limited amount of ecological diversity found among extant platyrrhines compared with the primates of other major biogeographical regions may well reflect, in part, the fact that this is a relatively young radiation. Whether these overall, similar rates of ecological diversification are the result of inherited features of life history such as generation time, basic ecological characteristics of the order, or even some methodological artifact, are topics well worth further examination. Comparison with rates and patterns of ecological diversification of other mammals on the same continents would be particularly interesting and useful.

The results above have demonstrated that the "island primate faunas" of Madagascar and South America have a relatively low correlation between ecological distance and phylogenetic distance among individual species as both show an explosive radiation of families rather than a slow accumulation over many epochs. It is certainly possible,

indeed likely, that many of the dates will be revised in future years. Nevertheless, assuming they are close to the real dates, they may offer some insight into global or regional events associated with these radiations. In neither case is the estimated timing of the major diversification coincident with what seems to be the initial colonization of the "island". For Madagascar, it seems as if the evolutionary divergence of the aye-aye preceded the appearance of other taxa by nearly 20 million years. Is this evidence for two colonizations of Madagascar? It is notable that the major radiation at 30 million years ago is roughly the time of both a dramatic cooling of earth temperatures and a major drop in sea level, both correlated with the advent of circumpolar currents in the Southern hemisphere and glaciation of Antarctica (Haq et al., 1987). Did the major radiation at 30 million years lead to the extinction of an earlier fauna? Answers to these questions will require better resolution of the divergence times for Malagasy primates and a Tertiary fossil record for that island.

For South America, the major radiations between 22 mya and 18 mya are, likewise, younger than the earliest appearance of anthropoids on that continent at about 25 mya (Fleagle et al., 1997). However, this timing correlates with a major period of Andean uplift (Pascual, 1984), an event that could well have led to both climatic and biogeographical disruption. Unfortunately the paleontological record of platyrrhine evolution is currently too widely distributed in time and space to reconstruct the biogeography of platyrrhine lineages with any detail (Fleagle et al., 1997).

For all faunas, there is a wide range of ecological distances between taxa with similar divergence times. The greatest ecological distances obviously reflect very distinct adaptive differences in phylogenetically distinct taxa. However, the instances of ecological similarity in distantly related taxa may be the result of either ecological convergence in separate lineages or primitive retentions. In this chapter it is not possible to sort out the two types of ecological similarity. However, more careful examination of individual cases should permit identification of these different processes of ecological similarity and perhaps some estimate of whether convergences are more common than primitive retentions in some faunas.

It is unfortunate that our current knowledge of the fossil record of any of the major continental regions inhabited by primates is not adequate to reconstruct details of the evolution and biogeography of the lineages comprising the extant communities in those regions. Indeed, it may never so complete. Nevertheless, the available fossil record and the results of recent molecular phylogenies are remarkably consistent in the broad outlines they provide of the major patterns and the timing of evolutionary radiations of primate lineages during the past 60 million years. When these data on primate phylogeny are combined with quantitative data on the ecological characteristics of living species in these four regions, several clear results emerge:

1. Ecological distance is positively correlated with phylogenetic distance. More distantly related taxa tend to be more different ecologically, for primates as a group and within individual faunas or regions. Similarly, more phylogenetically diverse faunas (those of Africa, Asia, and Madagascar) are also more ecologically diverse than the less phylogenetically diverse faunas of South America.

2. The average rate at which ecological diversity has increased as a function of phylogenetic divergence time between taxa is the same for all major geographical regions. Equal divergence times, on average, yield equal amounts of ecological distance between pairs of taxa in Africa, Asia, Madagascar or South America.

3. However, the correlation between phylogenetic distance and ecological distance for individual faunas reflects the different patterns of evolutionary diversification in the different biogeographical regions. In Africa and Asia where extant faunas are normally composed of a collection of lineages that have originated at successive times during the past 60 million years, there is a strong correlation between ecological distance and phylogenetic distance between pairs of species. However, on the "island continents" of Madagascar and South America where the living primate lineages seem to be largely the result of explosive evolutionary radiations during a relatively brief period of time, the correlation between ecological distance and phylogenetic distance is much lower.

In summary, both phylogenetic history and biogeography have played an important role in determining the patterns of ecological diversity of the extant primate faunas in different regions of the world.

REFERENCES

Andrews, P. (1981). Species diversity and diet in monkeys and apes during the Miocene. In *Aspects of Human Evolution*, ed. C. B. Stringer, pp. 25–61. London: Taylor & Francis.

Barry, J. C. (1987). The history and chronology of Siwalik cercopithecoids. *Human Evolution*, **2**, 47–58.

Beard, K. C. (1998a). A new genus of Tarsiidae (Mammalia: Primates) from the Middle Eocene of Shanxi Province, China, with notes on the historical biogeography of tarsiers. *Bulletin of the Carnegie Museum of Natural History* **34**, 260–77.

Beard, K. C. (1998b). Unmasking an Eocene primate enigma: the true identity of *Hoanghonius stehlinii*. *American Journal of Physical Anthropology*, Supplement **26**, 69.

Begun, D. R. (1997). A Eurasian origin of the Hominidae. *American Journal of Physical Anthropology*, Supplement **24**, 73–4.

Bown, T. M., Kraus, M. J., Wing, S. L., Fleagle, J. G., Tiffnay, B., Simons, E. L. & Vondra, C. F. (1982). The Fayum forest revisited. *Journal of Human Evolution*, **11**, 603–32.

Brown, J. H. & Lomolino, M. V. (1998). *Biogeography* (Second Edition) Sunderland, MA: Sinauer Associates.

Collura, R. V. & Stewart, C. B. (1995). Insertions and duplications of mtDNA in the nuclear genomes of Old World monkeys and hominoids. *Nature*, **378**, 485–9.

Collura, R. V. & Stewart, C.-B. (1996). Mitochondrial DNA Phylogeny of the colobine Old World monkeys. *Abstracts of the XVIth Congress of the International Primatology Society*, p. 727.

Delson, E. (1994) Evolutionary history of the colobine monkeys in paleoenvironmental perspective. In *Colobine Monkeys. Their Ecology, Behaviour and Evolution*, ed. A. G. Davies & J. F. Oates, pp. 11–44. Cambridge: Cambridge University Press.

Delson, E. (1996). The oldest monkeys in Asia. In *International Symposium: Evolution of Asian Primates*. Kyoto University Primate Research Institute. Aichi, Japan: Inuyama.

Delson, E. & Rosenberger, A. L. (1984). Are there any anthropoid primate living fossils? In *Living Fossils*, ed. N. Eldridge & S. M. Stanley, pp. 50–61. New York: Springer Verlag.

Disotell, T. R. (1996). The phylogeny of Old World monkeys. *Evolutionary Anthropology*, **5**, 18–24.

Fleagle, J. G. (1979). Size distributions of living and fossil primates. *Paleobiology*, **4**, 67–76.

Fleagle, J. G. (1984) Are there any fossil gibbons? In *The Lesser Apes. Evolutionary and Behavioural Biology*, ed. H . Preuschoft, D. J. Chivers, W. Y. Brockelman & N Creel, pp. 431–47. Edinburgh: Edinburgh University Press.

Fleagle, J. G. (1999). *Primate Adaptation and Evolution*. (Second Edition). San Diego: Academic Press.

Fleagle, J. G. & Kay, R. F. (1985). The Paleobiology of Catarrhines. In *Ancestors: The Hard Evidence*, ed. E. Delson, pp. 23–36. New York: Alan R. Liss.

Fleagle, J. G., Kay, R. F. & Anthony, M. R. L. (1997). Fossil new world monkeys. In *Vertebrate Paleontology in the Neotropics. The Miocene Fauna of La Venta Colombia*, ed. R. F. Kay, R. H. Madden, R. L. Cifelli & J. Flynn, pp. 473–95. Washington: Smithsonian Institution Press.

Fleagle, J. G. & Reed, K. E. (1996). Comparing primate communities: a multivariate approach. *Journal of Human Evolution*, 30, 489–510.

Fleagle, J. G. & Reed, K. E. (1997). Primate evolution. In *Encyclopedia of Cultural Anthropology*, ed. D. Levinson & M. Ember, pp. 1025–9. New York: Henry Holt and Company.

Ford, S. (1990). Platyrrhine evolution in the West Indies. *Journal of Human Evolution*, 19, 237–54.

Gebo, D. L., Dagosto, M., Beard, K. C. & Qi, T. (1988). The smallest primate. *American Journal of Physical Anthropology*, Suppl. 26, 86.

Godfrey, L. R., Jungers, W. L., Reed, K. E., Simons, E. L. & Chatrath, P. S. (1997). Subfossil lemurs. In *Natural Change and Human Impact in Madagascar*, ed. M. S. Goodman & B. D. Patterson, pp. 218–56. Washington: Smithsonian Institution Press.

Goodman, M., Porter, C. A., Czelusniak, J., Page, S. L., Schneider, H., Shoshani, J., Gunnell, G. & Groves, C. P. (1998). Toward a phylogenetic classification of primates based on DNA evidence complemented by fossil evidence. *Molecular Phylogenetic Evolution*, 9, 585–98.

Haq, B. U., Hardenbol, J. & Vail, P. R. (1987). Chronology of fluctuating sea levels since the Triassic. *Science* 235, 1156–67.

Hartwig, W. C. (1994). Patterns, puzzles, and perspectives on platyrrhine origins. In *Integrative Paths to the Past. Paleoanthropological Advances in Honor of F. Clark Howell*, ed. R. S. Corruccini & R. L. Ciochon, pp. 69–93. Englewood Cliffs, New Jersey: Prentice-Hall.

Hartwig, W. C. & Cartelle, C. (1996). A complete skeleton of the giant South American primate *Protopithecus*. *Nature* 381, 307–11.

Horovitz, I., & MacPhee, R. D. E. (1998). The Quaternary Cuban platyrrhine *Paralouatta varonai* and the origin of Antillean monkeys. *J. Human Evol.*

Horovitz, I. & Meyer, A. (1997). Evolutionary trends in the ecology of New World monkeys inferred from a combined, phylogenetic analysis of nuclear, mitochondrial, and morphological data. In *Molecular Evolution and Adaptive Radiation*. ed. T. J. Givnish & K. J. Sytsma, pp. 189–224. Cambridge: Cambridge University Press.

Ishida, H. & Pickford, M. (1997). A new late Miocene hominoid from Kenya: *Samburupithecus kiptalami* gen. et sp. nov. *Comptes Rendus D'Academie des Sciences (Paris). Sciences de la Terre et des Planets*, 326, 823–9.

Kappeler, P. M. & Heymann E. W. (1996). Nonconvergence in the evolution of primate life history and socio-ecology. *Biological Journal of the Linnean Society* 59, 297–326.

Kay, R. F. & Covert, H. H. (1984). Anatomy and behavior of extant primates. In *Food Acquisition and Processing in Primates*, ed. D. J. Chivers, B. A. Wood & A. Bilsborough, pp. 467–500. New York: Plenum.

Kay, R. F., Ross, C. & Williams, B. A. (1997). Anthropoid origins. *Science*, 275, 797–803.

MacPhee, R. D. E. & Jacobs L. L. (1986). *Nycticeboides simpsoni* and the morphology, adaptations, and relationships of Miocene Siwalik Lorisidae. *Contributions to Geology, University of Wyoming, Special Paper* 3, 131–61.

Martin, P. S. & Klein, R. G. (1984). *Quaternary Extinctions: A Prehistoric Revolution*. Tucson, AZ: University of Arizona Press.

Martin, R. D. (1993). Primate origins: plugging the gaps. *Nature* 363, 223–34.

Masters, J. C., Rayner, R. J., Ludewick, H., Zimmerman, E., Molez-Verriere, N., Vincent, F. & Nash, L. T. (1994). Phylogenetic relationships among the Galaginae as indicated by erythrocytic allozymes. *Primates* 35, 177–90.

Morales, J. C. & Melnick, D. J. (1998). Phylogenetic relationships of the macaques (Cercopithecidae: *Macaca*), as revealed by high resolution restriction site mapping of mitochondrial ribosomal genes. *Journal of Human Evolution* 34, 1–23.

Pascual, R. (1984). La succesion de las edades-mamifero, de los climas y del diastrofismo sudamericanos durante el Cenozoico: Fenómenos concurrentes. *Anales de la Academia Nacional de Ciencias Exacta, Fiscas y Naturales de Buenos Aires*, 36, 15–37.

Porter, C. A., Page, S. L., Czelusniak, J., Schneider, H.,

Schneider, M. P. C., Sampaio, I. & Goodman, M. (1997). Phylogeny and evolution of selected primates as determined by sequences of the e-globin locus and 5′ flanking regions. *International Journal of Primatology* 18, 261–95.

Potts, R. (1994). *Humankind's Descent: The Consequences of Ecological Instability*. NY: William Morrow.

Purvis, A. (1995). A composite estimate of primate phylogeny. *Philosophical Transactions of the Royal Society London*, **348**, 405–21.

Sarich, V. M. & Wilson, A. C. (1967). Immunological time scale for hominoid evolution. *Science*, **158**, 1200–3.

Sarich, V. M. & Cronin, J. E. (1980). South American mammal molecular systematics, evolutionary clocks, and continental drift. In *Evolutionary Biology of the New World Monkeys and Continental Drift*, eds. R. L. Ciochon & A. B. Chiarelli, pp. 399–421.

Schneider, H. & Rosenberger, A. L. (1996). Molecules, morphology and platyrrhine systematics. In *Adaptive Radiations of Neotropical Primates*, ed. M. A. Norconk, A. L. Rosenberger & P. A. Garber, pp. 3–19. New York: Plenum Press.

Schneider, H., Schneider, M. P. C., Sampaio, I., Harada, M. L., Stanhope, M., Czelusniak, J. & Goodman, M. (1993). Molecular phylogeny of the new world monkeys (Platyrrhini, Primates). *Molecular Phylogenetics and Evolution*, **2**, 225–42.

Schneider, H., Sampaio, I., Harada, M. L., Barroso, C. M. L., Schneider, M. P. C., Czelusniak, J. & Goodman, M. (1996). Molecular phylogeny of the New World monkeys (Platyrrhini, Primates) based on two unlinked nuclear genes: IRBP intron 1 and e-globin sequences. *American Journal of Physical Anthropology* **100**, 153–79.

Simons, E. L. & Rasmussen, T. (1994). A whole new world of ancestors: Eocene anthropoideans from Africa. *Evolutionary Anthropology* 3, 128–39.

Simpson, G. G. (1980). *Splendid Isolation*. New Haven: Yale University Press.

Stehli, F. G. & Webb, S. D. (1985). *The Great American Biotic Interchange*. NY: Plenum.

Stewart, C. B. & Disotell, T. R. (1998). Primate evolution – in and out of Africa. *Current Biology* 8, R582–R588.

Strait, S. G. (1988). Tooth use and the physical properties of food. *Evolutionary Anthropology*, **5**, 199–211.

Terborgh, J. (1992). *Diversity and the Tropical Rain Forest*, 242 pp. New York: W. H. Freeman.

Terborgh, J., & van Schaik, C. P. (1987). Convergence vs. nonconvergence in primate communities. In *Organization of Communities Past and Present*. ed. J. H. R. Gee & P. S. Giller, pp. 205–26. Oxford: Blackwell Scientific Publications.

Ungar, P. S. (1998). Dental allometry, morphology, and wear as evidence for diet in fossil primates. *Evolutionary Anthropology*, **6**, 205–17.

Van Schaik, C. P. & Kappeler, P. M. (1996). The social systems of gregarious lemurs: lack of convergence with anthropoids due to evolutionary disequilibrium? *Ethology* 102, 915–41.

Ward, S. (1997). The taxonomy and phylogenetic relationships of *Sivapithecus* revisited. In *Function, Phylogeny and Fossils: Miocene Hominoid Evolution and Adaptation*, ed. D. R. Begun, C. V. Ward & M. D. Rose, pp. 269–90. New York: Plenum Press.

Yoder, A. D. (1997). Back to the future: A synthesis of strepsirrhine systematics. *Evolutionary Anthropology* 6, 11–22.

7 · Population density of primates in communities: Differences in community structure

KAYE E. REED

INTRODUCTION

The studies of primate communities that have compared primate species' ecological characteristics have found major differences among Africa, the Neotropics, Asia, and Madagascar (Raemaekers *et al.*, 1980; Bourlière, 1985; Terborgh & van Schaik, 1987; Gautier-Hion, 1988; Ganzhorn, 1988, 1992; this volume; Terborgh, 1990; Fleagle & Reed, 1996; Kappeler & Heymann, 1996; McGraw, 1998). These studies have also shown ecological patterns within the same continental areas. However, most of these previous studies compare ecological differences among primate communities on different continents with attributes of individual species, i.e., contrasts in size, diet, and locomotor adaptations among individual species within communities. The unit of analysis is the *species*. Using these data, Fleagle & Reed (1996) showed that overall ecological space represented by ecological data occupied by primate species in communities were quite similar within continental areas, and were different between them. Thus, each primate species within each community held a particular position in ecological space (Hutchinson, 1978).

However, primates, for the most part, do not live individually. The density of primates, as well as diversity, presumably affects the size and shape of the ecological space that each community holds. For example, mammalian population density has been directly related to the size of an animal such that as animals get larger their population densities usually decrease (Fa & Purvis, 1997). It has been suggested that the scaling of this phenomenon is approximately the same for all mammalian herbivores (Damuth, 1981). Peters (1983) proposes that one of the most important reasons that population density falls with increasing body size is the constraint of food supply. That is, as plant

material is distributed relatively equally among populations of primary consumers, each species has access to the same amount of food. Thus, as larger animals eat more than smaller animals, their population densities are lower.

The interrelationship of body size, population density, food acquisition, and locomotor behavior has not yet been examined in primate communities. That is, comparisons among different primate communities have not analyzed species characteristics at the population level. In this chapter, I first examine how primate population density affects species diversity among primate communities. Second, I ask if body size patterns within and among continental areas change if density is considered, that is, what is the difference between species body size distribution and primate biomass distribution? Finally, I ask if the size and shape of ecological space that has previously been determined based on individual species attributes (Fleagle & Reed, 1996) is altered if population densities are considered. What resources do primates spend the most time acquiring? How does this vary among continental areas? To answer these questions, the unit of analysis is the *population* of each primate species, rather than the species.

PRIMATE COMMUNITY DATA

Primate community composition, population densities, and species body mass were collected from the literature for 15 primate communities in forested regions in Africa, Asia, South America, and Madagascar (Table 7.1, 7.3). Information on percentages of (1) time spent foraging for different dietary items, (2) time spent on different locomotor behaviors, and (3) activity pattern for eight communities were previously used in Fleagle & Reed (1996) and had also been acquired from published accounts. Primate community composition and species body mass were also collected from the literature for 49 primate communities in forested regions (Table 7.2). All primate body sizes are from Smith & Jungers (1997).

PRIMATE COMMUNITIES AND POPULATION DENSITY ANALYSES

Species diversity and population densities within and among continents

Species diversity indices are designed to depict the diversity of a community by a single number. Three indices were calculated for each of 15 communities: McIntosh U [$U = \sqrt{(\Sigma n_i^2)}$], McIntosh Evenness [$E = N\text{-}U/(N - (N/\sqrt{S}))$], and Simpson D ($D = \Sigma(n_i(n_i-1))/(N(N-1))$, where n_i is the populations density of each species, N is the total number of individuals, and S is the number of species.

While species richness is often considered as just the number of species, the McIntosh U index summarizes richness based on number of species and their abundances, in this case population density. This index gives a high degree of discrimination between communities (Magurran, 1981). The evenness (or equitability) index expresses the degree of equal distribution of species based on population density, thus the higher the index the more uniform the distribution of species. Finally, the Simpson D reflects the degree of dominance in the most common primates in the community. As this index decreases with increased diversity, the reciprocal is used to represent diversity (1/D).

Diversity indices for the 15 primate communities are listed in Table 7.3. In South America, numbers of species range from three at Guatopo to 13 at the Urucu River primate community. The McIntosh U index, however, shows relatively low diversity for all sites except Cocha Cashu in comparison with primate communities on other continents. South American community indices range from 44 to 100 (range = 56). Interestingly, the community with the fewest number of species, Guatopo, is not the lowest in diversity when population density is considered. All of these neotropical communities have higher evenness indices than communities in other regions. Thus, each species in each community is more equally represented. Finally, the Simpson Index reflects the evenness index in showing that most species are relatively common, but shows that Guatopo and Raleighvallen are either depauperate in

species number or have several rare species, respectively. Although neotropical primate communities are often considered quite species rich, the fairly low population diversity reflected in the McIntosh U parallels the low ecological diversity found in these primate communities (Fleagle & Reed 1996).

Asian communities show the least range in variation among communities in the McIntosh U index, from 56–90 (range = 34), but overall diversities are higher than those found in the Neotropics. The evenness indices show that many of the species are evenly distributed, but accompanied by a few rare species in each community. This accords with the Simpson Index which reveals that most of the species are in the mid-range of population densities for these communities. Thus there are one or two rare species, one dominant species, and most of the other primate species have roughly the same population densities in each community. Although species numbers ranging from three to ten have suggested low diversity for these communities (Caldecott 1986; Terborgh & van Schaik, 1987), they are actually more diverse when population densities are considered than are those of South America.

African communities have relatively high McIntosh U indices, ranging from 97–174 (range = 57), surpassing diversity in both Asian and South American communities, although the range in this variation is equal to that of South American communities. The evenness indices reflect a less even distribution of population densities, while the Simpson indices reveal that the dominance of species varies in these African communities. For example, Kibale does not have a particularly even distribution of primate densities, and while there is one dominant species, *Procolobus badius*, many of the other species have medium level densities and several others have very low densities. Therefore, the Simpson Index is fairly low. On the other hand, Makokou has many species that are dominant and only two species with low densities, thus the Simpson Index is relatively high.

Finally, Madagascar shows the highest McIntosh diversity indices ranging from 171–1257 (range = 1044), as well as the greatest range among communities. Even if Morondava, which has an extremely high index that reflects the very high population densities of five of the seven total species, is not included the variation in diversity still has the widest range (243). The evenness indices parallel those from Africa and Asia, showing that species populations are not as evenly distributed as in South America.

Table 7.1. *Focal primate communities listed by continental region with 28 variables used in text*

Continent	Locality	Species	MASS	PDEN	BMASS	%FR
Madagascar	Ranomafana	*Propithecus diadema*	6.100	30.55	186.36	47.00%
		Avahi laniger	1.175	68.00	79.90	25.00%
		Eulemur fulvus	2.215	60.00	132.90	75.00%
		Eulemur rubriventer	1.960	30.00	58.80	90.00%
		Varecia variegata	3.575	16.00	57.20	86.00%
		Hapalemur griseus	0.709	62.00	43.96	15.00%
		Hapalemur aureus	1.455	4.00	5.82	9.00%
		Hapalemur simus	1.725	7.00	12.08	15.00%
		Lepilemur microdon	0.970	3.00	2.91	0.00%
		Microcebus rufus	0.050	110.00	5.50	50.00%
		Cheirogaleus major	0.450	68.00	30.60	80.00%
		Daubentonia madagascariensis	2.555	3.00	7.67	50.00%
		TOTALS:		461.55	620.70	
	Morondava	*Microcebus murinus*	0.070	342.86	24.00	48.00%
		Cheirogaleus medius	0.200	240.00	48.00	85.00%
		Mizra coquerelli	0.330	30.00	9.90	25.00%
		Phaner furcifer	0.440	200.45	88.20	15.00%
		Lepilemur ruficaudus	0.769	360.00	276.84	25.00%
		Propithecus verreauxi	3.780	59.10	223.40	75.00%
		Eulemur fulvus	2.200	1061.00	2334.20	10.60%
		TOTALS:		2293.41	3004.54	
Asia	Kuala Lumpat	*Hylobates syndactylus*	10.900	4.50	49.05	44.00%
		Hylobates lar	5.640	6.10	34.40	58.00%
		Trachypithecus obscurus	7.080	31.00	219.48	42.00%
		Presbytis melalophus	6.543	74.00	484.18	64.30%
		Macaca fascicularis	4.475	39.00	174.53	58.00%
		Macaca nemestrina	8.850	0.50	4.43	87.70%
		Nycticebus coucang	0.653	10.00	6.53	80.00%
		TOTALS:		155.10	972.60	
	Ketambe	*Pongo pygmaeus*	56.750	3.00	170.25	55.00%
		Hylobates syndactylus	11.300	15.00	169.50	44.00%
		Hylobates lar	5.640	10.60	59.78	50.00%
		Presbytis thomasi	6.230	23.10	143.91	40.00%
		Macaca fascicularis	4.475	48.00	214.80	35.00%
		Macaca nemestrina	6.800	19.00	129.20	87.00%
		Nycticebus coucang	1.106	5.00	5.53	80.00%
		TOTALS:		123.70	892.97	
South America	Raleighvallen	*Saguinus midas*	0.545	23.47	12.79	68.75%
		Saimiri sciuerus	0.811	33.20	26.93	28.00%
		Pithecia pithecia	1.760	3.57	6.28	95.30%
		Chiropotes satanus	3.030	7.53	22.81	95.30%
		Cebus apella	3.085	13.30	41.03	53.90%

Table 7.1. (*cont.*)

%L	%FA	%AQ	%TQ	%CL	%LP	%BP	ACT	BFR	BL
53.00%	0.00%	0.00%	0.00%	10.00%	90.00%	0.00%	1	87.59	98.77
75.00%	0.00%	0.00%	0.00%	10.00%	90.00%	0.00%	3	19.98	59.93
25.00%	0.00%	50.00%	0.00%	0.00%	50.00%	0.00%	2	99.68	33.23
10.00%	0.00%	50.00%	0.00%	0.00%	50.00%	0.00%	2	52.92	5.88
14.00%	0.00%	100.00%	0.00%	0.00%	0.00%	0.00%	1	49.19	8.01
85.00%	0.0096	50.00%	0.00%	0.00%	50.00%	0.00%	1	6.59	37.36
91.00%	0.00%	50.00%	0.00%	0.00%	50.00%	0.00%	2	0.52	5.30
85.00%	0.00%	50.00%	0.00%	0.00%	50.00%	0.00%	2	1.81	10.26
100.00%	0.00%	0.00%	0.00%	20.00%	80.00%	0.00%	3	0.00	2.91
0.00%	50.00%	75.00%	0.00%	0.00%	25.00%	0.00%	3	2.75	0.00
0.00%	20.00%	100.00%	0.00%	0.00%	0.00%	0.00%	3	24.48	0.00
0.00%	50.00%	100.00%	0.00%	0.00%	0.00%	0.00%	3	3.83	0.00
								349.34	264.65
12.00%	40.00%	75.00%	0.00%	0.00%	25.00%	0.00%	3	11.52	2.88
5.00%	10.00%	100.00%	0.00%	0.00%	0.00%	0.00%	3	40.80	2.40
15.00%	60.00%	100.00%	0.00%	0.00%	0.00%	0.00%	3	2.48	1.49
70.00%	15.00%	50.00%	0.00%	0.00%	50.00%	0.00%	3	13.23	61.74
75.00%	25.00%	0.00%	0.00%	20.00%	80.00%	0.00%	3	69.21	207.63
25.00%	0.00%	0.00%	0.00%	10.00%	90.00%	0.00%	1	167.55	55.85
89.40%	0.00%	50.00%	0.00%	0.00%	50.00%	0.00%	2	247.43	2086.77
								552.22	2418.76
41.00%	15.00%	0.00%	0.00%	88.00%	6.00%	6.00%	1	21.58	20.11
29.00%	13.00%	0.00%	0.00%	77.00%	15.00%	8.00%	1	19.95	9.98
58.00%	0.00%	50.60%	0.00%	9.20%	40.20%	0.00%	1	92.18	127.30
35.70%	0.00%	20.50%	0.00%	12.00%	67.50%	0.00%	1	311.33	172.85
19.00%	23.00%	70.00%	27.00%	0.00%	3.00%	0.00%	1	101.22	33.16
11.90%	0.40%	20.00%	60.00%	20.00%	0.00%	0.00%	1	3.88	0.53
0.00%	20.00%	70.00%	0.00%	30.00%	0.00%	0.00%	3	5.22	0.00
								555.36	363.93
33.00%	12.00%	12.00%	0.00%	88.00%	0.00%	0.00%	1	93.64	56.18
43.00%	13.00%	0.00%	0.00%	88.00%	6.00%	6.00%	1	74.58	72.89
4.00%	46.00%	0.00%	0.00%	77.00%	15.00%	8.00%	1	29.89	2.39
56.00%	4.00%	20.50%	0.00%	12.00%	67.50%	0.00%	1	57.57	80.59
8.00%	57.00%	65.00%	10.00%	15.00%	10.00%	0.00%	1	75.18	17.18
12.60%	40.00%	20.00%	60.00%	20.00%	0.00%	0.00%	1	112.40	16.28
0.00%	20.00%	70.00%	0.00%	30.00%	0.00%	0.00%	3	4.42	0.00
								447.68	245.51
0.00%	31.25%	76.00%	0.00%	0.00%	24.00%	0.00%	1	8.79	0.00
0.00%	72.00%	75.00%	0.00%	3.00%	22.00%	0.00%	1	7.54	0.00
0.00%	4.70%	10.00%	0.00%	20.00%	70.00%	0.00%	1	5.99	0.00
0.00%	4.70%	80.00%	0.00%	2.00%	18.00%	0.00%	1	21.74	0.00
1.10%	45.00%	84.00%	0.00%	6.00%	10.00%	0.00%	1	22.12	0.45

Table 7.1. (*cont.*)

Continent	Locality	Species	BFA	BAQ	BTQ	BCL
Madagascar	Ranomafana	*Propithecus diadema*	0.00	0.00	0.00	18.64
		Avahi laniger	0.00	0.00	0.00	7.99
		Eulemur fulvus	0.00	66.45	0.00	0.00
		Eulemur rubriventer	0.00	29.40	0.00	0.00
		Varecia variegata	0.00	57.20	0.00	0.00
		Hapalemur griseus	0.00	21.98	0.00	0.00
		Hapalemur aureus	0.00	2.91	0.00	0.00
		Hapalemur simus	0.00	6.04	0.00	0.00
		Lepilemur microdon	0.00	0.00	0.00	0.58
		Microcebus rufus	2.75	4.13	0.00	0.00
		Cheirogaleus major	6.12	30.60	0.00	0.00
		Daubentonia madagascariensis	3.83	7.67	0.00	0.00
		TOTALS:	12.70	226.38	0.00	27.21
	Morondava	*Microcebus murinus*	9.60	18.00	0.00	0.00
		Cheirogaleus medius	4.80	48.00	0.00	0.00
		Mizra coquerelli	5.94	9.90	0.00	0.00
		Phaner furcifer	13.23	44.10	0.00	0.00
		Lepilemur ruficaudus	69.21	0.00	0.00	55.37
		Propithecus verreauxi	0.00	0.00	0.00	22.34
		Eulemur fulvus	0.00	1167.10	0.00	0.00
		TOTALS:	102.78	1287.10	0.00	77.71
Asia	Kuala Lumpat	*Hylobates syndactylus*	7.36	0.00	0.00	43.16
		Hylobates lar	4.47	0.00	0.00	26.49
		Trachypithecus obscurus	0.00	111.06	0.00	20.19
		Presbytis melalophus	0.00	99.26	0.00	58.10
		Macaca fascicularis	40.14	122.17	47.12	0.00
		Macaca nemestrina	0.02	0.89	2.66	0.89
		Nycticebus coucang	1.31	4.57	0.00	1.96
		TOTALS:	53.30	337.95	49.78	150.79
	Ketambe	*Pongo pygmaeus*	20.43	20.43	0.00	149.82
		Hylobates syndactylus	22.04	0.00	0.00	149.16
		Hylobates lar	27.50	0.00	0.00	46.03
		Presbytis thomasi	5.76	29.50	0.00	17.27
		Macaca fascicularis	122.44	139.62	21.48	32.22
		Macaca nemestrina	51.68	25.84	77.52	25.84
		Nycticebus coucang	1.11	3.87	0.00	1.66
		TOTALS:	250.96	219.26	99.00	422.00
South America	Raleighvallen	*Saguinus midas*	4.00	9.72	0.00	0.00
		Saimiri sciuerus	19.39	20.20	0.00	0.81
		Pithecia pithecia	0.30	0.63	0.00	1.26
		Chiropotes satanus	1.07	18.25	0.00	0.46
		Cebus apella	18.46	34.47	0.00	2.46

Table 7.1. (*cont.*)

BLP	BBP	BPFR	BPL	BPFA	BPAQ	BPTQ	BPCL	BPLP	BPBP
167.72	0.00	14.04%	15.84%	0.00%	0.00%	0.00%	2.99%	26.89%	0.00%
71.91	0.00	3.20%	9.61%	0.00%	0.00%	0.00%	1.28%	11.53%	0.00%
66.45	0.00	15.98%	5.33%	0.00%	10.65%	0.00%	0.00%	10.65%	0.00%
29.40	0.00	8.49%	0.94%	0.00%	4.71%	0.00%	0.00%	4.71%	0.00%
0.00	0.00	7.89%	1.28%	0.00%	9.17%	0.00%	0.00%	0.00%	0.00%
21.98	0.00	1.06%	5.99%	0.00%	3.52%	0.00%	0.00%	3.52%	0.00%
2.91	0.00	0.08%	0.85%	0.00%	0.47%	0.00%	0.00%	0.47%	0.00%
6.04	0.00	0.29%	1.65%	0.00%	0.97%	0.00%	0.00%	0.97%	0.00%
2.33	0.00	0.00%	0.47%	0.00%	0.00%	0.00%	0.09%	0.37%	0.00%
1.38	0.00	0.44%	0.00%	0.44%	0.66%	0.00%	0.00%	0.22%	0.00%
0.00	0.00	3.93%	0.00%	0.98%	4.91%	0.00%	0.00%	0.00%	0.00%
0.00	0.00	0.61%	0.00%	0.61%	1.23%	0.00%	0.00%	0.00%	0.00%
370.32	0.00	56.01%	41.96%	2.03%	36.29%	0.00%	4.36%	59.33%	0.00%
6.00	0.00	0.38%	0.10%	0.32%	0.60%	0.00%	0.00%	0.20%	0.00%
0.00	0.00	1.36%	0.80%	0.16%	1.60%	0.00%	0.00%	0.00%	0.00%
0.00	0.00	0.08%	0.50%	0.20%	0.33%	0.00%	0.00%	0.00%	0.00%
44.10	0.00	0.44%	2.05%	0.44%	1.47%	0.00%	0.00%	1.47%	0.00%
221.47	0.00	2.30%	6.91%	2.30%	0.00%	0.00%	1.84%	7.37%	0.00%
201.06	0.00	5.58%	1.86%	0.00%	0.00%	0.00%	0.74%	6.69%	0.00%
1167.10	0.00	8.24%	69.45%	0.00%	38.84%	0.00%	0.00%	38.84%	0.00%
1639.73	0.00	18.38%	81.67%	3.42%	42.84%	0.00%	2.58%	54.57%	0.00%
2.94	2.94	2.22%	2.07%	0.76%	0.00%	0.00%	4.44%	0.30%	0.30%
5.16	2.75	2.05%	1.03%	0.46%	0.00%	0.00%	2.72%	0.53%	0.28%
88.23	0.00	9.48%	13.09%	0.00%	11.42%	0.00%	2.08%	9.07%	0.00%
326.82	0.00	32.01%	17.77%	0.00%	10.21%	0.00%	5.97%	33.60%	0.00%
5.24	0.00	10.41%	3.41%	4.13%	12.56%	4.84%	0.00%	0.54%	0.00%
0.00	0.00	0.40%	0.05%	0.00%	0.09%	0.27%	0.09%	0.00%	0.00%
0.00	0.00	0.54%	0.00%	0.13%	0.47%	0.00%	0.20%	0.00%	0.00%
428.39	5.69	57.11%	37.42%	5.48%	34.74%	5.11%	15.50%	44.04%	0.58%
0.00	0.00	10.49%	6.29%	2.29%	2.29%	0.00%	16.78%	0.00%	0.00%
10.17	10.17	8.35%	8.16%	2.47%	0.00%	0.00%	16.70%	1.14%	1.14%
8.97	4.78	3.35%	0.27%	3.08%	0.00%	0.00%	5.16%	1.00%	0.54%
97.14	0.00	6.45%	9.03%	0.64%	3.30%	0.00%	1.93%	10.88%	0.00%
21.48	0.00	8.42%	1.92%	13.71%	15.64%	2.41%	3.61%	2.41%	0.00%
0.00	0.00	12.59%	1.82%	5.79%	2.89%	8.68%	2.89%	0.00%	0.00%
0.00	0.00	0.50%	0.00%	0.12%	2.43%	0.00%	0.19%	0.00%	0.00%
137.76	14.95	50.15%	27.49%	28.10%	26.55%	11.09%	47.26%	15.43%	1.68%
3.07	0.00	2.99%	0.00%	1.36%	3.30%	0.00%	0.00%	1.04%	0.00%
5.92	0.00	2.56%	0.00%	6.59%	6.86%	0.00%	0.27%	2.01%	0.00%
4.40	0.00	2.03%	0.00%	0.10%	0.21%	0.00%	0.43%	1.49%	0.00%
4.11	0.00	7.39%	0.00%	0.36%	6.20%	0.00%	0.15%	1.39%	0.00%
4.10	0.00	7.51%	0.15%	6.27%	11.71%	0.00%	0.84%	1.39%	0.00%

Table 7.1. (*cont.*)

Continent	Locality	Species	MASS	PDEN	BMASS	%FR
South America	Raleighvallen	*Alouatta seniculus*	5.950	17.00	101.15	41.40%
		Ateles paniscus	8.775	9.50	83.36	94.50%
		TOTALS:		107.57	294.35	
	Cocha Cashu	*Aotus trivirgatus*	0.775	23.33	18.08	80.00%
		Callicebus molloch	0.988	19.75	19.52	71.00%
		Saimiri scieurus	0.721	72.00	51.91	28.00%
		Cebus albifrons	2.735	30.00	82.05	65.00%
		Cebus apella	3.085	40.92	126.25	53.90%
		Alouatta seniculus	5.950	30.19	179.66	71.40%
		Ateles paniscus	8.775	24.95	218.95	94.50%
		Saguinus fuscicollis	0.351	12.00	4.21	52.00%
		Saguinus imperator	0.474	10.00	4.74	50.00%
		TOTALS:		263.14	705.37	
Africa	Kibale	*Pan troglodytes*	38.200	1.40	53.48	85.30%
		Colobus guereza	8.895	36.20	322.00	23.40%
		Procolobus badius	78.485	126.50	1073.35	22.80%
		Lophocebus albigena	7.135	2.30	16.41	66.10%
		Cercopithecus mitis	4.750	15.60	74.10	61.20%
		Cercopithecus ascanius	3.310	33.60	111.22	62.10%
		Cercopithecus l'hoesti	4.710	5.00	23.55	60.00%
		Papio anubis	19.200	0.01	0.19	72.20%
		Perodicticus potto	1.230	9.00	11.07	74.00%
		Galago senegalensis	0.213	87.00	18.53	10.00%
		Galagoides demidoff	0.062	65.00	4.00	20.00%
		TOTALS:		381.61	1707.90	
	Taï	*Cercopithecus diana*	4.550	16.80	76.45	87.60%
		Cercopithecus petaurista	3.650	29.30	106.95	87.20%
		Cercopithecus cambelli	3.600	15.00	54.00	84.80%
		Cercocebus atys	8.600	10.00	86.00	97.40%
		Procolobus badius	6.900	66.00	455.40	16.50%
		Colobus polykomos	9.100	23.50	213.85	46.90%
		Procolobus verus	4.450	21.00	93.45	5.50%
		Pan troglodytes	43.950	1.30	57.14	85.30%
		Perodicticus potto	0.833	8.35	6.95	69.00%
		Galagoides demidoff	0.062	58.33	3.59	20.00%
		TOTALS:		249.48	1153.78	

Table 7.1. (*cont.*)

%L	%FA	%AQ	%TQ	%CL	%LP	%BP	ACT	BFR	BL
58.60%	0.00%	80.00%	0.00%	16.00%	4.00%	0.00%	1	41.88	59.27
5.50%	0.00%	25.40%	0.00%	69.70%	4.20%	0.70%	1	78.77	4.58
								186.83	64.31
5.00%	15.00%	70.00%	0.00%	10.00%	20.00%	0.00%	3	14.47	0.90
26.00%	3.00%	60.00%	0.00%	10.00%	30.00%	0.00%	1	13.86	5.07
0.00%	72.00%	75.00%	0.00%	3.00%	22.00%	0.00%	1	14.54	0.00
0.00%	35.00%	70.00%	0.00%	10.00%	20.00%	0.00%	1	53.33	0.00
1.10%	45.00%	84.00%	0.00%	6.00%	10.00%	0.00%	1	68.05	1.39
28.60%	0.00%	80.00%	0.00%	16.00%	4.00%	0.00%	1	128.27	51.38
5.50%	0.00%	25.40%	0.00%	69.70%	4.20%	0.00%	1	206.91	12.04
0.00%	48.00%	48.00%	0.00%	19.00%	33.00%	0.00%	1	2.19	0.00
0.00%	50.00%	52.00%	0.00%	17.00%	31.00%	0.00%	1	2.37	0.00
								503.98	70.79
14.70%	0.00%	30.00%	50.00%	20.00%	0.00%	0.00%	1	45.62	7.86
76.60%	0.00%	50.00%	0.00%	0.00%	50.00%	0.00%	1	75.35	246.65
74.60%	2.60%	50.00%	0.00%	0.00%	50.00%	0.00%	1	244.72	800.72
7.90%	26.00%	70.00%	0.00%	30.00%	0.00%	0.00%	1	10.85	1.30
19.00%	19.80%	54.00%	0.00%	35.00%	11.00%	0.00%	1	45.35	14.08
16.10%	21.80%	39.00%	0.00%	45.00%	16.00%	0.00%	1	69.07	17.91
20.00%	20.00%	30.00%	40.00%	20.00%	10.00%	0.00%	1	14.13	4.71
7.80%	20.00%	30.00%	60.00%	5.00%	5.00%	0.00%	1	0.14	0.01
21.00%	5.00%	70.00%	0.00%	30.00%	0.00%	0.00%	3	8.19	2.32
50.00%	40.00%	20.00%	0.00%	0.00%	80.00%	0.00%	3	1.85	9.27
10.00%	70.00%	30.00%	0.00%	0.00%	70.00%	0.00%	3	0.80	0.40
								516.07	1105.23
7.70%	4.70%	70.10%	0.00%	19.40%	10.50%	0.00%	1	66.97	5.89
5.50%	7.30%	63.40%	0.00%	25.20%	11.40%	0.00%	1	93.26	5.88
0.00%	15.20%	80.30%	0.00%	14.50%	5.20%	0.00%	1	45.79	0.00
1.30%	1.30%	86.10%	0.00%	12.50%	1.40%	0.00%	1	83.76	1.12
83.50%	0.00%	65.20%	0.00%	17.00%	17.80%	0.00%	1	75.14	380.26
53.10%	0.00%	71.20%	0.00%	14.30%	14.50%	0.00%	1	100.30	113.55
94.50%	0.00%	67.60%	0.00%	12.00%	20.40%	0.00%	1	5.14	88.31
14.70%	0.00%	2.50%	84.10%	12.20%	0.00%	1.20%	1	48.74	8.40
21.00%	10.00%	70.00%	0.00%	30.00%	0.00%	0.00%	3	4.80	1.46
10.00%	70.00%	30.00%	0.00%	0.00%	70.00%	0.00%	3	0.72	0.36
								524.61	605.23

Table 7.1. (*cont.*)

Continent	Locality	Species	BFA	BAQ	BTQ	BCL
South America	Raleighvallen	*Alouatta seniculus*	0.00	80.92	0.00	16.18
		Ateles paniscus	0.00	21.17	0.00	58.10
		TOTALS:	43.22	185.36	0.00	79.27
	Cocha Cashu	*Aotus trivirgatus*	2.71	12.66	0.00	1.81
		Callicebus molloch	0.59	11.71	0.00	1.95
		Saimiri scieurus	37.38	38.93	0.00	1.56
		Cebus albifrons	28.72	57.44	0.00	8.21
		Cebus apella	56.81	106.05	0.00	7.57
		Alouatta seniculus	0.00	143.72	0.00	28.74
		Ateles paniscus	0.00	55.61	0.00	152.61
		Saguinus fuscicollis	2.02	2.02	0.00	0.80
		Saguinus imperator	2.37	2.46	0.00	0.81
		TOTALS:	130.60	430.60	0.00	204.06
Africa	Kibale	*Pan troglodytes*	0.00	16.04	26.74	10.70
		Colobus guereza	0.00	161.00	0.00	0.00
		Procolobus badius	27.91	536.67	0.00	0.00
		Lophocebus albigena	4.27	11.49	0.00	4.92
		Cercopithecus mitis	14.67	40.01	0.00	25.94
		Cercopithecus ascanius	24.25	43.38	0.00	50.05
		Cercopithecus l'hoesti	4.71	7.07	9.42	4.71
		Papio anubis	0.04	0.06	0.12	0.01
		Perodicticus potto	0.55	7.75	0.00	3.32
		Galago senegalensis	7.41	3.71	0.00	0.00
		Galagoides demidoff	2.80	1.20	0.00	0.00
		TOTALS:	86.61	828.38	36.28	99.65
	Taï	*Cercopithecus diana*	3.59	53.59	0.00	14.83
		Cercopithecus petaurista	7.81	67.80	0.00	26.95
		Cercopithecus cambelli	8.21	43.36	0.00	7.83
		Cercocebus atys	1.12	74.05	0.00	10.75
		Procolobus badius	0.00	296.92	0.00	77.42
		Colobus polykomos	0.00	152.26	0.00	30.58
		Procolobus verus	0.00	63.17	0.00	11.21
		Pan troglodytes	0.00	1.43	48.05	6.97
		Perodicticus potto	0.70	4.87	0.00	2.09
		Galagoides demidoff	2.51	1.08	0.00	0.00
		TOTALS:	23.94	758.53	48.05	188.63

Note: Key: MASS, species by mass in kilograms; PDEN, Population Density/km^2; BMASS, calculated by multiplying mean body weight (from Smith & Jungers, 1997) by population density, %FR (fruit), %L (leaves), %FA (fauna), estimated percentage of time spent foraging for these foods [in some cases reported percentages did not total to 100% so missing percentages were added to the fruit percentage (not over 5%)]; %AQ (Arboreal Quadrupedalism), %TQ (Terrestrial Quadrapedalism), %CL (Climbing), %LP (Leaping), %BP (Bipedality), estimated percentage of time spent on these various forms of locomotion; ACT, activity pattern (1: diurnal; 2: cathemeral, 3: nocturnal); BFR, BL,

Table 7.1. (*cont.*)

BLP	BBP	BPFR	BPL	BPFA	BPAQ	BPTQ	BPCL	BPLP	BPBP
4.05	0.00	14.23%	20.14%	0.00%	27.49%	0.00%	5.50%	1.37%	0.00%
3.50	0.58	26.76%	1.56%	0.00%	7.19%	0.00%	19.74%	1.19%	0.20%
29.15	0.58	63.47%	21.85%	14.68%	62.97%	0.00%	26.93%	9.88%	0.20%
3.62	0.00	2.05%	0.13%	0.38%	1.79%	0.00%	0.26%	0.51%	0.00%
5.86	0.00	1.96%	0.72%	0.08%	1.66%	0.00%	0.28%	0.83%	0.00%
11.42	0.00	2.06%	0.00%	5.30%	5.52%	0.00%	0.22%	1.62%	0.00%
16.41	0.00	7.56%	0.00%	4.07%	8.14%	0.00%	1.16%	2.33%	0.00%
12.62	0.00	9.65%	0.20%	8.05%	15.03%	0.00%	1.07%	1.79%	0.00%
7.19	0.00	18.19%	7.28%	0.00%	20.38%	0.00%	4.08%	1.02%	0.00%
9.20	1.53	29.33%	1.71%	0.00%	7.88%	0.00%	21.64%	1.30%	0.22%
1.39	0.00	0.31%	0.00%	0.29%	0.29%	0.00%	0.11%	0.20%	0.00%
1.47	0.00	0.34%	0.00%	0.34%	0.35%	0.00%	0.11%	0.21%	0.00%
69.18	1.53	71.45%	10.04%	18.51%	61.04%	0.00%	28.93%	9.81%	0.22%
0.00	0.00	2.67%	0.46%	0.00%	0.94%	1.57%	0.63%	0.00%	0.00%
161.00	0.00	4.41%	14.44%	0.00%	9.43%	0.00%	0.00%	9.43%	0.00%
536.67	0.00	14.33%	46.88%	1.63%	31.42%	0.00%	0.00%	31.42%	0.00%
0.00	0.00	0.64%	0.08%	0.25%	0.67%	0.00%	0.29%	0.00%	0.00%
8.15	0.00	2.66%	0.82%	0.86%	2.34%	0.00%	1.52%	0.48%	0.00%
17.80	0.00	4.04%	1.05%	1.42%	2.54%	0.00%	2.93%	1.04%	0.00%
2.36	0.00	0.83%	0.28%	0.28%	0.41%	0.55%	0.28%	0.14%	0.00%
0.01	0.00	0.01%	0.00%	0.00%	0.00%	0.01%	0.00%	0.00%	0.00%
0.00	0.00	0.48%	0.14%	0.03%	0.45%	0.00%	0.19%	0.00%	0.00%
14.82	0.00	0.11%	0.54%	0.43%	0.22%	0.00%	0.00%	0.87%	0.00%
2.80	0.00	0.05%	0.02%	0.16%	0.07%	0.00%	0.00%	0.16%	0.00%
743.61	0.00	30.22%	64.71%	5.06%	48.49%	2.13%	5.84%	43.54%	0.00%
8.03	0.00	5.80%	0.51%	0.31%	4.64%	0.00%	1.29%	0.70%	0.00%
12.19	0.00	8.08%	0.51%	0.68%	5.88%	0.00%	2.34%	1.06%	0.00%
2.81	0.00	3.97%	0.00%	0.71%	3.76%	0.00%	0.68%	0.24%	0.00%
1.20	0.00	7.26%	0.10%	0.10%	6.42%	0.00%	0.93%	0.10%	0.00%
81.06	0.00	6.51%	32.96%	0.00%	25.73%	0.00%	6.71%	7.03%	0.00%
31.01	0.00	8.69%	9.84%	0.00%	13.20%	0.00%	2.65%	2.69%	0.00%
19.06	0.00	0.45%	7.65%	0.00%	5.48%	0.00%	0.97%	1.65%	0.00%
0.00	0.69	4.22%	0.73%	0.00%	0.12%	4.16%	0.60%	0.00%	0.06%
0.00	0.00	0.42%	0.13%	0.06%	0.42%	0.00%	0.18%	0.00%	0.00%
2.51	0.00	0.06%	0.03%	0.22%	0.09%	0.00%	0.00%	0.22%	0.00%
157.87	0.69	45.46%	52.46%	2.08%	65.74%	4.16%	16.35%	13.69%	0.06%

BFA, BAQ, BTQ, BCL, BLP, BBP: biomass of each species allocated to each trophic and locomotor variable, i.e. total biomass of species multiplied by percentage of time spent on each variable (variable abbreviations as before); BPFR, BPL, BPFA, BPAQ, BPTQ, BPCL, BPLP, BPBP: the percentage of total community biomass supported by each trophic resource per species (variable abbreviations as before).

Table 7.2. *Body size distributions of primates in 49 communities with means for each continental area*

Code	Locality	Species	< 500 g (%)	501–1000 g (%)	1001–5000 g (%)	5001–10000 g (%)	> 10000 g (%)
1	Guinea Savanna, Nigeria	9	11.00	0.00	34.00	33.00	22.00
2	S Savanna Woodland, S Africa	5	20.00	0.00	30.00	30.00	20.00
3	Knysna Forest, S Africa	5	20.00	0.00	30.00	30.00	20.00
4	Serengeti NP, Tanzania	7	17.00	0.00	25.00	33.00	25.00
5	E Niger Forest, Nigeria	13	14.00	0.00	36.00	29.00	21.00
6	Cross River Forest, Nigeria	7	40.00	0.00	35.00	10.00	15.00
7	Ruwenzori UP, Uganda	10	20.00	0.00	20.00	20.00	40.00
8	Tiwai Island, Sierra Leone	10	10.00	0.00	50.00	30.00	10.00
9	Taï Forest, Ivory Coast	11	9.00	9.00	36.00	32.00	14.00
10	Kibale Forest, Uganda	11	18.00	0.00	36.36	27.27	18.18
11	W of Niger Forest, Nigeria	10	31.00	0.00	27.00	15.00	27.00
12	Ituri Forest, Dem Rep of Congo	12	7.00	0.00	23.00	53.00	17.00
13	Congo Basin, Equataire, DRC	11	0.00	0.00	46.00	34.00	20.00
14	Makokou Forest, Gabon	12	25.00	0.00	38.00	15.00	22.00
	African means		17.29	0.64	33.31	27.95	20.80
1	Horton Plains	3	33.00	0.00	0.00	67.00	0.00
2	Polonnaruwa	4	25.00	0.00	12.50	37.50	25.00
3	Pasoh	5	0.00	20.00	20.00	60.00	0.00
4	K Rompin 1	5	0.00	0.00	20.00	70.00	10.00
5	Ketambe, Sumatra	7	0.00	14.00	14.00	36.00	36.00
6	Kuala Lompat, Malaysia	7	0.00	14.00	14.00	50.00	22.00
7	Kutai	10	10.00	10.00	10.00	50.00	20.00
	Asian means		9.71	8.29	12.93	52.93	16.14
1	Perinet, Madagascar	10	18.00	27.00	46.00	9.00	0.00
2	Analamera, Madagascar	7	29.00	14.00	43.00	14.00	0.00
3	Andohahela, Madagascar	6	50.00	17.00	33.00	0.00	0.00
4	Ampijoroa, Madagascar	7	29.00	29.00	42.00	0.00	0.00
5	Ankarana, Madagascar	11	36.00	27.00	28.00	9.00	0.00
6	Berenty, Madagascar	6	33.00	17.00	50.00	0.00	0.00
7	Betampona, Madagascar	11	27.00	28.00	36.00	9.00	0.00
8	Beza-Mahafaly, Madagascar	5	40.00	20.00	40.00	0.00	0.00
9	Manombo, Madagascar	6	33.00	33.00	34.00	0.00	0.00
10	Manongarivo, Madagascar	8	37.50	25.00	37.50	0.00	0.00
11	Mantady, Madagascar	10	20.00	30.00	40.00	10.00	0.00
12	Montagne d'Ambre, Madagascar	7	43.00	14.00	43.00	0.00	0.00
13	Morondava, Madagascar	7	57.00	14.00	29.00	10.00	0.00
14	Namoroka, Madagascar	4	25.00	25.00	50.00	0.00	0.00
15	Ranomafana NP, Madagascar	12	16.60	8.30	66.70	8.30	0.00
16	Tsimanampetso, Madagascar	3	34.00	0.00	66.00	0.00	0.00
17	Tsingy de Bern, Madagascar	7	42.00	29.00	29.00	0.00	0.00
18	Verezanantsolo, Madagascar	9	33.00	11.00	34.00	22.00	0.00
	Malagasy means		33.51	20.46	41.51	4.52	0.00

Table 7.2. (*cont.*)

Code	Locality	Species	< 500 g (%)	501–1000 g (%)	1001–5000 g (%)	5001–10 000 g (%)	> 10 000 g (%)
1	Guatopo, Venezuela	3	0.00	0.00	33.30	66.70	0.00
2	Raleighvallen, Suriname	8	0.00	25.00	50.00	25.00	0.00
3	Rio Negro, Brazil	10	0.00	33.30	44.00	22.70	0.00
4	Rio Xingu, Brazil	9	11.00	23.00	44.00	22.00	0.00
5	RioNegro/Rio Japuro, Brazil	10	0.00	20.00	50.00	30.00	0.00
6	Cocha Cashu, Peru	10	30.00	25.00	25.00	20.00	0.00
7	Rio Roosevelt, Brazil	11	9.00	9.00	55.00	27.00	0.00
8	Upper Rio Joroa, Brazil	13	23.00	16.00	38.00	23.00	0.00
9	Urucu River, Brazil	13	16.30	16.40	51.00	16.30	0.00
10	Rio Javari, Brazil	14	14.00	14.00	50.00	22.00	0.00
	South American means		10.33	18.17	44.03	27.47	0.00

Table 7.3. *Species diversity indices for 15 primate communities*

Locality	No. of species	McIntosh U	McIntosh evenness	Simpson Dominance
Africa				
Makokou	12	130.1	0.875	7.14
Taï Forest	11	97	0.861	6.44
Ituri Forest	12	81.3	0.791	5.35
Kibale Forest	11	174.9	0.775	4.81
Asia				
Ketambe	7	59.8	0.831	4.41
Kutai	10	56.9	0.683	3.61
Kuala Lompat	7	90.1	0.730	3.23
South America				
Urucu River	13	49.7	0.918	9.34
Cocha Cashu	10	100.4	0.931	7.75
Raleighvallen	8	44.2	0.902	6.03
Guatopo	3	62.2	0.831	2.41
Madagascar				
Ranomafana	12	176	0.870	6.97
Perinet	11	171	0.883	5.86
Ampijoroa	7	413.6	0.804	4.013
Morondava	7	1214.6	0.783	3.57

The numerical representation of primate community population densities reflect, for the most part, the ecological space defined by these communities (Fleagle & Reed, 1996). The Malagasy primate communities reflected the largest distance from the centroid and the greatest taxonomic distance among the species, while the South American communities had the least. African primate communities are also more ecologically diverse than those found in Asia.

THE INFLUENCE OF BODY SIZE AND BIOMASS ON PRIMATE COMMUNITY STRUCTURE

Body size distributions within and among continental areas

To compare species body size distributions with biomass distributions, I first had to examine species body size patterns within and among continental regions by community. Therefore, males and females of each primate species from the 49 communities were placed into one of five body size categories: < 500 g (Category 1); 501–1000 g (Category 2); 1001–5000 g (Category 3); 5001–10 000 g (Category 4); and > 10 000 g (Category 5). Male and female mean masses were both used, so that if males were considerably larger than females they were sometimes placed in the next largest body size category. Percentages of primate species in each body size category were calculated for each community to make relative comparisons within and among continents (Table 7.2). For example, two primates in category 2 of a total community of five primate species would equal 20% (5 species × 2 sexes = 10 total). A Kruskal–Wallis test revealed that none of the within continent, community distributions of body size were significantly different from one another (Table 7.2; Fig. 7.1). The percentages of body masses for each community were then averaged so that a mean percentage of primates in each body size category for each continent was obtained (Fig. 7.2a).

An interesting pattern to emerge is that despite sharing almost identical families of living primates, Africa and Asia are quite different in the distribution of body sizes within continents (Fig. 7.2). African primates are more evenly distributed across body size categories 1, and 3–5, while Asian localities have the highest percentage of primate species in the 5 to 10 kg range (category 4). This size category includes most hylobatid, colobine, and macaque species; only lorises, tarsiers, and orangutans are outside of this body size range in Asia. Africa and South America share similar community body size percentages in categories 1 and 4. This reflects the body size similarities of lorisids and callitrichines, and cebid and *Cercopithecus* monkeys. African and Malagasy communities share a similarity in the relative percentages of the smallest primates in both communities, i.e., lorisids and cheirogaleids.

South American and Asian primate communities are similar only in body size category 2, again showing that percentages of loris and callitrichine species are similar in their respective communities. South America and Madagascar reflect similar distributions of body size categories 2 and 3, i.e., among cebids and lemurids. Communities from both regions are missing large-bodied primates, although that is a recent occurrence on Madagascar. In Madagascar, small-bodied cheirogaleid species equal an average of 30% of primates in those communities, surpassing relative abundance of callitrichine species in South American communities by 20%.

Thus, primate community structure based on body size is similar within continental regions, but is quite different among continental areas. Although there are some similarities in certain body size category percentages, continental regions have unique patterns of overall primate community body size structure.

Biomass distributions within and among continental areas

Primate biomass was calculated for the smaller data set (15 primate communities) and the percentages of biomass in each body size category were recorded. These percentages were then averaged for each category for each continent to provide mean biomass percentages in each body size category for each continental area. Figure 7.2 compares the differences in community composition of biomass between and among continents, as well as contrasting differences in biomass distribution with species body size distributions.

Malagasy communities are completely different from all other areas in relative biomass distribution (Fig. 7.2b). Greater than 70% of the primate community biomass in these communities is in Category 3 (1–5 kg), while in

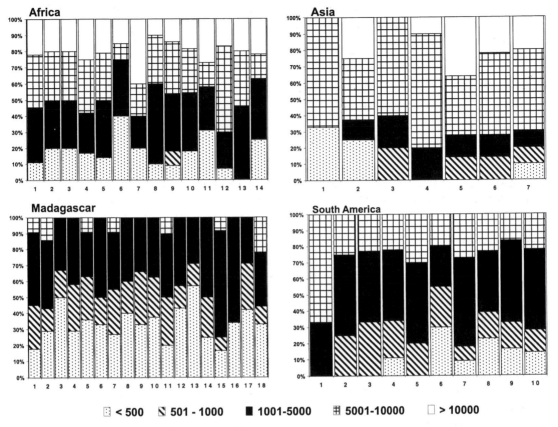

Fig. 7.1. Distribution of primate species body sizes in 49
communities. Key: Numbers are coded to Table 7.2.

Africa, Asia, and South America, much of the biomass
(55% or greater) is found in Category 4 (5–10 kg). Madagascar is also the only region in which biomass exceeds
2% in Category 1. Cheirogaleids are much more dense
within communities of Madagascar than their lorisid
counterparts are in Africa and Asia. Extant lemuroids fit
uniquely into Category 3 (1–5 kg), a fact that is evident
when biomass is compared with individual species body
sizes (Fig. 7.2).

In South America and Africa, the second highest distribution of relative biomass is found in Category 3, while in
Asia the second greatest percentages of biomass are found
in Category 5. Both African and Asian communities have
very little biomass in Category 2. While South American
primate species body size distributions in each community
are highest in Category 3 (cebids, pitheciines), Category 4

(atelines) has the greatest relative percentage of biomass.
Although African communities have a fairly even distribution of body sizes, the highest relative biomass category
is also in Category 4 (colobines, *Cercocebus*). Finally, Asian
communities have the highest percentage of species in
body size Category 4 (colobines), and this is reflected in the
relative percentages of biomass as well.

Thus, the differences between the community structures as represented by primate body sizes and primate
biomasses in communities are strikingly different within
continents. While the body size community structure
varies among continental areas, the structure represented
by primate biomasses is only extremely different between
Madagascar and all others. This may or may not be an artifact of recent extinctions of the very large Malagasy
lemurs. Most of these subfossils have been estimated to

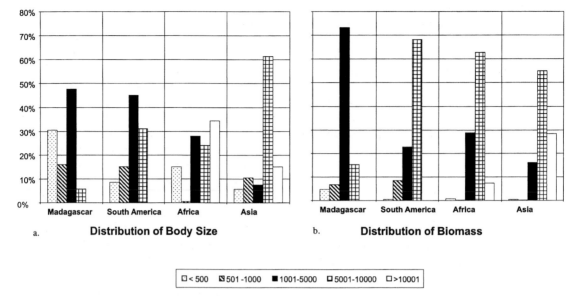

a. **Distribution of Body Size**   b. **Distribution of Biomass**

□ < 500   ◨ 501 -1000   ■ 1001-5000   ▨ 5001-10000   ▢ >10001

Fig. 7.2. Mean distribution of individual species body sizes compared to mean distributions of species biomass of primate communities on different continents.

weigh over 10 kg; if subfossil biomasses had the highest relative percentage in these communities, Madagascar would still differ from the other regions. The addition of these species and their densities, however, might shift the relative abundance of biomass so that Category 3 would have less contribution to the community structure. In any case, neither event would increase the relative biomass of Category 4 to be equivalent with primate communities on other continents.

Comparisons of body size and biomass within and among primate communities

Comparisons of body size among primates have been made at continental levels (Bourlière, 1985; Terborgh & van Schaik, 1987; Terborgh, 1992; Ganzhorn, chapter 8, this volume). However, this study concerns the patterning of primate body sizes within individual communities within and among continents. The distribution of individual primate species body mass within communities is quite similar intracontinentally, and different intercontinentally. To understand these patterns ecologically, more factors must be considered. Body size is also responsible

for what primates are able to consume (Kay, 1984). Thus among communities on different continents, we might correlate gross differences in the acquisition of trophic resources with dominant body size categories. About 45% of the species within South American and Malagasy communities are in the 1–5 kg range. These similarities among communities are only in body mass, however, as their trophic adaptations are quite different (Terborgh & van Schaik, 1987; Fleagle & Reed, 1996). In Asia, an average of 60% of the species in each community are in the 5–10 kg range. These species consist of folivore-frugivore colobines and frugivore-insectivore macaques. Finally, in Africa, there is no dominant body size category as primate body size is fairly evenly distributed. There does not appear to be a consistent pattern between dominant body size categories and trophic resources within primate communities on different continents.

Distributions of biomass in these same body size categories are more interesting, as three of the continental areas have similar patterns. First, Madagascar differs from all other continents. The dominant biomass category is 1–5 kg which includes most of the Lemuridae and some of the Indriidae. The biomass of these lemurs and indris contributes 70% to the total biomass of each primate community. In the other three continental regions, most of the biomass is found in the 5–10 kg category. For each continental area, these biomass categories include the most folivorous primates of each community. Thus, as has been noted in many previous studies, primate folivores have higher population densities than other primate trophic types, and this appears to contribute greatly to the biomass structural pattern consistent across continents, except for Madagascar which just has smaller folivores.

POPULATION DENSITY AND ECOLOGICAL SPACE WITHIN AND AMONG CONTINENTS

I have shown that there are differences in species diversity and body mass distributions when analyses based on individual species characteristics are compared with analyses including population densities of primates in communities. Therefore, I now address the question of whether or not the ecological space inhabited by primate species is altered if the primates are considered as a group rather than as individual species. In order to use population densities with ecological variables for comparisons, I restrict the analysis to the eight focal communities that were used in Fleagle & Reed (1996). Percentages of time spent on acquiring various food resources, and time spent in various locomotor postures were then multiplied by the biomass of each species. Biomass, rather than population density, was used so that ecological information contained in primate body sizes would also be included. The resultant biomass for each species population in each trophic and locomotor category was divided by total community biomass to arrive at percentage amounts that reflect the relative primate biomass that is spent on each type of trophic resource and locomotor category. For example, the biomass for *Hylobates lar* at Ketambe is 59.94 kg/km^2. This species eats 50% fruit, 4% leaves, and 46% fauna, thus gibbon biomass devoted to each of these trophic categories is 29.97 kg/km^2 in fruit, 2.4 kg/km^2 in leaves, and 27.57 kg/km^2 in fauna. Total primate biomass in each community was used to calculate the percentage of biomass supported by each trophic resource for each species. Thus, at Ketambe, the total primate biomass is 893.4 kg/km^2, and the *H. lar* population represents 3.35% of primate biomass that searches for fruit, 0.27% of primate biomass that searches for fruit, 0.27% of primate biomass that

feeds on leaves, and 3.09% of primate biomass time spent on foraging for fauna. Percentages of biomass per species in each ecological category were totaled to arrive at the total relative primate biomass utilizing each trophic resource and locomotor posture in each community (Table 7.1).

As various researchers report trophic resources differently, percentages of time spent on foodstuffs such as gums, seeds, and flowers were collapsed into the more common categories of either fruit or leaves. Previously, gums were added to fruit percentages (Fleagle & Reed, 1996), but as many gums are more structurally like foliage with high cellulose content (Hladik & Chivers, 1994), I have placed gums in the leaves category. Flowers and seeds are included in the fruit category, although it would be much better to include these trophic categories separately. Placing time spent on acquiring seeds and flowers in the fruit category allows more communities to be compared. The obvious disadvantage is that primates may spend more time eating seeds and less time eating flowers than they might spend on fruit acquisition (Janson & Chapman, chapter 14, this volume). When further data are available, it will be interesting to examine primate population ecological space with regard to time spent foraging for saps, gums, seeds, and flowers.

Percentage data in the literature about primate trophic and locomotor behaviors are usually available as time spent foraging for various foodstuffs and time spent on the locomotor activities moving toward these foodstuffs. This does not necessarily equate to percentages of food eaten, e.g., a medium sized primate could spend more time searching for fauna than fruit, but consume far less fauna (Janson, 1985; Janson & Chapman, chapter 14, this volume). Therefore, this analysis is technically showing the percentages of time that a primate population spends on acquiring these foods, both the type of food and the locomotor behaviors involved in getting to the food.

These percentages of ecological variables for each species population were examined in a principal coordinates analysis. PCO enables the description of representative ecological space encompassed by these primates in two-dimensional space (Fleagle & Reed, 1996). A 70×9 matrix based on 70 primate species and nine ecological variables was created. All percentages were arcsine transformed (Sokol & Rohlf, 1981). This multivariate technique enables the inter-correlation of primates with their variables to be more easily interpreted as a high percentage of

Table 7.4 *Percent of total variance for each of the nine factors in the principal coordinates analysis and loadings of the ecological variables on each factor*

Component	Eigenvalue	% of variance	Cumulative %
1	3.180	35.333	35.333
2	1.899	21.097	56.430
3	1.399	15.539	71.969
4	0.728	8.092	80.061
5	0.687	7.629	87.690
6	0.506	5.620	93.310
7	0.331	3.679	96.989
8	0.240	2.666	99.655
9	0.003	0.345	100.000

	Component								
	1	2	3	4	5	6	7	8	9
Fruit	0.887	0.144	0.003	−0.002	−0.009	0.274	−0.203	0.243	−0.008
Leaves	0.771	−0.450	−0.105	−0.169	0.267	−0.009	0.169	−0.220	−0.008
Arboreal quadruped	0.735	−0.286	0.376	0.194	−0.172	0.235	0.328	0.004	0.007
Leaping	0.697	−0.528	−0.008	0.005	0.284	−0.274	−0.234	0.104	0.008
Climbing	0.610	−0.560	−0.393	0.001	−0.003	0.236	−0.153	−0.277	0.006
Activity	−0.586	−0.416	−0.157	0.008	0.489	0.461	−0.002	0.002	0.0007
Bipedality	0.211	0.646	−0.576	0.00001	0.290	−0.105	0.271	0.196	0.001
Terrestrial quadruped	0.006	0.403	0.642	−0.590	0.264	0.005	0.00008	0.001	0.003
Fauna	0.100	0.499	0.562	0.552	0.322	−0.100	−0.004	−0.007	−0.002

the variation among species is often found on the first two principal coordinates axes (James & McCulloch, 1990). The positions of each population of species in each community were plotted within the overall ecological space described by the first two PCO axes.

The first factor accounts for 35.3% of the variance among species groups and is highly correlated with frugivory, folivory, arboreal quadrupedalism, and leaping (+) while activity pattern is the only negatively correlated variable. Factor 2, which accounts for 21.2% of the variance among species groups, is highly correlated with bipedality (+), climbing (+), leaping (−), and folivory (−). The third factor accounts for the final 15.5% of the variance with terrestrial quadrupedalism (+) loading most highly. This analysis is mostly based on the first two components which account for 56.4% of the variance among these factors (Table 7.4). This is roughly the same amount of variation as

reported for ecological variables based on species' ecological behaviors (Fleagle & Reed, 1996), but here the first factor accounts for 7% more of the variation.

Compared with overall phylogenetic patterns in ecological space when species are plotted without densities (Fleagle & Reed, 1996), the PCO results here are more phylogenetically overlapping. For example, hylobatids overlap with cercopithecines, colobines, atelines, and other hominoids, over a wider area of ecological space (Fig. 7.3), whereas they are fairly isolated in species-only analyses (Fleagle & Reed, 1996). Thus, the ecological space is defined by biomass and ecological variables such that primates with relatively low biomasses do not contribute as much to the defined ecological space. In general, primates with high population densities and medium range body masses contribute more to each ecological category than do small or large bodied primates with high and low population

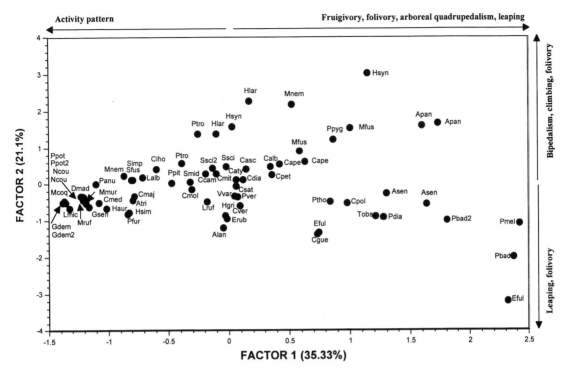

Fig. 7.3. Plot of all primate species on the first two factors of the principal coordinates analysis. Abbreviations based on the first letter of the genus followed by the first three letters of the species name.

densities respectively. While body size and ecological behavior are relatively conservative within phylogenetic groups (Fleagle & Reed, 1996), consideration of population densities of these primates alters this phylogenetic perspective. For example, *Lepilemur microdon* is positioned with other low density, nocturnal primates, but *L. ruficaudus* is found more centrally positioned as its density is 100 times that of *L. microdon*. As such it is contributing 6% of the total community biomass of Morondava on leaves, whereas *L. ruficaudus* is only contributing 0.47% to this effort at Ranomafana. In the analysis using species and ecological variables only, the two taxa were in relatively the same position (Fleagle & Reed, 1996). Therefore, ecological space is defined on both species' ecological characteristics and the effect of primate population density on those ecological variables. Species with higher population densities and body mass occupy the corners of ecological space, i.e., they expand the space more than lower density primates.

The shape of each community's ecological space, as defined by these parameters, is fairly similar overall. Various communities have subtle shape changes, but in general the polygons are all roughly triangular in shape. Figures 7.4 and 7.5 are based on plotting each community separately as defined by the first two PCO factors of total primate ecological space. The extremities of each triangular polygon depend upon how primate population density affects the ecological categories.

Ranomafana, despite the fact that there are almost twice the number of species, appears to have a smaller defined ecospace than does Morondava, with its high levels of population density, and thus biomass. Nevertheless, there is similarity between the Malagasy localities although Ranomafana is on the east coast and is a rainforest, while Morondava is on the west coast, and is a dry forest. First, they are the only localities that have no representation on the positive side of Factor 2. This is because there are no climbing (suspensory) or leaping faunivorous primates in Madagascar. All of the nocturnal and some cathemeral species are located on the negative side of Factor 1, not only are they nocturnal but these species are mostly small bodied and thus even with high population densities, the percentage of time spent on any foraging activity does not have as great an impact on the community as do those species with higher body masses and higher densities. However, Ranomafana is also the least triangular in shape of any community, and there are several primates in the center of the polygon. *Propithecus diadema* as a population is responsible for about 16% of total folivory, and the population of *Eulemur fulvus* accounts for 16% of total frugivory, thus pulling out the corners of the polygon. In contrast, at Morondava, *Eulemur fulvus* accounts for both 20% of time spent in pursuit of fruit and 58% of the total time spent on folivory for all species. *P. verreauxi*, despite having the next highest biomass does not come close to the impact that *E. fulvus* has on the ecological space of Morondava.

The Taï Forest is located in west Africa and Kibale Forest in east Africa, and the ecological shapes are somewhat different because of the biomass supported in the two different types of forests. Nevertheless, there are similarities. The ecological shapes are mostly triangular with *Procolobus badius* pulling out the folivory/frugivory/leaping corner in both communities. Thus, *P. badius* accounts for the highest percentage of both frugivory (14%) and folivory (47%) at Kibale. In the Taï forest, populations of *P. badius* are responsible for 33% of the folivory, but only 9% of the frugivory. *P. badius* at Taï appears higher along the *y*-axis than at Kibale because other primates in the Taï forest, such as *Pan troglodytes*, spend equivalent time on fruit. Taï Forest has the most primate populations in the center of the defined ecospace. These are mostly frugivorous *Cercopithecus* species that add protein to their diet with either small amounts of leaves or fauna. In both foraging time and biomass these *Cercopithecus* species are intermediate between the more dense colobines and the smaller prosimians.

Both of the Asian communities are triangular, but they do not have an overlapping shape. In fact, Kuala Lompat, in shape, is most similar to the Taï Forest in Africa. The corners of the Asian communities are held by different ecological, and thus phylogenetically different primate populations. The Asian communities are similar in the position of the space, however, and appear to hold more space above the 0 point of the *y*-axis than in African communities. Thus, the hylobatids pull this ecospace in the

direction of climbing (suspensory), faunivory, and frugivory. At Kuala Lompat, *Presbytis melalophus* has the highest biomass, and thus, as *P. badius* in Africa, is responsible for the highest percentages of total folivory (18%) and frugivory (32%). At Ketambe, *H. syndactylus*, *Macaca fascicularis* and *P. thomasi* are similar in percentages of total biomass spent on fruit and leaf acquisition and the entire right part of the polygon reflects this. The small bodied *Nycticebus coucang* is the third point in these polygons, having a low population density coupled with nocturnality.

Finally, South American ecospace is also roughly triangular in shape, but also occupies a separate region of overall space. Like Asian communities, South American communities occupy space in the suspensory/climbing, frugivorous section of the total primate area, although only *Ateles paniscus* pulls the space in this direction. In addition, they are the only communities in which the left point of the polygons are less than −1. This is partly due to *Aotus* being the only nocturnal primate. However, it also has to do with the relatively even population densities among most of the primate species. *A. paniscus* is therefore the heaviest, and as such has the highest biomass, while *P. pithecia* and all *Saguinus* species are the least dense in these communities. This means that as populations they do not influence the ecospace as much as the equally dense heavier primate species. Similar to cercopithecus in African communities, the smaller, frugivorous/faunivorous cebids are found in the central part of the triangles. The size of these ecospaces in two dimensions, however, is not any smaller than those of Madagascar. This indicates that if the South American communities are considered as primate populations, the apparent lack of ecological diversity (i.e., most neotropical primates are small bodied frugivores) is not reflected in the size of overall ecological space.

The primate populations in some of the communities are adhering to the outer reaches of the polygons, while in others there are several species directly in the center of the polygon. This is irrespective of the geographical location of the community such that Ranomafana, Taï, Ketambe, and Cocha Cashu all have centrally located species populations. In general these species are mostly in the 1–5 kg range in body size and have similar densities. In addition, they all process fruit in some combination with fauna or leaves and most are moving quadrupedally. The species in the centers appear more generalized in size, population

Table 7.5 *Three measures of dispersion among species populations in eight primate communities*

Community	Centroid distance (2 factors)	Euclidean distance (2 factors)	Euclidean distance (9 factors)
Ranomafana	0.698	1.09	3.55
Morondava	0.932	1.71	3.60
Kuala Lompat	0.918	2.09	4.31
Ketambe	1.210	1.95	6.01
Raleighvallen	0.571	1.35	3.84
Cocha Cashu	0.625	1.38	3.48
Kibale	0.719	1.62	2.93
Taï	0.580	1.38	2.72

density, diet and locomotor pattern, while the species at the periphery of the polygon have somewhat more specialized adaptations in higher concentrations. For example, hylobatids are the only primates in Asia that are not only suspensory, but also move bipedally and are found in relatively high densities. The smaller bodied primates in all communities for the most part have low biomasses and do not expand the primate ecological space beyond the lower left quadrant.

I used three methods to calculate the dispersion distance of these populations of species within communities because each method measures a different aspect of community dispersion (van Valkenburgh, 1988): (1) average distance of each species from the community centroid of the PCO factors; (2) average pairwise distance between species using the first two PCO factors; and (3) average pairwise distance between species based on all PCO factors (Table 7.5).

In average distance from the center of the community, Asian communities are more widely dispersed than any other communities. The lowest numbers in dispersion around the centroid (Taï, Cocha Cashu, and Raleighvallen) are found in communities where there are several species in the center of the ecological space. Thus, the more frugivorous primates cause a species clustering closer to the center of the ecospace.

The average ecological distance between each species as represented by two PCO factors gives a numerical representation of the plots in Figs. 7.4 and 7.5. South American communities define a somewhat larger ecological space than does the community of Ranomafana, and are roughly

equivalent to the ecospace defined by the Taï community. This is interesting because if ecological space is defined by the ecological attributes of individual species, then South American communities are tightly clustered because they are dominated by small, frugivorous–faunivorous primates (Fleagle & Reed, 1996). However, they are actually ecologically well-dispersed when biomass is considered because the large body sizes and density of the most folivorous primates add to the ecological space occupied by these primates.

Finally, the pairwise distance among all species in each community as represented by the complete analysis reveals that the Asian communities reflect the greatest distances among populations of species, followed by South American, Malagasy, and finally, African communities. In general this represents the contribution of folivores to community densities and biomass. The African communities have the smallest average distance because of numerous frugivores that are not particularly dense in these two communities. Asian communities occupy the largest ecospace, not only because of the density of folivorous primates, but because of the unique adaptations of hylobatids and *Pongo*.

Although this study is limited to eight primate communities, it suggests that the way primate communities are constructed is not only dependent upon their behavioral ecology (including body size), but also on the population density of each primate species. As all the primates in this study were plotted in common ecological space, the ways in which these communities overlap and differ are readily apparent (Fig. 7.6). The major differences among communities, both within and among continents, occur because, despite similar species, different primates at each community have higher densities, and thus consume more resources – no matter what those resources are.

Primate population density and ecospace discussion

Compared with other primate communities in this study, the ones from Madagascar hold a different type of ecological space. Malagasy communities are unique with their abundance of relatively small, folivorous primates, and the lack of faunivorous ones. Thus the ecological space is tightly confined, at least along Factor 2. In fact, if not for high densities of *E. fulvus* at Morondava and *P. diadema* at

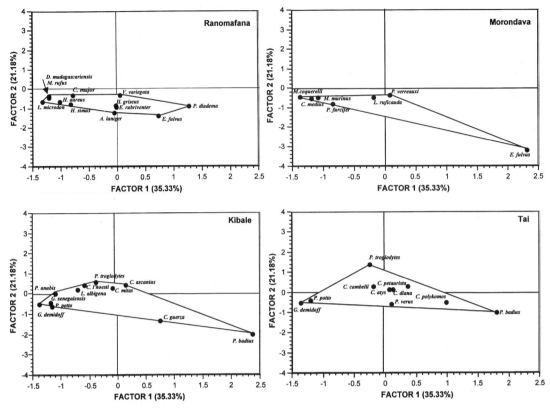

Fig. 7.4. Individual plots of Malagasy and African primate communities on the first two factors of the principal coordinates analysis.

Ranomafana, these Malagasy communities would occupy much smaller ecological space. The species found in the center of the ecospace of Ranomafana are positioned lower than the centrally located species of other, non-Malagasy communities. Again, this reflects the lack of more frugivorous primates on Madagascar. However, despite the absence of subfossil lemurs, the Malagasy communities' ecological areas are larger than those from Africa. While these large primates would add locomotor diversity to the community space, they likely subsisted on foliage and seeds, much like extant lemurs (Godfrey et al., 1997). As large body size usually indicates low population density (Fa & Purvis, 1997), it is possible that the ecological space held by the primate populations of Malagasy communities in the recent past would not be substantially larger with the inclusion of populations of larger, extinct species. Nevertheless, if these species could browse vegetation of lower

quality because of their larger body sizes, it is possible that they might have been denser than expected and the ecological space could be greatly expanded as it is with individual species' generated ecospace (see Godfrey et al., 1996).

Species in African communities are positioned more evenly around the origin of the eigenvectors with Cercopithecus and Cercocebus species in the central space. That is, medium sized, highly frugivorous, arboreal quadrupeds are located around the origin of the total ecospace held by primates. This may indicate that these primates are more generalized trophically compared with the primates on the periphery of the polygons. In Africa, those peripheral primates are (1) colobine species that spend time acquiring both fruit and leaves in high densities, (2) galagos that are nocturnal, faunivore/gumnivores with low biomass, and (3) large bodied, frugivorous chimps. Ganzhorn (chapter 8,

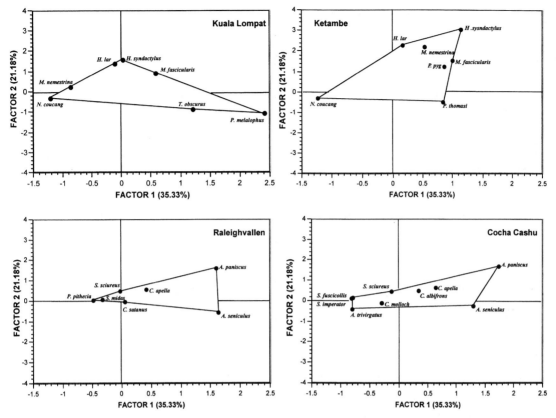

Fig. 7.5. Individual plots of Asian and South American primate communities on the first two factors of the principal coordinates analysis.

this volume) suggests that primate species are perhaps forced out to the periphery of ecological polygons when they are in competition with other species. However, in this analysis, the peripheral positions are determined by population densities rather than dietary adaptations.

Despite the fact that Asian communities have fewer numbers of species than other regions, the ecological space that they inhabit as populations is greater than all other communities. Thus, in spite of the lack of high species numbers or high diversity indices, Asian primate communities occupy greater ecological space than communities with higher species diversity. What is most interesting about this ecological space, however, is that there are no primates in the center of the ecological space. While peripheral species in African communities may avoid competition with frugivores, this does not appear to be the case

in Asian communities. Species are at the periphery because of the high densities of Asian colobine species combined with consumption of leaves and fruit. Hylobatids and macaque species are also generally far from the central ecospace as they supplement their fruit intake with fauna. While this behavior is also seen in some African cercopithecines, the densities of the Asian species are much greater.

South American communities have a larger ecospace when primate populations are included because although the densities of all primates are similar, the largest, semifolivorous species, such as *Aloutta*, are much heavier than the other species, resulting in pulling the ecospace out toward the folivore/frugivore/climbing side of Factor 1. The smaller cebids are still located in the more central position, and the callitrichines and pitheciines do not pull

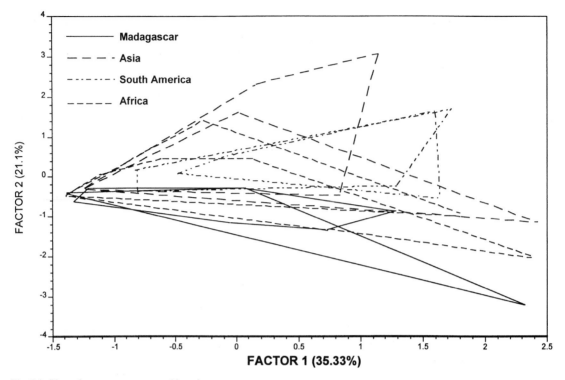

Fig. 7.6. Plots of ecospace encompassed by primate
communities on each continent on the first two factors of a
principal coordinates analysis.

the space as far to the right as do small species in other
communities. Thus, the major change in ecospace is due
to the large bodied atelines. Again, these species are far
from the centroid of the community, indicating a more
non-frugivorous diet and non-arboreal quadrupedal locomotor adaptations.

CONCLUSIONS

I proposed a series of questions at the beginning of this
chapter and I can attempt to answer some of them with the
data presented here.

(1) How does population density affect measures of
species diversity? In analyses that include the number of
species as the only measure of diversity, South American
primate communities are often considered the most
diverse having communities consisting of 7 to 14 species.
In the same vein, Asian communities are always considered the least diverse, with ten being the most primates in

any community. Likewise, Malagasy and African communities are just slightly less diverse than South American
ones. The addition of population densities, i.e., how those
species numbers are actually distributed in the community
changes the rank order of diversity. Thus Malagasy communities, because of the high densities of species have the
highest diversity values. South American communities have
some of the lowest values, probably because the populations of species, while distributed evenly among species,
are low. Asian communities also show fairly low diversity,
and are the least variable within their continental area.

(2) Are body size distributions of individual species
within communities the same as the distributions of biomass in those same communities? No. Body size distributions are fairly similar within continental areas and different among them. Biomass distributions are similar among
all continental areas except Madagascar. The biomass is
distributed such that the folivorous primates that range
from 5–10 kg on mainland areas and folivorous primates in

the 1–5 kg range on Madagascar incorporate the most biomass in their respective communities. Thus primate community structures based on biomass distributions are much more similar among continents than any other measure of community organization addressed here. The Malagasy community distribution of biomass is similar in that the biomass is held by folivores; the species are just smaller overall than folivores on the other continents.

(3) What part of the overall primate biomass in a community is supported by leaves, fruit, and fauna? What resources are being most utilized by each community? It depends upon the community. In general, relative biomass is greatest in more folivorous monkeys. So much so, that the relative biomass acquiring fruit by folivorous monkeys is often greater than the relative biomass acquiring fruit by frugivorous monkeys. Fauna supports the least amount of biomass in most communities.

(4) How does the ecological space occupied by individual species' attributes compare with space occupied by populations of species' trophic attributes? The ecospaces are different. Ecological space held by individual species is much more concordant between phylogeny and ecological characteristics (Fleagle & Reed, 1996) than space held by primate community populations. Ecological space displayed in this study, based on the addition of primate populations, suggests that if communities contain several, mostly frugivorous primates the overall ecological space is smaller than communities that have few of these primate species. This is because the frugivorous quadrupeds are located closer to one another in ecospace, thus reducing the average distance between all species. In communities with few to no frugivorous species, the primates are at the periphery of the polygons and have more distance between species.

Finally, this study has emphasized the importance of what primates do as a group, as opposed to individually. While the same species might exist within different communities, they spend time differently searching for foods, and densities vary for a variety of reasons, e.g., carrying capacity (Janson & Chapman, chapter 14, this volume), human hunting pressures (Peres, chapter 15, this volume), etc. Species at the perimeter of a polygon of ecological space are suggested to be those in competition (Ganzhorn, chapter 8, this volume). These species are different when their population densities are considered, and this may affect or contribute to interspecific competition. Both

types of studies, i.e., those of individual species and of populations of species, are important for identifying patterns in the structure of primate communities.

ACKNOWLEDGMENTS

I am grateful to the Wenner–Gren Foundation for providing support for the workshop from which this paper is derived. I thank Jorg Ganzhorn, Charles Janson, Louise Emmons, Peter Kappeler, and Colin Chapman for many interesting comments about population densities and biomass at the workshop. And words cannot express how much I appreciate the help of John Fleagle. This work was supported in part by a fellowship from the American Association of University Women.

REFERENCES

Bourlière, F. (1985). Primate communities: their structure and role in tropic ecosystems. *International Journal of Primatology*, **6**, 1–26.

Caldecott, J. O. (1986). An ecological and behavioral study of the pigtailed macaque. *Contributions to Primatology*, **21**, 1–259.

Damuth, J. (1981). Population density and body size in mammals. *Nature*, **290**, 699–700.

Fa, J. E. & Purvis, A. (1997). Body size, diet and population density in Afrotropical forest mammals: a comparison with neotropical species. *Journal of Animal Ecology*, **66**, 98–112.

Fleagle, J. G. & Reed, K. E. (1996). Comparing primate communities: a multivariate approach. *Journal of Human Evolution*, **30**, 489–510.

Ganzhorn, J. U. (1988). Food partitioning among Malagasy primates. *Oecologia (Berlin)*, **75**, 435–50.

Ganzhorn, J. U. (1992). Leaf chemistry and the biomass of folivorous primates in tropical forests. *Oecologia*, **91**, 540–7.

Gautier-Hion, A. (1988). Polyspecific associations among forest guenons: ecological, behavioral and evolutionary aspects. In *A Primate Radiation: Evolutionary Biology of the African Guenons*, eds. A. Gautier-Hion, F. Bourliere, J-P. Gautier & J. Kingdon, pp. 452–76. Cambridge: Cambridge University Press.

Godfrey, L., Jungers, W., Reed, K. E., Simons, E. L. & Chatrath, P. S. (1997). Primate subfossils: inferences about past and present primate community structure.

In *Natural Change and Human Impact in Madagascar*, ed. S. Goodman & B. Patterson, pp. 218–56. Washington: Smithsonian Institution.

Hladik, C.M. & Chivers, D. J. (1994) Foods and the digestive system. In *The Digestive System in Mammals: Food, Form, and Function*, eds. D. J.Chivers & P.L. Langer, pp. 65–73. Cambridge: Cambridge University Press.

Hutchinson, G. E. (1978). *An Introduction to Population Ecology*. New Haven: Yale University Press.

James, F. C. & McCulloch, C. E. (1990). Multivariate analysis in ecology and systematics: Panacea or pandora's box? *Annual Review of Ecology and Systematics*, **21**, 129–66.

Janson, C. H. (1985) Aggressive competition and individual food intake in wild brown capuchin monkeys. *Behavioral Ecology and Sociobiology* **18**, 125–38.

Kappeler, P. M. & Heymann, E. W. (1996). Nonconvergence in the evolution of primate life history socio-ecology. *Biological Journal of Linnean Society*, **59**, 297–326.

Kay, R. F. (1984). On the use of anatomical features to infer foraging behaviour in extinct primates. In *Adaptions for Foraging in Nonhuman Primates: Contributions to an Organismal Biology of Prosimians, Monkeys and Apes*, ed. P. S. Rodman & J. G. H. Cant, pp. 173–91. New York: Columbia University Press.

Magurran, A. E. (1981). *Ecological Diversity and Its Measurement*, 179 pp. Princeton: Princeton University Press.

McGraw, W. S. (1998). Comparitive locomotion and habitat use of six monkeys in the Taï Forest, Ivory Coast. *American Journal of Physical Anthropology*, **105**, 493–510.

Peters, R. H. (1983). *The Ecological Implications of Body Size*, 329 pp. Cambridge: Cambridge University Press.

Sokol, R. R. & Rohlf, F. J. (1981). *Biometry*, 859 pp. New York: W. H. Freeman & Co.

Terborgh, J. (1992). *Diversity and the Tropical Rain Forest*, 242 pp. New York: W. H. Freeman.

Terborgh, J. & van Schaik, C. P. (1987). Convergence *vs.* nonconvergence in primate communities. In *Organization of Communities, Past and Present*, eds. J. H. R. Gee & P. S. Giller, pp. 205—26. Oxford: Blackwell Scientific.

van Valkenburgh, B. (1988). Trophic diversity in past and present guilds of large predatory mammals. *Paleobiology*, **14**, 155–73.

8 · Body mass, competition and the structure of primate communities

JÖRG U. GANZHORN

INTRODUCTION

Understanding the processes influencing the distribution and abundance of organisms and their adaptations is a prime goal in ecology (Krebs, 1994). Although a community approach has been applied to primates for some time, most of the comparisons have been regionally restricted (Charles-Dominique, 1977; Struhsaker & Leland, 1979; Gautier-Hion, 1980; MacKinnon & MacKinnon, 1980; Mittermeier & van Roosmalen, 1981; Terborgh, 1983; Ganzhorn, 1989). A more global perspective of primate ecology was initiated in the 1980s with intercontinental comparisons of whole primate communities (Bourlière, 1985; Terborgh & van Schaik, 1987; Reed & Fleagle, 1995; Fleagle & Reed, 1996; Wright, 1997) and comparisons of primate and other mammalian radiations (Smith & Ganzhorn, 1996; Wright, 1996; Emmons, chapter 10, this volume).

In 1996, Fleagle and Reed used multivariate techniques to quantify and visualize a ten-dimensional niche space of primate species from eight different communities. The niche dimensions were based on body mass, activity cycle, locomotion and diet. According to this analysis, primate communities in the Old World show similar ecological diversity and occupy similar space in the ten-dimensional hypervolume, while the neotropical primate communities show lower overall diversity than the communities on other continents. Thus, the ecological space filled by primates in the neotropics is smaller than in other regions due to the lack of folivores, the lack of species with very large body mass, and the lack of a diverse set of nocturnal species among New World primates (Terborgh & van Schaik, 1987; Kappeler & Heymann, 1996; Wright, 1997).

A common feature of all communities analyzed by Fleagle & Reed (1996) is that the primate species of a given community are not dispersed uniformly in multivariate space, but, when reduced to a two-dimensional surface, they seem to be arranged at the periphery of a polygon, thus maximizing the space between them. Using the same methodological approach, a remarkably similar pattern as for primates has been described for dung beetles. There, the species most likely to compete were also somewhere at the periphery of a polygon within the multi-dimensional niche space whereas ephemeral species or species occuring at low densities were found to be distributed uniformly or randomly (Hanski & Cambefort, 1991). The former arrangement is to be expected if sympatric species not only compete within, but also between guilds, as suggested by an assembly rule developed by Fox (1987) and confirmed for lemurs of Madagascar (Ganzhorn, 1997; for discussion of potential artifacts due to confounding variables see, for example, Stone et al., 1996).

In their earlier paper, Reed & Fleagle (1995) showed that the number of primate species present at individual sites is positively correlated with mean annual rainfall in South America, Africa and Madagascar, but not in Asia. The rationale behind these positive correlations is, that increasing annual rainfall is correlated with floristic diversity and presumably with structural habitat complexity. This should result in increased niche diversity. Assuming that species compete over limited resources and that it is this type of competition which results in adaptive radiation, more niches should allow more different species to coexist in sympatry (see Kay & Madden, 1997, for a more extensive review of arguments). In fact, there is a close link between annual rainfall and tree species diversity in South America. Data for Africa and Asia are scant, but seem to fit the correlations found in the Neotropics (Gentry, 1988,

1993, 1995). For Madagascar, the argument could be taken one step further. There, annual rainfall is also correlated with tree species number. In addition, tree species richness and structural diversity are then positively correlated with the number of lemur species found in a given area, although areas of very high tree species richness seem to be avoided by lemurs, possibly due to the low density of suitable resources (Ganzhorn, 1994a; Ganzhorn et al., 1997). In South America, floristic richness, structural complexity, seasonality and fruit production are positively correlated with local tree species richness. But fruit production declines in areas with more than 2500 mm of rain per year while the other three variables remain fairly constant. The decline in fruit production is paralleled by a decline of the number of sympatric primate species at high levels of rainfall, but tree species diversity remains high. Thus, at high levels of rainfall the number of tree species and associated variables seem to be of subordinate importance to productivity (Kay et al., 1997). While this has been shown convincingly at high levels of rainfall, the relative importance of these confounded variables still has to be assessed at lower levels of annual precipitation, since as shown for lemurs (references as above) animals may modify or even reverse their reaction towards any given environmental factor depending on the environmental situation.

The question arising from the study of Reed and Fleagle is, then, why the Asian forest communities do not show the same correlation between annual rainfall and primate species numbers as do communities on other continents. Leaving phylogenetic constraints and historical biogeographic evolution under different environmental scenarios aside for the time being, possible explanations for the Asian peculiarities are:

(1)   a lack of correlation between annual rainfall and forest diversity, thus providing fewer potential niches for primate radiations;

(2)   a lack of primate niches, resulting in intensive competition over limited resources and thus reducing the number of potentially coexisting species;

(3)   a lack of competitive interaction between primate species, indicating that primate densities are below the potential carrying capacity of the habitat. This could be due to frequent disturbances which would not allow the consumers to reach the carrying capacity of the habitat. Since this idea can not be

tested, it will not be discussed below any further, even though it certainly has to be kept in mind. Alternatively, primates may not be able to reach their potential carrying capacity, because they are competitively excluded from appropriate resources by other animal taxa.

(4)   Thus, following hypothesis 3, the characteristics of the Asian primate communities could be a consequence of stronger competitive interactions of primates with other animal taxa in Asia than in other parts of the world.

In order to discriminate between the explanations outlined in hypotheses 1 through 4, it is crucial to identify whether competition is acting in primate communities or not. While it is basically impossible with descriptive studies to demonstrate competition, there are properties of community characteristics which are either consistent with or refute the idea of interspecific competition (Putman, 1994). Some of these properties will be discussed below.

POPULATION DENSITIES, BODY MASS AND DIET

Interspecific competition as an active force in structuring primate communities is only likely if population densities are oscillating around the carrying capacity of their habitats. Whether or not this is the case can be clarified through circumstantial evidence by the allometric relationships between population density and body mass, modified by trophic levels (Damuth, 1981, 1991, 1993; Peters, 1983; Robinson & Redford, 1986; Fa & Purvis, 1997; Reed, chapter 7, this volume). In mammals above about 100 g, species with higher body mass occur at lower densities than species with lower body mass (Brown, 1995; see Blackburn & Gaston, 1997 for a more detailed discussion). The slopes of the regression between logarithmically (to the base of 10) transformed data may vary between groups with different diets, but the generality of the relationship suggests that, on average, the upper densities of mammalian populations reflect physiological constraints of body mass and are adjusted by the carrying capacity of their habitat (for the time being, these allometries should be considered correlations rather than evolutionary causalities; see discussion by Witting, 1998; and Cates & Gittleman,

1998). Thus, deviations from the allometric scaling may indicate stochastic processes or depletion of resources by other consumers keeping the populations considered below the potential carrying capacity of their habitat. This line of argument will be used to examine whether the constraints acting upon Asian primates are different from those in other parts of the world.

Population densities

Population densities can be measured over large scales as crude densities or within specific habitats as ecological densities. Most surveys produce crude densities which represent averaged measurements obtained over large areas that also include habitats of varying suitability, including those that are not occupied by the species in question. Ecological densities represent the density of a species in suitable habitats. Based on empirical insights or rather on educated guesses, ecological densities can be calculated by multiplying crude densities by a factor of 2.8 (Eisenberg *et al.*, 1979; Prins & Reitsma, 1989).

Ecological and crude densities represent the two extremes of a continuum and are hard to define. If allometric relationships are to be assessed, it seems obvious that densities of any species have to be determined in the habitats where the species occurs. This is hampered by the lack of information about habitat patchiness, the scale at which patches are perceived by the animals, their actual habitat requirements and their performance in different habitats (Basset, 1995). Given our poor understanding of which factors are actually relevant for the distribution and abundance of primates (see Janson & Chapman, chapter 14, this volume), the data given for Asian and Malagasy primates were not transformed but rather the density estimates were taken as listed in the references considered, and averaged when more than one estimate was available per species. In contrast, the densities given by Robinson & Redford (1986) and Fa & Purvis (1997) had been transformed into ecological densities by these authors. For statistical analyses, this methodological discrepancy does not seem to be crucial. Since crude densities are transformed into ecological densities by multiplication by a simple factor, the distinction between ecological and crude densities has little effect on the statistical significance of abundance–body mass relationships (Blackburn & Gaston, 1997).

Body mass

Data on body mass was taken from Smith & Jungers (1997). Only body mass of females was used as female body mass may be more influenced by ecological and physiological demands of lactation while male body mass can be strongly influenced by sexual selection (Wrangham, 1980; van Schaik, 1996; Kappeler, 1996; Plavcan & van Schaik, 1997).

Dietary categories

Classification of primate diets based on food composition and behavioral data is a rather subjective undertaking (Oates, 1977; Janson & Chapman, chapter 14, this volume; Reed, chapter 7, this volume). In order to allow comparative analyses without too many assumptions, the very simple and simplistic classification into omnivores, frugivores (≥ 50% fruit in diet) and folivores (≥ 50% leaves in diet) applied earlier to lemur communities was also applied here (Ganzhorn, 1997). Omnivores here include the insectivores/omnivores and frugivores/omnivores of Robinson & Redford (1986) [R&R] and the faunivores and frugivores/faunivores of Kappeler & Heymann (1996) [K&H]. Frugivores include the frugivores/gramnivores and frugivores/herbivores of R&R and the frugivores and frugivores/folivores of K&H. Folivores represent the herbivores/browsers of R&R and the folivores of K&H.

The relationships of body mass, dietary preferences and primate population densities vary between regions (Table 8.1). In Madagascar, America, and, to a lesser extent, also in Africa, there are significant negative correlations between primate body mass and their population densities (Fig. 8.1). If dietary preferences are considered, the significance of the allometric relations increases in all regions but America. This may be a consequence of biologically too simplified assignment of the American species into dietary categories (see Janson & Chapman, chapter 14, this volume). Since the sample of Asian primates is influenced by biogeographic fragmentation and vicariance of ecospecies, the least square regression was repeated using only the primates of Borneo, which is comparable to Madagascar in area. But again, the regression between primate body mass and their densities is not significant according to least square regression ($P = 0.34$).

Using phylogenetic contrasts, the negative correlation

Table 8.1. *Relationships between body mass of female primates, population densities, dietary specializations (folivores, frugivores, faunivores/omnivores) and regional peculiarities (Madagascar, Africa, Asia, America) according to ANCOVA and least square regression of the data shown in Fig. 8.1*

ANCOVA				Least square regression	
Source of variation	d.f.	F	P	F	P
All regions					
Region	3	4.97	0.003		
Diet	2	5.24	0.006		
Body mass	1	23.54	< 0.001		
Model	6,134	14.13	< 0.001	33.57	< 0.001
$R^2 = 0.39$				$R^2 = 0.19$	
Madagascar					
Diet	2	1.82	0.183		
Body mass	1	10.28	0.004		
Model	3,24	4.37	0.014	8.89	0.006
$R^2 = 0.35$				$R^2 = 0.25$	
Africa					
Diet	1	19.64	< 0.001		
Body mass	1	10.28	< 0.001		
Model	2,28	12.37	< 0.001	3.11	0.089
$R^2 = 0.47$				$R^2 = 0.10$	
Asia					
Diet	2	2.28	0.115		
Body mass	1	5.26	0.027		
Model	3,39	2.04	0.124	1.47	0.232
$R^2 = 0.14$				$R^2 = 0.03$	
America					
Diet	2	0.06	0.804		
Body mass	1	3.96	0.054		
Model	3,36	3.75	0.033	7.62	0.009
$R^2 = 0.17$				$R^2 = 0.17$	

between densities and body mass remains significant only for the lemurs of Madagascar ($P = 0.01$), but becomes insignificant for all other regions considered. Analyses were based on the primate phylogeny presented by Purvis (1995) and the methodological approach of pairwise comparisons of the most closely related species proposed by Moller & Briskie (1995). Until the ongoing debate about the proper methods to be applied in the analyses of life history traits have been settled (Ricklefs & Starck, 1996), these discrepancies in the results have to be kept in mind so as not to attribute to the traditional approach of allometric relations more biological meaning than it actually deserves.

Another possibly confounding effect on population densities could be the number of sympatric primate species at a given site. One of the most comprehensive databases on primate densities compiled in standardized surveys that could be used to address this question, has been compiled by Peres (1997) on primate communities of the Amazon (see also Peres & Janson, chapter 3, this volume). I used the three most widely distributed primate species or ecospecies from this database to exemplify

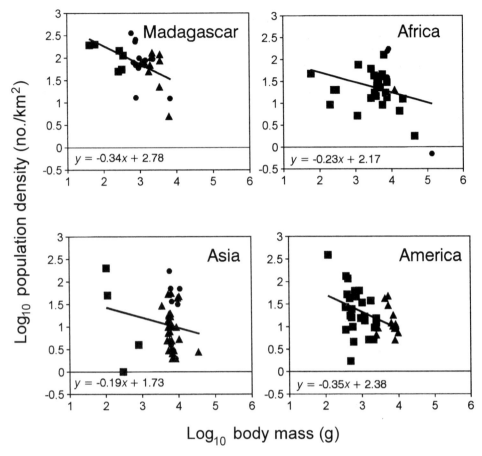

Fig. 8.1. Least square regressions between body mass (g) and mean population density (numbers per km²) for primates from Madagascar, Africa, Asia and America. Squares, omnivores; triangles, frugivores; dots, folivores. For statistics see Table 8.1. Estimates of populations densities were taken from: Madagascar: Charles-Dominique & Hladik (1971), Ganzhorn (1992, 1994b), Harcourt & Thornback (1990), Meyers (1993), Mutschler & Feistner (1995), Ganzhorn & Kappeler (1996), Sterling & Ramaroson (1996), Wright et al. (1997); Africa: Fa & Purvis (1997); Asia: MacKinnon (1986, 1987), MacKinnon & MacKinnon (1987), Davies (1994) and Lee (1995); America: Robinson & Redford (1986).

possible effects of the number of species per community on the population densities of these selected species. In Peres' database, ranging from 6 to 14 species, there was no effect of the number of coexisting primate species on the population densities of the three selected primate species (Table 8.2). The problem may be more pronounced in Asia where some insular sites might never have had a chance to assemble more species. But survey methods applied at different sites do not seem consistent enough to assess this question rigorously.

Using standard explanations for the allometric relation between body mass and population densities (discussed by Brown, 1995), these results indicate that primate population densities can approach some upper limits in a systematic way in Madagascar, Africa and the Americas, but to a much lesser extent in Asia. These upper limits may be set by physiological requirements of a given sized species and the habitat carrying capacity. Deviations from the allometric regression could be the result of spatial and temporal variation in habitat suitability, species specific adaptations and confounding effects of competing species.

Table 8.2. *Relation between the number of species present in a primate community and the densities (log$_{10}$) of primate species exemplified with neotropical primates, using data from Peres (1997). The general effect of species number was assessed in an analysis of covariance after taking into account effects of different vegetation types (terra firme and varzea) and species-specific effects of the primate species considered. The relation is exemplified using three primate species or ecospecies:* Saimiri *spp.,* Cebus apella *and* Alouatta seniculus

Variable	d.f.	F	P
Number of primate species per site	1	0.68	0.413
Vegetation type	1	10.74	0.002
Primate species	2	14.40	< 0.001
Model	4/43	17.48	< 0.001

Subsequently I will concentrate on signs of potential competition within the primate community and also between primates and other mammals. While this may be important it certainly is just one aspect contributing to the observed variability of communities within and between primate radiations. After all, even the most significant allometric regression and ANCOVA shown here capture no more than 47% of the variance in the data, thus leaving more than half of the variance unexplained.

BODY MASS RATIOS BETWEEN SPECIES WITH SIMILAR DIETS WITHIN A COMMUNITY

Indications of interspecific competition can be based on Hutchinson's rule that there are limits to the morphological similarity of coexisting species if they are competing over the same type of resources (Hutchinson, 1959; MacArthur, 1972; review by Brown, 1995). According to this idea, potentially competing species should differ at least by a factor of 1.3 to 1.4 in linear morphological traits and by a factor of 1.5 to 2 in body mass to allow stable coexistence (Emmons, 1980; Bowers & Brown, 1982). Certainly, body-mass ratios can only be used as an indication of species separation if the species in question are not segregated along some other dimensions (Emmons *et al.*, 1983). Although this "rule" and competition as the underlying force are still being debated vigorously, a comparative analysis of body mass ratios of coexisting species with similar diets may at least allow us to see whether size ratios of coexisting species differ between communities of different primate radiations, thus indicating different constraints on the evolution of body mass. The biological bases for the confirmation or rejection of the rule have then to be assessed separately.

Primate communities

Species composition of primate communities were compiled arbitrarily by searching through *Primate Conservation, Folia Primatologica, Primates, International Journal of Primatology* and the *American Journal of Primatology* and supplemented by other papers at hand. A total of 156 communities with three or more species were recorded. If possible, two communities were taken per number of species in a community for each region (Madagascar, Africa, Asia, and America). Care was taken that communities with the same number of species had different species composition, were geographically located as far apart as possible, and species composition had not been influenced by human activities recently (see Struhsaker, chapter 17, this volume).

The analyses of body mass ratios were restricted by the following conditions:

(1)   Except for Madagascar, only haplorhine primates were considered, because the distribution of the nocturnal strepsirhines is unclear: occurrence at a given site is often unknown, and sister-species may coexist unnoticed (Bearder *et al.*, 1995; Schwartz, 1996). Excluding strepsirhines from the analysis is likely to reduce the average body mass ratios for species in African and, to a lesser extent, in Asian primate communities. Since body mass ratios are only calculated for species within each dietary category, there would be few datapoints based on

strepsirhines. This would not change the overall picture and conclusions. Species with no ecological equivalents (such as *Daubentonia*) were not considered. Strepsirhines are included in all other analyses which do not require explicit knowledge of specific communities.

(2) Ratios were calculated with descending body mass only for species within each dietary category (within guilds); i.e. where competition should be most intense.

(3) Ratios were calculated for the species coexisting at a given locality, not for the continental species pool *in toto*. This eliminates vicariance effects of multiple inclusion of ecologically similar species in different places.

(4) Only communities with three to eight species were considered. Two communities were used per species number. This was done, first, to reduce violations of the assumptions for statistical analyses due to repeated occurrences of the same species even though this violation does not seem to be crucial (Bowers & Brown, 1982). Second, the datasets were reduced to match the Asian communities which contain eight primate species at the most.

(5) For the analysis of intercontinental variation in body mass ratios (Fig. 8.2), each pair of species was used only once. For looking at effects of the number of species per community on body mass ratios of the species within the community (Fig. 8.3), the same species pair could enter the analysis repeatedly, if the same species pair occurred in more than one community.

The medians of mass ratios are between 1.6 and 2.0 for sympatric species of the same guild in Madagascar, Africa and America, but the median is distinctly lower in Asian primate communities (Fig. 8.2; Kruskal–Wallis ANOVA: chi-square = 10.38, $P = 0.016$, d.f. = 3). Thus, the mass ratios of Asian primates do not match the pattern expected in communities in which interspecific competition between species with similar diets is mitigated by size differences (see also Reed, chapter 7, this volume). This result of body mass ratios is consistent with the previous analyses of abundance–body mass relations (Table 8.1). In these relations, there was no or only a marginally significant regression for Asian primates, indicating that competition

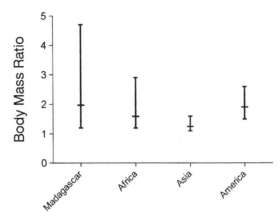

Fig. 8.2. Body mass ratios of pairs, with respect to body mass, adjacent species with similar diets for primate communities of three to eight species. Values indicated are medians and quartiles. Data on species composition was taken from: Madagascar: Meyers (1993), Sterling & Ramaroson (1996), Ganzhorn (1997); Africa: Gautier-Hion *et al.* (1980), Armstrong (1984), Bourlière (1985), Butynski (1985), Moore (1985), Carroll (1986), Harcourt & Nash (1986), Hart & Thomas (1986), Marsh (1986), Oates *et al.* (1990), McGraw (1994), Usongo & Fimbel (1995); Asia: MacKinnon (1986, 1987), MacKinnon & MacKinnon (1987), Tenaza (1987), Stanford (1988); America: Terborgh (1983), Glanz (1985), Stallings (1985), Soini (1986), Chapman *et al.* (1989), Rylands & Bernardes (1989), Cameron & Buchanan-Smith (1991/92), Peres (1997).

between primate species there is less intense than in America, Africa and Madagascar.

An important component of primate communities are the mixed-species associations known primarily from America and Africa (Gautier-Hion *et al.*, 1983; Terborgh, 1990; McGraw, 1994; Heymann, 1997; Noë & Bshary, 1997). In these associations competition avoidance seems to be based on mechanisms other than diverging body mass. The body-mass ratios of species occuring in these mixed-species troops will lower the average values of body-mass ratios, especially for America.

Further analyses of body mass ratios reveal more differences between primate community structure in different regions. In Africa and Asia, the body mass ratios do not vary systematically in relation to the number of species present in the community (Fig. 8.3; note that body mass ratios are log 2 transformed). However, in Madagascar and America body mass ratios decline significantly with increasing

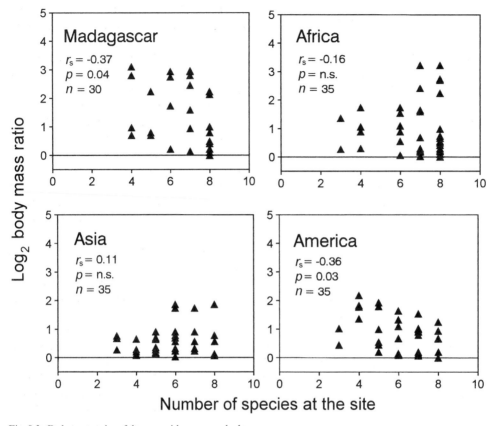

Fig. 8.3. Body mass ratios of the two, with respect to body mass, adjacent species with similar diets in relation to the number of species present in primate communities of three to eight species. Body mass ratios are log 2 transformed.

number of species in a community. This effect is even more pronounced once communities with more than eight species were considered and may indicate niche compression and increased species packing with increasing species number.

The data presented here are consistent with the analyses of Reed & Fleagle (1995) and Emmons (chapter 10, this volume). In Asian primate communities the number of species does not increase with the assumed increase in the availability of new niches. Also, in contrast to the situation in Madagascar, Africa and America, population densities of Asian primates show only a weak tendency to follow the general allometric laws of decreasing density with increasing body mass. And finally, body mass ratios of potentially competing species are smaller in Asian primate communities than anywhere else in the world.

All of these findings indicate that competition is an important component of primate communities in most parts of the world, except for Asia. However, this does not mean that Asian habitats are unlimited heaven for primates. In contrast, a more likely explanation would be that, though Asian primates compete among themselves (as suggested by the result of Fleagle & Reed (1996) that primate species in Asian communities seem to maximize the ecological space between them), they compete more heavily with non-primate taxa and that these non-primates limit the opportunities for the primates to adopt the "typical" interspecific interactions within their taxon. Thus, in order to understand primate community ecology we need to include non-primate taxa in our primatological studies.

COMPARISON OF PRIMATE AND NON-PRIMATE TAXA

Including non-primate taxa in the present analyses should allow us to describe two potential constraints imposed upon primates. If the distribution of primate body mass is influenced by interspecific competition with non-primate taxa, the two groups (primates and non-primates) should show complementary distributions of body mass. Alternatively, if primate body masses were constrained more by environmental conditions, such as the availability of ecological space (sensu Hanski & Cambefort, 1991; Fleagle & Reed, 1996; Godfrey et al., 1997), non-primate taxa should show similar body mass distributions as primates. In addition, if the different representation of dietary specializations of primates between continents were based on different resource availability, the same differences should also be apparent in other mammals.

Primates may compete with a variety of other taxa, including insects, reptiles and birds (Rockwood & Glander, 1979). This may have profound consequences for the evolution of primate adaptive radiations and their community structure (Sussman & Raven, 1978; Smith & Ganzhorn, 1996). However, other mammalian taxa may represent the most likely competitors as they share the feature of lactation. Nourishing the young with milk avoids the dietary problems associated with small body size of the growing individual encountered by some non-mammalian vertebrate taxa. In addition, primates are also more similar to other mammals than to birds and insects in their mode of locomotion, metabolism, digestive abilities, and body mass.

The analyses of (1) whether primate and non-primate taxa show complementary body mass distributions (thus indicating competition) or (2) whether the differences in body mass distributions and dietary preferences between continents are matched by similar differences in body mass of other mammalian taxa (thus indicating different resource limitations) were based on genera rather than on species to avoid redundancy due to species vicariance. Also, the taxonomic status of many small mammals is poorly known and it is highly likely that many species are still to be described (see Emmons, chapter 10, this volume). Analyses at the generic level may reduce this problem. If a genus contained species to be assigned to different dietary categories or belonging to different categories of body mass as defined in Fig. 8.4, the genus entered the analysis twice. The analysis was restricted to arboreal and terrestrial species occurring in forest habitats. True carnivores were not included, but more frugivorous taxa of the Canidae, Felidae, and Viverridae were considered. Since the statistical problem associated with vicariance seems most pronounced in insular Asia, the island of Borneo was used as a subsample of Asia's mammalian fauna. A total of 255 species groups were used in this analysis.

Figure 8.4 shows the distribution of body mass for primate and non-primate genera. In the Neotropics and Africa body mass of primate taxa does not differ from the distribution of body masses in non-primate groups (Kolmogorov–Smirnov tests: $P > 0.17$). On Borneo, primate genera are on average slightly heavier than other mammals (Kolmogorov–Smirnov test: chi-square = 7.41, d.f. = 2, $P < 0.05$). This difference is pronounced in Madagascar where only the bushpig reaches larger body mass than any of the extant lemurs (Kolmogorov–Smirnov test: chi-square = 18.78, d.f. = 2, $P < 0.001$).

The mammalian genera are evenly distributed among the three dietary categories in Madagascar (Fig. 8.5). This distribution deviates significantly from the other three regions (chi-square = 19.35, d.f. = 6, $P = 0.004$). The distribution of genera in different dietary categories is remarkably similar in South America, Africa and Borneo ($P > 0.5$) with many more omnivores and frugivores than folivores. This argues strongly against the idea that the differences in primate communities reported in intercontinental comparisons, such as the lack of folivorous primates in South America are due to different resource regimes (Terborgh & van Schaik, 1987; Kappeler & Heymann, 1996; Wright 1997), but rather a consequence of evolutionary history under different competitive regimes with non-primate taxa (Wright, 1997). However, despite the numerical argument illustrated in Figure 8.5, it remains unclear why folivorous primates are so poorly represented in the neotropics. None of the folivorous non-primate taxa reach densities that are even close to the densities of folivorous primates in the Old World.

SYNTHESIS

The present analysis has been performed under the assumption that competition is an active process affecting the structure of primate communities. This assumption

150   J. U. GANZHORN

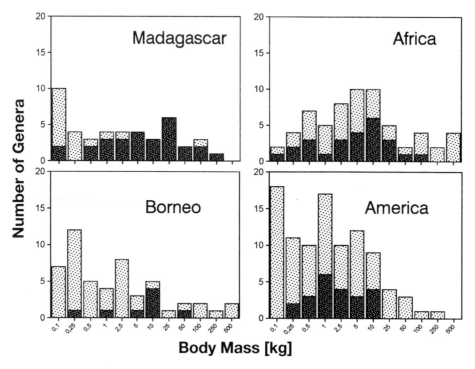

Fig. 8.4. Distribution of body mass (kg) in different primate
(black) and non-primate (stippled) groups based on genera. For
Madagascar subfossil taxa are included. In addition to the
primate references listed in Figs 8.1 and 8.2, data were taken
from: Madagascar: Ganzhorn *et al.* (1996), Goodman (1996),
Godfrey *et al.* (1997); Africa: Haltenorth & Diller (1977); Asia
(Borneo): Payne *et al.* (1985); America: Emmons & Feer (1990).

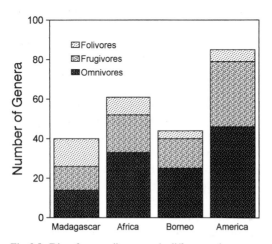

Fig. 8.5. Diet of mammalian genera in different regions.

was based on a sequence of plausibility arguments assem-
bled from different studies on different continents. First,
within regions and except for Asia, the number of primate
species in a community is correlated with annual rainfall.
Second, annual rainfall in turn is correlated with floristic
diversity at least in America and Madagascar. In concert,
the two correlations are consistent with the idea that
species compete over limited resources, that they avoid
competition by specializing and occupying distinct niches
and that higher habitat complexity provides more niches,
thus allowing more species to coexist. This sequence of
arguments has received support by modeling the assembly
of lemur communities with Fox's assembly rule (Ganzhorn, 1997). This rule assumes that species compete within
and between guilds. The highly ordered extant and subfossil lemur communities matched Fox's model almost
perfectly, thus adding to the circumstantial evidence that

Table 8.3. *Differences in the distribution of body mass between genera of tropical forest dwelling mammals (without flying and aquatic taxa) according to Kolmogorov–Smirnov two sample tests*

	Africa $n = 63$	Borneo $n = 52$	America $n = 96$
Madagascar $n = 44$	$D_{max} = 0.233$ $P > 0.05$	$D_{max} = 0.111$ $P > 0.05$	$D_{max} = 0.179$ $P > 0.05$
Africa		$D_{max} = 0.266$ $P < 0.05$	$D_{max} = 297$ $P < 0.005$
Borneo			$D_{max} = 0.076$ $P > 0.05$

Note: n, number of genera considered

interspecific competition is important for many primate communities.

Based on these results, four possible explanations were presented why primate communities in Asia do not show the same patterns seen in other regions. As it turned out, the original lines of arguments could not be confirmed if restricted to primates alone.

Lack of correlation between annual rainfall and forest (and thus niche) diversity

The diversity of tree species per unit area is lowest in Africa, followed by Madagascar, Asia and the Neotropics (Gentry, 1993; Abraham *et al.*, 1996; see also Barthlott *et al.*, 1996). Thus, Asian primate communities would be expected to have more primate species than Africa. Yet this is not true. However, if all mammalian taxa are considered, all of Africa's tropical forests have about the same number of taxa as the much smaller forested area of Borneo. Borneo, in turn, houses more mammalian genera than Madagascar (Fig. 8.5). These results support the idea that floral diversity is important for animal diversity in tropical forests, despite the findings that high tree species richness may coincide with low plant productivity and low numbers of primates at high levels of annual rainfall (Kay *et al.*, 1997). However, we should not expect linear correlations between environmental variables and primate community structure. After all, within a single contiguous forest lemurs were found more frequently in microhabitats of high tree species richness if the surrounding forest had few tree species, but the animals preferred forest parts with

relatively few tree species in a forest block of high tree species richness (Ganzhorn *et al.*, 1997). In conclusion, the low number of primate species in any given Asian community is unlikely to be the result of lacking niches.

Lack of primate niche diversification, resulting in intensive competition over limited resources and thus reducing the number of potentially coexisting species

Dietary diversity of Asian primates is higher than in any other region of the world (Kappeler & Heymann, 1996). This is also true for modes of locomotion and thus adaptations to structural habitat characteristics (Fleagle, 1999). Thus, per unit area, Asian forests offer at least as many options for adaptive radiations as tropical forests on other continents (review in Rosenzweig, 1995). However, fruit productivity varies widely between continents (Terborgh & van Schaik, 1987; Wright, 1997), and certainly plant productivity might play a major role in explaining why Asia is the "odd man out" (see Janson & Chapman, chapter 14, this volume). It has been suggested that Asian biodiversity and population densities of less mobile species are limited by the phenomenon of irregular mast fruiting and multiple year intervals between masts (Leighton & Leighton, 1983; review in Fleagle & Reed, 1996). Actually the relatively low biomass of primates (as shown in Fig. 8.1) might reflect these irregularities to some extent. But based on the literature survey presented here, the Asian mammalian fauna is at least as diverse or even more diverse than the mammalian fauna of other tropical forests

(Table 8.3). Also, in contrast to Central America, where fruit scarcity caused mass mortality among vertebrates (Foster, 1982), no signs of famine and related mortality could be documented in almost 20 years of observations in tropical South America and in Sumatra (van Schaik *et al.*, 1993). Thus, concentrating on plant productivity and primates alone seems to miss important aspects operating in Asian primate assemblages.

Lack of competitive interaction between primate species

According to Fleagle & Reed (1996), Asian primate species are as well separated within a multi-dimensional ecological space as are primates in other parts of the world, thus indicating competitive interactions between them. The data on body mass ratios are inconsistent with this result. The two aspects could be reconciled assuming that competition is mitigated by other means, such as differences in habitat use, dietary aspects, or temporal niche partitioning (MacKinnon & MacKinnon, 1980; Gautier-Hion *et al.*, 1983; Terborgh, 1983; Ganzhorn, 1989; Wright, 1989, 1996; Nickle & Heymann, 1996). Still, this does not explain why Asian primate densities do not follow the allometric abundance–body mass relationships described for the other regions. The most likely explanation for these discrepancies is presented below.

Consequence of strong interactions of primates with other mammalian taxa

Varying degrees of competitive interactions between primates and non-primate taxa could contribute to most of the phenomena listed above. At the beginning it has been argued that the negative relation between body mass and population density should be significant if the species in question fill their habitat to its carrying capacity (Fig. 8.1). Scatter around the regression line might indicate that other consumer species take their share and reduce the habitat's carrying capacity for given-sized primates to varying degrees. In Madagascar, where primates represent more than 60% of all mammal genera, the data fit the regression line best ($R^2 = 0.26$; Table 8.1). In Asia, where primates represent a minority of mammalian genera (but see Emmons, chapter 10, this volume), there is no significant relation between primate body mass and their densi-

ties, and the least square regression captures only 3% of the variance in the data. Thus, the goodness of fit between primate body mass and their expected abundance tends to increase with the percentage of primate taxa in a community. The macro-ecological approach seems sound, but the biological bases for these numbers and correlations remain obscure.

In their influential paper of 1987, Terborgh & van Schaik linked the differences in primate communities in different continents to differences in resource availability. According to their argument, the evolution of folivorous species should have been hampered by seasonally coinciding troughs in the production of leaves and fruits in the Neotropics. In addition to the different temporal pattern of food production on different continents, the Asian tropical forests stand out as their primary production is much lower than food production on other continents (Terborgh & van Schaik, 1987; review in Reed & Fleagle, 1996). However, on the genus level, dietary diversity among mammals is similar on all tropical continents (Fig. 8.5) and the high degree of dietary and morphological diversity in Asian primates indicates that there is a wide potential for adaptive radiation. But this diversity is not reflected in the representation of primates in Asian mammal communities to the same extent as in other parts of the world. Rather, the diversification has occurred in other mammalian taxa in this region. These non-primate taxa may occupy ecological space and thus might have not allowed intensive radiation towards different sized primates. Similarly, as outlined above, the lack of neotropical folivorous primates is unlikely to be a consequence of missing resources, but rather of the inability of primates to invade the folivorous niche already occupied by other taxa. Primates might then simply have arrived in the Neotropics too late to find enough niches for comprehensive ecological radiation (Wright, 1997). The idea that primates in the neotropics are restricted by non-primate taxa and can not invade new ecological space with increasing number of species is supported further by the finding that, if more species are added to the community, body mass ratios decline, indicating niche compression (Fig. 8.3). This is similar to the situation in Madagascar. There, in the absence of potential competitors, primates have radiated enormously and can occupy all the ecological space available. Yet the ecological space is finite and species packing also has to increase once more species are added to the community.

Finally, the characteristics of Asian primates fit best a scenario in which primates share resources with many other taxa. Therefore they compete less among themselves but more with other mammals (and possibly other taxa as well). This might be reflected in the lack of body mass ratios expected for competing taxa (Fig. 8.2) and the poor abundance–body mass relationship (Fig. 8.1).

The data and analyses presented here did not reveal a uniform picture about the evolution and actual constraints acting upon primate communities; it would have been naive to expect such a pattern. The main, but often neglected conclusion of this analysis is that communities in different regions show distinct patterns, but these patterns differ in various ways. Effects of other mammals (but certainly also of many other taxa not considered here) are part of the reason. Dietary diversity of all mammals is similar in all regions considered while dietary diversity of primates is not. This also suggests that non-primate taxa have profound effects on the evolution of primate communities through interspecific competition. Thus, the structure and evolution of primate communities can not be understood without considering non-primate taxa. These results indicate that evolutionary history under different competitive regimes with non-primate taxa rather than different resource regimes contribute to the variation in the composition and structure of primate communities.

ACKNOWLEDGMENTS

I would like to thank John Fleagle, Charles Janson and Kaye Reed for their invitation to participate in the workshop on "Primate communities" and the Wenner Gren Foundation and the Deutsche Forschungsgemeinschaft for financial support. Louise Emmons, John Fleagle, Eckhard Heymann, Charles Janson, Peter Kappeler, Carel van Schaik and Patricia Wright provided helpful comments on the manuscript.

REFERENCES

Abraham, J.-P., Rakotonirina, B., Randrianasolo, M., Ganzhorn, J. U., Jeannoda, V. & Leigh, E. G. (1996). Tree diversity on small plots in Madagascar: a preliminary review. *Revue d'Ecologie,* **51,** 93–116.

Armstrong, G (1984). *An Ecological Survey of Mamunta-Mayowoso Wildlife Reserve, Sierra Leone. Report to Lethbridge Community College.* Lethbridge, Alberta, Canada.

Barthlott, W., Lauer, W. & Placke, A. (1996). Global distribution of species diversity in vascular plants: towards a world map of phytodiversity. *Erdkunde,* **50,** 317–27.

Basset, A. (1995). Body size-related coexistence: an approach through allometric constraints on home-range use. *Ecology,* **76,** 1027–35.

Bearder, S. K., Honess, P. E. & Ambrose, L. (1995). Species recognition among galagos with special reference to mate recognition. In *Creatures of the Dark,* ed. L. Alterman, G. A. Doyle & M. K. Izard, pp. 331–52. New York: Plenum Press.

Blackburn, T. M. & Gaston, K. J. (1997). A critical assessment of the form of the interspecific relationship between abundance and body size in animals. *Journal of Animal Ecology,* **66,** 233–49.

Bourlière, F. (1985). Primate communities: their structure and role in tropical ecosystems. *International Journal of Primatology,* **6,** 1–26.

Bowers, M. A. & Brown, J. H. (1982). Body size and coexistence in desert rodents: chance or community structure? *Ecology,* **63,** 391–400.

Brown, J. H. (1995). *Macroecology.* Chicago: University of Chicago Press.

Butynski, T. M. (1985). Primates and their conservation in the Impenetrable (Bwindi) Forest, Uganda. *Primate Conservation,* **6,** 68–72.

Cameron, R., & Buchanan-Smith, H. (1991–1992). Primates of the Pando, Bolivia. *Primate Conservation,* **12–13,** 11–14.

Carroll, R. W. (1986). Status of the Lowland gorilla and other wildlife in the Dzanga-Sangha Region of southwestern Central African Republic. *Primate Conservation,* **7,** 38–41.

Cates, S. E. & Gittleman, J. L. (1998). Reply from S.E. Cates and J.L. Gittleman. *Trends in Ecology and Evolution,* **13,** 25.

Chapman, C. A., Chapman, L. & Glander, K. E. (1989). Primate populations in northwestern Costa Rica: potential for recovery. *Primate Conservation,* **10,** 37–44.

Charles-Dominique, P. (1977). *Ecology and Behaviour of Nocturnal Prosimians.* London: Duckworth.

Charles-Dominique, P. & Hladik, C. M. (1971). Le Lepilemur du sud de Madagascar: écologie, alimentation et vie sociale. *La Terre et la Vie,* **25,** 3–66.

Damuth, J. (1981). Population density and body size in mammals. *Nature*, **290**, 699–700.

Damuth, J. (1991). Of size and abundance. *Nature*, **351**, 268–269.

Damuth, J. (1993). Cope's rule, the island rule and the scaling of mammalian population density. *Nature*, **365**, 748–50.

Davies, A. G. (1994). Colobine populations. In *Colobine Monkeys: Their Ecology, Behaviour and Evolution*, ed. A. G. Davies & J. F. Oates, pp. 285–310. Cambridge: Cambridge University Press.

Eisenberg, J. F., O'Connell, M. & August, P. V. (1979). Density, productivity, and distribution of mammals in two Venezuelan habitats. In *Vertebrate Ecology in the Northern Neotropics*, ed. J. F. Eisenberg, pp. 187–207. Washington: Smithsonian Instition Press.

Emmons, L. H. (1980). Ecology and resource partitioning among nine species of African rain forest squirrels. *Ecological Monographs*, **50**, 31–54.

Emmons, L. H. & Feer, F. (1990). *Neotropical Rainforest Mammals*. Chicago: University of Chicago Press.

Emmons, L. H., Gautier-Hion, A. & Dubost, G. (1983). Community structure of the frugivorous-folivorous forest mammals of Gabon. *Journal of Zoology, London*, **199**, 209–22.

Fa, J. E. & Purvis, A. (1997). Body size, diet and population density in Afrotropical forest mammals: a comparison with neotropical species. *Journal of Animal Ecology*, **66**, 98–112.

Fleagle, J. G. (1999). *Primate Adaptation and Evolution*. London: Academic Press.

Fleagle, J. G. & Reed, K. E. (1996). Comparing primate communities: a multivariate approach. *Journal of Human Evolution*, **30**, 489–510.

Foster, R. B. (1982). Famine on Barro Colorado Island. In *The Ecology of a Tropical Forest*, ed. E. G. Leigh Jr., A. S. Stanley & D. M. Windsor, pp. 201–12. Washington, DC: Smithsonian Institution Press,.

Fox, B. J. (1987). Species assembly and the evolution of community structure. *Evolutionary Ecology*, **1**, 201–13.

Ganzhorn, J. U. (1989). Niche separation of the seven lemur species in the eastern rainforest of Madagascar. *Oecologia (Berlin)*, **79**, 279–86.

Ganzhorn, J. U. (1992). Leaf chemistry and the biomass of folivorous primates in tropical forests. *Oecologia (Berlin)*, **91**, 540–7.

Ganzhorn, J. U. (1994a). Lemurs as indicators for habitat change. In *Current Primatology: Ecology and Evolution, Volume I*, ed. B. Thierry, J. R. Anderson, J. J. Roeder & N. Herrenschmidt, pp. 51–56. Strasbourg: Université, Louis Pasteur.

Ganzhorn, J. U. (1994b). Les lémuriens. In *Recherches pour le Developpement: Inventaire biologique de la Forêt de Zombitse*, ed. S. M. Goodman & O. Langrand, Volume No Spécial, pp. 70–72. Antananarivo: Centre d'Information et de Documentation Scientifique et Technique.

Ganzhorn, J. U. (1997). Test of Fox's assembly rule for functional groups in lemur communities in Madagascar. *Journal of Zoology, London*, **241**, 533–42.

Ganzhorn, J. U. & Kappeler, P. M. (1996). Lemurs of the Kirindy Forest. In *Ecology and Economy of a Tropical Dry Forest in Madagascar*, ed. J. U. Ganzhorn & J.-P. Sorg, pp. 257–74, Primate Report, 46–1, Göttingen.

Ganzhorn, J. U., Malcomber, S., Andrianantoanina, O. & Goodman, S. M. (1997). Habitat characteristics and lemur species richness in Madagascar. *Biotropica*, **29**, 331–43.

Ganzhorn, J. U., Sommer, S., Abraham, J.-P., Ade, M., Raharivololona, B. M., Rakotavao, E. R., Rakotondrasoa, C. & Randriamarosoa, R. (1996). Mammals of the Kirindy Forest with special emphasis on *Hypogeomys antimena* and the effects of logging on the small mammal fauna. In *Ecology and Economy of a Tropical Dry Forest in Madagascar*, ed. J. U. Ganzhorn & J.-P. Sorg, pp. 215–232. Primate Report, 46–1, Göttingen.

Gautier-Hion, A. (1980). Seasonal variations of diet related to species and sex in a community of *Cercopithecus* monkeys. *Journal of Animal Ecology*, **49**, 237–69.

Gautier-Hion, A., Emmons, L. H. & Dubost, G. (1980). A comparison of the diets of three major groups of primary consumers of Gabon (primates, squirrels and ruminants). *Oecologia (Berlin)*, **45**, 182–9.

Gautier-Hion, Quris, A. R. & Gautier, J. P. (1983). Monospecific *vs* polyspecific life: a comparative study of foraging and antipredatory tactics in a community of *Cercopithecus* monkeys. *Behavioral Ecology and Sociobiology*, **12**, 325–35.

Gentry, A. H. (1988). Changes in plant community diversity and floristic composition on environmental and geographical gradients. *Annals of the Missouri Botanical Garden*, **75**, 1–35.

Gentry, A. H. (1993). Diversity and floristic composition of lowland tropical forest in Africa and South America. In *Relationships between Africa and South America*, ed. P. Goldblatt, pp. 500–47. New Haven: Yale University Press.

Gentry, A. H. (1995). Diversity and floristic composition of neotropical dry forests. In *Seasonally Dry Tropical Forests*, ed. S. H. Bullock, H. A. Mooney & E. Medina, pp. 146–94. Cambridge: Cambridge University Press.

Glanz, W. E. (1985). The terrestrial mammal fauna of Barro Colorado Island: census and long-term changes. In *The Ecology of a Tropical Forest: Seasonal Rhythms and Long-term Changes,* ed. E. G. Leigh Jr., A. S. Rand & D. M. Windsor, pp. 455–68. Washington, DC: Smithsonian Institution Press.

Godfrey, L. R., Jungers, W. L., Reed, K. E., Simons, E. L. & Chatrath, P. S.. (1997). Inferences about past and present primate communities in Madagascar. In *Natural Change and Human Impact in Madagascar,* ed. S. M. Goodman & B. D. Patterson, pp. 218–256. Washington, DC: Smithsonian Institution Press.

Goodman, S. M. (1996). A floral and faunal inventory of the eastern Slopes of the Réserve Naturelle Intégrale d'Andringitra, Madagascar: with reference to elevational variation. *Fieldiana: Zoology, New Series Edition,* 85. Chicago: Field Museum of Natural History.

Haltenorth, T. & Diller, H. (1977). *Säugetiere Afrikas und Madagaskars.* BLV München: Verlagsgesellschaft.

Hanski, I. & Cambefort, Y. (1991). Resource partitioning. In *Dung Beetle Ecology,* ed. I. Hanski & Y. Cambefort, pp. 330–49. Princeton: Princeton University Press.

Harcourt, C. S. & Nash, L. T. (1986). Social organization of galagos in Kenyan coastal forests.1: *Galago zanzibaricus. American Journal of Primatology,* **10**, 339–55.

Harcourt, C. & Thornback, J. (1990). *Lemurs of Madagascar and the Comoros.* Gland: IUCN.

Hart, J. A. & Thomas, S. (1986). The Ituri Forest of Zaire: primate diversity and prospects for conservation. *Primate Conservation,* **7**, 42–4.

Heymann, E. W. (1997). The relationship between body size and mixed-species troops of Tamarins (*Saguinus* spp.). *Folia Primatologica,* **68**, 287–95.

Hutchinson, G. E. (1959). Homage to Santa Rosalia, or why are there so many kinds of animals? *American Naturalist,* **93**, 137–59.

Kappeler, P. M. (1996). Intrasexual selection and phylogenetic constraints in the evolution of sexual canine dimorphism in strepsirhine primates. *Journal of Evolutionary Biology,* **9**, 43–65.

Kappeler, P. M. & Heymann, E. W. (1996). Nonconvergence in the evolution of primate life history and socio-ecology. *Biological Journal Linnean Society,* **59**, 297–326.

Kay, R. F. & Madden, R. H. (1997). Mammals and rainfall: paleoecology of the middle Mioene at La Venta (Columbia, South America). *Journal of Human Evolution,* **32**, 161–99.

Kay, R. F., Madden, R. H., van Schaik, C. & Higdon, D. (1997). Primate species richness is determined by plant productivity: implications for conservation. *Proceedings of the National Academy of Sciences USA,* **94**, 13023–7.

Krebs, C. J. (1994*). Ecology: The Experimental Analysis of Distribution and Abundance,* 4th Edition. New York: Harper Collins College Publishers.

Lee, R. J. (1995). Population survey of the Crested Black Macaque (*Macaca nigra*) at Manembonembo Nature Reserve in North Sulawesi, Indonesia. *Primate Conservation,* **16**, 63–5.

Leighton, M. & Leighton, D. R. (1983). Vertebrate responses to fruiting seasonality within a Bornean rain forest. In *Tropical Rain Forest: Ecology and Management,* ed. S. L. Sutton, T. C. Whitmore & A. C. Chadwick, pp. 181–196. Oxford: Blackwell Scientific.

MacArthur, R. H. (1972). *Geographical Ecology.* Englewood Cliffs, New Jersey: Prentice-Hall.

MacKinnon, J. R. & MacKinnon, K. S. (1980). Niche differentiation in a primate community. In *Malayan Forest Primates,* ed. D. J. Chivers, pp. 167–90. New York: Plenum Press.

MacKinnon, J. & MacKinnon, K. (1987). Conservation status of the primates of the Indo-Chinese subregion. *Primate Conservation,* **8**, 187–95.

MacKinnon, K. (1986). The conservation status of nonhuman primates in Indonesia. In *Primates: The Road to Self-sustaining Populations,* ed. K. Benirschke, pp. 99–126. New York, NY: Springer.

MacKinnon, K. (1987). Conservation status of primates in Malaysia with special reference to Indonesia. *Primate Conservation,* **8**, 175–86.

Marsh, C. (1986). A resurvey of Tana River primates and their habitat. *Primate Conservation,* **7**, 72–82.

McGraw, S. (1994). Census, habitat preference, and polyspecific associations of six monkeys in the Lomako Forest, Zaire. *American Journal of Primatology,* **34**, 295–307.

Meyers, D. M. (1993). *The effects of resource seasonality on behavior and reproduction in the Golden-Crowned Sifaka* (Propithecus tattersalli, *SIMONS, 1988) in three Malagasy forests.* Ph.D. Dissertation. Duke University.

Mittermeier, R. A. & van Roosmalen, M. G. M. (1981). Preliminary observations on habitat utilization and diet in eight Surinam monkeys. *Folia Primatologica,* **36**, 1–39.

Moller, A. P. & Briskie, J. V. (1995). Extra-pair paternity, sperm competition and the evolution of testis size in birds. *Behavioral Ecology and Sociobiology,* **36**, 357–65.

Moore, J. (1985). Chimpanzee survey in Mali, West Africa. *Primate Conservation,* **6**, 59–63.

Mutschler, T. & Feistner, A. T. C. (1995). Conservation status and distribution of the Alaotran Gentle Lemur *Hapalemur griseus alaotrensis*. *Oryx*, **29**, 267–74.

Nickle, D. A. & Heymann, E. W. (1996). Predation on Orthoptera and other orders of insects by tamarin monkeys, *Saguinus mystax* and *Saguinus fuscicollis nigrifrons* (Primates: Callitrichidae), in north-eastern Peru. *Journal of Zoology, London*, **239**, 799–819.

Noë, R. & Bshary, R. (1997). The formation of red colobus – diana monkey associations under predation pressure from chimpanzees. *Proceedings of the Royal Society of London B*, **264**, 253–9.

Oates, J. F. (1977). The Guereza and its food. In *Primate Ecology*, ed. T. H. Clutton-Brock, pp. 276–321. London: Academic Press.

Oates, J. F., Whitesides, G. H., Davies, A. G., Waterman, P. G., Green, S. M., Dasilva, G. L. & Mole, S. (1990). Determinants of variation in tropical forest primate biomass: new evidence from West Africa. *Ecology*, **71**, 328–43.

Payne, J., Francis, C. M. & Phillipps, K. (1985). *A Field Guide to the Mammals of Borneo*. WWF Malaysia. Kuala Lumpur: The Sabah Society.

Peres, C. A. (1997). Primate community structure at twenty western Amazonian flooded and unflooded forests. *Journal of Tropical Ecology*, **13**, 381–405.

Peters, R. H. (1983). *The Ecological Implications of Body Size*. Cambridge: Cambridge University Press.

Plavcan, J. M. & van Schaik, C. P. (1997). Intrasexual competition and body weight dimorphism in anthropoid primates. *American Journal of Physical Anthropology*, **103**, 37–68.

Prins, H. H. T. & Reitsma, J. M. (1989). Mammalian biomass in an African equatorial rain forest. *Journal of Animal Ecology*, **58**, 851–61.

Purvis, A. (1995). A comparative estimate of primate phylogeny. *Philosophical Transactions of the Royal Society London B*, **348**, 405–21.

Putman, P. J. (1994). *Community Ecology*. London: Chapman & Hall.

Reed, K. E. & Fleagle, J. G. (1995). Geographic and climatic control of primate diversity. *Proceedings of the National Academy of Sciences USA*, **92**, 7874–6.

Ricklefs, R. E. & Starck, J. M. (1996). Applications of phylogenetically independent contrasts: a mixed progress report. *Oikos*, **77**, 167–72.

Robinson, J. G. & Redford, K. H.. (1986). Body size, diet, and population density of neotropical forest mammals. *American Naturalist*, **128**, 665–80.

Rockwood, L. L. & Glander, K. E. (1979). Howling monkeys and leaf-cutting ants: comparative foraging in a tropical deciduous forest. *Biotropica*, **11**, 1–10.

Rosenzweig, M. L. (1995). *Species Diversity in Space and Time*. Cambridge: Cambridge University Press.

Rylands, A. B. & Bernardes, A. T. (1989). Two priority regions for primate conservation in the Brazilian Amazon. *Primate Conservation*, **10**, 56–62.

Schwartz, J. (1996). *Pseudopotto martini*: a new genus and species of extant lorisiform primate. *Anthropological Papers of the American Museum of Natural History*, **78**, 1–14.

Smith, A. P. & Ganzhorn, J. U. (1996). Convergence and divergence in community structure and dietary adaptation in Australian possums and gliders and Malagasy lemurs. *Australian Journal of Ecology*, **21**, 31–46.

Smith, R. J. & Jungers, W. L. (1997). Body mass in comparative primatology. *Journal of Human Evolution*, **32**, 523–59.

Soini, P. (1986). A synecological study of a primate community in the Pacaya-Samiria National Reserve, Peru. *Primate Conservation*, **7**, 63–71.

Stallings, J. R. (1985). Distribution and status of primates in Paraguay. *Primate Conservation*, **6**, 51–58.

Stanford, C. B. (1988). Ecology of the Capped Langur and Phayre's Leaf Monkey in Bangladesh. *Primate Conservation*, **9**, 125–8.

Sterling, E. J. & Ramaroson, M. G. (1996). Rapid assessment of primate fauna of the eastern slopes of the RNI d'Andringitra, Madagascar. In *A floral and Faunal Inventory of the Eastern Side of the Reserve Naturelle Intégrale d'Andringitra, Madagascar: With Reference to Elevational Variation, Fieldiana: Zoology Edition, Volume 85*, ed. S. M. Goodman, pp. 293–305. Chicago: Field Museum Natural History.

Stone, L., Dayan, T. & Simberloff, D. (1996). Community-wide assembly patterns unmasked: the importance of species' differing geographic ranges. *American Naturalist*, **148**, 997–1015.

Struhsaker, T. T. & Leland L. (1979). Socioecology of five sympatric monkey species in the Kibale Forest, Uganda. *Advances in the Study of Behavior*, **9**, 159–228.

Sussman, R. W. & Raven, P. H. (1978). Pollination by lemurs and marsupials: An archaic coevolved system. *Science*, **200**, 731–6.

Tenaza, R. (1987). The status of primates and their habitats in the Pagai Islands, Indonesia. *Primate Conservation*, **8**, 104–10.

Terborgh, J. (1983). *Five New World Primates*. Princeton, New Jersey, USA: Princeton University Press.

Terborgh, J. (1990). Mixed flocks and polyspecific associations: costs and benefits of mixed groups to birds and monkeys. *American Journal of Primatology*, **21**, 87–100.

Terborgh, J. & van Schaik, C. P. (1987). Convergence vs. nonconvergence in primate communities. In *Organization of Communities*, ed. J. H. R. Gee & P. S. Giller, pp. 205–26. Oxford: Blackwell Scientific.

Usongo, L. & Fimbel, C. (1995). Preliminary survey of arboreal primates in Lobeke Forest Reserve, south-east Cameroon. *African Primates*, **1**, 46–48.

van Schaik, C. P. (1996). Social evolution in primates: the role of ecological factors and male behaviour. *Proceedings of the British Academy*, **88**, 9–31.

van Schaik, C. P., Terborgh, J. W. & Wright, S. J. (1993). The phenology of tropical forests: Adaptive significance and consequences for primary consumers. *Annual Review Ecology and Systematics*, **24**, 353–377.

Witting, L. (1998). Body mass allometries caused by physiologial or ecological constraints? *Trends in Ecology and Evolution*, **13**, 25.

Wrangham, R. W. (1980). An ecological model of female-bonded primate groups. *Behaviour*, **75**, 262–300.

Wright, P. C. (1989). The nocturnal primate niche in the New World. *Journal of Human Evolution*, **18**, 635–58.

Wright, P. C. (1996). The Neotropical primate adaptation to nocturnality: Feeding in the night (*Aotus nigriceps and A. azarae*). In *Adaptive Radiations of Neotropical Primates*, ed. M. Norconk, A. Rosenberger & P. Garber, pp.369–382. New York: Plenum Press.

Wright, P. C. (1997). Behavioral and ecological comparisons of Neotropical and Malagasy primates. In *New World Primates: Ecology, Evolution, and Behavior*, ed. W. G. Kinzey, pp. 127–41. New York: Aldine de Gruyter.

Wright, P. C., Heckscher, S. K. & Dunham, A. F. (1997). Predation on Milne-Edward's Sifaka (*Propithecus diadema edwardsi*) by the Fossa (*Cryptoprocta ferox*) in the rain forest of southeastern Madagascar. *Folia Primatologica*, **68**, 34–43.

9 · Convergence and divergence in primate social systems

PETER M. KAPPELER

INTRODUCTION

Current socio-ecological theory assumes that social systems are the result of adaptations, and thus subject to evolutionary modification by selection (Crook & Gartlan, 1966; Emlen & Oring, 1977; Wrangham, 1987; Ims, 1988). Social systems are not direct targets of selection, however, because they represent the outcome of individual behavioral interactions and strategies (Hinde, 1976). The underlying behavior of individuals towards conspecifics is thought to be largely shaped by ecological factors, such as the distribution of risks and resources in the environment (Wrangham, 1980; van Schaik & van Hooff, 1983; Terborgh & Janson, 1986; van Schaik, 1989), as well as the resulting social boundary conditions (Janson, 1986; van Schaik, 1996). Behavior constitutes, therefore, the crucial interface between individuals and their environment (Terborgh, 1992). Individuals decide at the behavioral level whether they lead a solitary life or form permanent groups, and which group size and composition is optimal under a given set of ecological conditions. An animal's behavioral decisions at this level are therefore a target of selection where convergences among distantly related taxa as a result of similar selection pressures can be expected.

Among mammals, primates exhibit an extraordinary level of diversity in social systems (Smuts *et al.*, 1987), rivaled perhaps only by that found among marsupials and carnivores (Gittleman, 1989; Strahan, 1995). They display all fundamental types of mammalian grouping patterns (Clutton-Brock, 1989) and show a variety of bonding patterns, especially in gregarious taxa (van Schaik, 1989). Thus, the evolution of convergences is not principally constrained by primate-specific traits.

For the purpose of an analysis of convergence in social systems, the living primates can be divided into four major groups with potentially important variation in taxonomic diversity. Africa and Asia contain heterogeneous communities of strepsirhines and haplorhines, whereas Madagascar and the New World have been colonized by either a strepsirhine or haplorhine radiation, respectively, in the absence of members of the other group (Martin, 1990; Fleagle, 1998). Thus, any study of similarities and differences among these four groups must consider these potentially confounding effects of historical biogeography in order to distinguish between homology and convergence.

Because this and similar analyses focus on variation within an order, similarities due to common ancestry, i.e. homologies, are potentially likely and may lead to unwarranted conclusions about convergence. This concern seems negligible in this particular case, however, because the traits under investigation appear to be very flexible in evolutionary times. This point is nicely illustrated by the variation in social systems exhibited by the apes (Short, 1979; Wrangham, 1987). The solitary orangutan, the pair-living gibbons, the single-male, multi-female groups of gorillas, the male-bonded, multi-male, multi-female groups of chimpanzees and the female-bonded groups of bonobos represent almost all currently recognized primate social systems within a clade that shared the last common ancestor about 18 million years ago (Purvis, 1995). Because lemurs have been isolated on Madagascar for more than 45 million years, platyrrhines in America for about 40 million years and the major groups of African and Asian cercopithecoids have been separated for about 14 million years (Purvis, 1995), all four primate communities should have been separated long enough for their members to develop social adaptations independently.

A second important assumption for studying convergence concerns the similarity of ecological pressures faced by primates in different regions. Because the majority of taxa are confined to tropical and subtropical regions dominated by virtually the same plant families, this may create a range of broadly similar habitats in each region (Richard, 1985; Terborgh & van Schaik, 1987; Gentry, 1988; Mack, 1993). It should be noted, however, that more specific habitat characteristics may vary considerably and affect fundamental aspects of socio-ecology of the respective primates (Terborgh & van Schaik, 1987). These differences among regions or similar variation at smaller scales are still poorly documented, however, and so are their consequences for primates (but see Reed & Fleagle, 1995; Fleagle & Reed, 1996).

With these caveats in mind, I would like to address two main topics. First, did primates in Africa, Asia, America and Madagascar evolve the same diversity of social systems? To examine this question, I will compare primate communities of the four regions at the coarsest level, i.e., by examining the presence and proportion of taxa representing the three fundamental types of social organization: solitary, pair-living and group-living species.

In solitary species, adult males and females spend the majority of their activity alone, occupying individual home ranges that may overlap within and between the sexes to various extents (Bearder, 1987; Kappeler, 1997). Examples include the genera *Galago*, *Perodicticus*, *Microcebus* and *Pongo*. In pair-living species, one adult male and one adult female are permanently associated and coordinate their activities in a joint home range, as for example in the genera *Indri*, *Callicebus* and *Hylobates* (van Schaik & Dunbar, 1990). Group-living species are characterized by the permanent association of more than two adult males and females, be they closely (e.g., *Cercopithecus* spp., *Alouatta* spp., *Macaca* spp.) or more loosely (e.g., *Ateles* spp., *Eulemur* spp., *Pan*) associated in space and/or time.

Second, I will compare the size, composition and bonding patterns among group-living taxa in more detail because the interactions among these variables are responsible for much of the total variation in primate social systems among regions. Some attempts at describing and explaining variation in primate social systems at these levels have already been made (Terborgh & van Schaik, 1987; Wrangham, 1987), but were based on smaller data sets and did not explicitly compare these four biogeographic regions.

DETERMINANTS OF SOCIAL SYSTEMS

In addition to ecological factors, such as predation and the distribution and abundance of resources, basic life history traits, such as body size, activity, pattern of infant care, which in turn may all be closely dependent upon phylogenetic factors, have been identified as important determinants of social systems (Clutton-Brock & Harvey, 1977; Wrangham 1980; van Schaik, 1983; Terborgh & Janson, 1986; van Schaik & Kappeler, 1993; Di Fiore & Rendall, 1994). I will therefore begin by summarizing variation among primate communities in these traits, before focusing on the various aspects of their social systems. More detailed analyses and discussion, based on a data set with information for more than 160 species, representing 58 of 59 extant genera, can be found in Kappeler & Heymann (1996). Additional information for the present analyses was taken from Rowe (1996).

First, because of phylogenetic inertia, one expects less diversity among species within genera than among genera within families, etc (Harvey & Pagel, 1991). It is therefore possible that potential differences among the four primate communities in any trait are partly due to differences in their taxonomic diversity. To investigate this possibility, one can obtain a crude measure of taxonomic diversity for each group by counting the number of taxa at each level (Table 9.1). Because the number of families and genera, where most phylogenetic inertia is located, is remarkably similar among the four groups, and the proportion of families, genera and species does not differ significantly among the four regions (Kappeler & Heymann, 1996), it is unlikely that the following and other similar analyses are biased by taxonomic artifacts.

Second, there are several qualitative and quantitative differences in body size among the four primate communities. As has been noted previously, neotropical primates are on average smaller than those of other regions, partly because no living species exceeds 10 kg (Terborgh & van Schaik, 1987; Peres, 1994; Kappeler & Heymann, 1996); only a few recently discovered Neotropical fossil taxa were apparently heavier than 20 kg (Hartwig, 1995). In addition, the body mass frequency distributions of primates in the four regions are significantly different from each other, except for the Africa–Madagascar comparison, indicating that the proportion of species in different size ranges varies among regions (Kappeler & Heymann, 1996). Thus, differences

Table 9.1. *Taxonomic diversity of primates in major biogeographic areas*

	Number of			
	Families	Genera	Species	Subspecies
Africa	4	18	64	180
Asia	5	12	62	182
America	2	16	77	165
Madagascar	5	14	32	51
Total	16	60	235	578

Note: See Kappeler & Heymann (1996) for details.

in body size may affect variation in social systems across regions, especially those related to grouping patterns and group size (see Terborgh & Janson, 1986).

Third, the fundamental pattern of early infant care varies among primates and has important consequences for social organization. Females in species that park their newborn young in nests, tree holes or protected parts of the vegetation are not permanently associated with adult males (van Schaik & Kappeler, 1997). *Varecia variegata* provides the only exception to this rule. Primate females that carry their dependent young from birth on with them, in contrast, are permanently associated with at least one adult male, *Pongo pygmaeus* being the only exception to this other rule. The Neotropics is the only region with only carrying females (or males; but see Soini, 1988); all other regions contain "parkers" and "carriers". This life history constraint obviously limits the degrees of freedom available to primate females as far as decisions about their grouping patterns are concerned, and must therefore be considered in the respective comparisons (see below).

Finally, comparison of categorical activity frequency distributions revealed significant variation among the four regions in the proportions of diurnal and nocturnal species (Kappeler & Heymann, 1996). This heterogeneity is caused by a significantly larger proportion of nocturnal species in Madagascar, which is also home to all but one of the cath-emeral primates. Because nocturnal activity appears to be incompatible with group-living, these differences among regions have to be considered in an analysis of social systems, as well.

The possible causes of these differences among regions and potential interactions among these variables have been

discussed before (Kappeler & Heymann, 1996), as well as in other contributions to this volume, and will therefore not be considered further here.

CONVERGENCE IN SOCIAL ORGANIZATION AND STRUCTURE

A species' social organization is defined by the number and sex of regularly associated conspecifics (Struhsaker, 1969). The following analyses revealed that the four regions differ qualitatively and quantitatively in the proportion of primates with different types of social organization (Kappeler & Heymann, 1996). On a qualitative level, America is the only region without solitary primates and Africa the only region without consistently pair-living ones. A more quantitative comparison, counting only genera, indicated that the four regions also differ in the proportion of solitary, pair- and group-living taxa (Fig. 9.1), but, because of statistical problems due to the large number of cells with small expected values, this should only be taken as indicating a trend.

Interestingly, the vast majority of genera does not include species with different types of social organization. The genus *Tarsius* is the only one containing solitary and pair-living species, whereas the lemur genera *Hapalemur*, *Eulemur* and *Varecia*, the Neotropical *Cebuella*, *Saguinus* and *Leontopithecus*, as well as the Asian genera *Simias* and *Presbytis*, are distinct in containing pair- and group-living

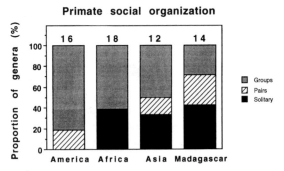

Fig. 9.1. The proportion of primate genera in four major biogeographic regions with different types of social organization. Genera with variation in social organization were counted twice. The proportion of solitary, pair- and group-living genera differs significantly across regions ($\chi^2 = 17.05$, 6 d.f., $P = 0.009$, but see text).

Table 9.2. *Evolutionary transitions among primate social systems in the four major biogeographic areas. Transitions between solitary (S), pair-living (P) and group-living (G) taxa are shown; transitions between different types of social organization are summarized (—) separately*

	S→P	S→G	S→S	P→G	P→S	P→P	G→S	G→P	G→G	—
Madagascar	2	1	10	1	—	1	—	3	9	7
Africa	—	1	11	—	—	—	—	—	16	1
Asia	1	—	1	—	—	—	1	3	6	5
America	—	—	—	—	—	1	—	3	15	3
Total	3	2	31	1	—	2	1	9	46	16

Note: See Fig. 2 for reference.

species (or populations). Thus, the small amount of variation at this low taxonomic level indicates a potentially profound phylogenetic effect on social organization (Di Fiore & Rendall, 1994; Rendall & Di Fiore, 1995).

Because differences in grouping pattern among regions may therefore be partly due to either founder effects or phylogenetic inertia, in addition to the determinants discussed above, I performed several phylogenetic analyses to explore these interrelationships. To examine potential founder effects, it is necessary to identify ancestral character states for the relevant taxa and to obtain information on the likelihood of transitions among social systems. I addressed these questions with a parsimony-based phylogenetic reconstruction (Maddison & Maddison, 1992) of social organization, mapped onto a composite primate phylogeny (Purvis, 1995).

This reconstruction (Fig. 9.2) revealed that the ancestral lemur was solitary and that the ancestral New World primate was group-living. The ancestral African and Asian anthropoids were group-living, but lorises and tarsiers contribute solitary ancestral taxa to these two regions, respectively. Using a more recent alternative phylogeny for New World primates (Schneider *et al.*, 1996) did not affect these conclusions (Fig. 9.3); only the ancestral state of the Pitheciinae could not be reconstructed unequivocally.

The six possible transitions, from solitary to pair- or group-living, from pair- to group-living and vice versa occurred with unequal frequencies (Table 9.2). Group-living evolved once in the ancestral anthropoid, presumably in Africa, and twice among lemurs. Changes from group- to pair-living occurred most frequently, both in general and in each region separately, and much more frequently than the reverse transition. The transition back to a solitary life occurred only once, in orangutans. Thus, the lack of solitary primates in the New World may be due to a constraint that makes this particular transition very difficult.

Group size

The primates of the four regions clearly differ in their diversity of group sizes (Kappeler & Heymann, 1996). At the qualitative level, Africa is the region with the greatest diversity. Asia lacks species that form very large groups, i.e., those with more than 50 or so members, even though some species may form large aggregations (e.g., *Rhinopithecus* spp.). The average size of lemur groups is even more reduced; species averaging more than 15–20 animals per group are lacking entirely in Madagascar. Species that form very large groups are rare in the New World, as well. All pairwise comparisons revealed that only the frequency distribution of group sizes of Malagasy primates differs significantly from those of all other regions. Additional comparisons among genera revealed that, at the generic level, Asian, and especially Malagasy primates live on average in smaller groups than their African and Neotropical relatives, after controlling for differences in body size (Janson & Goldsmith, 1995; Kappeler & Heymann, 1996).

Group structure

The majority of primates live in permanent groups containing more than two adults. The number and ratio of

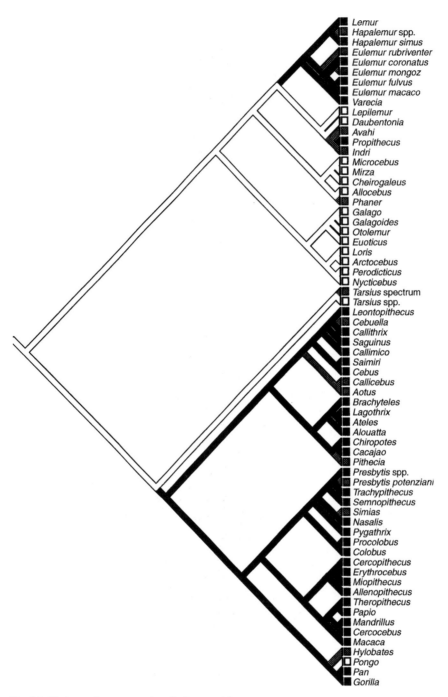

Fig. 9.2. Phylogenetic reconstruction of primate social organization. Using parsimony, the character state for various ancestral primate taxa is reconstructed on a composite primate phylogeny (Purvis, 1995). White bars: solitary taxa, stippled: pair-living taxa, black: group-living taxa.

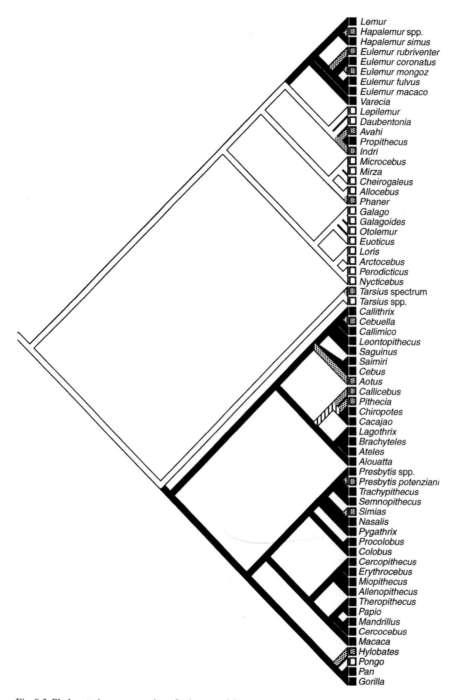

Fig. 9.3 Phylogenetic reconstruction of primate social organization, using an alternative phylogeny for New World primates (Schneider *et al.*, 1996). White bars: solitary taxa, stippled: pair-living taxa, black: group-living taxa.

Table 9.3. *Frequency of different group structures among group-living primates in the four major biogeographic areas*

	1M	MM	PA	MM–P	PA–P	MM–PA	1M–MM	VAR
Madagascar	—	5	—	1	—	—	2	1
Africa	20	19	—	1	—	—	8	—
Asia	13	14	—	1	—	—	8	1
America	—	25	—	2	1	6	7	3
Total	33	63	—	5	1	6	25	5

Note: The following categories are used: one-male, multi-female groups (1M), multi-male, multi-female groups (MM), polyandrous one-female, multi-male groups (PA), some combinations of two categories, and variable species (VAR) exhibiting more than two categories (including pairs, P). For the first three categories, only species with "little variation" were considered, i.e., >80% of censused groups fall within that category.

adult males and females, which I define as group structure, is highly variable among these group-living species. As a result, much of the existing variation in social systems, as well as important qualitative differences with consequences for mating systems, for example, are included in this category, even though these descriptions of group structure do not always allow to infer the actual mating system (Cords, 1987). Specifically, groups consisting of one male and several females, several males and one female, and so-called multi-male, multi-female groups have been lumped for the previous analyses in this category. To obtain a finer resolution, I will focus on this aspect of variation among the group-living species next by analyzing distributions of modal group structure (Table 9.3).

Variation in group composition among group-living primates has two striking aspects. First, taxa representing some categories are lacking entirely in certain regions. Specifically, species that invariably consist of one adult male and several adult females ("harem groups") are absent in Madagascar and the New World. In contrast, over 30% of group-living Old World primates live primarily in such one-male groups. If this category is expanded to include those species that form both single- and multi-male groups, even about 55% of group-living African and Asian primates can be said to form one-male groups on a regular basis. Moreover, species that regularly form so-called polyandrous groups consisting of one female and two or more males are restricted to the New World. There is much intraspecific variation in space and time around this theme, however, resulting in the frequent occurrence of multi-male groups and pairs in the same population

(Kinzey & Cunningham, 1994). Only some groups of the lemur genera *Hapalemur, Eulemur* and *Propithecus* were observed to exhibit similar variation, but in most cases nothing is known about their history or stability (Kappeler, 1997).

Second, in all four regions, there is a proportion of 20–30% of the respective species with considerable intraspecific variation in group structure. The most common category in all regions is that containing both single-male and multi-male groups. Pairs and multi-male groups within the same species are also found in all regions. A few species, from the genera *Propithecus, Eulemur, Hapalemur, Callimico, Callithrix, Leontopithecus* and *Simias*, show even variation among more than two categories, including pairs.

These patterns of variation in group structure at the species level are less constrained by phylogenetic effects than variation in social organization because the majority of genera of group-living primates in each region is not homogeneous with respect to the type of group structure displayed by its members (Table 9.4). That is to say that closely-related sister species are more likely to have different group structures than to be uniform in this respect. Thus, once groups are formed, they are very unlikely to change into other types of social organization during speciation events, whereas group structure can be quite flexible at this level.

At the level of spatio-temporal cohesion of groups, some interesting convergences can be discerned. Members of the genera *Eulemur, Varecia, Ateles, Chiropotes, Lagothrix, Brachyteles* and *Pan* live in groups that split up in subgroups of various size and composition, including pairs,

Table 9.4. *Frequency of group-living primate genera in the four major biogeographic areas that are homogeneous for a category of group structure. This tabulation includes monotypic genera and all species with genera for which sufficiently detailed information is available*

	1M	MM	VAR	not homogeneous
Madagascar	–	1	1	3
Africa	2	3	1	5
Asia	1	–	2	3
America	–	6	2	5
Total	3	10	6	16

Note: 1M, one-male, multi-female groups; MM, multi-male, multi-female groups; VAR, variable species exhibiting more than two categories (including pairs, P).

for variable periods of time. These similarities and their possible ecological causes have already been discussed in detail elsewhere (Chapman, 1990*a*; Chapman *et al.*, 1995; Kappeler, 1997). It is noticeable that species with such a fission-fusion system are apparently lacking in Asia, where only some *Macaca* and large *Trachypithecus* groups may split up into subgroups occasionally (van Schaik & van Noordwijk, 1986; Newton & Dunbar, 1994).

A final parallel has evolved between some species of *Papio*, *Theropithecus*, *Colobus* and *Pygathrix*, where several basic social units, mostly consisting of one male and several females, may join to form larger associations (Stammbach, 1987; Newton & Dunbar, 1994). Such a type of organization is found neither in Madagascar, nor in the New World.

Social structure

Finally, group-living primates within grouping types also differ in social structure, defined as the distribution of social bonds of distinct qualities among group members (Hinde, 1976). The strongest social bonds may develop within or between the sexes. The quality of bonds within and between the sexes depend primarily on the interaction between feeding competition and degree of relatedness, and can be inferred from data on affinitive, affiliative, agonistic and cooperative behavior (van Schaik, 1989). In this

last section, I will attempt a crude comparison among primates of the four regions at this level.

Four types of bonding structures can be recognized in group-living primates (Sterck *et al.*, 1997). In so-called female-bonded species, these bonds develop between females that remain in their natal groups, whereas in non-female-bonded species female social relationships may be weakly differentiated and more pronounced between females and the usually single resident male (Wrangham, 1980; van Schaik, 1989). These females may also leave their natal group. In a few so-called male-bonded species, males are philopatric and develop strong bonds among each other (Struhsaker, 1980; Wrangham, 1987). Finally, individual male-female pairs may exhibit the most salient bonds in taxa in which both sexes seem to emigrate from their natal group (Goldizen, 1987; van Schaik & Kappeler, 1993).

Because these classifications require detailed long-term studies, there is not enough information for a well-grounded quantitative comparison across regions. Nevertheless, some tentative conclusions about similarities and differences among regions seem possible. Female-bonded and non-female-bonded group-living primates can be found in Africa (e.g., *Papio* spp. and *Theropithecus* vs. *P. hamadryas* and *Gorilla*, respectively), Asia (*Macaca* and *Semnopithecus* vs. *Presbytis*) and America (*Cebus* and *Saimiri boliviensis* vs. *S. oerstedi* and *Alouatta*), whereas none of the Malagasy lemurs appear to be female-bonded (see Kappeler, 1997). Similarly, male-bonded species occur in Africa (e.g., *Pan*, *Procolobus badius*) and America (*Ateles*, *Brachyteles*), whereas these categories are absent in Madagascar and Asia. Salient male-female bonds, finally, have been described in lemurs (*Eulemur*, *Varecia*) and some New World monkeys (*Saguinus*, *Leontopithecus*), but are rare in the Old World (some *Papio* and *Macaca*).

DISCUSSION

These crude comparisons of primate social systems across major communities revealed more differences than similarities among them. Similarities in social systems are most noticeable in the structure of group-living species, where parallels among fission-fusion species, for example, offer an opportunity to identify ecological bases of convergences (Chapman, 1990*b*; Chapman *et al.*, 1995). Parallels in bonding patterns, which are expected to be closely related to the nature of feeding competition (Wrangham, 1980; van Schaik, 1989), can be found among anthropoids, whereas the limited available information does not indicate the presence of the typical male- or female-bonded species among gregarious lemurs (Kappeler, 1997). At a similar level of organization, pair-living members of different communities developed striking similarities in behavioral mechanisms employed to maintain their social organization (Pollock, 1986; Cowlishaw, 1992). Thus, there is much less convergence in various aspects of social systems than may be expected based on the ubiquity of apparently broadly similar selection pressures. This lack of convergence may therefore indicate that the link between ecology and behavior is less direct than previously thought, and that other, internal, factors, such as life history traits, influence aspects of social systems, as well.

Common ancestry and historical biogeography shape primate social systems importantly. Poorly understood phylogenetic constraints are associated with patterns of variation among regions at various levels (see also Janson & Goldsmith, 1995). This observation complements earlier reports of phylogenetic clustering in ecological multivariate space (Fleagle & Reed, 1996) and the importance of phylogeny in shaping patterns of coexistence within local communities (Houle, 1997) in all four areas. Illumination of the functional bases of these constraints should be a major goal for future studies of convergence in primate evolution.

Given the differences in potential determinants of social systems among these four communities, it is perhaps not surprising that the four regions display little convergence in continent-wide patterns of social organization. Most interesting in this respect is the lack of certain categories of social organization in some regions, as for example solitary species in the New World or pair-living ones in Africa. The former may ultimately be due to a historical accident and the (life history?) constraints hampering a subsequent transition back to a solitary life style (Wright, 1989; van Schaik & Kappeler, 1993; Dunbar, 1995a). In this context it would be interesting to know more about interspecific variation in life history and social organization among tarsiers, who may have had diurnal, and thus most likely nonsolitary ancestors (Martin, 1990). The lack of pair-living taxa in Africa may ultimately have social or ecological reasons, but it does not present a theoretical problem because the associated monogamous mating system is not expected to be favored among mammals as a result of the large asymmetry between the sexes in their potential reproductive rates (Clutton-Brock & Parker, 1992; van Schaik & Dunbar, 1990).

Differences among regions in the frequency distributions of group structure of the respective primate species may be due to ecological factors or to intrinsic life history characteristics. For example, the fact that groups consisting of one reproductive female and several males are confined to Neotropical callitrichines has been attributed to the preponderance of twin births in these taxa and the resulting need for paternal care for the young (Dunbar, 1995a). The virtual lack of one-male groups in Madagascar and America, on the other hand, may be due to the fact that groups of females in these regions cannot be economically defended by a single male. This inability, in turn, may be due to the, on average, large number of females associated for ecological reasons in the New World (Robinson & Janson, 1987), the marked synchrony of reproduction among the small number of females in lemur groups (Richard & Dewar, 1991), or the lack of primarily terrestrial species in these regions.

Lemurs display the most idiosyncrasies in grouping patterns for which a combination of factors may be responsible. First, the small proportion of group-living species may be an artifact of the extinction of the subfossil lemurs, if they were non-solitary, which is likely, given their large body size (Kappeler & Heymann, 1996). Although the largest subfossil lemurs may have been solitary, the majority of them were in the size range of extant group-living primates.

Second, the relative small size of lemur groups may be due to a number of factors. For example, small food patches that are rapidly depleted or patch characteristics, such as low density and scattered distribution, that incur high travel costs should favor small groups (Janson & Goldsmith, 1995; Overdorff, 1996). Moreover, small group size may also be the result of relaxed predation pressure (Janson & Goldsmith, 1995). A number of recent studies indicated, however, that extant lemurs are under a constant threat of predation from raptors and viverrids (Goodman et al., 1993; Goodman, 1994a,b; Rasoloarison et al., 1995), so that larger groups would be expected.

Group size in primates may also be limited by the amount of social information individuals can process. This ability may ultimately be constrained by neocortex size.

Dunbar (1992; 1995*b*) presented analyses that confirmed the predicted positive correlation between neocortex size and group size. More recent analyses controlling for phylogenetic and allometric effects revealed, however, that neocortex and group size are not correlated in lemurs and other strepsirhines (Barton, 1996).

Finally, it has recently been proposed that most of the unusual social and morphological features of lemurs, including small group size, may not reflect stable adaptations, but rather the result of an evolutionary disequilibrium following recent ecological changes in Madagascar (van Schaik & Kappeler, 1996). The recent extinction of several lemurs and large eagles (Godfrey *et al.,* 1990; Goodman, 1994*a*; Goodman & Rakotozafy, 1995), in particular, may have allowed previously nocturnal lemurs to become diurnal. This hypothesis is supported by the occurrence of many adaptations to nocturnal activity in extant cathemeral and diurnal species (Martin, 1990; van Schaik & Kappeler, 1996). As is well known, nocturnal activity appears to constrain group size considerably (Terborgh & Janson, 1986); several other lines of evidence indicate indeed that the living diurnal lemurs or their immediate ancestors were pair-living (van Schaik & Kappeler, 1993; 1996).

Thus, there are largely untested hypotheses involving ecological mechanisms and adaptations that may explain these social idiosyncrasies of lemurs, indicating that aspects of non-convergence may be due to ecological factors. Other contributors to this volume discuss in more detail ecological differences in phenological cycles, plant diversity, fruit size, tree size and density among various regions that should affect all aspects of primate communities and societies. Even if the majority of primate habitats were characterized by similar abiotic features, they may differ in aspects of community structure that affect primate adaptations. Competition with other taxa, such as arboreal mammals and frugivorous birds, and differences in predation pressure, in particular, may also influence the availability of particular niches for primates. We should therefore conduct future studies of convergence as ecologists, rather than primatologists.

ACKNOWLEDGMENTS

I thank John Fleagle, Charles Janson, and Kaye Reed for their invitation to participate in the Primate Communities workshop, the Wenner Gren Foundation and the Deutsche Forschungsgemeinschaft for their financial support, and Eckhard Heymann, Jörg Ganzhorn, John Fleagle, Charles Janson, and Carel van Schaik for many stimulating discussions and comments on this manuscript.

REFERENCES

Barton, R. (1996). Neocortex size and behavioural ecology in primates. *Proceedings of the Royal Society of London* B, **263**, 173–7.

Bearder, S. K. (1987). Lorises, bushbabies, and tarsiers: diverse societies in solitary foragers. In *Primate Societies*, ed. B. B. Smuts, D. L. Cheney, R. M. Seyfarth, R. W. Wrangham & T. T. Struhsaker, pp. 11–24. Chicago: University of Chicago Press.

Chapman, C. A. (1990*a*). Ecological constraints on group size in three species of neotropical primates. *Folia Primatologica*, **55**, 1–9.

Chapman, C. A. (1990*b*). Association patterns of spider monkeys: the influence of ecology and sex on social organization. *Behavioural Ecology and Sociobiology*, **26**, 409–14.

Chapman, C. A., Wrangham, R. W. & Chapman, L. (1995). Ecological constraints on group size: an analysis of spider monkey and chimpanzee subgroups. *Behavioural Ecology and Sociobiology*, **36**, 59–70.

Clutton-Brock, T. H. (1989). Mammalian mating systems. *Proceedings of the Royal Society London* B **236**, 339–372.

Clutton-Brock, T. H. & Harvey, P. H. (1977). Primate ecology and social organization. *Journal of Zoology*, London **183**, 1–39.

Clutton-Brock, T. H. & Parker, G. A. (1991). Potential reproductive rates and the operation of sexual selection. *Quarterly Review of Biology*, **67**, 437–56.

Cords, M. (1987). Forest guenons and Patas monkeys: male-male competition in one-male groups. In *Primate Societies*, ed. B. B. Smuts, D. L. Cheney, R. M. Seyfarth, R. W. Wrangham & T. T. Struhsaker, pp. 98–111. Chicago: University of Chicago Press.

Cowlishaw, G. (1992). Song function in gibbons. *Behaviour*, **121**, 131–153.

Crook, J. H. & Gartlan, J. C. (1966). Evolution of primate societies. *Nature* **210**, 1200–3.

Di Fiore, A. & Rendall, D. (1994). Evolution of social organization: a reappraisal for primates by using phylogenetic methods. *Proceedings of the National Academy of Sciences USA*, **91**, 9941–9945.

Dunbar, R. (1992). Neocortex size as a constraint on group size in primates. *Journal of Human Evolution*, **20**, 469–93.

Dunbar, R. (1995a). The mating system of callitrichid primates: I. Conditions for the coevolution of pair bonding and twinning. *Animal Behaviour*, **50**, 1057–70.

Dunbar, R. (1995b). Neocortex size and group size in primates: a test of the hypothesis. *Journal of Human Evolution*, **28**, 287–96.

Emlen, S. T. & Oring, L. W. (1977). Ecology, sexual selection, and the evolution of mating systems. *Science*, **197**, 215–23.

Fleagle, J. G. (1998). *Primate Adaptation and Evolution*. New York: Academic Press.

Fleagle, J. G. & Reed, K. E. (1996). Comparing primate communities: a multivariate approach. *Journal of Human Evolution*, **30**, 489–510.

Gentry, A. (1988). Changes in plant community diversity and floristic composition of environmental and geographical gradients. *Annals of Missouri Botanical Garden*, **75**, 1–34.

Gittleman, J. (1989). *Carnivore Behavior, Ecology and Evolution*. Ithaca: Cornell University Press.

Godfrey, L. R., Sutherland, M. R., Petto, A. J. & Boy, D. S. (1990). Size, space, and adaptation in some subfossil lemurs from Madagascar. *American Journal of Physical Anthropology*, **81**, 45–66.

Goldizen, A. W. (1987). Tamarins and marmosets: communal care of offspring. In *Primate Societies*, ed. B. B. Smuts, D. L. Cheney, R. M. Seyfarth, R. W. Wrangham & T. T. Struhsaker, pp. 34–43. Chicago: University of Chicago Press.

Goodman, S. M. (1994a). Description of a new species of subfossil eagle from Madagascar: *Stephanoaetus* (Aves: Falconiformes) from the deposits of Ampasambazimba. *Proceedings of the Biological Society of Washington*, **107**, 421–8.

Goodman, S. M. (1994b). The enigma of anti-predator behavior in lemurs: evidence of a large extinct eagle on Madagascar. *International Journal of Primatology*, **15**, 129–34.

Goodman, S. M. & Rakotozafy, L. (1995). Evidence for the existence of two species of *Aquila* on Madagascar during the Quarternary. *Geobios* **28**, 241–6.

Goodman, S. M., O'Connor, S. & Langrand, O. (1993). A review of predation on lemurs: implications for the evolution of social behavior in small, nocturnal primates. In *Lemur Social Systems and Their Ecological Basis*, ed. P. M. Kappeler & J. U. Ganzhorn, pp. 51–66. New York: Plenum Press.

Hartwig, W. (1995). A giant New World monkey from the Pleistocene of Brazil. *Journal of Human Evolution*, **28**, 189–95.

Harvey, P. &. Pagel, M (1991). *The Comparative Method in Evolutionary Biology*. Oxford: Oxford University Press.

Hinde, R. A. (1976). Interactions, relationships and social structure. *Man* **11**, 1–17.

Houle, A. (1997). The role of phylogeny and behavioral competition in the evolution of coexistence among primates. *Canadian Journal of Zoology*, **75**, 827–46.

Ims, R. (1988). Spatial clumping of sexually receptive females induces space sharing among male voles. *Nature,* **335**, 541–3.

Janson, C. H. (1986). The mating system as a determinant of social evolution in capuchin monkeys (*Cebus*). In *Proceedings of the Xth International Congress of Primatology* ed. J. Else & P.C. Lee, pp. 169–79. Cambridge: Cambridge University Press.

Janson, C. H. & Goldsmith, M. (1995). Predicting group size in primates: foraging costs and predation risks. *Behavioural Ecology*, **6**, 326–36.

Kappeler, P. M. (1997). Determinants of primate social organization: comparative evidence and new insights from Malagasy lemurs. *Biological Reviews,* **72**, 111–51.

Kappeler, P. M. & Heymann, E. W. (1996). Nonconvergence in the evolution of primate life history and socio-ecology. *Biological Journal of the Linnean Society*, **59**, 297–326.

Kinzey, W. G. & Cunningham, E. P. (1994). Variability in platyrrhine social organization. *American Journal of Primatology*, **34**, 185–98.

Mack, A. (1993). The sizes of vertebrate-dispersed fruits: a neotropical-paleotropical comparison. *American Naturalist*, **142**, 840–56.

Maddison, W. & Maddison, D. (1992). *MacClade: Analysis of Phylogeny and Character Evolution*. Sunderland: Sinauer Press.

Martin, R. D. (1990). *Primate Origins and Evolution*. London: Chapman & Hall.

Newton, P. & Dunbar, R. (1994). Colobine monkey societies. In *Colobine Monkeys: Their Ecology, Behaviour and Evolution,* ed. A. Davies & J. Oates, pp. 311–46. Cambridge: Cambridge University Press.

Overdorff, D. (1996). Ecological correlates to social structure in two lemur species in Madagascar. *American Journal of Physical Anthropology*, **100**, 487–506.

Peres, C. (1994). Which are the largest New World monkeys? *Journal of Human Evolution*, **26**, 245–9.

Pollock, J. I. (1986). The song of the Indris (*Indri indri*, Primates: Lemuroidea): natural history, form and function. *International Journal of Primatology*, 7, 225–64.

Purvis, A. (1995). A composite estimate of primate phylogeny. *Philosophical Transactions of the Royal Society of London B*, 348, 405–21.

Rasoloarison, R., Rasolonandrasana, B., Ganzhorn, J. & Goodman, S. (1995). Predation on vertebrates in the Kirindy forest, Western Madagascar. *Ecotropica*, 1, 59–65.

Reed, K. E. & Fleagle, J. G. (1995). Geographic and climatic control of primate diversity. *Proceedings of the National Academy of Sciences USA*, 92, 7874–6.

Rendall, D. & Di Fiore, A. (1995). The road less traveled: phylogenetic perspectives in primatology. *Evolutionary Anthropology*, 43–52.

Richard, A. F. (1985). *Primates in Nature*. New York:Freeman.

Richard, A. F. & Dewar, R. E. (1991). Lemur ecology. *Annual Review of Ecology and Systematics*, 22, 145–75.

Robinson, J. G. & Janson, C. H. (1987). Capuchins, squirrel monkeys, and atelines: socioecological convergence with Old World primates. In *Primate Societies*, ed. B. B. Smuts, D. L. Cheney, R. M. Seyfarth, R. W. Wrangham & T. T. Struhsaker, pp. 69–82. Chicago: Chicago University Press.

Rowe, N. (1996). *The Pictorial Guide to the Living Primates*. East Hampton, NY: Pogonias Press.

Schneider, H., Sampaio, I., Harada, M. L., Barroso, C. M. L., Schneider, M. P. C., Czelusniak, J. & Goodman, M. (1996). Molecular phylogeny of the New World monkeys (Platyrrhini, Primates) based on two unlinked nuclear genes: IRBP Intron 1 and e-globin sequences. *American Journal of Anthropology*, 100, 153–79.

Short, R. V. (1979). Sexual selection and its component parts, somatic and genital selection, as illustrated by man and the great apes. *Advances in the Study of Behaviour*, 9, 131–58.

Smuts, B. B., Cheney, D. L., Seyfarth, R. M., Wrangham, R. W. & Struhsaker, T. T. (ed.) (1987). *Primate Societies*. Chicago: Chicago University Press.

Soini, P. (1988). The pygmy marmoset, genus *Cebuella*. In *Ecology and Behavior of Neotropical Primates, Vol.2.*, ed. R. A. Mittermeier, A. B. Rylands, A. F. Coimbra-Filho & G. A. B. da Fonseca, pp. 79–129. WWF: Washington.

Stammbach, E. (1987). Desert, forest and montane baboons: multi-level societies. In *Primate Societies*, ed. B. B. Smuts, D. L. Cheney, R. M. Seyfarth, R. W. Wrangham & T. T. Struhsaker, pp. 112–120. Chicago: University of Chicago Press.

Sterck, E. H. M., Watts, D. P. & van Schaik, C. P. (1997). The evolution of female social relationships in nonhuman primates. *Behavioural Ecology and Sociobiology*, 41, 291–310.

Strahan, R. (1995). *The Mammals of Australia*. Chatswood, NSW: Reed Books.

Struhsaker, T. T. (1969). Correlates of ecology and social organization among African cercopithecines. *Folia Primatologica*, 11, 80–118.

Struhsaker, T. T. (1980). Comparison of the behavior and ecology of red colobus and redtail monkeys in the Kibale Forest, Uganda. *African Journal of Ecology*, 18, 33–51.

Terborgh, J. (1992). *Diversity and the Tropical Rainforest*. New York: Scientific American Library.

Terborgh, J. & Janson, C. H. (1986). The socioecology of primate groups. *Annual Review of Ecology and Systematics*, 17, 111–35.

Terborgh, J. & van Schaik, C. P. (1987). Convergence vs. nonconvergence in primate communities. In *Organisation of Communities*, ed. J. Gee & P. Giller, pp. 205–26. Oxford: Blackwell Scientific.

van Schaik, C. P. (1983). Why are diurnal primates living in groups? *Behaviour*, 87, 120–44.

van Schaik, C. P. (1989). The ecology of social relationships amongst female primates. In *Comparative Socioecology*, ed. V. Standen & R. A. Foley, pp. 195–218. Oxford: Blackwell Scientific.

van Schaik, C. P. (1996). Social evolution in primates: the role of ecological factors and male behaviour. *Proceedings of the British Academy*, 88, 9–31.

van Schaik, C. P. & Dunbar, R. I. M. (1990). The evolution of monogamy in large primates: a new hypothesis and some crucial tests. *Behaviour*, 115, 30–62.

van Schaik, C. P. & Kappeler, P. M. (1993). Life history, activity period and lemur social systems. In *Lemur Social Systems and Their Ecological Basis*, ed. P. M. Kappeler & J. U. Ganzhorn, pp. 241–60. New York: Plenum Press.

van Schaik, C. P. & Kappeler, P. M. (1996). The social systems of gregarious lemurs: lack of convergence with anthropoids due to evolutionary disequilibrium? *Ethology*, 102, 915–41.

van Schaik, C. P. & Kappeler, P. M. (1997). Infanticide risk and the evolution of male-female association in primates. *Proceedings of the Royal Society of London B*, 264, 1687–94.

van Schaik, C. P. & van Hooff, J. A. R. A. M. (1983). On the ultimate causes of primate social systems. *Behaviour*, 85, 91–117.

van Schaik, C. P. & van Noordwijk, M. A. (1986). The hidden costs of sociality: intra-group variation in feeding strategies in Sumatran long-tailed macaques (*Macaca fascicularis*). *Behaviour*, **99**, 296–315.

Wrangham, R. W. (1980). An ecological model of female-bonded primate groups. *Behaviour*, **75**, 262–300.

Wrangham, R. W. (1987). Evolution of social structure. In *Primate Societies*, ed. B. B. Smuts, D. L. Cheney, R. M. Seyfarth, R. W. Wrangham & T. T. Struhsaker, pp. 282–97. Chicago: University of Chicago Press.

Wright, P. C. (1989). The nocturnal primate niche in the New World. *Journal of Human Evolution*, **18**, 635–58.

10 • Of mice and monkeys: Primates as predictors of mammal community richness

LOUISE H. EMMONS

INTRODUCTION

Understanding the patterns of community structure is one of the eternal, basic goals of ecology, while measuring and monitoring diversity is a necessity for conservation biology. The fundamental data set for both of these is the simple species list. Every chapter in this book is directly or indirectly based on site-specific primate lists. Complete primate lists for localities number in dozens for every major biogeographic region, but despite the obvious ecological importance of mammals, the complete composition of few, if any, tropical forest mammal faunas is known. This knowledge vacuum is a consequence not only of the intrinsic difficulty of sampling species which cannot be seen and are not readily captured, but also of large temporal fluctuations in small mammal densities that result in rare species often being undetectable except during periods of peak population. Of all forest mammals, primates are the most easily identified and most quickly inventoried and monitored. In this chapter I explore the question of how primate species richness relates to the species richness of mammals in other taxa at given localities; how these patterns vary among different continents; on what scale patterns occur; and whether primates are good indicators of mammalian species richness. I conclude with comments on the possible underlying causes of some of the patterns discovered.

RATIONALE

Mammal species are difficult to inventory, but some (animals not captured by standard methods, such as shrews, small insectivorous marsupials, high-flying bats, etc.) are much more difficult than others. For this reason, 'complete' inventories of the most difficult habitat, tropical rainforest, may not exist. Of all mammals, primates are the most easily inventoried, being diurnal, large, conspicuous, and noisy. Visual transect surveys of primates in the neotropics usually can achieve 50% inventory of all species present within 16 census hours, and complete inventory within 50 hours or often fewer (Fig. 10.1; Peres, chapter 15, this volume). When primates are collected as specimens during mammal inventories, an asymptote is usually reached for primates after 10 to 20 collecting days, but for rodents this takes about 40 days or much longer (Fig. 10.2). Several years or even decades of repeated sampling were needed to obtain nearly complete lists of mammals for rich localities, such as Makokou, Gabon or Cocha Cashu, Peru. Because such sampling has been done for few localities worldwide and is unlikely to ever comprise more than a tiny handful of sites and cover an inconsequential fraction of the landscape, it is of importance to know if inventories of the most easily censused 10% of the fauna, primates, are useful as predictors of the greater, uninventoried segments of an assemblage. If they are, this might also tell us something about the ecological characteristics of primate communities.

SITE LISTS AND METHODS

The mammal lists used are from the published literature and from a few unpublished reports (Table 10.1). The geographical coverage of a particular species list varies from a few square km in single sites, to hundreds of square km in large regions (Nigeria), but they are restricted to faunas drawn from chiefly forest habitats. Lowland forest mammals in general have geographic ranges large enough that good site lists are nearly identical to lists from much larger

Table 10.1. *Mammal inventory lists*

Locality	Latitude[a]	Rainfall, mm	Primates	Native Rodents	Marsupials	Insectivore/ Tenrecs/ Shrews	Carnivores
Neotropics							
BCI, Panama*	9 N	2600	4	14	6		13
La Selva, Costa Rica*	10 N	3962	4	16	5		14
Arataye, French Guiana*	4 N	2750	7	21	9		11
Kartabo, Guyana*	6 N	2550	6	20	7		13
Guatopo, Venezuela	10 N	2100	3	18	8		13
Imataca, Venezuela	6 N	1500	3	22	9		16
Cunucunuma, Venezuela*	4 N		7	11	8		7
Balta, Peru*	10 S		10	24	11		15
Cuzco Amazonico, Peru*	13 S	2387	7	22	9		11
Cocha Cashu, Peru*	12 S	2000	13	27	12		14
Huampami, Peru	4 S		5	19	9		14
MCSE, Brazil*	2 S	2186	6	17	9		8
Rio Doce, Brazil	20 S	1480	4	11	9		
Xingu, Brazil*	4 S		7	23	8		[2]
Calilegua (<600 m), Argentina	24 S	2000	1	8	3		9
Africa							
Makokou, Gabon	1 N	1724	16	33		12	10
Lopé, Gabon	0 N	1506	14				12
Kongana, CAR	3 N	1457	11	22		16	10
Maïko, Congo	0 S		12	26		10	18
Okapis, Congo	1 N	1750	15	31			14
E. Nigeria	6 N	1600	14	17		6	6
W. Nigeria	7 N	1600	8	18		5	4
Madagascar							
Manantantely	25 S		5	3		6	2
Nahampoana	25 S		5	2		5	2
Ranomafana 1000 m	21 S	2600	12	6		8	5
Andringitra 1200 m	22 S		9	6		8	2
Andringitra 800 m	22 S		9	7		6	4
Anjanaharibe-Sud 1260 m	14 S		8	7		8	
Vohibasia	22 S	800	8	1		6	
Andranomay 1300 m	18 S	1204	9	6		10	
Ambohitanteley 1450 m	18 S	1680	4	1		6	
Asia							
Danum Valley, Borneo	5 N	2822	8	30		2	22
Mt. Kinabalu lowland, Borneo	6 N		7	35		9	14
Kuala Lompat, Malaysia	4 N	2120	6	23			6
Pasoh Forest Reserve, Malaysia	3 N	2000	7	29		2	14
Ketambe, Sumatra	4 N	2229	7	13			14
Wilpattu, Ceylon	8 N	1000	4				11

Note: [a]Latitude rounded to the nearest degree
*Asterisks mark localities for which expected faunas were calculated in Voss & Emmons (1996).

Table 10.1. (*cont.*)

Locality	Large Ungulates	Bats	Total non-bats	Total non-bat non-primates	Total mammals	Reference
Neotropics						
BCI, Panama*	5	64	49	45	113	Voss & Emmons, 1996
La Selva, Costa Rica*	5	65	52	54	117	Voss & Emmons, 1996
Arataye, French Guiana*	5	61	61	54	122	Voss & Emmons, 1996
Kartabo, Guyana*	5		60	46		Voss & Emmons, 1996
Guatopo, Venezuela	5	60		52	115	Ochoa et al.,1995
Imataca, Venezuela	6	78	66	63	144	Ochoa 1995
Cunucunuma, Venezuela*	3	50	43	63	93	Voss & Emmons, 1996
Balta, Peru*	4	56	74	64	130	Voss & Emmons, 1996
Cuzco Amazonico, Peru*	4	44	59	52	103	Voss & Emmons 1996
Cocha Cashu, Peru*	5	60	79	66	139	Voss & Emmons, 1996
Huampami, Peru	4	24	51	63	75	Patton et al., 1982 & pers comm.
MCSE, Brazil*	5		53	48	53	Voss & Emmons, 1996
Rio Doce, Brazil				47		Stallings et al., 1991; Fonseca & Kierulff, 1988
Xingu, Brazil*	3	47	48	41	95	Voss & Emmons 1996
Calilegua (<600 m), Argentina	2	15	23	22	22	Heinonen & Bosso, 1994
Africa						
Makokou, Gabon	13	34	89	73	124	Anon., 1979; unpubl. data
Lopé, Gabon	13					White, 1994
Kongana, CAR	13		77	66		Ray, 1996
Maïko, Congo	14	26	84	74	110	Colyn, 1986
Okapis, Congo	15		83	68		Hart & Bengana, 1996; Dudu et al. unpubl. data
E. Nigeria	10	23	55	49	78	Happold, 1985
W. Nigeria	11	22	48	43	70	Happold, 1985
Madagascar						
Manantantely		4	16	11	20	Creighton, unpubl. data
Nahampoana		4	14	9	18	Creighton, unpubl. data
Ranomafana 1000 m			30	18		Ryan et al., In Press
Andringitra 1200 m		2	25	16	27	Goodman, 1996
Andringitra 800 m		3	27	18	30	Goodman, 1996
Anjanaharibe-Sud 1260 m						Goodman, 1998
Vohibasia						Langrand & Goodman, 1997
Andranomay 1300 m						Rakotondravony & Goodman, 1998
Ambohitanteley 1450 m						Rakotondravony & Goodman, 1998
Asia						
Danum Valley, Borneo	9	28	81	73	109	Anon., 1993
Mt. Kinabalu lowland, Borneo	4	12	72	65	84	Lim & Muul, 1978
Kuala Lompat, Malaysia	8	18	46	40	64	Medway & Wells, 1971
Pasoh Forest Reserve, Malaysia	8	26	65	58	91	Kemper, 1988; Francis, 1990
Ketambe, Sumatra	6		46	39		C. van Schaik, in litt
Wilpattu, Ceylon	7					Eisenberg & Lockhart 1972

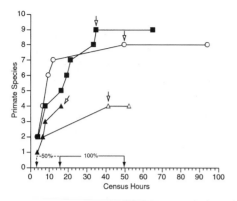

Fig. 10.1. The number of primate species recorded with time during transect surveys by L. H. Emmons. Squares, Cocha Cashu, Peru 1978; open circles, Cocha Cashu 1983; closed triangles, Barro Colorado Island, Panama; open triangles, MCSE reserves, Brazil. Arrows indicate times when all resident, diurnal species had been seen (a species present on the trail system at Cocha Cashu in 1978 was not found there in 1983).

areas of the same habitat. In larger geographic samples, only faunas from forests at uniform elevations were considered as units. For Madagascar, faunas of different elevational zones in the same altitudinal transect are in one case listed as separate data sets. This could present a problem of independence of the data, but any two lists taken from the same biogeographic region, such as the whole of equatorial Africa, have a similar problem, and for the analyses of interest here, the more individual faunal samples that can be included, the more enlightening the results will be, so similarity of faunal pools was not an argument for excluding data sets.

The null hypothesis for this chapter is that the number of primate species at any given locality is not predictive of the number of mammal species of other orders at that site. The major problem in trying to examine this question was finding, for each continent or island, lists of non-primate mammals (NPM) from a range of localities with different numbers of primate species. At the outset, I can state with

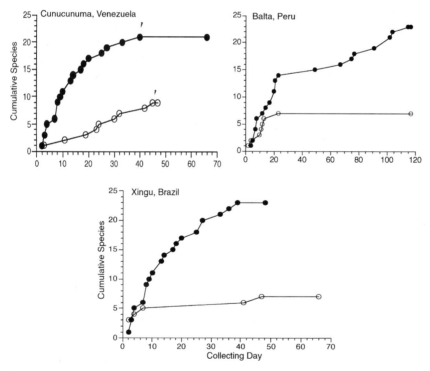

Fig. 10.2. Species accumulation curves for neotropical mammal inventories (after data compiled in Voss & Emmons, 1996). Closed circles, rodents; open circles, primates.

great confidence that the NPM lists for continental localities are, with few exceptions, significantly incomplete (Voss & Emmons, 1996). Not only that, but there is wide variation in incompleteness, depending on the order of mammals concerned, with bats, small insectivores and small marsupials being the most poorly known (not captured by standard methods), and large mammals such as artiodactyla and carnivores being the best sampled. Moreover, recent work in the Amazon basin of Western Brazil (Patton *et al.*, pers. comm.) and in Madagascar (Goodman, 1996) has shown that as much as 10% of the small terrestrial mammal fauna is still undescribed. This may be true of many other regions as well. For each site the mammal species list is broken into orders or functional groups (large ungulates), and I have included or excluded orders in the analyses depending on whether or not the lists seem to be reasonably complete (Table 10.1). This permits the use of good, but partial data sets. For example, Steven Goodman's recent pitfall sampling of shrew tenrecs in Madagascar has uncovered rich faunas where standard trapping failed to do so, but there are no convincing bat lists for Madagascar sites with other inventory data. For Asian localities, few mammal lists were complete enough to make valid comparisons of total faunas: few lists included more than a few of the many expected species of squirrels, shrews, and small mice (Asian surveys often rely on locally made mesh rat traps which leak tiny mammals). Even fewer Asian lists included creditable bat faunas, because methods used on other continents fail to capture equivalent percentages of Asian bat communities (Francis, 1994). The total number of non-flying mammal species is used below as a baseline measure of mammal species richness, not because bats have less importance, but because many more data sets are available without, than with, bats. Despite the weaknesses of individual lists or parts of lists, many are excellent and the results of years of effort.

Another difficulty arises from the lack of an adequate geographical representation among the lists from certain regions. For example, no good mammal inventory lists have been published for the equatorial neotropical localities with highest primate numbers. As will be seen below, this makes it difficult to discern or evaluate some underlying biogeographical patterns.

The analyses below can be viewed as a heuristic exercise to demonstrate tendencies and trends rather than firm conclusions. Because the list data are incomplete, and each

list is incomplete to a different extent and for different taxa, and because geographic coverage is skewed, I have avoided statistical tests because assumptions about comparable samples are violated, and statistics might lend an air of confidence or validation beyond the level warranted by the data. Furthermore, the number of data sets used is so small, that addition or subtraction of a single point that falls outside of the others could change the slopes of the regressions. Most graphs were drawn with DeltaGraph™ 4.0.

TOTAL MAMMAL FAUNAS

When the total known non-flying mammal faunas of all of the sites are plotted against the numbers of primates, there is a clear trend among African and Malagasy faunas for increasing numbers of primates to be accompanied by increasing numbers of other mammal species (Fig. 10.3). However, for Asia (Fig. 10.3), there is no evident relationship between primates and other mammals. There is an inherent problem with the Asian data sets used, in that they all have almost the same numbers of primate species, such that they cannot show how other taxa vary with changes in primate richness. This evenness of primate richness in Asia may be a distinctive feature of the fauna. Asian localities fall within the cloud of neotropical sites (Fig. 10.3), but if the Asian species lists are about equally complete, primate richness does not tend to increase much with richness of NPM in Asia, in contrast to the patterns for other regions.

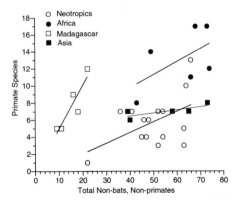

Fig. 10.3. The total numbers of non-flying, non-primate species related to total primate species.

Among neotropical sites, there is a strong positive relationship between primates and other mammals, but only when there are six or more primates in the fauna. Where primate richness is low, (3 to 5 spp.) there is a broad scatter of total mammal richness, but all values are in the direction of higher than expected mammal richness, if a line drawn through the richer sites is considered as the expected fauna. This could be due to either a lower than expected number of primates (see Peres, chapter 15 and Struhsaker, chapter 17, this volume), or to the addition of non-primate elements from other habitats (savanna, montane, dry forest) at sites where primates are reaching their environmental limits.

The data for the sites with few primates (Table 10.1) confirms this supposition. Two sites, Cuzco Amazonico and Huampami, were hunted out of large primates and have altered faunas. Three more, Rio Doce, BCI and La Selva, are in the Atlantic forest of Brazil or Central America, where the faunal pool differs (they perhaps belong on a separate analysis), while the final two, Imataca and Guatopo, are partly in dry forest or savanna habitat. All of these sites are outside of the Amazon Basin (but so are some other Guianan sites). The linear relationships derived from the data (Total other mammals = $a \times$ primates + b) were: Neotropics, TM = 1.9P + 3.9; Africa, TM = 2.5P + 3.1; Madagascar, TM = 1.7P + 1.4.

To evaluate the possible errors due to incomplete faunal data sets for the neotropical sites, we can use the expected total primates and expected total other mammal species for ten of the same sites, as determined in Voss & Emmons (1996) from the known geographic range of each species in the fauna (Table 10.1). This shows a remarkable, almost perfect relationship between expected numbers of primates, and the expected numbers of other, non-flying mammals (Fig. 10.4). The regression line generated from the expected numbers (TM = 2.3p +2.8) is quite close to that obtained with the real data, despite the poor fit of the latter (Figs. 10.3 and 10.4). On this basis, neotropical primate richness appears to be a very good predictor of the overall richness of other terrestrial (non-bat) mammals, at least throughout the range of lowland faunas included in those ten sites, but the real inventory data currently available often fail to show this, because either the inventory lists are incomplete or faunas have been altered by man. For the neotropics, at least, generating a list of expected primates may therefore yield a better prediction of the

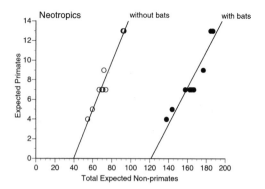

Fig. 10.4. Expected total other mammals species in relation to expected total primate species for 10 neotropical localities (data from Voss & Emmons, 1996).

number of species in the balance of the fauna than do actual lists (see Ganzhorn, in press, for a similar situation in Madagascar).

INDIVIDUAL FAUNAL COMPONENTS

Do all mammalian taxa show parallel trends in species richness? They can be examined individually by order and by region.

For bats, there are few acceptable data sets outside of the neotropics, and even within that region, few are near completion. In the New World, the correspondence between bat and primate richness is weak (Fig 10.5). This is due to extremely high bat richness in the primate-poor Guyana region of the northern neotropics, from Venezuela to French Guiana. The few data sets for Africa seem to show a more predictable positive association between bat and primate species richness, but there are insufficient data for much confidence (Fig 10.5). Recent biogeographic studies of West African bats by Fahr (1996, and pers. comm.) show that based on their ranges, many more bat species can be expected at equatorial African sites than are currently known on any lists. Two of the four points for Asia could be said to lie on the same regression as for Africa (Fig. 10.5), but in reality the four Asian points show a doubling of the bat fauna with virtually no difference in primate numbers. The poorest Asian bat fauna (Mt. Kinabalu) may be an especially incomplete list. Overall, the much larger bat fauna of the neotropics is striking, as is the similarity in size of Paleotropic inventories.

Rodents, in contrast, have been better sampled world-

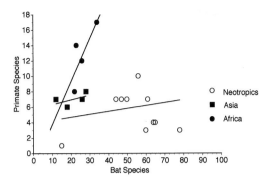

Fig. 10.5. The numbers of bat species in relation to the numbers of primate species.

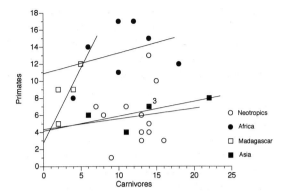

Fig. 10.7. The numbers of carnivore species in relation to the number of primate species.

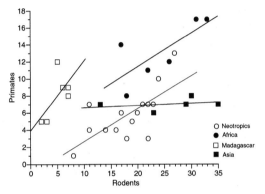

Fig. 10.6. The numbers of rodent species in relation to the number of primate species, introduced rodents excluded.

wide. Moreover, the way in which they are inventoried tends to be more uniform, so that lists can be compared more reliably. For the neotropics, Africa, and Madagascar, the number of rodent species increases regularly with the numbers of primates (Fig. 10.6), while for Asia, again, tripling of the rodents is not associated with any increase in primates among these sites. The regressions for the continents do not form a continuum, but the curve for Africa lies above that for the neotropics, because Africa evidently has higher relative and maximum numbers of both primates and rodents. Madagascar is radically different, with high primate, and very low rodent richness, such that primates outnumber rodents at a given locality. Some Asian localities lie among the neotropical points, but others show the opposite tendency from Madagascar: many rodents for relatively few primates.

Along with Primates, Carnivora is the taxon of maximum global focus for mammal conservation. Their diversity relationships with primate faunas are therefore of particular interest. Carnivore richness shows virtually no positive association with primate richness for either Africa or the neotropics; but for Madagascar, there is a strong trend for more carnivores to be found with increasing primate numbers (Fig 10.7). Asia may show a slight trend, but again for this region, almost tripling of the carnivores from six to 22 species is accompanied by an almost inconsequential change from only six to eight species of primates.

The large ungulates (elephants, Perissodactyla, Artiodactyla) show no association with primate species richness in the neotropics, but in Africa there is a definite positive trend, while Asian sites are ambiguous, although some sites seem to lie on the same curve as African sites (Fig. 10.8). Africa is outstanding for its high ungulate richness and the neotropics for its poverty, while Asia is sandwiched between. Interestingly, Asia falls on a continuum with Africa, as if the factors which control primate and ungulate species numbers in the two regions covary in the same way: a relation not seen for the other taxa compared.

The final speciose orders are the small insectivorous mammals: marsupials in the neotropics, Insectivora such as shrews in the Old World, and tenrecs in Madagascar. All of these groups include a number of tiny species that are difficult to capture with standard methods, or rare, so they are almost always under-represented in inventories, except when special methods such as pitfall traps are used to capture them. A dramatic illustration of this was shown by Ray (1996), who captured only two species of shrews

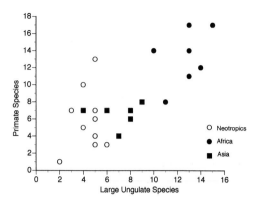

Fig. 10.8. The numbers of large ungulate species (Artiodactyla, Perissodactyla, Elephants) in relation to the number of primate species.

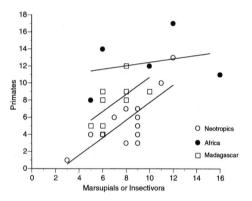

Fig. 10.9. The numbers of marsupial (neotropics), insectivores (Africa), and/or tenrec (Madagascar, Africa) species in relation to the number of primate species.

among 1957 trapped small mammals, but found 16 species, including one or two undescribed taxa, in the feces of small carnivores at the same site. This site is the far right point among the African set (for this reason little confidence can be placed in the other African points, and Asia has been excluded for lack of any satisfactory data). Likewise, recent work in Brazil by Patton *et al.* (1997) and in French Guiana by Voss & Simmons (pers. comm.) has uncovered higher marsupial numbers than shown in Fig. 10.9. Nevertheless, for all regions, there are strong positive associations between this sector of the fauna and primate species numbers (Fig. 10.9). Species richnesses cluster in a similar group for all geographic regions, but much more inventory

and taxonomic work is needed before we can have confidence in the insectivore picture. The true curve will likely shift much to the right for all areas.

To use again the expected faunas for the ten neotropical sites to evaluate errors due to incomplete sampling (Fig. 10.10A), we can see that marsupials have a nearly perfect correspondence with primates, rodents a very good one, but carnivores almost none. For rodents and marsupials, the expected maximum numbers (33,16 spp.) in western Amazonia are approximately double the minimum expected numbers (17,7 spp.) from Central America. But for carnivores, the little total variation in expected numbers (11–16) does not track primate richness. Bat species show a strong tendency to increase in association with primate species (Fig. 10.10B) especially at lower bat and primate richnesses, but the association fails at highest richnesses of primates. It appears that for biogeographical reasons, highest bat diversities occur in the Guyana region, where primate richness is low; while highest primate diversity occurs near the equator in Western Amazonia, where bat richness seems lower. It is important to repeat that none of these mammal inventories has been carried out in the region of Amazonia where mammal species richness is expected to be highest, and thus the correlation between bats and primates could improve with fuller data.

DISCUSSION

Continental patterns

Primates are the only order of mammals for which there are sufficient data on complete tropical forest communities to undertake the kinds of valuable comparative analyses seen throughout this book. For other orders of mammals, the few available incomplete lists from widely scattered localities, and the sketchy information on species ecology, form an inadequate basis on which to either examine biogeographic, evolutionary, genetic, and ecological questions, or make biologically based conservation decisions. The analyses presented in this chapter suffer from the same problem, but the excellent fit of expected mammal numbers to expected primates for ten neotropical sites suggests that many of the relationships suggested in the current data sets will turn out to be much stronger when the database is improved.

In an early comparison of mammal communities in

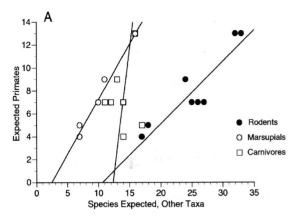

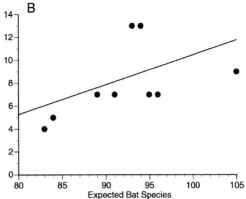

Fig. 10.10. The numbers of expected primate species in 10 neotropical localities in relation to: (a) the expected number of rodent, marsupial, and carnivore species; and (b) the expected bat species (data from Voss & Emmons, 1996).

Amazonia, Emmons (1984) observed that when faunas varied in species richness from place to place in Amazonia, only certain taxa exhibited variation, while others were almost invariant. The richness of primates, small rodents, marsupials, and to a lesser extent small carnivores varies, while that of large rodents, large carnivores, lagomorphs, ungulates (tapir, deer, and peccaries), and xenarthrans (anteaters and armadillos) were virtually constant. A recent more detailed analysis, with better species lists, confirms this earlier suggestion, and adds bat faunas to the list of more variable taxa. In general, speciose orders tend to show the most geographic variation in species richness.

In the neotropics, primate richness seems to be a good predictor of the richness of rodents and marsupials, but only weakly of bats and carnivores, and not at all for large ungulates. In Africa, all taxa seem to increase with primate richness, although the insectivore data are poor, probably from poor sampling. Likewise, in Madagascar all sampled taxa generally increase in richness with primates, so that it seems likely that primates are good predictors of virtually all of the non-flying mammal fauna (for bats there are insufficient data for analysis). In Asia, in contrast, a doubling or tripling in the numbers of species of other mammals is associated with only a weak, or no increase in primate species. Are these continental differences meaningful? If so, what underlies them?

Two global phenomena are in general linked to gradients in species richness. Latitude is the best known, and the

latitudinal gradient in species diversity of mammals is well established (Kaufman, 1995; Eeley & Lawes, chapter 12, this volume). If the sites listed are arranged by latitude and richness of primates or total mammals, some curious patterns appear (Figs. 10.11 and 10.12). The African sites show an almost perfect correspondence between latitude and mammal species, but in the Neotropics the correspondence is quite poor, while in Asia, it actually appears reversed, with more species known from the higher latitude sites. For primates alone, all three regions and Madagascar show the 'expected' inverse relationship between primates and latitude, but for Madagascar and the neotropics the association is very weak (Peres & Janson, chapter 3, this volume). In Madagascar it is probable that humans have altered the faunas and habitats such that current inventories do not well reflect the primate communities as they naturally evolved (Ganzhorn, in press), in which case it is futile to speculate about why they do not follow current environmental rules. For Asia, the reverse total species gradient may be due to a combination of land-mass (island biogeography) effects and poor inventory samples, because the latitudinal gradient for primates alone is as it should be. Nevertheless, the difference between the gradients for all mammals and primates may be real, and reflect the poor correspondence in Asia generally between primate richness and that of other mammals. The wide range in species numbers between 5 and 10 degrees latitude in the New World is a biogeographical

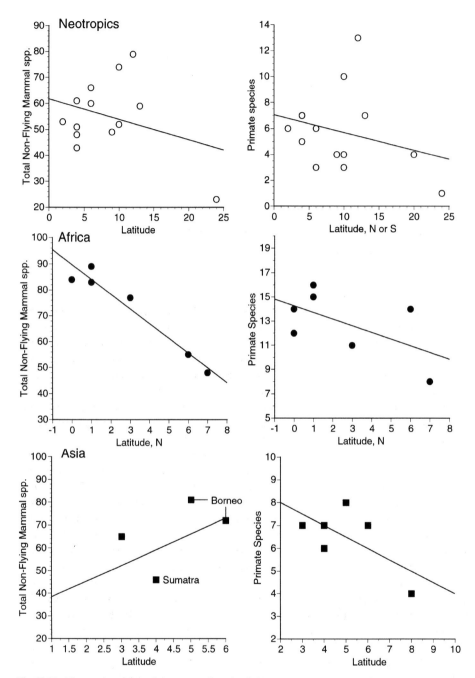

Fig. 10.11. The numbers of non-flying mammal species (left figures) and primate species (right figures) in a locality as a function of latitude, in degrees north or south of the equator.

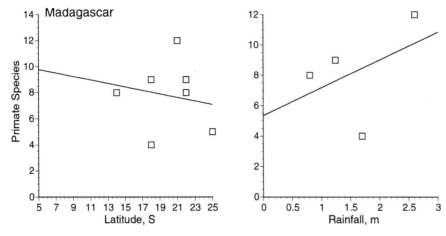

Fig. 10.12. The number of primate species as a function of latitude (left figure), or rainfall (right figure), for Madagascar localities.

phenomenon: the whole Guyana region is species-poor (except in bats), despite low latitude, so that historical or environmental effects override the latitudinal gradient. As mentioned above, and as with the Asian data, the localities for which good neotropical inventories are available are not drawn from the potentially richest sites.

The second global correlate of species richness, rainfall (Reed & Fleagle, 1995; Gentry, 1988), is of course causally related to latitude, but because of local landmass effects, there is enormous rainfall variation within any given parallel. A perhaps more important aspect of rainfall, its distribution in time (seasonality), is very tightly linked to latitude (see Chapman et al., chapter 1, this volume). The variation of species richness with rainfall among our sites (Figs. 10.12 and 10.13) again shows curious effects. Asia, which had 'reverse' latitude gradient, here shows tight positive correlation of species richness with rainfall, suggesting that rainfall effects might override the latitudinal gradient. Africa, which had tight latitude dependence for total mammals, shows much weaker gradients with rainfall (but note that the range of differences in rainfall between the sites is small). Madagascar likewise shows a definite increase in primates with rainfall, as also shown by Ganzhorn et al. (1997), and Reed & Fleagle (1995). In sharp contrast, the neotropics shows a striking, complete absence of relationship between rainfall and species richness (see Peres & Janson, chapter 3, this volume).

If the ten neotropical sites are plotted with their

expected faunas (Fig. 10.14), a remarkable negative, 'reverse' association with rainfall appears, and moreover, a 'reverse' positive association of species with latitude is seen in this data set. For latitude, I believe this is a false picture caused by the location of mammal inventories in peripheral sites; and for rainfall it is likewise because the highest rainfall sites are peripheral coastal areas at high latitudes in Central America, Coastal Brazil and the Guyanas, where species richness is low.

From a much larger database of sites with primate inventories, Reed & Fleagle (1995) drew similar species/rainfall curves for the same four geographic areas. Their results differed from those presented here for two regions: for the neotropics, they showed a fairly strong, positive correlation with rainfall, while for Asia, there was virtually none. These enigmatic differences may be simple sampling problems, from poor geographic coverage among sites with good mammal inventory, but they merit closer study (see also Kay et al., 1997; Peres & Janson, chapter 3, this volume).

A conclusion from the two relationships between faunas and latitude or rainfall is that with the single exception of the anomalous, reverse positive trend of latitude and total fauna in Asia (Fig. 10.11), primate species richness patterns mirror the patterns of the whole mammal faunas for the same sets of sites. The factors or constraints which globally control primate and total mammal numbers therefore march together at scales of both individual sites and continents. Over 20 hypotheses have been proposed to

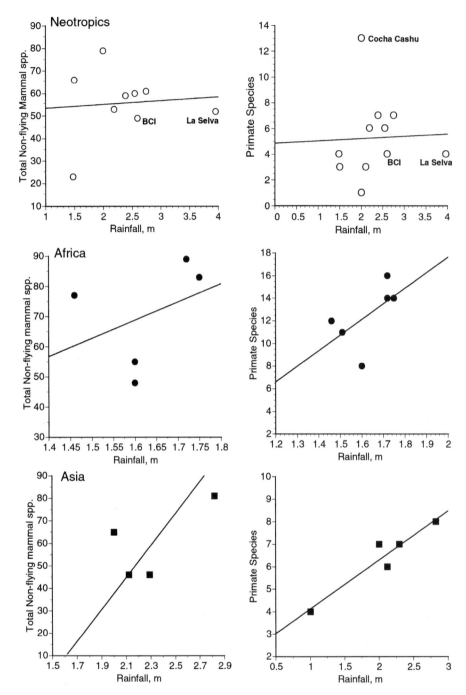

Fig. 10.13. The numbers of non-flying mammal species (left figures) and primate species (right figures) in a locality as a function of annual rainfall in meters.

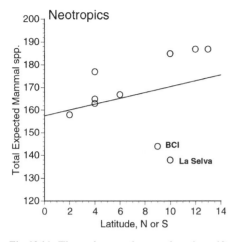
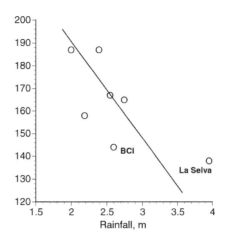

Fig. 10.14. The total expected mammal species at 10 neotropical localities as a function of latitude (left figure) or annual rainfall (right figure).

explain the gradients of species richness with latitude (Kaufman, 1995; Eeley & Lawes, chapter 12, this volume), and the gradient with rainfall generally can be included as part of the same phenomenon. This global question of proximate factors is outside the domain of this chapter (see Eeley & Lawes, chapter 12, this volume); but we can further explore the behavior of individual taxa with relation to primates.

The neotropical taxa which increase most clearly with primates are small-bodied, partially insectivorous gramnivore/frugivores (rodents) and faunivore/frugivores (marsupials). Rodents also eat some buds and leaves, as do primates. The arthropods in the diets of most rainforest members in all three taxa are gleaned from surfaces. If species richness as well as density is resource linked, then it seems probable that as fruit and arthropod food resources increase such that more primates are supported, conditions will likewise favor other frugivore–insectivores (see also Janson & Chapman, chapter 14, this volume). As noted by Emmons (1984) and Peres & Janson (chapter 3, this volume), the primates which show the most presence/absence variation in the neotropics are mainly the smaller ones: callithrichines, *Callimico*, *Callicebus*, and *Saimiri*, and to these we can now add the pitheciines. These small primates (< 3 kg), may more closely approach rodents and marsupials in resource use than do larger primates, such that areas which lack one group, lack the other. Small body size

may confer increased susceptibility to seasonal resource changes or deficits, and/or confer a competitive disadvantage with larger species (Emmons, 1984). The very curious absence of association in Asia between primate and other mammal richnesses (and rainfall; Reed & Fleagle, 1995) may be partly, if not wholly, due to the large body sizes and absence of geographically variable small species among Asian primates (Terborgh & van Schaik, 1987; but see Ganzhorn, chapter 8, this volume). Conversely, the apparently tight association between species numbers of primates and those of all other taxa in Madagascar could be a function of the small body sizes of Malagasy lemurs.

The neotropics has extremely low ungulate richness (six sympatric, maximum) and all Amazonian sites have the same expected number (five). Africa, in contrast, has a diversified fauna of from 10 to 16 species per forest site. The Bovidae in particular has an important radiation of forest species (up to nine sympatric), the majority of which are primarily frugivores. Again, a speciose frugivorous taxon shows an increase in richness with that of primates. The Asian sites fall exactly half-way between Africa and Amazonia in ungulate richness (four to nine species, Fig. 10.8).

For Carnivora, a different pattern emerges: both Africa and the neotropics have diverse forest faunas (8–18 spp.) generally in the same range (10–15 spp.), but although speciose, this order has virtually no association with primate numbers in those regions. This may be because carnivores have virtually no latitudinal gradient in the Americas, at least, from the equator to the Canadian border (Emmons *et al.*, 1984). As they are secondary consumers, the species richness of their prey is not important, only its abundance, size and location, so meat-eaters do not increase in richness with latitude or rainfall. Asia has the world's richest forest carnivore faunas, with as many as 22 co-occurring species, and in this region there seems to be a weak trend for increasing primate numbers with richer carnivore faunas; but as before, the low variability in primate faunas makes it difficult to validate any associations. The high carnivore diversity in Asia is due to rich felid and mustelid faunas (carnivorous), but also to largely frugivorous viverrid species: of the 22 carnivora known from Danum Valley, seven are civets that eat fruit. This group of carnivores might be expected to track primate richness most closely. Madagascar is the only one of the four regions where primate and carnivore numbers have a tight positive association. The native carnivore fauna of Madagascar is entirely composed of an ecologically diverse radiation of carnivorous Viverridae that feed on vertebrates and invertebrates. Some may eat occasional fruit, but none has the arboreal frugivore role seen in African and Asian palm civets and New World olingos and kinkajous (Procyonidae), which share fruit resources with monkeys. One can speculate again that small body sizes may make Malagasy carnivores more susceptible to environmental variation, but a more probable explanation is that Malagasy carnivore inventories are incomplete and/or that as with the primates, humans have altered their distributions, to produce an artificial gradient in species richness dependent on the size, location, and history of forest reserves.

The regional differences in the patterns of how primate richness relate to the rest of mammal assemblages are largely due to differences in taxon composition resulting from historical biogeography (e.g. whether or not there are bovids, viverrids, or marsupials). Because of such effects, use of primates as faunal indicators must be tailored to individual regional patterns.

Asia in all of these analyses, as in those of Reed & Fleagle (1995) and Ganzhorn (chapter 8, this volume), is 'the odd man out.' It has an unusually low range of primate richness values and little variation from site to site among the data sets used. It is not clear whether this phenomenon is real, or an artifact of sampling site location. Africa and the New World have large contiguous landmasses, with species arrays that are displayed across a continuous environmental gradient (a reason that Africa has such a perfect latitude/species curve). Asia is now broken into disconnected fragmentary islands with discontinuous species pools (Borneo, Sumatra, Sri Lanka, the Malay Peninsula), perhaps none of which is large enough to maintain high primate richness (Reed & Fleagle, 1995). It would be valuable to gather data sets from the Indian subcontinent for comparison (see Gupta & Chivers, chapter 2, this volume). In Asia large increases in species numbers of other mammalian taxa are not accompanied by equivalent increases in primate species, as they are in other regions. That richnesses of other mammalian taxa in Asia are as high as those in other regions, and much more variable than those of primates, suggests that the resource base in Asia is not more uniform from site to site than it is in other regions, and that land masses and forests are large enough to sustain high diversities of other taxa. I have suggested above that body size may be a factor in Asian insensitivity to variation, but body size differences themselves require a different level of explanation (Terborgh & van Schaik, 1987; Reed, chapter 7, this volume).

The scale of patterns

Mammal communities vary considerably on local scales, largely among the same habitat gradients which affect primates (Peres & Janson, chapter 3, Peres, chapter 15 and Janson & Chapman, chapter 14, this volume); especially between floodplain and interfluvial terra firme formations, and between dense, sunlight-exposed 'edge' or high disturbance vegetations, and closed canopy, open understory, forest interior habitats. All of the inventory lists used here are broad surveys of all local habitat types combined. Because their distributions and habitat requirements are better known than those of other forest mammals, primates might be used as habitat indicators for other mammals on local scales, but there is not much information on other mammals with which to establish a baseline.

The largest mammal species tend to have the largest geographic ranges: both among primates and among other taxa. The species which show the most geographic variation in presence/absence, are also those with the most β diversity of species, and these tend to be small-bodied. These three factors (range size, patchiness, speciation) are doubtless causally related. For example, in the neotropics, callithrichids are the most variable primates on all counts, as are murid and echimyid rodents, while *Ateles*, *Alouatta* and the largest rodents (capybara, paca) are the least variable and therefore the least informative. The same five large ungulates and four largest felids, the largest porcupine, and all the forest armadillos and anteaters, among others, potentially occur everywhere in Amazonia, and are thus of no use as indicators of either faunal diversity or of special assemblages with endemism. Large species are the best indicators of hunting pressure (Peres, chapter 15 and Struhsaker, chapter 17, this volume) but if conservation of biodiversity is the goal, the focus should be on the distribution of small species, and on the geographic scale on which changes in these occur.

The geographical delimitations of ranges of primates and other mammals have often been taken a priori to coincide with major river barriers (Ayres & Clutton-Brock, 1992). Patton *et al.* (1997) have recently found strong genetic evidence from rodents and marsupials that species-level differentiation, with groups of concordant species, can occur within currently undivided blocks of forest, presumably as a result of past geological vicariant events. To map and understand species differentiation and distribution, surveys must address a finer scale than large interfluvial basins. Fortunately, primate richness seems to be a good indicator of richness of the variable taxa of small mammals. It remains to be seen whether patterns of primate endemism also mirror those of other forest mammals. For rodents, marsupials, insectivores, and bats, exact knowledge of both geographic ranges and often of the systematics is currently inadequate to evaluate this.

Conclusions

In the absence of other information, and except in Asia, the number of primate species at a site relative to the number found at other places in the same zoogeographic region can be used as an indicator of the likely richness of several other mammalian orders, although certain orders fail to correlate with primate richness in certain regions. Small-bodied forest frugivores and insectivores seem to show the most positive association with primate richness. However, there is much scatter among the data, and exact numbers of non-primate species cannot be predicted from the number of primates, only general trends. A better database may eliminate much of the scatter seen in the stronger associations presented here, and for the neotropics, an estimate of the expected primate fauna for a locality seems to be an excellent indicator of the expected species in the balance of the fauna. The global paucity of adequate mammal inventories is a serious problem for conservation, because without knowledge of where species are, nothing can be done to plan land use for their conservation. There is an urgent need for many, and better, well vouchered inventories, both to uncover the many undescribed mammals, and to understand the biogeography of all mammalian taxa. It is clearly important to carefully document primate faunas as well as those of more cryptic species. Unfortunately, the importance of this basic knowledge is unrecognized by most funding agencies and many scientists.

The information presented here strongly suggests that for maximum impact, conservation should focus not only on single species or large species, but on diversity within and between communities. A current conservation fallacy is that protecting large charismatic umbrella species such as jaguars or elephants will automatically conserve the less glamorous taxa as well. In fact, jaguars can be protected almost anywhere in all of tropical America, chimps or elephants in many parts of Africa, whereas small rare species require conservation of specific locations. A strategy for protection of forest areas of highest primate species richness, and those with different species assemblages, is the most likely to favor the largest number of other mammal species from the most geographically variable families and genera. The large glamour species are always found where each small endemic forest species occurs, but not vice-versa.

Future research needs include acquiring complete, vouchered mammal inventories of key areas – globally, but especially for Asian and African sites, and analyzing features of scale, between local habitats, and among small and larger local regions.

SUMMARY OF RESULTS

1. Primates can be more rapidly and completely inventoried than any other similarly speciose mammalian taxon.
2. In the neotropics, Africa, and Madagascar, the total number of other non-volant mammal species in forested localities increases with increasing primate richness, but the fit of actual data sets to the regressions is not tight

except for Madagascar. Due to the low range of primate richnesses in our sample of Asian sites, no satisfactory association between primates and other species numbers can be seen for Asia, where large increases in species richness of other mammals seem not to be accompanied by increasing primate richness, although there are a few slight positive trends.

3. For the neotropics, analysis of expected local numbers of species suggests that deviations from a nearly perfect linear relationship between primate species numbers and that of non-primate mammals are largely due to incompleteness of the baseline inventories, suggesting that primate richness may be an excellent indicator of general mammal richness in Amazonia.

4. When individual orders of mammals are considered, for Africa, bats, rodents, ungulates, and insectivores increase with primate species richness, but carnivores show little correlation; for the neotropics, rodents and marsupials increase regularly with primates, bats more poorly, and carnivores and ungulates show no correlation; while for Madagascar alone, all taxa (rodents, tenrecs, carnivores and bats) show strong positive association with primate species numbers. Little can be said from our data sets about Asia, but in general, increases in richness of other mammalian taxa are not there associated with increases in primate richness, but nor can they be ruled out.

5. In the absence of other inventory information, the species richness of primates at a locality is a useful indicator of the probable richness of the non-primate mammals, but such prognoses should be applied on a taxon by taxon and case by case basis, and should not be construed as a substitute for trying to acquire as complete a mammal inventory as possible.

ACKNOWLEDGMENTS

Many thanks to Steve Goodman, John Hart and Carel van Schaik for making available unpublished data sets in manuscripts and reports. I am grateful to Colin Chapman, John Fleagle, Jörg Ganzhorn, Charles Janson, and Daphne Onderdonk for improving the manuscript with their good suggestions. The Smithsonian Institution, Division of Mammals, generously provided the author with a place to work.

REFERENCES

Anon. (1979). Liste des vertébrés du bassin de L'Ivindo (République Gabonaise) – poissons exceptés. *Unpub. report*: 42 pp.

Anon. (1993). *Danum Valley Conservation Area: A Checklist of Vertebrates.* Kota Kinabalu: Innoprise Corporation.

Ayres, J. M. & Clutton-Brock, T. H. (1992). River boundaries and species range size in Amazonian primates. *American Naturalist*, **140**, 531–7.

Colyn, M. (1986). Les mammifères de forêt ombrophile entre les riveres Tshopo et Maiko (Région du Haut-Zaire). *Bulletin de l'Instut Royal des Sciencess Naturelles de Belgique: Biologie*, **56**, 21–26.

Creighton, G. K. (1992). Faunal Study: Madagascar Minerals Project, Final Report: Madagascar Minerals Project. Unpublished report: QIT Fer et Titane.

Dudu, A., Katuala, G. B. & Hart, J. A. (ca. 1995). The Small Mammal Fauna of the Réserve de Faune à Okapis, Ituri Forest, Zaire, with a special emphasis on the muroid rodents. Unpublished report of the Wildlife Conservation Society.

Eisenberg, J. F. & Lockhart, M. (1972). An ecological reconnaisance of Wilpattu National Park, Ceylon. *Smithsonian Contributions to Zoology*, **101**, 118 pp.

Emmons, L.H. (1984). Geographic variation in densities and diversities of non-flying mammals in Amazonia. *Biotropica*, **16**, 210–22.

Emmons, L.H., Gautier-Hion, A. & Dubost, G. (1983). Community structure of the frugivorous-folivorous forest mammals of Gabon. *Journal of Zoology, London*, **199**, 209–22.

Fahr, Jakob (1996). *Die Chiroptera der Elfenbeinküste (unter Beruckschtigung des westafrikanischen Raumes): Taxonomie, Habitatpräferenzen und Lebensgemeinshaften.* Diplomarbeit, Lehrstuhl fur Tierökologie und Tropenbiologie, Julius-Maximilians-Universität, Würtzburg.

Fonseca, G. A. B. da & Kierulff, M.C.M. (1988). Biology and natural history of Brazillian Atlantic forest small mammals. *Bulletin of the Florida State Museum, Biological Sciences*, **34**, 99–152.

Francis, C. M. (1990). Trophic structure of bat communities in the understory of lowland dipterocarp rainforest in Malaysia. *Journal of Tropical Ecology*, **6**, 421–31.

Francis, C.M. (1994). Vertical stratification of fruit bats (Pteropodidae) in lowland dipterocarp rainforest in Malaysia. *Journal of Tropical Ecology*, **10**, 523–30.

Ganzhorn, J. U. (in press). Lemurs as indicators for

assessing biodiversity in forest ecosystems of Madagascar: Why it does not work. In *Aspects of Biodiversity in Ecosystems – Analysis of Different Complexity Levels in Theory and Practice*, ed. A Kratochwil. Dordrecht: Kluwer Academic.

Ganzhorn, J. U., Malcomber, S., Andrianantoanina, O. & Goodman, S.M. (1997). Habitat characteristics and lemur species richness in Madagascar. *Biotropica*, **29**, 321–43.

Gautier-Hion, A., Duplantier, J. M., Quris, R., Feer, F., Sourd, C., Decoux, J. P., Dubost, G., Emmons, L. H., Erard, C., Hecketsweiler, P., Moungazi, A., Roussilhon, C. & Thiollay, J.-M. (1985). Fruit characters as a basis of fruit choice and seed dispersal by a tropical vertebrate community. *Oecologia*, **65**, 324–37.

Gautier-Hion, A., Emmons, L. H. & Dubost, G. (1980). A comparison of the diets of three major groups of primary consumers of Gabon (primates, squirrels, ruminants). *Oecologia*, **45**, 182–9.

Gentry, A. (1988). Changes in plant community diversity and floristic composition on environmental and geographical gradients. *Annals of the Missouri Botanical Garden*, **75**, 1–34.

Goodman, S. M. (ed.) (1996). A floral and faunal inventory of the eastern slopes of the Réserve Naturelle Intégrale d'Andringitra, Madagascar: with reference to elevational variation. *Fieldiana Zoology, New Series*. Vol. 85.

Happold, D. C. D. (1985). Geographic ecology of Nigerian mammals. *Musée Royal de L'Afrique Centrale Tervurven, Belgique, Sciences Zoologiques*, **246**, 4–50.

Hart, J. A. & Bengana, F. L. (1996). La Faune Mammalienne connue actuellement et supposé être dans la réserve à Okapis. Centre de Formation et de Recherche en Conservations Forestière, Unpublished report.

Heinonen, S. & Bosso, A. (1994). Nuevos aportes para el conocimiento de la mastofauna del Parque Nacional Calilegua (Provincia de Jujuy, Argentina). *Mastozoologia Neotropical*, **1**, 51–60.

Kaufman, D. M. (1995). Diversity of New World mammals: universality of the latitudinal gradients of species and bauplans. *Journal of Mammalogy*, **76**, 322–34.

Kay, R. F., Madden, R. H., van Schaik, C. & Higdon, D. (1997). Primate species richness is determined by plant productivity: implications for conservation. *Proceedings of the National Academy of Sciences, USA*, **94**, 13023–7.

Kemper, C. (1998). The mammals of Pasoh Forest Reserve, Penninsular Malaysia. *Malayan Nature Journal*, **42**, 1–19.

Langrand, O. & Goodman, S. M. (1997). Inventaire biologique: forêt de Vohibasia et d'Isoky-Vohimena. *Recherches pour le Developpment, Série Sciences Biologiques*, no. 12.

Lim, B. L. & Muul, I. (1978). Small mammals. In *Kinabalu, Summit of Borneo*, ed. M. Luping, W. Chin & R. E. Dingley, pp. 403–57. Kota Kinabalu: Sabah Society.

Medway, Lord & Wells, D.R. (1971). Diversity and density of birds and mammals at Kuala Lompat, Pahang. *Malayan Nature Journal*, **24**, 238–47.

Ochoa, J. G. (1995). Los Mamíferos de la region de Imataca, Venezuela. *Acta Científica Venezolana, Zoologia*, **46**, 272–87.

Ochoa, J. G. Aguilera, M. & Soriano, P. (1995). Los mamíferos del Parque Nacional Guatopo (Venezuela): lista actualizada y estudio comunitario. *Acta Científica Venezolana, Zoologia*, **46**, 174–87.

Pacheco, V., Patterson, B. D., Patton, J. L., Emmons, L. H., Solari, S. & Ascorra, C. (1993). List of mammal species known to occur in Manu Biosphere Reserve, Perú. *Publicaciones del Museo de Historia Natural Universidad Nacional Mayor de San Marcos, Serie A Zoologica*, **44**, 1–12.

Patton, J. L., Berlin, B. & Berlin, E.A. (1982). Aboriginal perspectives of a mammal community in Amazonian Perú: knowledge and utilization patterns among the Aguaruna Jivaro. In *Mammalian biology in South America*, ed. M. A. Mares & H. H. Genoways, Pymatuning Symposium in Ecology **6**, 111–28.

Patton, J. L. & Reig, O. A. (1989). Genetic differentiation among echimyid rodents, with emphasis on spiny rats, geus *Proechimys*. In *Advances in Neotropical Mammalogy*, ed. K. H. Redford & J. F. Eisenberg, pp. 75–96. Gainesville: Sandhill Crane Press.

Patton, J. L., da Silva, M. N., Lara, M. C. & Mustrangi, M. A. (1997). Diversity, differentiation, and the historical biogeography of non-volant small mammals. In *Tropical Forest Remnants: Ecology, Management, and Conservation of Fragmented Communities*, ed. W. F. Laurence & R. O. Bierregaard, Jr., pp. 455–65. Chicago: University of Chicago Press.

Payne, J., Francis, C. M. & Phillips, K. (1985). *A Field Guide to the Mammals of Borneo*. Kuala Lumpur: The Sabah Society.

Rakotondravony, D. & Goodman, S. (1998). Inventaire biologique Forêt d'Andranomay, Anjozorobe. *Recherches pour le Développment, Série Sciences Biologiques*, no. 13.

Ray, J. C. (1996). *Resource Use Patterns Among Mongooses and Other Carnivores in a Central African Rain Forest*. PhD, Gainsville, University of Florida.

Reed, K. E. & Fleagle, J. G. (1995). Geographic and climatic control of primate diversity. *Proceedings of the National Academy of Science*, **92**, 7874–6.

Ryan, J. M., Emmons, L. H., Raholimavo, E. & Creighton, G. K. (in press). Non-primate mammals of Parc National de Ranomafana. In *Biodiversity research in the Rain Forest of Madagascar Ranomafana National Park*, ed. P. Wright. Covelo, CA: Island Press.

Skalli, A. & Dubost, G. (1986). Comparaison de deux faunes de mamifères tropicaux par l'analyse des caractères de leur mode de vie. *Les Cahiers de l'analyse des Données*, **11**, 403–40.

Stallings, J. R., da Fonseca, G. A. B., de Souzì Pinto, L. P., de Souza Aguiar, L. M. & Lima Sábato, E. (1991). Mamíferos do Parque Florestal Estadual do Rio Doce, Minas Gerais, Brasil. *Revista Brasileria Zoologia*, **7**, 663–77.

Terborgh, J. & van Schaik, C. P. (1987). Convergence vs. nonconvergence in primate communities. In *Organization of Communities: Past and Present*, ed. J. H. R. Gee & P.S. Giller, pp. 205–26. Oxford: Blackwell Scientific.

Voss, R. S. & Emmons, L. H. (1996). Mammalian diversity in Neotropical lowland rainforests: a preliminary assessment. *Bulletin of the American Museum of Natural History* no. 230.

White, L. J. T. (1994). Biomass of rain forest mammals in the Lopé Reserve, Gabon. *Journal of Animal Ecology*, **63**, 499–512.

11 • Comparing communities

JOHN G. FLEAGLE, CHARLES H. JANSON AND KAYE E. REED

Following upon the chapters in the first section of this volume, which provided geographically restricted overviews of a single biogeographical region, the chapters in this section have provided broad comparisons between the primates of different regions. Fleagle & Reed (chapter 6) review the fossil record of primate evolution over the past 60 million years and examine the relationship between quantitative measures of ecological distance and phylogenetic divergence times within and between communities in the four major geographical regions. Their results indicate that the rate of ecological divergence between pairs of taxa is constant for primates of all regions, but that the strength of the correlation varies depending upon the biogeographical history of the region. They find a major distinction between the extant faunas of Africa and Asia, which contain elements from many different radiations from the past 60 million years and show a high correlation between divergence time and ecological distance, and the faunas of South America and Madagascar that are the result of more temporally restricted explosive radiations.

In contrast to most of the chapters in this volume which compare communities in terms of the component species, Reed's chapter 7 compares modern communities in the four geographical areas from the perspective of how much primate biomass is supported by different types of resources. Comparing the body masses of primates in different regions, she finds Asian communities stand out in having a very restricted size range compared with primates of the other regions.

Ganzhorn (chapter 8) compares patterns of body mass among the primate communities of different biogeographical regions to examine the extent to which there is evidence for competition among primate species within communities. In his comparisons Asia is again the unusual region in

showing unusually high overlap in body size among potentially competing species.

Kappeler (chapter 9) compares social organization in the primates of different regions and finds little evidence of convergence in social systems of primates of different regions despite evidence of considerable variability within continents. In contrast with other regions, South America is poor in nocturnal and solitary species, while group-living Malagasy primates show very different patterns of social organization from those found in Africa and Asia. Although potentially due to ecological factors, it seems likely that many of these patterns reflect phylogenetic constraints or artifacts.

Emmons (chapter 10) examines the relationship between primate diversity and that of the total mammalian fauna and finds that there is a very good correlation once one corrects for likely sampling problems for other groups of mammals. This result is important for biodiversity estimates since primates are more easily surveyed than other mammals because of their size and diurnal behavior. The most obvious outlier in her analysis is Asia which shows remarkably little variation in the diversity of the primate fauna. Emmons finds both primate and mammalian diversity showing similar correlations with latitude and rainfall, except for Asia. Throughout her analyses Asia is the unusual region, in large part due to the relatively low and relatively invariant primate diversity. She notes that at sites in Africa, South America and Madagascar it is the smaller frugivorous species that generate a high diversity, but these species are generally missing from Asia.

Overall, there are two repeated themes that emerge from this section. One is that biogeographical history has clearly played a role in the patterns of ecological and behavioral diversity seen among the extant primate communities.

The main contrasts are the shorter time for extant primates to radiate on South America and the unique isolated radiation of lemurs on Madagascar in the absence of anthropoids and many other mammalian competitors. The other observation that appears repeatedly in these chapters is the low α diversity and uniform body sizes of Asian primates compared with those of other regions. Numerous suggestions have been put forth to explain the unusual pattern of primate species in Asia, including species—areas effects, due to the fragmented geography that have limited species diversity relative to larger continental areas, low reliable fruit productivity on a year to year basis in association with the common pattern of masting fruit cycles, or greater competition from small, sympatric frugivorous carnivores and rodents. At present there is no clear answer.

12 • Large-scale patterns of species richness and species range size in anthropoid primates

HARRIET A.C. EELEY AND MICHAEL J. LAWES

INTRODUCTION

Anthropoid primates are found throughout the tropical regions of Africa, South and Central America, India and south-east Asia, extending into subtropical regions in Africa and South America, and in east Asia into temperate latitudes as far north as Japan (Wolfheim, 1983; Napier & Napier, 1985; Richard, 1985; Fleagle, 1999). Throughout this range, they occupy a variety of habitats. However, given that roughly 90% of all primate species world-wide are restricted to tropical forest habitats (Mittermeier, 1988), the distribution of the majority is best described by the distribution of the forest biome. Anthropoid primates have a relatively young evolutionary history. In the Old World, they have undergone two major radiations during the Miocene epoch (22–5 mya): hominoids and cercopithecoids, including both colobines and cercopithecines (baboons, macaques, mangabeys, geladas, and guenons: Richard, 1985; Hamilton, 1988; Leakey, 1988; Fleagle, 1999). The evolutionary history of the New World anthropoids is less well recorded, but most modern groups are recognizable from the Miocene (Fleagle, 1999) and several extant genera are thought to have diversified during the Plio-Pleistocene (5 mya–10000 BP; e.g., callitrichines Kinzey, 1982; see also Fleagle & Reed, chapter 6, this volume). These radiations are associated with major environmental changes and the current distribution of anthropoids is to a large extent a consequence of the climatic events that took place during the late Quaternary. In a comparatively short time, anthropoid primates have become a relatively diverse group and comprise an important component of the mammalian community particularly within the forest biome.

Within this broad range, there are a number of clear biogeographical patterns that describe the distribution of anthropoid primates at a large scale. The recent development of macroecological theory provides a platform from which these might be explored and underlying processes examined. Macroecology emphasizes the "non-experimental, statistical investigation of the relationships between the dynamics and interactions of species populations that have typically been studied on small scales by ecologists, and the processes of speciation, extinction, and expansion and contraction of ranges that have been investigated at much larger scales by biogeographers, palaeontologists and macroevolutionists" (Brown, 1995). By bridging geographical and local spatial scales and ecological and evolutionary temporal frames, macroecology explains patterns of abundance, distribution and diversity. A number of biogeographical patterns have been highlighted, describing the gross distribution of taxa world-wide. First, there is the latitudinal decline in species richness: species richness decreases with increasing distance from the equator, a pattern that has long been recognized (Wallace, 1876, 1878) and is widely observed (Fischer, 1960; Pianka, 1983; Brown, 1988; Gaston & Williams, 1996). Second, among a number of groups a latitudinal gradient in species range size has been identified mirroring the species richness gradient (Stevens, 1989; Pagel et al., 1991; Letcher & Harvey, 1994): species range size, on average, increases with increasing distance from the equator. However, although this pattern has been termed 'Rapoport's rule' (Stevens, 1989) it is by no means universal (Gaston et al., 1998), and indeed has been most often described among taxa within the northern temperate latitudes (Rhode, 1996). Species richness and range size gradients have also been observed among some taxa in association with both longitude and altitude (Pagel et al., 1991; Stevens, 1992; Smith et al.,

1994). Lastly, the frequency distribution of range size among many taxa is highly skewed, with the majority of species occupying relatively small ranges (Brown *et al.*, 1996; Gaston, 1996). Such patterns reflect the interplay of various ecological and evolutionary processes, including patterns of speciation, extinction and dispersal associated with environmental history and the distribution of biotic and physical barriers, phylogenetic interrelationships, species level characteristics and the play-offs between niche breadth, abundance and body size.

Here, we question how the patterns observed among anthropoid primates conform to these general biogeographical patterns, and examine some of the underlying mechanisms. In doing so we provide a broader perspective to more localized community-based studies. Given their primarily tropical distribution, and the degree to which they are associated with the forest biome, anthropoid primates provide an interesting contrast to most taxa in which general broad-scale patterns have been observed. They are also an appropriate group for investigation as the current distribution of species is relatively well defined and their evolutionary history relatively well understood. The discussion is restricted to the anthropoid primates of the continental landmasses of Africa and South and Central America. Given their shared evolutionary history and broadly similar biogeographical conditions, these two regions form a useful comparison. Not included in this review, however, is any discussion of the anthropoids of south-east Asia as the processes influencing biogeographical patterns are unique to this region and dominated by island effects (Cracraft, 1994). This is reflected in the relatively low diversity of south-east Asian primate communities (Reed & Fleagle, 1995; Fleagle & Reed, 1996).

AN OVERVIEW OF ANTHROPOID DISTRIBUTION IN SOUTH AMERICA AND AFRICA

New World anthropoid primates (platyrrhines) are commonly divided into five or more relatively old subfamilies placed in two families, Cebidae and Atelidae. They are distributed across Central and South America, from southern Mexico in the north to southern Brazil and northern Argentina in the south (Fig. 12.1a) (Wolfheim, 1983; Napier & Napier, 1985). All New World anthropoids are primarily forest dwelling, and their distribution at the broad-scale is well described by that of the forest biome (Fig. 12.1a). Callitrichines range throughout the Amazon basin to the foothills of the Andes in the west, and south to the tropical forests of coastal south-eastern Brazil which are inhabited by the lion tamarin (*Leontopithecus rosalia*) and common marmoset (*Callithrix jacchus*) (Moynihan, 1976; Wolfheim, 1983; Napier & Napier, 1985). The group's most northerly representative, the crested tamarin (*Saguinus oedipus*), extends into Central America as far as the Costa Rica/Panama border. Other subfamilies occupy a more extensive range; howler monkeys (genus *Alouatta*) being the most widely distributed of all platyrrhine genera and occurring from southern Mexico to approximately 27 °S in Argentina (Eisenberg, 1979). Spider and woolly monkeys (subfamily *Atelinae*), capuchins (genus *Cebus*) and the douroucouli (*Aotus*) are also widely distributed, while most other New World monkeys are found mainly in the Amazon basin (Moynihan, 1976; Eisenberg, 1979; Wolfheim, 1983; Napier & Napier, 1985; Peres & Janson, chapter 3, this volume).

Although 85% of African anthropoids (catarrhines) are primarily restricted to forest habitats, those groups whose members also occupy woodland and savanna habitats have the greatest distribution within the continent (Fig. 12.1b). Two families are represented among the African anthropoids: the apes (Pongidae) and colobine and cercopithecine monkeys (Cercopithecidae), with the latter being divided into two relatively speciose tribes, the Papionini and Cercopithecini (Groves, 1989). The Papionini is the most widely distributed of these, represented in forest habitats by mandrills and drills (genus *Mandrillus*) and mangabeys (*Cercocebus* and *Lophocebus*), in woodland, savanna and subdesert regions by baboons (genus *Papio*), and in the high altitude grasslands of Ethiopia by geladas (*Theropithecus gelada*). The tribe extends across sub-Saharan Africa from Senegal west to the Arabian Peninsula and south to the Cape of Good Hope, and in the Atlas Mountains north of the Sahara it is represented by the barbary macaque (*Macaca sylvanus*) (Wolfheim, 1983; Napier & Napier, 1985; Kingdon, 1997). Cercopithecini are also widely distributed south of the Sahara, with the patas monkey (*Erythrocebus patas*) occupying wooded and savanna habitats from Senegal to western Ethiopia, and the vervet monkey (*Cercopithecus aethiops* species group) ranging beyond this south to South Africa (Wolfheim, 1983; Napier & Napier, 1985; Lernould, 1988; Kingdon, 1997).

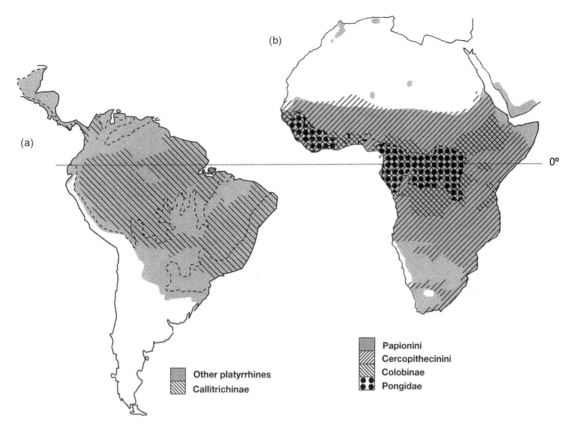

Fig. 12.1. The distribution of anthropoid primates in (a) South America, and (b) Africa (after Wolfheim, 1983; Lernould, 1988). On both continents, the majority of anthropoids occupy tropical forest habitats. The forest biome is outlined by a dashed line [- - - - -] and in South America includes seasonal forest (after Richard, 1985).

The majority of cercopithecines, however, are restricted to the main forest blocks of west and central Africa and are typified by the forest dwelling guenons (genus *Cercopithecus*). Colobine monkeys too occupy forest habitats from Senegal in the west though central Africa, ranging into Ethiopia, Kenya and Tanzania along montane and riparian forest belts (Wolfheim, 1983; Napier & Napier, 1985; Oates *et al.*, 1994; Kingdon, 1997). Similarly, the African apes are primarily found in the forests of central Africa, east to the mountains of Zaire and Rwanda, with the chimpanzee (*Pan troglodytes*) inhabiting wooded savanna southeast to Tanzania (Wolfheim, 1983; Napier & Napier, 1985; Kingdon, 1997).

PATTERNS OF SPECIES RICHNESS AND RANGE SIZE

Africa

A number of large-scale biogeographical patterns may be described among the anthropoid primates of Africa, the most striking of which is a clear latitudinal gradient in species richness (Fig. 12.2a; Cowlishaw & Hacker, 1997; Eeley & Foley, in press). Here African catarrhines conform to a near universal pattern (Brown, 1995; Gaston & Williams, 1996), discernible at various taxonomic levels and across many different groups (for example, mammals: McCoy & Connor, 1980; Willig & Selcer, 1989; Pagel *et al.*, 1991; reptiles: Schall & Pianka, 1978; Pianka, 1983; insects: Cushman *et al.*, 1993; trees: Currie & Paquin, 1987). Catarrhine species richness is concentrated more

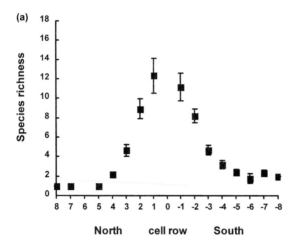

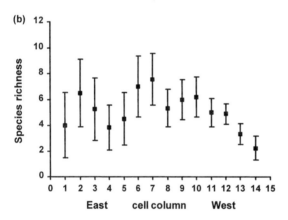

Fig. 12.2. The species richness of African anthropoid primates (a) decreases with increasing distance from the equator (r = −0.802, P < 0.001, n = 102), and (b) shows no significant relationship with longitude across the continent (r = −0.016, NS, n = 102). Species richness was determined in 102 250 000 km² cells comprising the African continent. Mean species richness and standard error for cell rows north and south of the equator, and columns east to west across the continent are shown (reproduced from Eeley & Foley, in press, *Biodiversity and Conservation*, with kind permission from Kluwer Academic Publishers).

specifically in the equatorial regions of central and west Africa (Figs. 12.3a and 12.4), a pattern that is reflected at the community level. This lowland forest area incorporates six regional primate communities identified by Oates (1996); namely the Upper Guinea, southern Nigeria,

Cameroon, western equatorial Africa, Congo Basin, and eastern Zaire communities. The communities of western equatorial Africa and eastern Zaire are recognized as particularly species rich, with the former comprising at least 20 primate species (Oates, 1996; Chapman *et al.*, chapter 1, this volume). Although catarrhine species richness at the broad-scale appears to decrease from the west to the east of the continent, and in particular from the western rift valley eastward, the gross pattern is a latitudinal one and there is no consistent longitudinal trend (Fig. 12.2b).

The pattern of anthropoid species richness clearly reflects the distribution of major vegetation types within the African continent, with a strong gradient of decreasing species richness from wetter to more arid biomes (Grubb, 1982). The central equatorial region of Africa is one of the richest areas in the world in terms of primate species and also has some of the most extensive forest cover (Mittermeier, 1988). This oversimplification belies the complexity of the mosaic of habitats that comprise the African continent, but it is sufficient to note that the pattern of primate diversity at the broad-scale is strongly associated with the forest biome (see Figs. 12.1b and 12.4). There is also a sharp species turnover across the forest/savanna boundary, the general habitat affiliation of species being closely tied to their phylogenetic relationship (Grubb, 1982). For example, the genera *Cercopithecus*, *Cercocebus* and *Colobus* are for the most part characteristic of the forest biome, while *Erythrocebus* and *Papio* typify savanna habitats.

The geographical distribution of anthropoid range sizes across the African continent shows a latitudinal gradient mirroring that of species richness. Species range sizes (extent of occurrence *sensu* Gaston, 1991) on average increase with increasing distance from the equator (Fig. 12.5). In doing so African catarrhines appear to conform to Rapoport's rule (Stevens, 1989). As originally described, Rapoport's rule states that, on average, the latitudinal range of species increases with increasing latitude (Stevens, 1989), although a similar latitudinal gradient may be recognized in relation to broader areal measures of geographical range size (for example, Pagel *et al.*, 1991; Lecher & Harvey, 1994; Brown, 1995; Blackburn & Gaston, 1996; Hughes *et al.*, 1996; Gaston *et al.*, 1998). Looking more closely, however, support for Rapoport's rule among African primates is mixed. Spatial autocorrelation may bias the strength of the above relationship (Rhode *et al.*, 1993; Letcher & Harvey, 1994), although the trend remains once this has been taken

(a)

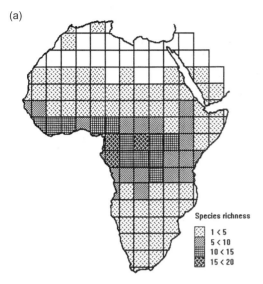

(b)

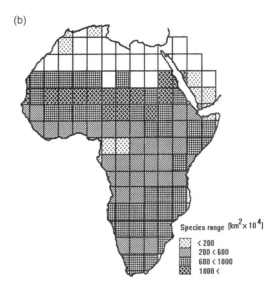

Fig. 12.3. The broad-scale distribution of (a) species richness and (b) mean species range size of African anthropoids (cell size 250000 km²). High species richness and low average range size is focused particularly in western equatorial Africa (reproduced from Eeley & Foley, in press, *Biodiversity and Conservation*, with kind permission from Kluwer Academic Publishers).

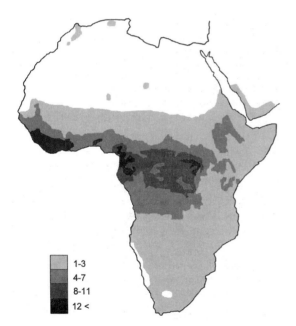

Fig. 12.4. The distribution of anthropoid species richness across Africa (after Wolfheim, 1983; Lernould, 1988). Species richness is concentrated in the forested central and western equatorial regions, and gradients of richness are steepest in the vicinity of elevational barriers (compare with Fig. 12.12).

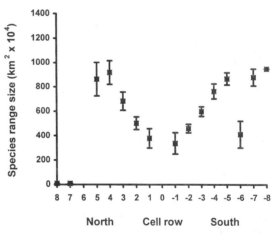

Fig. 12.5. Species range size of African anthropoids increases with increasing distance from the equator ($r = 0.279$, $P = 0.004$, $n = 108$). Average range size was determined in 102 250000 km² cells comprising the African continent. Mean and standard error for cell rows north and south of the equator are shown (reproduced from Eeley & Foley, in press, *Biodiversity and Conservation*, with kind permission from Kluwer Academic Publishers). Average range size is reduced at cell row –6 as the presence of the Kalahari and Namib deserts at this latitude means fewer cells are occupied by widespread species such as *Cercopithecus aethiops* and *C. mitis*.

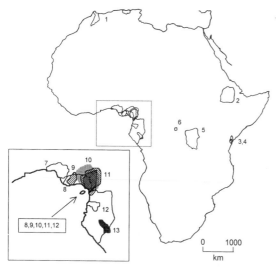

Fig. 12.6. The distribution of microareal anthropoids in Africa, i.e. those comprising the lowest quartile range size of the 51 species recognized (Rapoport, 1982; after Wolfheim, 1983; Lernould, 1988). Microareal species are found in association with areas of high relief, and cluster in western equatorial Africa where the Adamawa Highlands lie in close juxtaposition with coastal and riverine barriers (compare with Fig. 12.12). This region is one of marked endemism, characterized both by high species richness and, on average, low range size, and influences broad-scale biogeographical patterns in Africa. Species labeled: (1) *Macaca sylvanus*, (2) *Theropithecus gelada*, (3) *Cercocebus galeritus*, (4) *Procolobus ruformitratus*, (5) *Cercopithecus hamlyni*, (6) *C. salongo*, (7) *C. erythrogaster*, (8) *Procolobus preussi*, (9) *Cercopithecus erythrotis*, (10) *Mandrillus leucophaeus*, (11) *Cercopithecus preussi*, (12) *Colobus satanas*, (13) *Cercopithecus solatus*.

into account (Eeley & Foley, in press). In contrast, in an inter-specific analysis which included both catarrhine and strepsirhine primates and removed both spatial and phylogenetic autocorrelation, Cowlishaw & Hacker (1997) find no overall correlation between species' latitudinal range and latitude. However, the relationship to the north and south of the equator is different: south of the equator the latitudinal range of species does increase with latitude but north of the equator it does not (Cowlishaw & Hacker, 1997). Exploring this pattern further, Cowlishaw & Hacker find a strong correlation between species' latitudinal range and climatic variability (primarily seasonality of rainfall) both north and south of the equator, a relationship that

subsumes that between latitudinal range and latitude *per se*. They suggest therefore that while species' latitudinal extent is determined by adaptation to climatic variation, this in turn does not always directly parallel latitude. The increased width of the African continent north of the equator may explain why there is a stronger overall latitudinal gradient when range size is measured as extent of occurrence rather than as latitudinal extent. Species occupying savanna habitats north of the equator, such as the patas monkey and olive baboon (*Papio anubis*) have a broad longitudinal range.

There is a general association between the primary habitat type occupied by primates and the extent of their geographical range (Grubb, 1982; Oates, 1996). Among African catarrhines, the average range size of species primarily occupying non-forest habitats is approximately five times that of forest dwelling species. Thus, the broad-scale distribution of species range size reflects that of species richness, and is lowest in the forest regions of western equatorial Africa (Fig. 12.3b). The high level of specific and subspecific endemism in this area, which includes the West Cameroon, Rio Muni and Ogooué Centres of Endemism (Grubb, 1990), is well recorded (Grubb, 1990; Kingdon, 1990; Oates, 1996), and clearly illustrated by mapping the distribution of those species occupying the most restricted ranges, or 'microareal' species (Rapoport, 1982) (Fig. 12.6). The distribution of microareal species is of interest in that, where local diversity is high, species richness is often swelled by an increase in the number of restricted-range species (Gaston, 1994; Gaston & Williams, 1996; but see Prendergast *et al.*, 1993). Average species range size scales negatively with species richness (Fig. 12.7), a relationship that is independent of the broad-scale effects of latitude (and longitude) (Eeley & Foley, in press), and areas of particularly high primate species richness across Africa (richness 'hotspots') also tend to be rich in species with restricted ranges (Hacker *et al.*, 1998).

However, although the species-rich areas of western equatorial Africa are characterized by species occupying, on average, relatively small ranges, regions of small average range size are also located in the extreme north and east of the continent, in areas of low species richness (Fig. 12.3b). Two broad trends can be identified among the mainly non-forest species which inhabit these low richness/higher latitude regions. While most such areas are occupied by a few wide-ranging species such as the

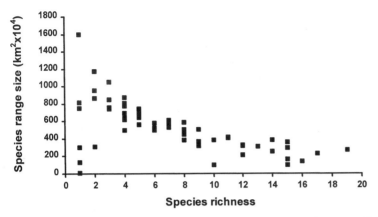

Fig. 12.7. There is a negative correlation between the species richness of African anthropoids and mean range size (r = −0.401, P < 0.001, n = 102) (reproduced from Eeley & Foley, in press, *Biodiversity and Conservation*, with kind permission from Kluwer Academic Publishers). In areas of high species richness, numbers are swelled by an increase in restricted-range species. Areas of very low species richness show considerable variation in terms of species range size, and may be occupied either by widespread species (such as *Erythrocebus patas*) or species occupying only a limited area (such as *Theropithecus gelada*).

vervet monkey, the patas monkey and the olive baboon, some, for example in North Africa and the Ethiopian highlands, are occupied by restricted-range species, the barbary macaque and the hamadryas baboon (*Papio hamadryas*), respectively. Accordingly, Hacker *et al.* (1998) describe the geographical distribution of areas characterized by restricted-range species as patchy, and note that these do not always coincide with areas of high species richness.

South America

The broad-scale geographical distribution of species richness and range size among South American primates show patterns similar to those observed in Africa. First, species richness shows a strong latitudinal gradient (Fig. 12.8a). Thirty-seven primate species occur within 5° of the equator, declining to four species at around 25° S and 19 species 10° N (Ruggiero, 1994; Peres & Janson, chapter 3, this volume). Species richness declines further into Central America, with only three species occurring as far as 18° north (Wolfheim, 1983). Species richness is highest in the western Amazonian basin (Fig. 12.9). Second, the geographical distribution of species range size (latitudinal extent) mirrors the latitudinal pattern of species richness and, on average, the latitudinal range of species increases with increasing distance from the equator (Fig. 12.8b;

Ruggiero, 1994). Although the strength of this relationship may be biased by the effects of spatial autocorrelation (Rhode *et al.*, 1993; Letcher & Harvey, 1994), it is supported by Ruggiero's observation that, as in Africa, microareal species tend to occur in equatorial regions (Fig. 12.10). To our knowledge, however, the latitudinal gradient in species range size has not been explored separately to the north and south of the equator, or in relation to total range size measured as extent of occurrence. It might be expected, for example, that in Central America while latitudinal range may continue to increase northwards (unimpeded by any major habitat change until approximately 17° north), the greatly reduced width of the continent would restrict total range size. However, observations among other taxa where total range size and latitudinal extent tend to co-vary (Gaston *et al.*, 1998), and a general correlation between the latitudinal and longitudinal extent of species ranges (Brown, 1995), might militate against this proposal. The latter patterns remain to be explored among anthropoid primates. Third, combining the two broadscale patterns there is a negative relationship between species density and average species range size (latitudinal extent) (Ruggiero, 1994). Finally, as in Africa, these patterns are strongly linked to the distribution of the forest biome, with high primate diversity and high levels of endemism occurring particularly in regions with extensive forest cover (Mittermeier, 1988).

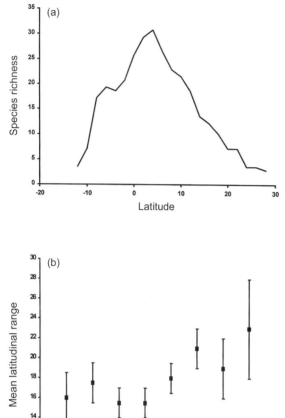

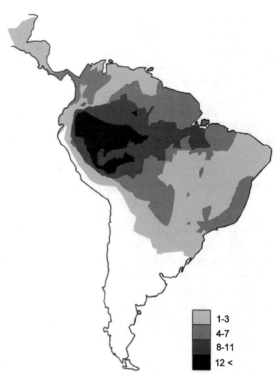

Fig. 12.9. The distribution of anthropoid species richness across South America (after Wolfheim, 1983). Species richness is concentrated in the forested equatorial regions, especially in the upper Amazon Basin, and gradients of richness are steepest in the vicinity of the Andes mountain range (compare with Fig. 12.12).

Fig. 12.8. Among South American anthropoids there is a latitudinal gradient in both (a) species richness ($Rs = -0.810$, $P = 0.001$) and (b) range size, represented by mean latitudinal extent ($Rs = 0.880$, $P = 0.02$) (redrawn with permission from Ruggiero, 1994, Journal of Biogeography, Blackwell Science Ltd.).

PROCESSES SHAPING BROAD-SCALE PATTERNS OF SPECIES RICHNESS AND RANGE SIZE

Alternative explanations for latitudinal gradients in species richness and range size

The latitudinal species richness gradient has been a major focus of biogeographical study since the pioneering work of Wallace (1876, 1878; also Dobzhansky, 1950; Fischer,

1960; Simpson, 1964; Wilson, 1974). In the search for an explanation of this general zoogeographic trend over a dozen hypotheses have been proposed (Stevens, 1989). These are separable into two groups. The first comprises ecological processes taking place at the local-scale and includes, for example, variation in primary productivity, habitat heterogeneity and structural complexity, interspecific competition, predation, present-day climatic variation and disturbance regimes (Hutchinson, 1959; MacArthur, 1965, 1972; MacArthur et al., 1966; Pianka, 1967; Connell, 1978; Hubbell, 1979, 1980). The second group operates at the regional scale and over evolutionary time, and includes geomorphological and paleoclimatic change, time since disturbance, evolutionary history, and patterns of speciation and extinction (Fischer, 1960; Haffer, 1969; Rosenzweig, 1975; Stanley, 1979; Prance, 1982). These

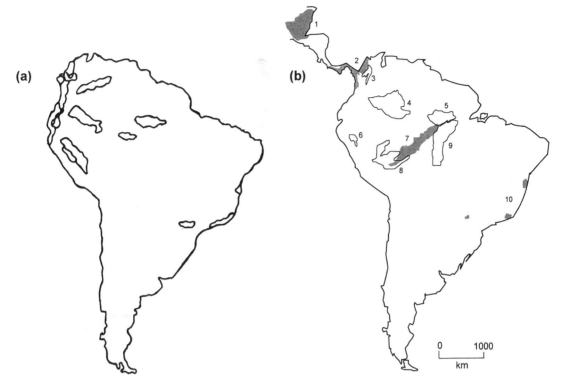

Fig. 12.10. The distribution of microareal anthropoids in South America: (a) redrawn with permission from Ruggiero (1994); (b) this analysis: (1) *Alouatta pigra*, (2) *Saguinus oedipus*, (3) *S. leucopus*, (4) *S. inustus*, (5) *S. bicolor*, (6) *Lagothrix flavicauda*, (7) *Saguinus labiatus*, (8) *S. imperator*, (9) *Callithrix humeralifer*, (10) *Leontopithecus rosalia* (after Wolfheim, 1983). Differences between the two patterns presented are likely due to differences in the taxonomy adopted, in distributional information, and the inclusion in our analysis of Central America. This highlights some of the problems of working with macroecological data (see Brown, 1995). Both patterns demonstrate that microareal species are mainly found in equatorial regions, which are characterized both by low average range size and high species richness (see text). However, while Ruggiero shows that most microareal species cluster around the northern Andes, our analysis suggests that the majority are found within the Amazon Basin with no clear clustering. Both elevational and riverine barriers are therefore likely to be important in limiting the distribution of microareal platyrrhines.

postulated mechanisms have been reviewed extensively (Simpson, 1964; MacArthur, 1972; Brown & Gibson, 1983; Brown, 1988; Rosenzweig, 1992, 1995; Cox & Moore, 1993), but their relative merits remain contentious (Gaston & Williams, 1996).

One mechanism proposed to account for the high diversity of tropical regions that has received considerable attention in the recent ecological and biogeographical literature is the Rapoport rescue hypothesis (Stevens, 1989). Stevens (1989) described Rapoport's rule for a variety of taxa in North America and suggested an underlying functional relationship between this and the latitudinal gradient in species richness. He argued that broad environmental tolerance is selectively advantageous at higher latitudes, in response to greater seasonal climatic variation, and consequently species will be able to occupy larger geographical areas, while species at lower latitudes in general have a narrower environmental tolerance and more restricted microhabitat requirements (the environmental variability hypothesis). Given equal dispersal abilities of species in all regions (a premise which has been questioned, e.g., Huston, 1994; Gaston *et al.*, 1998), Stevens went on to argue that local species richness at low latitudes may be increased relative to that at high latitudes by an increased 'mass effect' (Schmida & Wilson, 1985), or 'rescue effect' (Brown & Kodric-Brown, 1977), i.e., individuals occurring outside their range in areas where they may survive but in which they are unable to maintain a population (Stevens, 1989).

While latitudinal species range size gradients have been identified among a number of higher taxa, in particular in northern temperate latitudes (for example, mammals: Stevens, 1989; Pagel *et al.*, 1991; Letcher & Harvey, 1994; reptiles and amphibians: Stevens, 1989; and fish: Stevens, 1989; Rhode *et al.*, 1993), at lower taxonomic levels groups vary considerably in their adherence to Rapoport's rule (France, 1992; Ricklefs & Latham, 1992; Rhode *et al.*, 1993; Roy *et al.* 1994; Ruggiero, 1994). Indeed, the status of the latitudinal gradient in species range size as a general ecological 'rule' has been questioned (Gaston *et al.*, 1998). Primates are unusual in the degree to which they appear to follow this broad biogeographical pattern: of seven different mammalian orders examined in South America only primates, carnivores and bats support Rapoport's rule (Ruggiero, 1994). At first sight the Rapoport rescue effect might, therefore, appear to account for the high diversity of anthropoid primates in equatorial regions of Africa and South America. However, evidence that primates do not consistently show parallel latitudinal species richness and species range size gradients argues against the mechanism (Cowlishaw & Hacker, 1997). Furthermore, accepting the Rapoport rescue effect as a driving mechanism for the increased species richness of low latitudes, would mean adopting an essentially locally based, ecological process to account for regional patterns. Increasing evidence suggests that this is unlikely, rather the opposite is more generally true (Ricklefs, 1987, 1989; Cornell & Lawton, 1992; Cornell, 1993; Ricklefs & Schluter, 1993*a*; Cracraft, 1994; Gaston, 1994; Gaston & Williams, 1996).

There has been considerable debate surrounding both Rapoport's rule and the Rapoport rescue effect (Rhode *et al.*, 1993; Colwell & Hurtt, 1994; Roy *et al.*, 1994; Rhode, 1996; Price *et al.* 1997; Gaston *et al.*, 1998). Stevens (1989) proposed the environmental variability hypothesis to explain the broad geographical ranges of species at higher latitudes (see above), however a number of alternative mechanisms have also been suggested to underlie the latitudinal gradient in species range size (Brown, 1995; Gaston *et al.*, 1998). Brown (1995) offers five, not mutually exclusive, hypotheses from the pattern occurring in North America: (1) the shape of the continent, species ranges get smaller towards the south as the tapering of the continent limits the physical space over which they may be distributed; (2) an increase in biotic interactions curtails the range size of species at lower latitudes where local diversity is higher; (3) dispersal to and recolonization from 'sink' habitats is increased at lower latitudes, where environmental conditions vary less in time and space, allowing species with very small ranges to persist; (4) differential extinction and colonization in association with glacial/interglacial cycles at higher latitudes have removed species with narrow requirements and restricted geographical ranges (see also Price *et al.*, 1997); (5) based on Janzen (1967), elevational barriers are more severe at low latitudes, where lowland species are subject to less seasonal climatic variation, promoting the formation and persistence of narrowly endemic species. A sixth hypothesis, that the pattern is related to the extent of biogeographical provinces and species turnover associated with boundaries between major biomes has also been proposed (Roy *et al.*, 1994; Smith *et al.*, 1994; Blackburn & Gaston, 1996; Rhode, 1996; Gaston *et al.*, 1998).

Given the shape of Africa and South America, both of which reach their widest extent at or near the equator, Brown's first hypothesis is implausible in terms of anthropoid primates. Indeed, Brown also rejects the first hypothesis for North America as the latitudinal trend in decreasing range size is observed before the width of the continent tapers significantly. Of the remaining hypotheses, two and three are essentially local processes, while hypotheses four, five and six occur within a broader geographical and temporal framework. Stevens' original climatic variability hypothesis spans both ecological and historical scales. An insightful method for distinguishing the relative importance of processes occurring at the local and the regional scale has been developed recently in the field of community ecology, and is explored below. Hypothesis four may not apply directly to the tropical and subtropical latitudes inhabited by anthropoid primates, but there is considerable evidence that here too major environmental change occurred during the late Quaternary which could have shaped present-day patterns of distribution and diversity among primates (see below). Cowlishaw & Hacker's (1997) evidence indicates that adaptation to climatic variability is key to the latitudinal range of African primates, but, as latitude is not always a good surrogate of climatic variability, patterns of species range size do not always underlie the latitudinal gradient in species richness. In short, even where primates do conform to Rapoport's rule, it is important to divorce this latitudinal gradient in species range size from the Rapoport rescue effect linking it with the latitudinal gradient in species richness, and to investigate alternative explanatory processes for both broad-scale patterns.

Regional vs. local processes

In describing geographical patterns of diversity, primatologists have traditionally emphasized the utility of either historical or ecological processes. The former have generally been invoked in consideration of regional zoogeographic patterns and the distribution of endemic taxa (Kinzey, 1982; Rogers et al., 1982; Hamilton, 1988; Colyn, 1991; Colyn et al., 1991; Grubb, 1982, 1990), while proponents of ecological processes have sought to explain variation in diversity between different assemblages (Emmons, 1984; Bourlière, 1985; Terborgh & van Schaik, 1987). In this approach, primatologists have followed a traditional dichotomy between biogeographers and ecologists (Ricklefs, 1989). In general, ecologists have emphasised the role of locally-based, ecological processes, such as predation, competition or the limits of the physical environment, in determining patterns of local diversity and community structure (Ricklefs, 1987, 1989; see Janson & Peres, chapter 3 and Peres, chapter 15, this volume).

Recently, however, increased attention has been given to the role of regional processes, particularly species formation, geographical dispersal and climate change, in shaping local communities (Ricklefs & Schluter, 1993b; Balmford, 1996; Caley & Schluter, 1997). Ricklefs (1987) points out that if local diversity is deterministically related to local processes, local communities should converge under similar physical conditions. This often fails to occur. Increasing evidence suggests that, on the contrary, many local communities are unsaturated and local diversity is instead dependent upon regional diversity and regional processes (Ricklefs, 1987, 1989; Cornell & Lawton, 1992; Ricklefs & Schluter, 1993b; Caley & Schluter, 1997). It is possible to test this assumption by examining the diversity of local communities in relation to the richness of the regions within which they lie. If local processes determine local richness, then local richness will reach a

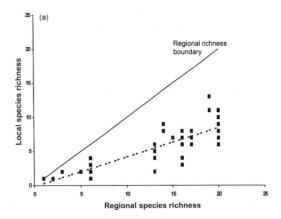

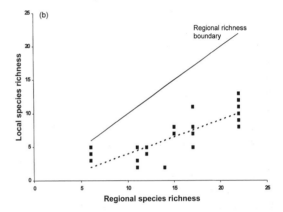

Fig. 12.11. Local species richness scales with regional richness for both (a) African and (b) South American anthropoids. These relationships are best described by linear functions (Africa: $y = -0.16 + 0.43x$, $F_{1,41} = 68.52$, $P < 0.001$, $R^2 = 0.63$; South America: $y = -1.02 + 0.50x$, $F_{1,31} = 86.22$, $P < 0.001$, $R^2 = 0.74$). Local communities are not saturated but instead are proportionally enriched from the regional pool. Regional richness is freed of local constraint and determined primarily by regional processes.

limit that is independent of regional richness (Ricklefs, 1987, 1989; Cornell & Lawton, 1992; Cornell, 1993; Cresswell et al., 1995).

A preliminary analysis suggests that local communities of anthropoid primates in Africa and South America are not saturated (Fig. 12.11; Lawes & Eeley, in prep.). On both continents, local richness fails to asymptote with increasing regional richness. Instead, increasing local richness is best described by a linear relationship with regional richness. If primate communities are not saturated then the role of regional processes in shaping broad-scale patterns of diversity, and in structuring local communities, cannot be ignored. While local (or alpha) diversity reflects the interplay of both current environmental conditions, which enable many species to persist, and past conditions that have enabled these species to accumulate (Brown, 1995), it is the latter that are more important with regard to anthropoid primates. Local communities are proportionally enriched from the regional pool and their diversity is limited by the processes that determine the richness of the latter, broadly speciation and dispersal (Ricklefs, 1989), and regional richness is independent of local processes. What then are the regional and historical processes that might be driving both broad-scale and local patterns?

Evolutionary and biogeographical history

With the majority of anthropoid primates being restricted to tropical forest habitats, present-day patterns of species richness and species range size have been strongly influenced by the historical and regional processes occurring within this biome (Grubb, 1982, 1990; Kinzey, 1982; Hamilton, 1988; Kingdon, 1990). Since the introduction of the concept by Haffer (1969; and Moreau, 1966, 1969), refuge theory has received a great deal of attention as a driving force for the evolution of species in the forested tropics during the late Quaternary (for example, Vanzolini, 1973; Prance, 1982; Whitmore & Prance, 1987; Lynch, 1988; Terborgh, 1992; Richards, 1996). In brief, refuge theory suggests that the climatic fluctuations of the Pleistocene period, which were characterized at high latitudes by the expansion and contraction of glacial ice sheets, were reflected at lower latitudes by successive periods of cooler, drier and warmer, wetter conditions, and the associated contraction and re-expansion of tropical forest (Hamilton, 1976, 1989; Haffer, 1982; Livingstone, 1982; Van der Hammen, 1982, 1991; Hamilton & Taylor, 1991). The present-day distribution of many groups in the tropical regions of both Africa and South America are thought to be the outcome of evolution and speciation acting within this framework (for example, birds: Diamond & Hamilton, 1980; Crowe & Crowe, 1982; Pearson, 1982; Haffer, 1987; butterflies: Turner, 1982; Brown, 1987; frogs: Duellman, 1982; Kingdon, 1990; mammals: Grubb, 1982). However, the debate surrounding refuge theory, the patterns, processes and precise evolutionary mechanisms involved,

remain far from resolved (Endler, 1982; Grubb, 1982; Connor, 1986; Gentry, 1986; Mayr & O'Hara, 1986; Colinvaux, 1987, 1989; Cracraft & Prum, 1988; Lynch, 1988).

Historical processes associated with the expansion and contraction of both Afromontane and lowland forests during the Pliocene (1.9–5.4 mya) and Pleistocene (10 000 BP –1.9 mya), have been advanced a number of times to explain the current pattern of primate distribution (Africa: Rogers et al., 1982; Hamilton, 1988; Oates, 1988, 1996; Kingdon, 1990, Lawes, 1990; Colyn, 1988, 1991; Colyn et al., 1991; Grubb, 1990; Chapman et al., chapter 1, this volume: South America: Kinzey, 1982). If refuge theory is to be accepted as an explanation of the broad-scale patterns described here, then the cycles of forest expansion and contraction should coincide with periods of primate speciation and radiation. There is evidence that this is so, for example, in South America among the marmosets and tamarins (Callitrichinae), in which differentiation is thought to have occurred during the late Quaternary (Hershkovitz, 1977; Kinzey, 1982). In Africa, similarly, the forest-adapted genus Cercopithecus first appeared in the fossil record around 2.9 million years ago (Leakey, 1988) and has undergone a major radiation in the last million years (Hamilton, 1988; Oates, 1988; Kingdon, 1990). During the past two million years there have been roughly 20 cycles of forest expansion and contraction (Hamilton, 1988), providing ample opportunity for isolation and divergence of populations and speciation within the forest dwelling primate community. Cercopithecus is by far the largest and most diverse genus of anthropoid primates in Africa, and the evolutionary history of this taxon has shaped many of the present-day broad-scale patterns.

The lowland forest region of equatorial Africa is illustrative of the complex pattern of diversity and endemism thought to result from evolution within a framework of forest refugia. Here four main refugia are traditionally recognized, based on the distribution of endemic species and subspecies of a number of taxa (including primates); one either side of the Zaire river basin, the West Central (West Lower Guinea or Cameroon/Gabon) and East Central (East Lower Guinea or Central) refugia, and two smaller refugia in west Africa, in Liberia and Ghana (Endler, 1982; Grubb, 1982; Mayr & O'Hara, 1986; Hamilton 1988; Kingdon, 1990). Grubb (1990) points out that for primates these major forest refugia may have been subdivided by riverine barriers (see below), resulting in a greater total number of refugia than for non-primate taxa. More recently, a fifth major fluvial refuge has been proposed in the Zaire river basin (Colyn, 1988, 1991; Colyn et al., 1991). Up to 24 species groups of primates currently inhabit the lowland forest region of Africa, each comprising a number of allopatric taxa and showing a high turnover between sister taxa associated with different communities thought to derive from these refugia (Grubb, 1990; Oates, 1996). However, the geographical patterns shown by these different species groups vary, and are best described on a taxon by taxon basis (Rahm, 1970; Grubb, 1982, 1990; Hamilton, 1988; Lernould, 1988; Oates, 1988; Colyn, 1988, 1991; Colyn et al., 1991). The *Cercopithecus cephus* species group serves here as an illustrative example (Lernould, 1988; Grubb, 1990; Oates, 1996). Within this group, *Cercopithecus petaurista* is part of the Upper Guinea community, derived from the Liberian refuge, and *C. erythrogaster* and *C. erythrotis sclateri* part of the southern Nigeria community, derived from both the Ghanaian refuge and West Central refuge. Two further subspecies of *C. erythrotis* (*C. e. erythrotis* and *C. e. camerunensis*) are endemic to the Cameroon community, while the western equatorial community includes *Cercopithecus cephus*; both communities deriving from the West Central refuge. *Cercopithecus ascanius whitesidei* and *C. a. katangae* occur within the Congo Basin which likely derives from the Zaire river refuge, while *C. a. schmidti* forms part of the Eastern Zaire community which is suggested to comprise an overlap of species which have dispersed both from the latter and montane refugia associated with the central Rift Valley to the east (Colyn, 1988; Colyn et al., 1991).

The importance of barriers

Barriers are essentially an interpretation of the edge of a taxon's range, generally in terms of either direct or indirect ecophysiological processes (Rosen, 1988). By restricting the dispersal of individuals, barriers limit the distribution of taxa. Both physical and biotic barriers are identified: physical barriers relate to geographical features such as the boundary between land and sea, rivers, and topography. Biotic barriers relate, for example, to changes in vegetation or the distribution of parasitic species (Rapoport, 1982; Myers & Giller, 1988; Brown et al., 1996). Barriers are proposed either to split a species' range, providing the opportunity for allopatric speciation to occur by preventing

gene-flow between the resultant vicariant populations, or to limit the outward spread of a taxon from its evolutionary center of origin, for example a forest 'refuge' (Myers & Giller, 1988). Major barriers, that simultaneously curtail many taxa, may influence patterns of species diversity and endemism both locally and regionally.

The distribution of major vegetation types, associated with continent-wide climatic variation, has already been observed to have a fundamental influence on the broad-scale pattern of anthropoid distribution and diversity in Africa and South America. As most primate species are restricted to forest habitats, the boundary between the savanna and forest biomes is particularly important (compare Figs. 12.1, 12.4 and 12.9). The distribution of major biomes and the associated turnover in species (hypothesis six, above) thus plays a role in determining the gross patterns described among anthropoids. In Africa, the latitudinal pattern of species richness is also strongly influenced by the Sahara desert region, which limits the northward distribution of savanna species such as the patas monkey and the olive baboon. On a smaller scale, the Dahomey Gap in West Africa, a region of low rainfall and dry savanna vegetation separating the Upper and Lower Guinea forest (Fig. 12.12), is a zoogeographic barrier which limits the distribution of a number of forest primates such as *Cercopithecus erythrogaster* to the east and *C. diana roloway* and *Procolobus badius waldroni* to the west (Oates, 1988). The expansion of both the Dahomey Gap and the Baoulé-V, a similar area in central Ivory Coast where the forest/savanna mosaic comes to within 100 m of the coast, was probably responsible for separating primate populations in west African forest refugia during the climatic changes of the Pleistocene (Oates, 1988). In South America, the fact that primate distribution is primarily limited by the distribution of the forest biome is illustrated by the high levels of specific and subspecific endemism among the primates of the isolated Atlantic forests of coastal south-eastern Brazil (Kinzey, 1982; Mittermeier, 1988).

Riverine barriers too restrict the distribution of many anthropoid species (Africa: Rahm, 1970; Oates, 1988; Colyn, 1988, 1991; Grubb, 1990; Colyn et al., 1991; South America: Kinzey, 1982; Ayres & Clutton-Brock, 1992), particularly those that are forest dwelling and/or of relatively small body size, such as *Cercopithecus* among African anthropoids and *Saguinus* and *Callicebus* in South America

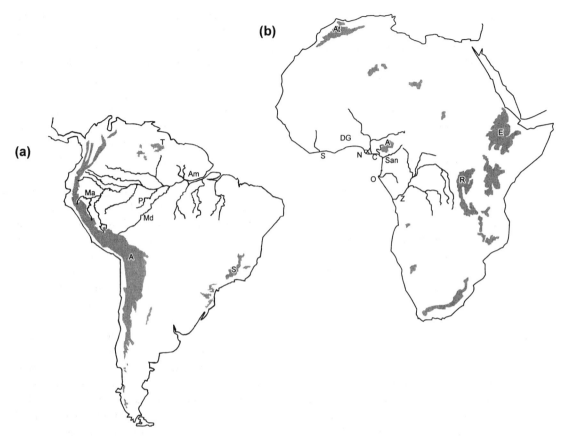

Fig. 12.12. Major topographical features and biogeographical barriers within the forest biome in (a) South America, (b) Africa. Features mentioned in the text are labeled. South America: (A) Andes mountains, (Am) Amazon, (Ma) Marañon, (P) Purus, (Md) Madeira, (T) Tapajós. (S) Sierra da Mantiqueira; Africa: (At) Atlas mountains, (E) Ethiopian highlands, (R) Ruwenzoris, (A) Adamawa highlands, (S) Sassandra, (N) Niger, (C) Cross, (San) Sanaga, (O) Ogooué, (Z) Zaire, (DG) Dahomey Gap.

(Ayres & Clutton-Brock, 1992; Eeley, 1994a). The density of barriers provided by the exceptionally large river systems that dominate the lowland forest regions of equatorial Africa and South America (Fig. 12.12) contributes to the high diversity and generally smaller ranges of anthropoids in these areas (Figs. 12.4, 12.6, 12.9 and 12.10) by limiting the geographical spread of species and restricting gene flow between populations. The effectiveness of river barriers, however, varies with their size, flow and geographical location (Ayres & Clutton-Brock, 1992; Grubb, 1990). Slow, meandering rivers provide less of a barrier to primate populations than larger and faster flowing rivers, while those rivers that lie completely within the forest block may be more easily circumvented. In Africa, the Sanaga, Zaire, Kasai, Sassandra, and Niger rivers, in particular, are recognised as important barriers to primate distribution, limiting up to 69 specific and subspecific taxa (Grubb, 1990; see also Chapman et al., chapter 1, this volume). The variation in climate and vegetation associated with rivers, and the concentration of human populations along their banks, also contributes to their effectiveness as distributional barriers (Oates, 1988; Ayres & Clutton-Brock, 1992).

However, while rivers separate many lowland forest communities, at least limited dispersal occurs across most rivers, enabling some gene flow to occur between sister taxa inhabiting opposite banks (Oates, 1988; Grubb, 1990; Ayres & Clutton-Brock, 1992). For example, an area of overlap and hybridization between *C. erythrotis camerunensis* and *C. cephus cephus* occurs around the headwaters of the Sanaga river in Africa (Grubb, 1990). The importance of rivers in terms of promoting vicariant evolution has therefore been questioned (Kinzey, 1982; Oates, 1988), and river barriers may primarily serve to maintain the patterns of distribution that are the legacy of evolutionary events taking place within a Pleistocene framework of forest contraction and expansion. In areas where there is a high density of rivers, however, these may have had a more fundamental effect by influencing the pattern of distribution of forest refugia. In such regions, it is argued, forest contraction is more likely to have lead to the separation of a number of local populations and, consequently, to a greater diversity of taxa than in areas where river density is low (Grubb, 1990; Ayres & Clutton-Brock, 1992). In Africa, such a situation may have occurred within the major fluvial refuge of the Zaire River basin and within the West Central refuge (Colyn, 1988; Grubb, 1990).

Two points should be made in connection with elevational barriers. The first is that regions of higher altitude may have served as a focus for forest refugia during periods of increased aridity in the tropical lowlands, as it is argued that here rainfall remained high and conditions conducive to forest persistence (e.g. Haffer, 1982; Whitmore & Prance, 1987). The second is that the distribution of species may be limited at high altitudes, the effect being mediated through associated climatic and vegetational change. Thus, while species richness may be expected to be low at high altitudes it might be highest in some areas of intermediate elevation. The importance of elevational barriers to the gross patterns described is highlighted both by the pattern of species richness and the distribution of microareal species. Gradients of species richness tend to be steeper in the vicinity of higher elevation areas (compare Figs. 12.4, 12.9 and 12.12). The pattern is most clearly shown in South America where there is a rapid decline in richness over a relatively short distance abutting the Andes mountains, especially in equatorial latitudes, providing support for Brown's fifth hypothesis (above). The Andean mountain chain presents a major barrier to primate distribution. There is no equivalent in Africa, although relatively steep species richness gradients are found around the Ruwenzoris and the Adamawa Highlands. The distribution of microareal species in South America is complex, and two different patterns are presented based on different distributional information (Fig. 12.10). Ruggiero (1994) shows the majority of restricted-range species to be found on the flanks of the northern Andes, with most lying to the west of the Andes between the mountain chain and the coast. Interestingly, a similar pattern occurs in south-eastern Brasil, where the Atlantic forests, rich in endemic species and subspecies (Mittermeier, 1988), again lie between the coast and an area of high relief, the Serra da Mantiqueira. While our own representation of microareal platyrrhines also indicates that the Andes may limit some species, particularly where South and Central America join, most are found within the Amazon Basin where their distribution is limited at least in part by tributaries such as the Tapajós, Madeira, Purus and Marañon. It is likely, therefore, that both elevational and riverine barriers play a significant role in limiting the distribution of the most restricted range species in South America. The association between high endemicity and the juxtaposition of coastal, topographic and riverine barriers is clearly apparent in Africa, where the majority of microareal species cluster in the western equatorial region between the Adamawa Highlands, the coast and the Niger, Cross and Sanaga rivers (Figs. 12.4 and 12.12). It seems likely that anthropoid populations which diverged within highland refugia and close to additional physical barriers were limited in the area within which they could expand their range once environmental conditions became conducive for them to do so. In South America, for example, platyrrhine species dispersing from the Chocó refuge to the west of the Andes in Colombia and Equador (Whitmore & Prance, 1987) are limited within a relatively short distance by the coast to one side and the mountains to the other. Likewise, the dispersal of species from refugia along the eastern Andean foothills, such as the Napo refuge, may have been curtailed both by the mountains and by the network of Amazon river tributaries, and a similar pattern is observed in Africa in association with the Cameroon refuge.

ECOLOGICAL AND PHYLOGENETIC DETERMINANTS OF RANGE SIZE

The frequency distribution of species range size

The frequency distribution of geographical range size among anthropoid primates in both Africa and South

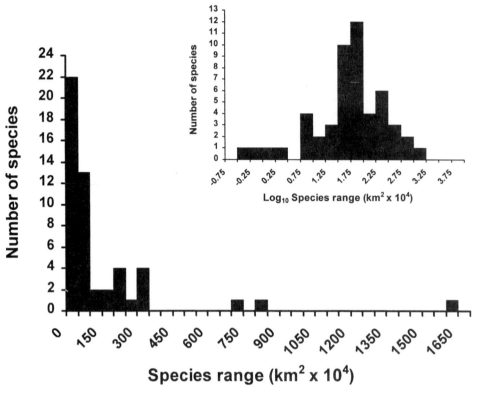

Fig. 12.13. The frequency distribution of species range size across 51 African anthropoid primates is highly right-skewed; median range size is 63.8 km² × 10⁴ and mean range size 151.62 km² × 10⁴. Inset shows the frequency distribution when range size is log$_{10}$ transformed; range size approximates to a lognormal distribution.

America follows a classic "hollow curve" (Gaston, 1996); that is, it is highly right-skewed and approximates to a log-normal distribution (Fig. 12.13; Ruggiero, 1994; Eeley & Foley, in press). The majority of anthropoids thus occupy relatively small ranges and few are widely distributed. For example, in Africa while mean range size is approximately 1.5 million km², this is biased by such savanna dwelling species as the patas monkey, the olive baboon and the vervet monkey species group, and median range size is about a third of this, 640 000 km². In this, anthropoid primates conform to a general pattern that was first noted by Willis (1922) and has seen a revival of interest in a number of recent macroecological studies (Brown, 1984, 1995; Gaston, 1990, 1994, 1996; Brown *et al.*, 1996). Lognormal range size frequency distributions have been described for many taxa (including mammals: Pagel *et al.*, 1991; Letcher & Harvey, 1994; Ruggiero, 1994; birds: Gaston, 1994;

Brown, 1995; and insects: Gaston, 1990, 1994). Willis (1922) attributed this pattern to species history; young species occupying smaller ranges, and ranges increasing over time. It may also be put down to the random, stochastic processes that underlie many biological phenomena (Rapoport, 1982; Gaston, 1994; Ruggiero, 1994). However, exploring the distribution of species range size more explicitly, Gaston (1994; 1996) suggests the pattern results instead from the interplay of temporal variation in species range sizes, evolutionary patterns of speciation, and ecological limitations (as well as various methodological artifacts).

Niche breadth and range size

Brown (1995) notes that one of the implications of the 'hollow curve' frequency distribution of species range size is that related species differ widely in their ecological requirements and niche breadth. Both abiotic and biotic factors interact to limit a species' distribution, influencing local population dynamics such that a species' range is curtailed at the point where the sum of emigration and death outweighs the sum of immigration and birth (Caughley et al., 1988; Gaston, 1990; Brown, 1995). Species that exploit a variety of resources (i.e., those that have a broad ecological niche, sensu Hutchinson, 1957) are likely to become both more locally abundant and more widespread (McNaughton & Wolf, 1970; Brown, 1984, 1995; Brown & Maurer, 1987). A positive relationship between species range size and niche breadth (such as environmental tolerance or habitat characteristics) is therefore expected, and has been observed among a variety of taxa (for example, mammals: Glazier, 1980; Pagel et al., 1991; marine bivalves: Jackson, 1974). Although Gaston (1994) argues that niche 'pattern' and resource availability may be at least as important as niche breadth in influencing abundance and range size patterns.

One approach to assessing environmental tolerance is to measure the variation in climatic conditions experienced by a species within its range. Thus, among African primates Cowlishaw & Hacker (1997) have shown a significant positive relationship between species range size and the seasonality of rainfall measured at the range center. South American primates exhibit a similar strong relationship between the annual range in temperature extremes and species range size (Ruggiero, 1994). Among African anthropoids, species range size is also related to habitat and dietary niche breadth (Fig. 12.14). Species that occupy a large geographical area, e.g., the vervet monkey species group, olive baboon, chimpanzee and guereza (Colobus guereza) generally exploit a greater variety of habitat and food types than congeneric species occupying smaller ranges. This is reflected at the broad-scale by the observation that regions of high species richness and low average range size are also characterized by a narrow average habitat and dietary breadth (i.e., increased habitat and dietary specificity) among the resident species (Fig. 12.15; Eeley & Foley, in press). However, the circularity of the relationship between the size of a species' geographical range and its niche breadth cannot be ignored. Does a species occupy a large range because of its broad niche, or is the latter a necessary consequence of the former and the spatial heterogeneity of the environment (Williamson, 1988)?

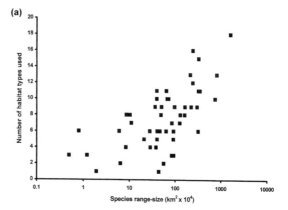

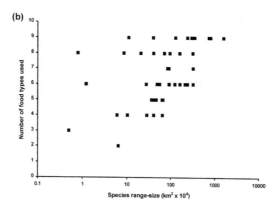

Fig. 12.14. Species range sizes increase with increasing niche breadth in terms of (a) the number of habitats used (r = 0.618, P < 0.001, n = 50) and (b) the number of food types used (r = 438, P < 0.01, n = 45) across African anthropoids (see Appendix 12.1 for categorization of habitat and food types) (reproduced from Eeley & Foley, in press, *Biodiversity and Conservation*, with kind permission from Kluwer Academic Publishers).

This point has been made most frequently when considering the relationship between the number of habitat types a species occupies and its range size, and the direction of these relationships remains ambiguous. It should also be borne in mind that niche breadth may relate to (1) the range of conditions tolerated by individuals, (2) within population variation (e.g., differences between males and females), and (3) variation across populations within the species geographical range (Brown, 1995), the relative significance of which, in terms of species range size, remains to be elucidated among anthropoid primates.

(a)

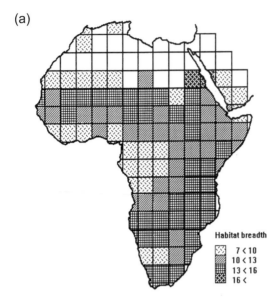

Habitat breadth
- 7 < 10
- 10 < 13
- 13 < 16
- 16 <

(b)

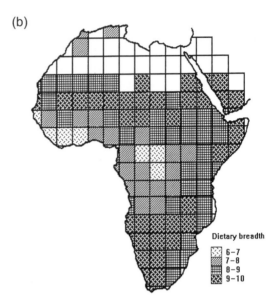

Dietary breadth
- 6−7
- 7−8
- 8−9
- 9−10

Fig. 12.15. The geographical distribution of (a) average habitat niche breadth, and (b) average dietary niche breadth of anthropoid primates across Africa (cell size 250 000 km^2) (reproduced from Eeley & Foley, in press, *Biodiversity and* *Conservation*, with kind permission from Kluwer Academic Publishers). Species are on average more specialized in central and western Africa, in regions where species richness is high and range sizes on average are low (see Fig. 12.3).

Abundance, body size and range size

It is argued that those species which occupy a broad niche and are geographically widespread will also be able to maintain a higher abundance across their range than more specialized species; i.e., that the "Jack-of-all-trades is the master of all" (Brown, 1984, 1995; Brown & Maurer, 1987; Gaston & Lawton, 1988, 1990; Gaston, 1990, 1994; Lawton, 1993; Lawton *et al.*, 1994; also Hanski, 1982). As Brown (1995) points out, this goes against the common view of a trade-off between abundance and range size, based on the allometric constraints of resource utilization. A positive relationship between range size and abundance among closely related species (i.e., within genera or sub-families) was first recorded by Darwin (1859), and has been observed among several taxa (birds: Ricklefs & Cox, 1978; Lawton, 1993; insects: Gaston & Lawton, 1990; also Gaston, 1994; Brown, 1995; Gaston *et al.*, 1997). However, the strength of the abundance/range size relationship may be confounded both by sampling biases, the ranges of species which are locally rare (less abundant) tending to be underestimated, and how well the 'reference habitat' (within which local abundance is measured) represents the

geographical region of interest (Brown, 1984; Gaston, 1990, 1994; Gaston & Lawton, 1990; Lawton, 1993). Brown (1995) notes that the pattern is more complex where larger and less closely related groups are concerned. Both he and Gaston (1994) suggest that when range size is measured over the broad-scale and across larger taxa, the inter-specific relationship between range size and abundance may be better described as triangular than strictly linear. Maximum density increases with increasing range size and, while there are no species that occupy a small range and have a high density, widespread species may occur at either a high or low density (Gaston, 1994) (Fig. 12.16a).

African anthropoids conform to this general pattern; the relationship between species range size and abundance (measured as average population density) is roughly triangular (Fig. 12.17a). Widespread species occur at both high and low densities, while species which occupy relatively small ranges, such as *Cercopithecus erythrogaster* and *Macaca sylvanus*, occur at lower densities. There are three exceptions to this general pattern. The Tana River mangabey and red colobus monkeys (*Cercocebus galeritus* and *Procolobus ruformitratus*) occur at notably high densities for

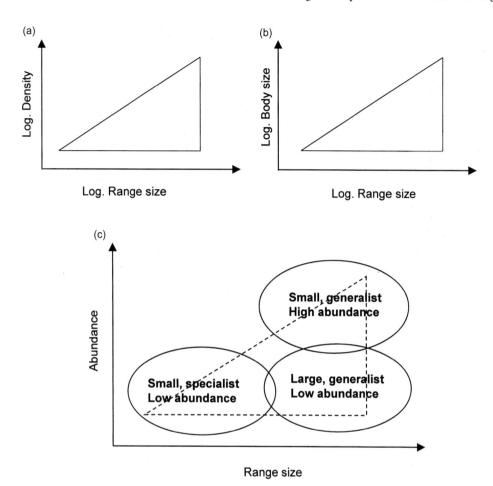

Fig. 12.16. The general, theoretical relationship between species range size and (a) abundance (density), (b) body size (after Gaston, 1994); (c) combines these two relationships with the general positive correlation between range size and niche breadth, to identify some expected species characteristics in relation to range size.

their range size, and the red colobus (*Procolobus badius*) is also unusual in occurring at exceptionally high densities in some regions such as the Kibale Forest (Struhsaker, 1975; Struhsaker & Leyland, 1987). However, only one species, *Cercopithecus erythrogaster*, falls into the lowest quartile for both species range size and abundance; an average of 3% over the group and a figure which is far less than that found among other taxa (between 14 and 25%) (Gaston, 1994). This may be due to a lack of information on the population density of a number of the most restricted-range species, and which were therefore excluded from this preliminary analysis.

Associated with the relationship between species range size and abundance is another, more complex, pattern, the relationship between body size and range size. Brown & Maurer (1987, 1989; Brown, 1995; also Damuth, 1981) argue that large species are constrained by their energetic requirements to have larger individual home ranges and therefore occur at lower population densities than smaller species. As a consequence, they reason, large species that occupy only a small geographical range will have a higher probability of extinction than those that are widespread. Both the influence of body size on dispersal ability and the tendency for smaller species to be more specialized serve to reinforce this pattern (Brown & Maurer, 1989; Gaston, 1990). Gaston & Lawton (1988) suggest that, given the

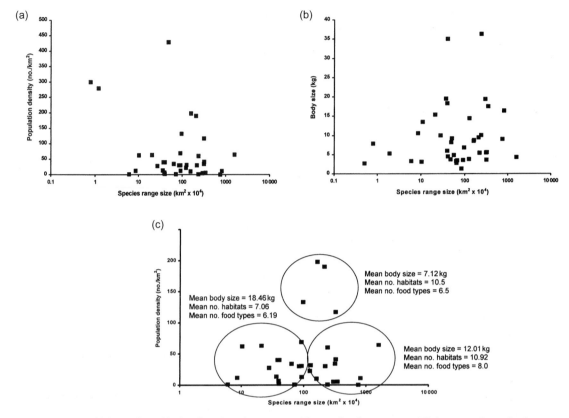

Fig. 12.17. Among African anthropoids there is a triangular relationship between species range size and both (a) abundance, measured as population density ($n = 34$), and (b) body size ($n = 40$). Species occupying small ranges have a low abundance, while widespread species exist at both high and low density (the three outliers are identified in the text). Small species occupy either small or large ranges, while larger species tend to be widespread. The first relationship is redrawn (c), with the three outliers removed, to show how species habitat and dietary niche breadth characteristics vary with these relationships, and the expected pattern identified in Fig. 12.16c (see text for discussion and Appendix 12.1 for categorization of habitat and food types).

general inverse relationship between body size and intrinsic rate of population increase, the opposite pattern might also occur and smaller species may occupy larger ranges. They argue that on colonizing a site, smaller species may attain a higher population density more rapidly than larger species and consequently would be less susceptible to stochastic extinction and would become widespread. However, supporting evidence for either pattern is limited (Glazier, 1980; Gaston & Lawton, 1988; Gaston, 1994; Lawton et al., 1994). Gaston (1994) suggests that the relationship between range size and body size, like that between range size and abundance, might be triangular: those species which occupy small ranges having small body size, while widespread species may be either small or large (Fig. 12.16b). Such a pattern has been described for example among New World birds (Blackburn & Gaston, 1996; see Peres & Janson, chapter 3, this volume). The African anthropoids again provide evidence of such a pattern. In general, larger species occur over large geographical areas, while small species may either be widespread or relatively range restricted (Fig. 12.17b).

Is it possible then to combine the relationship between range size and abundance, expected on the basis of niche breadth, with that between body size and range size, mediated through energetic constraint, dispersal ability or intrinsic rate of population increase? In general, small specialist species might be expected to occupy small ranges and occur at a lower abundance, small generalist species might be relatively abundant and occupy large ranges, while large species will be widespread and generalist but occur at a lower abundance (Fig. 12.16c). Large species occupying small ranges will rapidly become extinct. This is supported by Gaston's observation that evidence for a general abundance–body size relationship is mixed: while large species tend to be rare (in terms of their abundance), rare species do not fall into a particular size class and consequently small species may be either abundant or rare (Gaston, 1994). Again, such a pattern appears to occur among the African anthropoid primates (Fig. 12.17c), although in detail it is more complex and the degree of specialization is better shown in terms of habitat breadth than dietary breadth. Widespread species are, on average, more generalist than species occupying smaller ranges, while among wider-ranging species those which are generally more abundant such as *Cercocebus torquatus*, *Colobus angolensis* and *Cercopithecus ascanius*, are on average smaller than those of low abundance, such as *Papio anubis*, *P. cynocephalus* and *Pan troglodytes*. However, there is considerable variation within this pattern, and species occupying the small range size /low abundance category are, unexpectedly, on average the largest. Although this category does include relatively small species, such as *Cercopithecus erythrogaster*, *C. petaurista* and *Miopithecus talapoin*, average body size is biased by the presence of species such as *Macaca sylvanus*, *Theropithecus gelada* and *Gorilla gorilla*. These are either paleoendemic species (see below) or, as anticipated above, in danger of extinction.

Phylogenetic relationship and temporal dynamics

Range size, like body size, abundance, and niche breadth, is a species characteristic and a phylogenetic bias is therefore expected in all the relationships described above (Harvey & Pagel, 1991; Lawton, 1993; Gaston 1994, 1996). Among catarrhine primates in general, the maximum variance in species range size is attributable to the genus level of the taxonomic hierarchy (Eeley, 1994b). For the purposes of analysis, genera may therefore be considered relatively free of phylogenetic constraint (Harvey & Pagel, 1991), and any residual variance associated with other taxonomic levels may be ignored. Across genera comprising

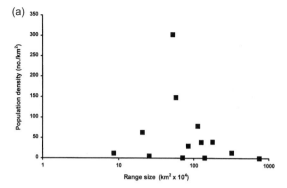

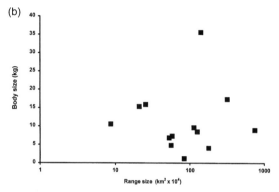

Fig. 12.18. Similar triangular relationships are observed between mean species range size and both (a) mean abundance (population density), and (b) mean body size across genera of African anthropoids (n = 13) as were observed across species (Fig. 12.17).

the African anthropoids the same roughly triangular relationships between range size, abundance and body size are observed as at the species level (Fig. 12.18). Similarly, at the genus level average species range size continues to increase with average habitat niche breadth (Fig. 12.19). These relationships are therefore real, but species' level characteristics are primarily determined by the genera to which they belong.

Gaston (1996) suggests that the observed lognormal frequency distribution of species range size represents a "slice across the temporal trajectories of the geographic ranges of the species in an assemblage". He also points out, however, that there is no commonly accepted model for the temporal change in species range size from speciation to extinction. Instead, two alternative views predominate; the first suggests considerable dynamism in species range

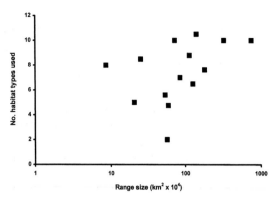

Fig. 12.19. Genus-average species range size increases with increasing average habitat niche breadth (see Appendix 12.1 for a categorization of habitat types), showing a similar pattern at the genus level to that observed across species of African anthropoids (Fig. 12.14a).

size over relatively long time scales (thousands to millions of years), while the second argues for relative stasis (Gaston, 1994). If range size is dynamic, and species respond to long-term variation in, for example, climate or the evolution and distribution of organisms with which they interact (including competitors, predators and parasites) then they would likely follow independent trajectories in terms of their range size. It would then be difficult to formulate any general model (Gaston, 1994).

On the other hand, there is evidence that species range size is relatively static over long time periods (Jablonski, 1987; Ricklefs & Latham, 1992; Branch, 1985; Brown *et al.*, 1996). Models based on such evidence predict that periods of increase and decrease in range size (at speciation and to extinction, respectively) are relatively brief (Lawton, 1993; Gaston, 1994). If this is the case range size may be correlated between related, extant species and between ancestor and descendant taxa (Gaston, 1994). This would account for the genus level fidelity of species-range size observed among African anthropoids. Reconciling the two views is difficult. However, the fluctuations in range size which are observed, for example, in response to climate change (Hengeveld, 1990) mostly occur within constraints set by higher-level phylogenetic characteristics, such as body size, life-history strategy, or habitat and dietary characteristics. As Ricklefs & Latham (1992) observe, "If adaptations influencing ecological and geographical extent are conservative, the distribution of sister

taxa in geographically isolated but climatically and ecologically similar regions should vary in parallel as they respond to shifts in global climate patterns." This would explain the similar overall patterns observed in Africa and South America as forest catarrhines and platyrrhines respond similarly to changing environmental conditions over the long term. Differences within the catarrhine primates, for example between the genera *Papio* and *Cercopithecus*, arise where taxa have come to occupy different climatic and ecological biomes.

The observable differences between genera may relate in part also to a temporal trajectory at this level. Species occupying relatively small ranges belong either to young, radiating genera (and therefore are neoendemic species) or to relatively old genera (and are paleoendemics, or relict species). Within the African anthropoids, the many *Cercopithecus* species characterize the former, while the gelada and the barbary macaque typify the latter. The dominance of African anthropoids by *Cercopithecus* species may thus shape the overall range size frequency distribution both because of the general ecological characteristics of this genus, and as a result of its relative youth. This contributes to an understanding of the differences observed in terms of species ranges size between areas characterized in general by low species-range size. Low range size is accompanied by high richness in areas inhabited by younger genera, for example central and west Africa, while areas inhabited by relict species tend to be marginal and are characterized both by low range size and low species richness, for example the Ethiopian highlands and the Atlas mountains.

SYNTHESIS

The distribution of anthropoid primates across the continents of Africa and South America shows a number of consistent, broad-scale, biogeographical patterns in terms of species richness and species range size. These patterns are examined in the light of recent macroecological theory, and the putative processes that shape the regional distribution and diversity of primate species are explored.

(1)   Anthropoid species richness is concentrated in equatorial regions and declines with increasing latitude.

(2)   Species range size, in contrast, appears to increase with latitude. Equatorial species generally occupy

small ranges, while species inhabiting higher latitudes are, on average, more widespread.

(3) The frequency distribution of species range size is highly skewed, with the majority of anthropoids occupying relatively small ranges and, in accordance with these patterns, there is a general negative relationship between species richness and average species range size.

(4) The distribution and species richness of anthropoid primates is best described in relation to the distribution of major vegetation types. The majority of anthropoids are forest dwelling and high species richness and generally small range size is associated with the forest biome. The relatively few species that inhabit woodland and savanna environments tend to be more widespread.

(5) Among anthropoid primates in Africa and South America, regional, historical processes play the primary role in shaping broad-scale patterns of diversity, and patterns of anthropoid species richness and species range size fundamentally reflect their evolutionary history within the forest biome.

(6) Refuge theory describes the process of evolution within a framework of forest contraction and expansion resulting from climatic change during the Pleistocene, and provides an explanation for the high species richness and endemism observed among forest primates today. Resulting patterns may be maintained and reinforced by major physical and biotic barriers that limit dispersal; most importantly changes in vegetation type and topographical, coastal and riverine barriers.

(7) Species-specific characteristics such as niche-breadth, abundance and body size combine to determine individual species range size. In general, species that have a broad niche-breadth are more widespread than are more specialized species, and maintain a greater abundance across their range. Large species also are generally widespread, although small species may occupy both small and large areas. Such characteristics govern how taxa are limited by distributional barriers, and how they respond to regional, evolutionary processes, shaping continent-wide patterns of distribution.

(8) The broad-scale patterns of diversity are strongly related to phylogenetic relationships within the anthropoid lineage, and general trends may be shaped by a few speciose taxa, such as the relatively young genus *Cercopithecus* in Africa. Notable exceptions to the general patterns are paleoendemic or relict species, such as the gelada, which tend to occupy small ranges in more marginal areas.

(9) The fact that the majority of primates occupy relatively small ranges and the general negative correlation between species richness and average range size has implications for conservation. For instance, these findings suggest that a relatively small environmental change may threaten a relatively large number of species (Pagel *et al.*, 1991). This is particularly true as regions of high species richness and high endemism are associated closely with the forest biome, a habitat under constant threat.

Like its subject, this review is broad-ranging, and has only touched the surface of a field that is itself rapidly evolving. Some general trends have been identified which contribute to an understanding of the broad-scale patterns of diversity among anthropoid primates, the interrelationships between species richness, species range size and ecological characteristics, and an appreciation of the regional processes that underlie these patterns.

ACKNOWLEDGMENTS

Many thanks to John Fleagle, Charles Janson, Kevin Gaston, Christophe Soligo and Guy Cowlishaw for their valuable comments on an earlier draft of this manuscript. Harriet Eeley was supported during this time by a postdoctoral fellowship from the Foundation for Research Development (FRD), South Africa, and Michael Lawes by an FRD research grant.

APPENDIX 12.1

The number of habitat types occupied and the number of food types used by 51 African anthropoid primates was determined on the basis of an extensive literature review (Eeley, 1994*b*). Twenty-four habitat categories of habitats and nine food types were identified.

Habitat types

1. Primary tropical forest, rainforest, mature lowland moist forest
2. Secondary forest, regenerating forest, disturbed forest
3. Gallery forest, riverine or riparian forest
4. Swamp forest, seasonally flooded forest
5. Montane forest, cloud forest
6. Bamboo forest
7. Semi-deciduous forest, mixed evergreen and deciduous forest
8. Deciduous forest
9. Coastal forest
10. Forest edge, forest pockets
11. Tropical woodland
12. Temperate woodland
13. Savanna woodland, deciduous woodland
14. Savanna grassland
15. Montane grassland
16. Lowland grassland
17. Marsh, reedbeds, flooded grassland
18. Semi-arid steppe, subdesert
19. Scrubland, bushland
20. Dry forest, thorn forest
21. Desert
22. Rocks and outcrops
23. Cultivated land, plantations, gardens
24. Urban areas, towns

Food types

1. Mature leaves
2. Young leaves, shoots, buds
3. Fruit
4. Seeds
5. Roots, bulbs, tubers
6. Gum, bark
7. Grass
8. Meat, eggs
9. Insects

REFERENCES

Ayres, J. M. & Clutton-Brock, T. H. (1992). River boundaries and species range size in Amazonian primates. *American Naturalist*, **40**, 531–7.

Balmford, A. (1996). Extinction filters and current resilience: the significance of past selection pressures for conservation biology. *TREE*, **11**, 193–6.

Blackburn, T. & Gaston, K. J. (1996). Spatial patterns in the geographic range sizes of bird species in the New World. *Philosophical Transactions of the Royal Society London, B*, **351**, 897–912.

Bourlière, F. (1985). Primate communities: their structure and role in tropical ecosystems. *International Journal of Primatology*, **6**, 1–26.

Branch, G. M. (1985). Competition: its role in ecology and evolution in intertidal communities. In *Species and Speciation*, ed. E. S. Vrba, pp. 97–104. Transvaal Museum Monograph No. 4. Pretoria: Transvaal Museum.

Brown, J. H. (1984). On the relationship between abundance and distribution of species. *American Naturalist*, **124**, 255–79.

Brown, J. H. (1988). Species diversity. In *Analytical Biogeography: An Integrated Approach to the Study of Animal and Plant Distributions*, ed. A. A. Myers & P. S. Giller, pp. 57–89. New York: Chapman & Hall.

Brown, J. H. (1995). *Macroecology*. Chicago: Chicago University Press.

Brown, J. H. & Gibson, A. C. (1983). *Biogeography*. St. Louis, Missouri: C. V. Mosby.

Brown, J. H. & Kodric-Brown, A. (1977). Turnover rates in insular biogeography, effect of immigration on extinction. *Ecology*, **58**, 445–9.

Brown, J. H. & Maurer, B. A. (1987). Evolution of species assemblages: effects of energetic constraints and species dynamics on the diversification of the North American avifauna. *American Naturalist*, **130**, 1–17.

Brown, J. H. & Maurer, B. A. (1989). Macroecology: the division of food and space among species on continents. *Science*, **243**, 1145–50.

Brown, J. H., Stevens, G. C. & Kaufman, D. M. (1996). The geographic range: size, shape, boundaries and internal structure. *Annual Review Ecology and Systematics*, **27**, 597–623.

Brown, K. S. Jr. (1987). Biogeography and evolution in Neotropical butterflies. In *Biogeography and Quaternary History in Tropical America*, ed. T. C. Whitmore & G. T. Prance, pp. 66–104. Oxford: Clarendon Press.

Caley, M. J. & Schluter, D. (1997). The relationship

between local and regional diversity. *Ecology*, **78**, 70–80.

Caughley, G., Grice, D., Barker, R. & Brown, B. (1988). The edge of the range. *Journal of Animal Ecology*, **57**, 771–85.

Colinvaux, P. A. (1987). Amazon diversity in light of the paleoecological record. *Quaternary Science Reviews*, **6**, 93–114.

Colinvaux, P. A. (1989). The past and future Amazon. *Scientific American*, **May**, 68–74.

Colwell, R. K. and Hurtt, G. C. (1994). Nonbiological gradients in species richness and a spurious Rapoport effect. *American Naturalist*, **144**, 570–95.

Colyn, M. M. (1988). Distribution of guenons in the Zaire-Lomani river system. In *A Primate Radiation: Evolutionary Biology of the African Guenons*, ed. A. Gautier-Hion, F. Bourlière, J.-P. Gautier & J. Kingdon, pp. 104–124. Cambridge: Cambridge University Press.

Colyn, M. M. (1991). L'Importance zoogéographique du bassin du fleuve Zaïre pour la spéciation: le cas des primates simiens. *Musee Royal de L'Afrique Centrale Tervuren, Belgique, Annales Sciences Zoologiques* Volume 264.

Colyn, M. M., Gautier-Hion, A. & Verheyen, W. (1991). A re-appraisal of palaeoenvironmental history in Central Africa: evidence for a major fluvial refuge in the Zaire Basin. *Journal of Biogeography*, **18**, 403–7.

Connell, J. H. (1978). Diversity in tropical rain forests and coral reefs. *Science*, **199**, 1302–10.

Connor, E. F. (1986). The role of Pleistocene forest refugia in the evolution and biogeography of tropical biotas. *TREE*, **1**, 165–9.

Cornell, H. V. (1993). Unsaturated patterns in species assemblages: the role of regional processes in setting local species richness. In *Species Diversity in Ecological Communities*, ed. R. E. Ricklefs & D. Schluter, pp 243–252. Chicago: Chicago University Press.

Cornell, H. V. & Lawton, J. H. (1992). Species richness, local and regional processes, and limits to the richness of ecological communities: a theoretical perspective. *Journal of Animal Ecology*, **61**, 1–12.

Cowlishaw, G. & Hacker, J. E. (1997). Distribution, diversity, and latitude in African primates. *American Naturalist*, **150**, 505–12.

Cox, C. B. & Moore, P. D. (1993). *Biogeography: An Ecological and Evolutionary Approach*, 5th edn. Oxford: Blackwell Scientific.

Cracraft, J. (1994). Species diversity, biogeography, and the evolution of biotas. *American Zoologist*, **34**, 33–47.

Cracraft, J. & Prum, R. O. (1988). Patterns and process of diversification: speciation and historical congruence in some neotropical birds. *Evolution*, **42**, 603–20.

Cresswell, J. E., Vidal-Martinez, V. M. & Crichton, N. J. (1995). The investigation of saturation in the species richness of communities: some comments on methodology. *Oikos*, **72**, 301–4.

Crowe, T. M. & Crowe, A. A. (1982). Patterns of distribution, diversity and endemism in Afrotropical birds. *Journal of Zoology, London, B.*, **198**, 417–42.

Currie, D. J. & Paquin, V. (1987). Large-scale biogeographical patterns of species-richness of trees. *Nature*, **329**, 326–7.

Cushman, J. H., Lawton, J. H. & Manly, B. F. J. (1993). Latitudinal patterns in European ant assemblages: variation in species richness and body size. *Oecologia*, **95**, 30–7.

Damuth, J. (1981). Population density and body size in mammals. *Nature*, **290**, 699–700.

Darwin, C. (1859). *On the Origin of Species*. London: John Murray.

Diamond, A. W. & Hamilton, A. C. (1980). The distribution of forest passerine birds and Quaternary climatic change in tropical Africa. *Journal of Zoology, London, B.* **191**, 379–402.

Dobzhansky, T. (1950). Evolution in the tropics. *American Scientist*, **38**, 209–21.

Duellman, W. E. (1982). Quaternary climate-ecological fluctuations in the lowland tropics: frogs and forests. In *Biological Diversification in the Tropics*, ed. G. T. Prance, pp. 389–402. New York: Columbia University Press.

Eeley, H. A. C. (1994*a*). A digital method for the analysis of primate range boundaries. In *Current Primatology. Selected Proceedings of the XIVth Congress of the International Primatological Society* Vol. 1., ed. B. Thierry, J. R. Anderson, J. J. Roeder & N. Herrenschmidt, pp. 123–32. Strasbourg.

Eeley, H. A. C. (1994*b*). *Ecological and Evolutionary Patterns of Primate Species Area*. Unpublished PhD. Thesis. University of Cambridge, Cambridge.

Eeley, H. A. C. & Foley, R. A. (In press). Species richness, species range size and ecological specialisation among African primates: geographical patterns and conservation implications. *Biodiversity and Conservation*.

Eisenberg, J. F. (1979). Habitat, economy and society: some correlations and hypotheses for the Neotropical primates. In *Primate Ecology and Human Origins*, ed. I. S. Bernstein & E. O. Smith, pp. 215–62. New York: Garland STPM Press.

Emmons, L. H. (1984). Geographic variation in densities and diversities of non-flying mammals in Amazonia. *Biotropica*, **16**, 210–22.

Endler, J. A. (1982). Pleistocene forest refuges: fact or fancy? In *Biological Diversity in the Tropics*, ed. G. T. Prance, pp. 641–57. New York: Columbia University Press.

Fischer, A. G. (1960). Latitudinal variations in organic diversity. *Evolution*, **14**, 64–81.

Fleagle, J. G. (1999). *Primate Adaptation and Evolution*. San Diego, California: Academic Press.

Fleagle, J. G. & Reed, K. E. (1996). Comparing primate communities: a multivariate approach. *Journal of Human Evolution*, **30**, 489–510.

France, R. (1992). The North American latitudinal gradient in species richness and geographical range of freshwater crayfish and amphipods. *American Naturalist*, **139**, 342–54.

Gaston, K. J. (1990). Patterns in the geographical ranges of species. *Biological Review*, **65**, 105–29.

Gaston, K. J. (1991). How large is a species' geographic range? *Oikos*, **61**, 434–8.

Gaston, K. J. (1994). *Rarity*. London: Chapman & Hall.

Gaston, K. J. (1996). Species-range-size distributions: patterns, mechanisms and implications. *TREE*, **11**, 197–201.

Gaston, K. J. & Lawton, J. H. (1988). Patterns in the distribution and abundance of insect populations. *Nature*, **311**, 709–12.

Gaston, K. J. & Lawton, J. H. (1990). Effects of scale and habitat on the ralationship between regional and local abundance. *Oikos*, **58**, 329–35.

Gaston, K. J. & Williams, P. H. (1996). Spatial patterns in taxonomic diversity. In *Biodiversity: A Biology of Numbers and Difference*, ed. K. J. Gaston, pp. 202–29. Oxford: Blackwell Science.

Gaston, K. J., Blackburn T. M. & Lawton, J. H. (1997). Interspecific abundance and range size relationships: an appraisal of mechanisms. *Journal of Animal Ecology*, **66**, 579–601.

Gaston, K. J., Blackburn, T. M. & Spicer, J. I. (1998). Rapoport's rule: time for an epitaph? *TREE*, **13**, 70–4.

Gentry, A. H. (1986). Endemism in tropical versus temperate plant communities. In *Conservation Biology*, ed. M. E. Soulé, pp. 153–81. Sunderland, Mass.: Sinauer Associates.

Glazier, D. S. (1980). Ecological shifts and the evolution of geographically restricted species of North American *Peromyscus* (mice). *Journal of Biogeography*, **7**, 63–83.

Groves, C. P. (1989). *A Theory of Human and Primate Evolution*. Oxford: Clarendon Press.

Grubb, P. (1982). Refuges and dispersal in the speciation of African forest mammals. In *Biological Diversity in the Tropics*, ed. G. T. Prance, pp. 537–53. New York: Columbia University Press.

Grubb, P. (1990). Primate geography in the Afro-tropical forest biome. In *Vertebrates in the Tropics*, ed. G. Peters & R. Hutterer, pp. 187–214. Bonn: Museum Alexander Koenig.

Hacker, J. E., Cowlishaw, G. & Williams, P. H. (1998). Patterns of African primate diversity and their evaluation for the selection of conservation areas. *Biological Conservation*, **84**, 251–62.

Haffer, J. (1969) Speciation in Amazonian forest birds. *Science*, **165**, 131–7.

Haffer, J. (1982). General aspects of the Refuge Theory. In *Biological Diversification in the Tropics*, ed. G. T. Prance, pp. 6–24. New York: Columbia University Press.

Haffer, J. (1987). Biogeography of Neotropical birds. In *Biogeography and Quarternary History in Tropical America*, ed. T. C. Whitmore & G. T. Prance, pp. 105–50. Oxford: Clarendon Press.

Hamilton, A. C. (1976). The significance of patterns of distribution shown by forest plants and animals in tropical Africa for the reconstruction of upper Pleistocene palaeoenvironments, a review. *Palaeoecology of Africa*, **9**, 63–97.

Hamilton, A. C. (1988). Guenon evolution and forest history. In *A Primate Radiation: Evolutionary Biology of the African Guenons*, ed. A. Gautier-Hion, F. Bourlière, J.-P. Gautier & J. Kingdon, pp. 13–34. Cambridge: Cambridge University Press.

Hamilton, A. C. (1989). African forests. In *Rain Forest Ecosystems: Biogeographical and Ecological Studies*, ed. H. Leith & M. J. A. Werger, pp. 155–82. Ecosystems of the World 14B. Vol. 2. Amsterdam: Elsevier.

Hamilton, A. C. & Taylor, D. (1991). History of climate and forests in tropical Africa during the last 8 million years. *Climatic Change*, **19**, 65–78.

Hanski, I. (1982). Dynamics of regional distribution: the core and satellite species hypothesis. *Oikos*, **38**, 210–21.

Harvey, P. H. & Pagel, M. D. (1991). *The Comparative Method in Evolutionary Ecology*. Oxford: Oxford University Press.

Hengeveld, R. (1990). *Dynamic Biogeography*. Cambridge: Cambridge University Press.

Hershkovitz, P. (1977). *Living New World Monkeys (Platyrrhini)*. Vol. 1. Chicago: Chicago University Press.

Hubbell, S. P. (1979). Tree dispersion, abundance and

diversity in a dry tropical forest. *Science, 203,* 1299–309.

Hubbell, S. P. (1980). Seed predation and coexistence of tree species in tropical forests. *Oikos,* **35**, 214–29.

Hughes, L., Cawsey, E. M. & Westoby, M. (1996). Geographic and climatic range sizes of Australian eucalypts and a test of Rapoport's rule. *Global Ecology and Biogeography Letters,* **5**, 128–42.

Huston, M. A. (1994). *Biological Diversity. The Coexistence of Species on Changing Landscapes.* Cambridge: Cambridge University Press.

Hutchinson, G. E. (1957). Concluding Remarks. *Cold Spring Harbour Symposium on Quaternay Biology,* **22**, 415–27.

Hutchinson, G. E. (1959). Homage to Santa Rosalia, or 'Why are there so many kinds of animals?' *American Naturalist,* **93**, 145–59.

Jablonski, D. (1987). Heritability at the species level: analysis of geographic ranges of Cretaceous mollusks. *Science,* **238**, 360–3.

Jackson, J. B. C. (1974). Biogeographic consequences of eurytopy and stenotopy among marine bivalves and their evolutionary significance. *American Naturalist,* **108**, 541–60.

Janzen, D. H. (1967). Why mountain passes are higher in the tropics. *American Naturalist,* **101**, 233–43.

Kingdon, J. (1990). *Island Africa. The Evolution of Africa's Rare Animals and Plants.* London: Collins.

Kingdon, J. (1997). *The Kingdon Field Guide to African Mammals.* San Diego: Academic Press.

Kinzey, W. G. (1982). Distribution of primates and forest refuges. In *Biological Diversity in the Tropics,* ed. G. T. Prance, pp. 455–82. New York: Columbia University Press.

Lawes, M. J. (1990). The distribution of the samango monkey (*Cercopithecus mitis erythrarchus* Peters, 1852 and *Cercopithecus mitis labiatus* I. Geoffroy, 1843) and forest history in southern Africa. *Journal of Biogeography,* **17**, 669–80.

Lawton, J. H. (1993). Range, population abundance and conservation. *TREE,* **8**, 409–13.

Lawton, J. H., Nee, S., Letcher, A. J. & Harvey, P. H. (1994). Animal distributions: patterns and processes. In *Large-scale Ecology and Conservation Biology,* ed. P. J. Edwards, R. M. May & N. R. Webb, pp. 7–12. Oxford: Blackwell Scientific.

Leakey, M. (1988). Fossil evidence for the evolution of the guenons. In *A Primate Radiation: Evolutionary Biology of the African Guenons,* ed. A. Gautier-Hion, F. Bourlière, J.-P. Gautier & J. Kingdon, pp. 54–78. Cambridge: Cambridge University Press.

Lernould, J.-M. (1988). Classification and geographical distribution of guenons: a review. In *A Primate Radiation: Evolutionary Biology of the African Guenons,* ed. A. Gautier-Hion, F. Bourlière, J.-P. Gautier & J. Kingdon, pp. 54–78. Cambridge: Cambridge University Press.

Letcher, A. J. & Harvey, P. H. (1994). Variation in geographical range size among mammals of the Palearctic. *American Naturalist,* **144**, 30–42.

Livingstone, D. A. (1982). Quaternary geography of Africa and the refuge theory. In *Biological Diversity in the Tropics,* ed. G. T. Prance, pp. 523–36. New York: Columbia University Press.

Lynch, J. D. (1988). Refugia. In *Analytical Biogeography: An Integrated Approach to the Study of Animal and Plant Distributions,* ed. A. A. Myers & P. S. Giller, pp. 311–42. London: Chapman & Hall.

MacArthur, R. H. (1965). Patterns of species diversity. *Biological Review,* **40**, 510–33.

MacArthur, R. H. (1972). *Geographical Ecology. Patterns in the Distribution of Species.* Princeton, New Jersey: Harper and Row.

MacArthur, R. H., Recher, H. & Cody, M. (1966). On the relation between habitat selection and species diversity. *American Naturalist,* **100**, 319–22.

Mayr, E. & O'Hara, R. J. (1986). The biogeographic evidence supporting the Pleistocene forest refuge hypothesis. *Evolution,* **40**, 55–67.

McCoy, E. D. & Connor, E. F. (1980). Latitudinal gradients in the species diversity of North American mammals. *Evolution,* **34**, 193–203.

McNaughton, S. J. and Wolf, L. L. (1970). Dominance and the niche in ecological systems. *Science,* **167**, 131–9.

Mittermeier, R. A. (1988). Primate diversity and the tropical forest: case studies from Brazil and Madagascar and the importance of the megadiversity countries. In *Biodiversity,* ed. E. O. Wilson & F. M. Peter, pp. 145–54. Washington, DC: National Academy.

Moreau, R. E. (1966). *The Bird Faunas of Africa and its Islands.* London: Academic Press.

Moreau, R. E. (1969). Climatic changes and the distribution of forest vertebrates in West Africa. *Journal of Zoology, London,* **158**, 39–61.

Moynihan, M. (1976). *The New World Primates: Adaptive Radiation and the Evolution of Social Behaviour, Languages, and Intelligence.* Princeton, N.J: Princeton University Press.

Myers, A. A. & Giller, P. S. (1988). *Analytical Biogeography: An Integrated Approach to the Study of Animal and Plant Distributions.* London: Chapman & Hall.

Napier, J. R. & Napier, P. H. (1985). *The Natural History of the Primates*. London: British Museum (Natural History).

Oates, J. F. (1988). The distribution of *Cercopithecus* monkeys in West African forests. In *A Primate Radiation: Evolutionary Biology of the African Guenons*, ed. A. Gautier-Hion, F. Bourlière, J.-P. Gautier & J. Kingdon, pp. 79–103. Cambridge: Cambridge University Press.

Oates, J. F. (1996). *African Primates*. Gland, Switzerland and Cambridge, UK: IUCN.

Oates, J. F., Davies, A. G. & Delson, E. (1994). The diversity of living colobines. In *Colobine Monkeys: Their Ecology, Behaviour and Evolution*, ed. A. G. Davies and J. F. Oates, pp. 45–73. Cambridge: Cambridge University Press.

Pagel, M. D., May, R. M. & Collie, A. R. (1991). Ecological aspects of the geographical distribution and diversity of mammalian species. *American Naturalist*, **137**, 791–815.

Pearson, D. L. (1982). Historical factors and bird species richness. In *Biological Diversity in the Tropics*, ed. G. T. Prance, pp. 441–452. New York: Columbia University Press.

Pianka, E. R. (1967). On lizard species diversity: North American flatland deserts. *Ecology*, **48**, 331–51.

Pianka, E. R. (1983). *Evolutionary Ecology*, 3rd edn. New York: Harper & Row.

Prance, G. T. (ed.) (1982). *Biological Diversity in the Tropics*. New York: Columbia University Press.

Prendergast, J. R., Quinn, R. M., Lawton, J. H., Eversham, B. C. & Gibbons, D. W. (1993). Rare species, the coincidence of diversity hotspots and conservation strategies. *Nature*, **365**, 335–7.

Price, T. D., Helbig, A. J. & Richman, A. D. (1997). Evolution of breeding distributions in the Old World leaf warblers (Genus *Phylloscopus*). *Evolution*, **51**, 552–61.

Rahm, U. H. (1970). Ecology, zoogeography and systematics of some African forest monkeys. In *Old World Monkeys*, ed. J. R. Napier & P. H. Napier, pp. 589–626. London: Academic Press.

Rapoport, E. H. (1982). *Areography. Geographical Strategies of Species*. Oxford: Pergamon Press.

Reed, K. E. & Fleagle, J. G. (1995). Geographic and climatic control of primate diversity. *Proceedings of the National Academy of Sciences USA*, **92**, 7874–6.

Rhode, K. (1996). Rapoport's Rule is a local phenomenon and cannot explain latitudinal gradients in species diversity. *Biodiversity Letters*, **3**, 10–13.

Rhode, K., Heap, M. & Heap, D. (1993). Rapoport's rule does not apply to marine teleosts and cannot explain latitudinal gradients in species richness. *American Naturalist*, **142**, 1–16.

Richard, A. F. (1985). *Primates in Nature*. New York: W. H. Freeman and Co.

Richards, P. W. (1996). *The Tropical Rain Forest: An Ecological Study*, 2nd edn. Cambridge: Cambridge University Press.

Ricklefs, R. E. (1987). Community diversity: relative roles of local and regional processes. *Science*, **235**, 167–171.

Ricklefs, R. E. (1989). Speciation and diversity: the integration of local and regional processes. In *Speciation and its Consequences*, ed. D. Otte & J. A. Endler, pp. 599–622. Sunderland, Mass.: Sinauer Associates.

Ricklefs, R. E. & Cox, G. W. (1978). Stage of taxon cycle, habitat distribution, and population density in the avifauna of the West Indies. *American Naturalist*, **112**, 875–95.

Ricklefs, R. E. & Latham, R. E. (1992). Intercontinental correlation of geographical ranges suggests stasis in ecological traits of relict genera of temperate perennial herbs. *American Naturalist*, **139**, 1305–21.

Ricklefs, R. E. & Schluter, D. (1993a). Species diversity: regional and historical influences. In *Species Diversity in Ecological Communities*, ed. R. E. Ricklefs & D. Schluter, pp. 350–63. Chicago: University of Chicago Press.

Ricklefs, R. E. & Schluter, D (eds.) (1993b). *Species Diversity in Ecological Communities*. Chicago: University of Chicago Press.

Rogers, W. A., Owen, C. F. & Homewood, K. M. (1982). Biogeography of East African forest mammals. *Journal of Biogeography*, **9**, 41–54.

Rosen, B. R. (1988). Biogeographic patterns: a perceptual overview. In *Analytical Biogeography*, ed. A. A. Myers & P. S. Giller, pp. 23–53. London: Chapman and Hall.

Rosenzweig, M. L. (1975). On continental steady states of species diversity. In *Ecology and Evolution of Communities*, ed. M. L. Cody & J. M. Diamond, pp. 121–40. Cambridge, Mass: Harvard University Press.

Rosenzweig, M. L. (1992). Species diversity gradients: we know more and less than we thought. *Journal of Mammalology*, **73**, 715–30.

Rosenzweig, M. L. (1995). *Species Diversity in Space and Time*. Cambridge: Cambridge University Press.

Roy, K., Jablonski, D. & Valentine, J. W. (1994). Eastern Pacific molluscan provinces and latitudinal diversity gradient: No evidence for "Rapoport's rule". *Proceedings of the National Academy of Sciences, USA*, **91**, 8871–4.

Ruggiero, A. (1994). Latitudinal correlates of the sizes of mammalian geographical ranges in South America. *Journal of Biogeography*, **21**, 545–58.

Schall, J. J. & Pianka, E. R. (1978). Geographical trends in numbers of species. *Science*, **201**, 679–86.

Schmida, A. & Wilson, M. V. (1985). Biological determinants of species diversity. *Journal of Biogeography*, **12**, 1–20.

Simpson, G. G. (1964). Species density of North American recent mammals. *Systematic Zoology*, **13**, 57–73.

Smith, F. D. M., May, R. M. & Harvey, P. H. (1994). Geographical ranges of Australian mammals. *Journal of Animal Ecology*, **63**, 441–50.

Stanley, S. M. (1979). *Macroevolution: Pattern and Process*. San Francisco: W. H. Freeman and Co.

Stevens, G. C. (1989). The latitudinal gradient in geographical range: how many species coexist in the tropics. *American Naturalist*, **133**, 240–56.

Stevens, G. C. (1992). The elevational gradient in altitudinal range: an extension of Rapoport's latitudinal rule to altitude. *American Naturalist*, **140**, 893–911.

Struhsaker, T. T. (1975). *The Red Colobus Monkey*. Chicago: University of Chicago Press.

Struhsaker, T. T. & Leyland, L. (1987). Colobines: infanticide by adult males. In *Primate Societies*, ed. B. B. Smuts, D. L. Cheney, R. M. Seyfarth, R. W. Wrangham & T. T. Struhsaker, pp. 83–97. Chicago: University of Chicago Press.

Terborgh, J. (1992). *Diversity and the Tropical Rainforest*. New York: Scientific American Library.

Terborgh, J. & Van Schaik, C. P. (1987). Convergence vs. nonconvergence in primate communities. In *Organisation of Communities*, ed. J. H. R. Gee & P. S. Giller, pp. 205–26. Oxford: Blackwell Scientific.

Turner, J. R. G. (1982). How do refuges produce biological diversity? Allopatry and parapatry, extinction and gene flow in mimetic butterflies. In *Biological Diversification in the Tropics*, ed. G. T. Prance, pp. 309–35. New York: Columbia University Press.

Van der Hammen, T. (1982). Paleoecology of Tropical South America. In *Biological Diversification in the Tropics*, ed. G. T. Prance, pp. 60–77. New York: Columbia University Press.

Van der Hammen, T. (1991). Palaeoecological background: neotropics. *Climate Change*, **19**, 37–47.

Vanzolini, P. E. (1973). Paleoclimates, relief, and species multiplication in equatorial forests. In *Tropical Forest Ecosystems in Africa and South America: A Comparative Review*, ed. B. J. Meggers, E. S. Ayensu & W. D. Duckworth, pp. 255–8. Washington DC: Smithsonian Institute Press.

Wallace, A. R. (1876). *The Geographical Distribution of Animals: With a Study of the Relations of Living and Extinct Faunas as Elucidating the Past Changes of the Earth's Surface*. London: Macmillan.

Wallace, A. R. (1878). *Tropical Nature and Other Essays*. London and New York: Macmillan.

Whitmore, T. C. & Prance, G. T. (1987). *Biogeography and Quaternary History in Tropical America*. Oxford: Clarendon Press.

Williamson, M. (1988). Relationship of species number to area, distance and other variables. In *Analytical Biogeography: An Integrated Approach to the Study of Animal and Plant Distributions*, ed. A. A. Myers & P. S. Giller, pp. 91–115. New York: Chapman & Hall.

Willig, M. R. & Selcer, K. W. (1989). Bat species density gradients in the New World: a statistical assessment. *Journal of Biogeography*, **16**, 189–195.

Willis, J. C. (1922). *Age and Area*. Cambridge: Cambridge University Press.

Wilson, J. W. (1974). Analytical zoogeography of North American mammals. *Evolution*, **28**, 124–40.

Wolfheim, J. H. (1983). *Primates of the World: Distribution, Abundance and Conservation*. Seattle and London: University of Washington Press.

13 • The recent evolutionary past of primate communities: Likely environmental impacts during the past three millennia

CAROLINE TUTIN AND LEE WHITE

INTRODUCTION

Primate communities in the tropical rainforests of Africa, South America, and south-east Asia vary within and between areas in the number of species that co-exist, the biomass of the community, and the population densities of the component species. Some of this variability relates to broad ecological factors such as floral diversity and composition (which defines potential diet), and climate (which defines seasonal patterns of availability of many foods), but some is not easily related to present-day conditions and historical factors are frequently invoked (Butynski, 1990; Oates *et al.*, 1990). 'Deep' evolutionary history defines the primate species assemblage of each continent and of geographically distinct regions within each continent (see Fleagle & Reed, chapter 6, this volume), but enigmas remain in understanding local patterns of distributions or differences in population densities. In this chapter we consider four kinds of historical event that may have had lasting impacts on primate communities. We restrict our considerations to the past three millennia, which represent approximately 200 to 1000 generations for most primate species.

The categories of events known, or likely, to have affected primate populations in the recent past are: climate change; human activities; outbreaks of disease; and natural disasters. In the following sections we consider whether or not evidence exists to suggest that some variations observed in present-day primate species and communities are a lasting result of these historical factors (see also Struhsaker, chapter 17, this volume). Technological development has intensified the impact of human activities on primates over the past 50 to 100 years. Specific impacts of this recent period are addressed in other chapters in this volume (see Peres, chapter 15 and Wright & Jernvall, chapter 18) and are not examined in detail here.

The primate community of the Lopé Reserve in central Gabon is used to illustrate many points because, at this site, multi-disciplinary research over the past 15 years has accumulated extensive data not only on the primate species of the present-day community, but also on vegetation history and archeology of the region.

CLIMATE CHANGE

Past climate change will have affected rainforest primates in two ways: firstly by changing the amount of habitat available; and secondly, by disrupting seasonal patterns of availability of fruit foods.

Forest cover

The extent of closed canopy forest in Africa has varied considerably in recent times (see Maley, 1996). In periods of forest retreat, fragmentation of large forest areas into blocks or refuges, surrounded by savanna dominated habitats, will have led to genetic isolation of populations of forest-dependent species.

Compared to the well-documented history of the climate of temperate zones, many details of the extent of change in past climates of tropical regions remain obscure. However, it is well established that the extent of tropical rainforest cover has fluctuated extensively in the past in the Neotropics, south-east Asia and in Africa (Whitmore & Prance, 1987; Maley, 1996; Houghton, 1995) .

To generalize for the recent past in equatorial Africa: forest cover expanded from a minimum 18 000 to 12 000 years before present (B.P.), reached a maximum 10 000 to

Table 13.1. *Chronology of climate history and human occupation of the Lopé area in central Gabon, over the past 18 000 years*

Years B.P.	Probable climate change	Impact on forest habitat	Key events in chronology of human occupation
18 000	Drier and cooler than present-day[a]	Forests shrink progressively ending as montane and riverine refuges in savanna landscape	Middle stone age culture, hunter-gatherers
12 000	Similar to present-day	Forests expansion starts	
5000		Forest cover maximal, scattered savanna refuges	Bantu migrations from north begin, bringing agriculture and pottery
2800	More seasonal, probably drier and warmer	Forests shrink, savanna expands	Bananas arrive. Large increase in human population
2000	Similar to present-day	Forests start to expand	Iron technology
1400		Rapid forest expansion	Dramatic population crash
800		Status quo, forest expansion blocked by regular savanna fires	Ancestors of present-day people arrive

Note: [a]Present-day climate: 1550 mm rainfall, 3-month dry season which is cloudy and the coolest time of year (see Tutin & Fernandez, 1993a)

5000 B.P., was reduced again, but to a lesser extent between 3000 to 2000 B.P., and is presently expanding (Livingstone, 1975; Maley, 1996, 1997) . The major vegetation changes in the Congo Basin can be related to shifts in weather patterns (see Table 13.1), which were in turn linked to changes in temperature of the waters off the Atlantic coast of west Africa (Maley, 1997). Between about 3700 to 3000 B.P., cooling of coastal waters resulted in climatic conditions similar to those experienced today, with a long, cool cloudy dry season between June and September. From about 2800 to 2000 B.P. sea temperatures were warmer and the cool dry season was replaced by sunny conditions. These would have subjected plants to increased physiological stress, especially as overall annual rainfall may have been reduced. Forests retreated in many parts of Africa during this period, and it was probably at this time that the Dahomey gap opened up.

Phenological patterns

Some present-day tree species respond to specific weather-related cues to synchronize their reproduction. For example, in Gabon the low temperatures of the cool dry season (July and August are the coldest months of the year)

trigger flowering in a group of tree species (Tutin & Fernandez, 1993a). In years when dry season temperatures remain unusually warm a suite of species fail to flower, and hence to set fruit, resulting in periods of fruit scarcity of up to eight consecutive months. While not all plant species require external cues to trigger synchronized flowering, weather patterns influence plant reproduction in general ways leading to marked seasonal patterns of fruit availability in most primate habitats. Relatively minor changes in climate could have led to extended periods of food scarcity for frugivores by disrupting fruiting cycles (Tutin & White, 1998). Unusually warm dry seasons at Lopé are particularly disruptive of fruiting patterns today. Similar occasional deviations from normal climate can affect fruit production in other tropical areas, for example, on Barro Colorado Island where if rain falls during the dry season fruit crop failures cause famine that result in mortality of frugivores (Foster, 1982a). Hence, past periods of climate change (such as 2800 to 2000 years ago in the Congo Basin when the months from June to September were warm and sunny) are likely to have led to immediate disruption of flowering and fruiting patterns.

Impact of past climate change on primates

The extent of the short and long-term impact of climate change on primate species will depend on factors such as their mobility and dietary flexibility. Since these vary between species, community structure will have changed and competitive interactions between species will have altered when fruiting patterns were disrupted, as a result of climate change, and when forest gave way to savanna. Primate communities of present-day rainforests are more species rich than those in habitats dominated by woodland or by savanna but comparisons of biomass are hindered by a lack of data from savanna habitats.

Few primate fossils from the recent past have been found as soils in the tropics are acidic. The present-day distributions and phylogenetic relationships of species and subspecies within diverse groups such as cercopithecines, colobines and callitrichines, reflect major forest fragmentations of the past that allowed speciation to occur during periods of reproductive isolation. Most of these events pre-date the period under consideration here, but the reduction in forest habitat in west-central Africa 3000 to 2000 B.P., will have exerted selective pressures different to those operating today. In the absence of fossils, a comparison of how species within present-day primate communities with access to both forest and savanna make use of these contrasting habitats opens a "window" on past environments.

At Lopé, in central Gabon, the present-day landscape is dominated by tropical rain forest, but islands of ancient savanna-dominated habitat persist (Oslisly et al., 1996). The vegetation mosaic at Lopé has been shaped by past climate change, but fires of human origin have, in the past, as now, played a role in conserving savanna habitat during climatic phases when forest vegetation was expanding (see below, "Human Activities" and summary in Table 13.1). Two broad types of continuous forest are recognized which differ in structure and plant species composition: Marantaceae Forest (representing the most recent period of forest expansion) and Closed Canopy Forest (believed to have persisted during the climatic change of 3000 to 2000 B.P.). White & Tutin (in press) and White & Abernethy (1997) discuss details of vegetation history and dynamics at Lopé. Within the savannas there are small areas of forest: gallery forests that grow along permanent watercourses and isolated blocks of forest, or 'bosquets'.

Table 13.2. *Comparison of biomass (kg/km²) in forest and savanna ecotones in the Lopé Reserve, Gabon*

Species	Marantaceae forest[a]	Forest fragments[b]
Gorilla g. gorilla	66.4	37.4
Pan t. troglodytes	21.3	38.1
Mandrillus sphinx	11.4	221.0
Colobus satanas	107.9	65.0
Lophocebus albigena	43.1	33.9
Cercopithecus nictitans	71.2	135.3
Cercopithecus pogonias	12.4	2.1
Cercopithecus cephus	10.6	61.9
Total primates	344.3	594.7
Other large and medium-sized diurnal mammals	4389.8	5418.2
Total biomass	4734.1	6012.9

Note: [a]Sites 1 and 5; White, 1994
[b]Tutin et al., 1997a

The present-day savanna habitat at Lopé may not be identical to that of the past, but the persistence of some gallery or riverine forests during periods when continuous forest cover shrank seems likely and is supported by botanical and zoological data from other parts of tropical Africa (Colyn et al., 1991; Rietkerk et al., 1995). Thus the present-day savannas probably resemble the habitat that was much more extensive in the recent past at Lopé (see Fig. 13.1).

The primate community at Lopé includes eight diurnal species: *Gorilla g. gorilla, Pan t. troglodytes, Mandrillus sphinx, Colobus satanas, Lophocebus albigena, Cercopithecus nictitans, C. pogonias* and *C. cephus*. All are considered as "forest" species and no savanna specialists occur. Primate biomass was measured in each of the two major forest types at Lopé from repeated 5 km transect censuses (White, 1994a); and in forest fragments within the savanna zone by sweep censuses (Tutin et al., 1997a). Table 13.2 compares primate biomass in two contrasting habitats: forest fragments within the savannas and one of the major forest types, Marantaceae Forest, which surrounds the present-day Lopé savannas. Of the eight species, four were found to live at greater densities in continuous forest than in fragments and vice versa. Total biomass was greater in

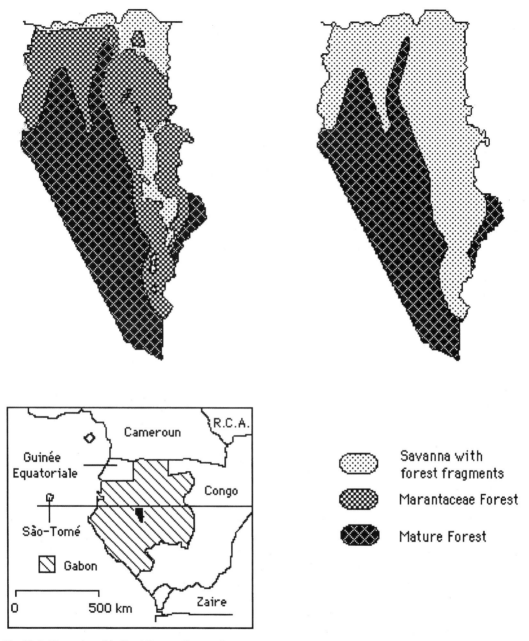

Fig. 13. 1. Vegetation of the Lopé Reserve: Present-day
distribution of major habitats, on the left; likely distribution of
habitats 1400 years ago, on the right.

the fragments than in continuous forest but much of this difference came from the very high figure for mandrill biomass in forest fragments. The four species that adapt well to the fragmented habitat fall into two distinct categories of "users": residents and transients. Groups of two species of guenon (*C. nictitans* and *cephus*) are resident within certain fragments where they occupy home ranges 8 to 25 times smaller than those of conspecific groups in the neighboring continuous forest. In contrast, chimpanzees and mandrills are transient visitors and move across open savanna to use the fragmented habitat intensively either on a regular seasonal basis (mandrills), or at times when fruit is abundant (chimpanzees) (Rogers *et al.*, 1996; Tutin *et al.*, 1997a). Gorillas, mangabeys and colobus are also transients, but visit fragments less frequently. *Cercopithecus pogonias*, which resembles the other two guenons in body size, diet and social organization, almost never visits the fragmented habitat for reasons that remain obscure.

It is clear from these data that the recent evolutionary history of the eight members of the present-day primate community at Lopé will have differed. As recently as 1400 years ago, the savanna dominated habitat was much more extensive than it is today and forest habitat was proportionally reduced (see Figure 13.1). Distances between continuous forest and most of the forest fragments of this period will have been great due to the extended area of savanna dominated habitat. Two species of guenon, *C. nictitans* and *C. cephus*, are likely to have persisted in small home ranges within gallery forests and in bosquets as small as about 5 hectares. Our data suggest that chimpanzees and mandrills are also likely to have adapted to life in the fragmented habitat, but the extremely large home ranges (120 to 560 km^2) of chimpanzees in savanna dominated habitats in drier parts of Africa (Suzuki, 1969; Izawa, 1970; Kano, 1972; Baldwin *et al.*, 1982), suggest that they would have required large areas of habitat. In contrast, gorillas, mangabeys, black colobus and crowned guenons, the species that make only rare visits to present-day fragments, are likely to have been restricted to the shrunken area of continuous forest.

At Kibale Forest, Uganda, as at Lopé, not all members of the forest primate community use adjacent fragmented habitat (Onderdonk, 1998). Some parallels with the Lopé study exist but differences between the two sites also emerged. Table 13.3 summarizes probable differences in community structure in fragmented and continuous forest

habitats at Lopé and Kibale in terms of species' ability to persist in fragmented forest. This comparison shows that both within and between sites, patterns of use by primates of forest fragments do not correlate consistently with traits such as diet, body size, group size, or home-range size.

The structural and botanical differences between forest fragments embedded in savanna and continuous forest are great and have major implications for primates in terms of food availability and vulnerability to predators. For all species vulnerability to predators is high when crossing savannas as all depend on trees for evasion of terrestrial predators (leopard, golden cat and humans) and they are more exposed to birds of prey. For species able to reside within a single fragment (*C. nictitans* & *cephus* at Lopé; *Colobus guereza* at Kibale) predator pressure may not differ significantly from that experienced in continuous forest, but behavioral observations suggest that primates feel vulnerable when they move between fragments, or between continuous forest and fragments. All primates at Lopé show extreme vigilance when crossing savannas and select the shortest routes, which they cross rapidly in tight formation often "supervized" by adult males (SEGC unpublished data).

At Lopé, botanical species composition differs between fragments and continuous forest (Tutin *et al.*, 1997a). Not only does the limited area of the former mean that diversity is less, but also, some plant species are restricted to one or other of the habitats (White & Abernethy, 1997). All eight primate species at Lopé have diets that are dominated by fruit (seven species) or seeds (one species) and all have diverse diets including 46 to 220 different foods (Tutin *et al.*, 1997b). Seasonal variation in diets is great and availability of fruit and seeds varies both within and between years (Tutin *et al.*, 1991; White, 1994b). The reduced diversity and number of plants in fragments means that resident groups of monkeys will be exposed more often to periods of food shortage, than those in continuous forest. More data are required to determine the exact habitat requirements of primate species and, given the increasing speed of forest fragmentation through human action, such research is a conservation priority.

Overall, less food is available in fragments than in continuous forest, but there are illuminating exceptions. Preferred fruit from rare species, or that of fig trees that fruit asynchronously, is occasionally abundant in fragments and

Table 13.3. *Summary of patterns of use of fragmented forest habitat by members of continuous forest primate communities at Lopé and Kibale*

Species	Lopé	Kibale
Gorilla gorilla	Rare transient	X
Pan troglodytes	Frequent transient	Frequent transient
Mandrillus sphinx	Frequent transient	X
Papio anubis[a]	X	Frequent transient
Colobus satanas	Rare transient	X
C. guereza	X	Resident
C. badius	X	Rare resident
Lophocebus albigena	Rare transient	Absent
Cercopithecus nictitans	Resident	X
C. mitis	X	Absent
C. pogonias	Absent	X
C. l'hoesti	X	Absent
C. cephus	Resident	X
C. ascanius	X	Frequent transient
C. aethiops[b]	X	Frequent transient
No. species in community:		
Continuous forest	8	7
Fragmented forest	4	6

Note: [a]Species unlikely to persist in continuous forest without access to savanna;
Rare transient, makes occasional visits from adjacent continuous forest; Frequent transient, makes frequent visits to fragments taken to indicate ability to persist in fragmented habitat but not within a single fragment; Resident, species able to live in a single fragment; Absent, species not recorded in fragments; X, species not present at site. See text for details.
Source: Based on Tutin *et al.*, (1997*a*) for Lopé and Onderdonk (1998) for Kibale.

may attract transient visitors, particularly mangabeys, chimpanzees and gorillas that spend hours, or days, within a small forest fragment. At such times resident groups are faced with intense competition from the larger species. In contrast, fruit crops of common species sometimes persist for weeks or even months longer in fragments than in continuous forest, allowing resident monkeys to benefit from the reduced levels of feeding competition (Tutin, 1999). Thus, defining habitat requirements of frugivores is not easy as both crop size and levels of competition vary in ways that are difficult to predict. When preferred foods are scarce, alternative foods are eaten (Tutin *et al.*, 1997*b*), but the ability to move out of the fragment is probably essential in cases of extreme resource depletion.

In summary, patterns of primate residency in fragments appear to be limited by food availability (determined by

plant species diversity that in turn is influenced by fragment size). At Lopé, fragments of >5 ha can support a resident group of *C. cephus* or *C. nictitans* while at Kibale, groups of black and white colobus can live in even smaller fragments, probably because of their folivorous diet. Other primate species do not reside in fragments, presumably because there is insufficient food to sustain them year-round. Increased vulnerability to predators while crossing savannas seems likely to be a major factor in determining patterns of transient visits to fragments. (Physiological costs may also be imposed by exposure to the high temperature of unshaded savannas.) Costs are impossible to quantify as data on natural predation of primates are scarce but if the cost is equally high for all species, the observed differences in visiting frequency may be related to the benefits which, in turn, are defined by dietary needs and food preferences. These vary between species and also within species over time, tracking community-wide changes in fruit and seed availability. For chimpanzees, who appear to have the strongest dependency, or preference, for fruit over other foods among the Lopé primates (Tutin et al., 1997b), the benefits of exploiting scattered ephemeral crops in fragments may outweigh increased predation risk incurred while crossing savanna. Other factors such as body size, group size, and social structure enter into the cost-benefit equation. For individual gorillas, predation risk may be reduced by the defensive behavior of silverback males, but benefits may also be reduced by high levels of intraspecific feeding competition if the amount of fruit in the fragment is limited.

Mandrills are an interesting case as they use the fragmented habitat at Lopé on a seasonal basis with visits concentrated in the long dry season when fruit is generally scarce. In this habitat mandrills travel in hordes frequently numbering 700 (and up to 1300 individuals in exceptional circumstances). They alternate rapid travel with periods of temporary residence of 1 to 3 days within selected fragments. Mandrills remain an enigmatic species as their nomadic exploitation of vast home ranges makes them impossible subjects of classic primate field study (Lahm, 1986). Information obtained opportunistically over 13 years at Lopé shows them to have an eclectic diet with a preference for fruit and seeds (Rogers et al., 1996). It is tempting to speculate that their unusual form of social organization is primarily adapted to a fragmented habitat and evolved in past times when savannas were extensive in west-central Africa. Individuals within the huge hordes sweeping into fragments and across savanna gaps, may gain sufficient advantages from predator protection and in inter-specific feeding competition to outweigh the high costs of feeding competition imposed by the size of the group. Similarly, the small size and cryptic behavior of the smallest guenon at Lopé, *Cercopithecus cephus* could have evolved as adaptations to fragmented habitats where edge-to-area ratio is high.

Although evidence is circumstantial there are indications that some aspects of morphology and behavior of the four species that use the present-day fragmented habitat at Lopé regularly, may have evolved in past times when these habitats were much more extensive. However, the lack of consistent correlations between species-specific factors and the ability of forest primate species to persist in fragmented habitats (as shown in Table 13.3) warns against generalizations.

While the comparison of present-day patterns of use of the two contrasting ecotones at Lopé is instructive, it does not exactly mirror past events. Forest shrinkage due to changing climate (such as occurred 3000 B.P., see Maley, 1997) is likely to have involved an initial period when many trees stopped flowering (Tutin & Fernandez, 1993a), reducing the availability of fruit. Trees would have then progressively disappeared, either dying as a result of physiological stress or vanishing gradually because seeds were no longer produced. Faced with an environment no longer providing sufficient food, primates would have had three options: to move to a viable habitat; to adapt to a different diet; or to die. Paleontological studies show that species did move large distances to "track" suitable habitat during periods of major climatic change in Europe (Butzer, 1972). Movement seems the most probable response of animals such as the primate species at Lopé, although some degree of dietary adaptation is also likely and some deaths inevitable. In a scenario of climate induced vegetation change, the likelihood of finding suitable areas of empty habitat are low and there is little doubt that these periods led to intensified selection as competition for food increased.

HUMAN ACTIVITIES

Human impacts likely to have exerted selective pressures on primates in their recent evolutionary past include competition for food, hunting and modification by humans of the shared habitat. Below, we distinguish between indirect impacts of human action on primates (i.e. habitat modification) and direct impacts (competition and hunting). We again use the Lopé area as a case study as archeological and primatological research have been conducted for more than a decade at the same site. Some aspects of past scenarios can be reconstructed from data on interactions between present-day sympatric primates, and clues of past patterns of behavior of humans come from archeological and ethnological research (Vansina, 1990; Hladik et al., 1993; Shaw et al., 1993; Hather, 1994; Stahl, 1995a; Sutton, 1996) and, while largely speculative, a few generalizations emerge.

The history of human use of tropical rainforest habitats is poorly known when compared to other vegetation zones as archeologists are faced with enormous methodological problems in what, for them is a 'hostile' working environment (Zeidler, 1995): locating sites is difficult, as are investigating, dating, and interpreting finds. Great progress has been made recently in developing novel techniques (e.g., recovery and identification of microscopic remains of plants and animals and identification of charcoal to species-specific level) which add to information provided by shards of pottery and vestiges of iron smelting. A multidisciplinary approach, integrating linguistic, ethnological and palynological data with the artifacts left by the humans of the past, has allowed reconstruction of patterns of human occupation of some parts of the rainforests of west-central Africa and of the neotropics during the past five millennia (see papers in Shaw, 1993; Stahl, 1995a; Sutton, 1996). Much debate remains about the timing of crucial events such as the emergence and development of agriculture and the invention of iron-smelting technology, but it is abundantly clear that humans were living in rainforests throughout the period considered in this chapter.

At Lopé, in central Gabon, the primate community of eight diurnal species (see Table 13.2) has co-existed with humans for a very long time (Tutin & Oslisly, 1995). Archeological research has documented a long history of human occupation dating back at least 350 000 years (Oslisly & Peyrot, 1992). Artifacts and radiocarbon dating give a picture of the past three millennia. Occupation of central Gabon was not constant and at least three cultural groups arrived successively from the north bringing increasingly sophisticated technologies (Oslisly, 1996). A human population crash occurred at about 1400 B.P. and the Lopé area (and indeed much of the interior of central-west Africa) was sparsely occupied by humans until the arrival of the ancestors of present-day groups around 800 B.P. (de Maret, 1996; Oslisly, in press). Key events in this chronology are summarized in Table 13.1. Below, we examine likely direct and indirect impacts of humans on the primate community using information from Lopé and, where possible, from other sites and continents.

Direct impacts of humans on primates

Competition

The general scenario of human use of the Congo Basin forest block of west-central Africa 5000 years ago is believed to have been occupation by scattered groups of hunter-gatherers within the forest and of fishermen-gatherers along the coast and larger rivers (Vansina, 1990; Eggert, 1996), although some argue, based on present-day lifestyles of hunter-gatherers, that humans could not have survived in these forests prior to agriculture (Hart & Hart, 1986; Bailey et al., 1989). Starting at about 5000 B.P. and continuing until 1500 B.P. the "Bantu migrations" brought waves of immigrants into this huge rainforest block from the savanna dominated habitats further north. These people brought pottery into the Congo Basin and probably had already begun to farm, clearing fields for the cultivation of yams and sowing seeds of useful plants at their settlement sites. The impulsion for some, or all, of the migrations may have been related to climate-induced changes in vegetation that simultaneously made life more difficult in the northern savanna zone and easier in the forests (de Maret et al., 1987; Schwartz, 1992; de Maret, 1996).

Examination of numerous human settlement sites dating from the past three millennia show that at Lopé, many were situated on hill tops in the savanna ecotone (see Oslisly, 1993 for details of site names and dating), although the existence of human settlement sites within forests has recently been established (Oslisly & White, 1996). It is difficult to recreate the diet of humans living at Lopé in past times, but they undoubtedly depended heavily on wild plant foods. Large sweet fruit, nuts, tubers and large seeds are all likely to have been important foods even for farmers. Plant species diversity is much greater in forest than in savanna and it seems likely that many of the important wild plant foods of humans came then, as now, from the forest. Archeological evidence shows that large nuts and succulent fruit such as *Canarium schweinfurthii*, were carried by humans from the forest to savanna settlement sites. Nuts can be stored for many months and the kernels of species such as *Coula edulis*, *Panda oleosa* and *Elaeis guineensis* are rich in fat and protein. Stone anvils, with characteristic depressions resulting from repeated use, are associated with many of the human settlement sites at Lopé. The kernels of these large nuts are major foods of chimpanzees in West Africa who use wooden clubs and stone tools to crack the hard shells (Sugiyama & Koman, 1979; Boesch & Boesch, 1990). Chimpanzee populations in the rest of Africa have not acquired this technology (McGrew, 1992), despite those at Lopé having access both to a large number of potential tools (McGrew et al., 1997) and long exposure to nut-cracking humans.

Competition with other species for fruit and nuts is likely to have been intense, principally because humans were gathering rather than eating *in situ* and would have wished to collect large quantities in order to share and/or store part of the harvest. The greater intelligence of humans would have allowed analysis of competition and the formation of resource-defense strategies. For large crops of desirable fruit or nuts, it seems likely that humans would have invented ways of reducing feeding competition. Barriers, fire and noise may have been sufficient to deter primates, but pit-traps and snares may have been used to reinforce defenses. For omnivorous humans, a concentration of desirable fruit was not only a resource, but also "bait" that could be exploited to attract "meat." Thus forest-dwelling humans may have evolved trapping skills to simultaneously protect shared plant foods and acquire animal foods: the ultimate in exclusion-competition is "to have your cake and eat it too."

Hunting

Apart from the regular peaceful polyspecific associations formed by the arboreal monkeys, the usual pattern of interspecific interaction among primates is mutual avoidance. Aggressive interactions between the sympatric species at Lopé are limited to occasional predation by chimpanzees on arboreal monkeys (Tutin & Fernandez, 1993*b*). Humans are more likely to have initiated aggressive interactions in disputes over shared plant foods or, like chimpanzees, by actively hunting primates for meat. The large terrestrial species will have been major competitors of humans for fruit and will also have been more susceptible to harassment. Gorillas and to a lesser extent chimpanzees and mandrills are capable of intimidating displays and the apes will occasionally attack when threatened by humans (Sabater Pi, 1966), making them formidable opponents. Really effective weapons (spears and arrows with metal tips) appeared in the human arsenal at Lopé around 2500 B.P., but stone tools in combination with pit-traps and snares would have allowed predation from an early date, as suggested by the discovery of old gorilla and chimpanzee bones associated with human artifacts dated to 5000 B.P. in a cave in Cameroun (de Maret *et al.*, 1987).

Hunting arboreal species would have been more difficult prior to the arrival of firearms. Stahl (1995*b*) described the varied challenges faced by past human hunters in trop-

ical forests and showed that specific hunting technologies evolved for different groups of mammals in the Neotropics. As a hunter could not carry his whole arsenal of weapons, decisions about prey choice were made at the start of a hunt and are likely to have been guided by local conditions of relative abundance, ease of capture, prey behavior and taste preferences of the hunter. Primates are often the most numerous non-flying mammalian occupants of tropical forest ecosystems, but the majority are arboreal and relatively small. In the Neotropics, the absence of terrestrial primates perhaps contributed to the development of varied and effective methods of hunting arboreal species prior to firearms (Mittermeier, 1987). In the Ituri Forest, Zaire, nets are the principal method of hunting and even bow hunters preferentially target terrestrial species, devoting little time to hunting arboreal primates because of lower success rates (Wilkie & Curran, 1991). Throughout the Congo Basin, terrestrial primates are likely to have suffered higher levels of human predation, particularly after the introduction of dogs some 1000 years ago (Clutton-Brock, 1996). In present-day east Africa, Madagascar and parts of Asia primates are seldom hunted because of cultural taboos (Mittermeier, 1987), but the antiquity of such traditions is unknown. Hunting probably contributed to the extinction of some of the species of giant lemur following the arrival of humans in Madagascar some 1500–2000 years ago (Sussman *et al.*, 1985).

The exact impact of past hunting of primates by humans is difficult to assess. It seems likely that terrestrial species were more vulnerable than arboreal ones but, in the Neotropics, humans did devise effective ways to hunt arboreal species. Apart from the case of Madagascar, extermination of primate populations by hunting seems unlikely to have occurred prior to the arrival of guns. Some aspects of behavior of terrestrial African primates may have been influenced by their long history of being harassed and hunted by humans. Possible examples include the intimidation displays of males and the engrained suspicion of humans that makes habituation of the apes of the Congo Basin so difficult. The large group size of mandrills and the protective behavior shown by the large males may have been advantageous in encounters with humans in past times, but it should not be forgotten that large carnivores are also largely terrestrial and will have exerted similar selective pressures.

Indirect impacts of humans on primates

Humans more than any other species have a capacity to modify their environment. While hunting has undoubtedly affected primates over the past millennia, the impact of humans on the vegetation of the shared habitats has had far-reaching indirect effects on primates. Fires and forest-clearance for agriculture, alone, or in combination have left indelible marks on tropical landscapes. In addition, for a very long time humans have moved themselves and other species around the world. The domestication of certain plant species and the subsequent spread of major crops into different regions and continents were factors with enduring impacts on tropical forests (see below). Humans also moved some primate species long distances resulting in the establishment of exotic populations on certain islands (e.g., *Cercopithecus aethiops* in Barbados, *C. mona* on Sao Tomé and *Macaca fasicularis* on Mauritius).

Fire

Fire is known to have played a role in the dynamics of all tropical forest areas of the world in past times as witnessed by layers of charcoal found in South America (Sanford *et al.*, 1985), Africa (Swaine, 1992; Hart *et al.*, 1996), south-east Asia (Maloney, 1994) and Australia (Kershaw, 1992). While some natural fires may have occurred from lightning strikes, fires of human origin were much more common and widespread. Fire does not spread in present-day undisturbed rain forests (Tutin *et al.*, 1996), but may do so in times of drought, particularly when forest structure has been modified by human activities (Mackie, 1984). Large-scale fires will kill less mobile primate species and have a long-lasting indirect impact by destroying food plants.

Farming

Forest clearance for agriculture has had a profound impact on the tropical rainforest habitats of primates over the whole period covered in this chapter. Recent research has shown widespread and large-scale modification of tropical forests in South America, south-east Asia and Africa. In Africa, a picture of significant human impact on forests is emerging wherever palynologists, ecologists and archeologists team up. At Lopé, in central Gabon, numerous neolithic and iron age archeological sites have been located

deep in the forest, in association with extensive charcoal deposits, suggesting that much of the forest of this area has been altered by human activities (Oslisly & White, 1996). In Okomu, in south-western Nigeria, Jones (1956) documented extensive remains of pottery and charcoal (mostly oil palm kernels) in a layer starting at 25 cm below the modern soil surface, in what were considered some of the finest pristine mahogany forests in the British Empire (Richards, 1996). This layer has recently been dated to 700 to 750 B.P. suggesting that the forests of Okomu are in fact an old secondary formation (L. J. T. White & J. F. Oates, unpublished data). In northern Congo, Fay (1997) reports finding large numbers of oil palm nuts in forests where *Elaeis guineensis* is currently restricted to human habitations. Carbon dates of 100 palm nuts fell between 2400 to 900 B.P., peaking between 1800 to 1700 years B.P. and then declining sharply. Given the tens of thousands of palm nuts and the extent of their distribution, Fay concludes that much of northern Congo, south-east Cameroun and south-west Central African Republic was cultivated up to about 1600 years B.P. and that a population crash resulted in widespread forest regeneration. Similar results have been obtained in south-west Cameroun where *Elaeis guineensis* pollen in lake sediments rose from 2730 B.P. to a peak around 1600 B.P. and then declined (e.g., Reynaud-Ferrera *et al.*, 1996). In Ituri, eastern Zaire, evidence of extensive forest fires, at least some of which are thought to have been related to humans, have been reported (Hart *et al.*, 1996). Here almost 70% of soil pits contained charcoal at depths up to 80 cm and radiocarbon ages were between 130 to 4190 B.P., mostly in the last 2000 years.

The widespread impact of humans on the forest in the last three millennia is probably related to the rapid revolution in tropical farming produced by domestication of wild plants and the spread of these crops within and between continents. The history of domestication and spread of major crops such as rice, maize and bananas can be traced by the study of phytoliths (characteristically shaped inorganic crystals) at habitation sites and in soil (Pearsall, 1994). Native crops were being grown on all continents by 3000 B.P.: yams in African forests; bananas and rice in south-east Asia; and maize in the Neotropics. The date of arrival of bananas in Africa is not known precisely, but genetic evidence and present-day distributions show three major arrivals with the earliest reaching the east coast about 2500 to 3000 years ago (de Langhe *et al.*, 1996) and by 1500–2000 B.P., plantain bananas were the major food of humans in the Congo Basin. The spread of farming and introduction of exotic crops led to population increase that in turn, increased the amount of forest cleared for plantations. Vansina (1990) believes that forest clearance in the Congo Basin did not depend on iron tools, but was achieved by ring-barking trees with polished stone axes and subsequent burning. Indeed he argues in the context of human settlement in the region, that "the impact of metallurgy was on the whole perhaps smaller than that of the humble banana" (p. 61). The high yield of bananas allowed farming cultures to spread deeper into forests where they absorbed hunter-gatherers into their own culture or into dependent relationships.

The poor soils of the rainforests meant that farming was itinerant and settlements moved on when the land was exhausted. Thus large areas of forest were modified and regeneration within old plantations led to an increased heterogeneity of vegetation. This kind of disturbance increased the food supply of some primates but is likely to have had differential effects on species within communities (Fimbel, 1994). In marginal areas, and especially when fire was common, forest cover will have been reduced and forest blocks fragmented. The creation of secondary, regenerating, forests favors some species in some habitats, as in the more modern example of selective logging, which by increasing plant species diversity in monodominant forests in Uganda, allows some primate species to live at increased population densities (Plumptre & Reynolds, 1994). However, differences exist in responses to habitat disturbance both between primate species and between habitats (Johns & Skorupa, 1987; Fimbel, 1994; White & Tutin, in press). In savanna dominated areas, human activities can lead indirectly to limited reforestation. At Lopé, poles of *Millettia* used to build houses will often re-sprout. Once the village is abandoned, the growth of these shrubs shades out grasses such that fire is excluded, allowing seedlings of a variety of woody species to grow and create isolated forest patches which gradually expand.

Human impacts on vegetation have been diverse and, whilst not always deleterious to forest primates, it is certain that negative effects have far outweighed positive ones during the past three millennia.

Active hunting may not have been intense prior to the arrival of firearms, but many animals were drawn to plantations and terrestrial species, particularly, will have been caught in the intricate traps and snares placed to defend crops. These processes will have intensified over time with human population growth and the arrival of new crops, such as manioc and maize which reached central Africa from the Neotropics in the 17th century. However, overall, human population density in tropical forest areas probably remained quite low, being limited by endemic diseases such as malaria and yellow fever. In addition, there were periodic population crashes which may have come, in part, from contact with exotic infectious diseases for which no immunity existed being introduced via increased mobility of humans.

DISEASE

Long-term data suggest that disease causes about 50% of deaths in wild populations of primates (Cheney et al., 1981; Goodall, 1986; Nishida et al., 1990). Other major causes of death are predation and injury. While minor epidemics of infectious disease that kill several members of the same group have been recorded at intervals, natural epidemics leading to widespread mortality across populations of wild primates appear to be rare. However, the possibility of infectious diseases having decimated populations in past millennia exists.

Serological studies show that primates are hosts to a range of viruses, bacteria and parasites (Leininger et al., 1978; Cousins, 1984; Blakeslee et al., 1987; Tsujimoto et al., 1988; Eberle & Hilliard, 1989; Peeters et al., 1989; Davies et al., 1991). The majority of evidence suggests that most native viruses of primates are non-pathogenic for their natural hosts, but for infections such as the SIV retroviruses of African primates, non-pathogeneity may be an evolved state achieved after a period when infected individuals died. Evidence from Taï Forest shows that chimpanzees, like humans, usually die if exposed to the filovirus Ebola (Le Guenno et al., 1995). Primates in general are susceptible to human illness as illustrated by a number of epidemics recorded in habituated communities of chimpanzees where the likely cause of infection was from people (Hosaka, 1995); the infection of humans and mountain gorillas by some of the same intestinal parasites (Ashford et al., 1990); and a case of transmission of monkey pox from wild chimpanzees to humans (Mutombo et al., 1983).

A review of the evolutionary ecology of diseases (Sheldon

& Read, 1997) makes a useful distinction between pathogens transmitted between individuals by sexual vs. non-sexual (airborne or physical contact) means, the former being generally less virulent. The greatest risk of primates encountering new, potentially dangerous pathogens is through cross-species infection. Viruses and bacteria can "jump" between closely related species and primates that kill and eat vertebrates are exposed to this danger more than strictly vegetarian species, although some risk of cross-species infection may come from associating in cohesive polyspecific associations.

A range of behaviors in primates function to minimize disease transmission within groups, for example, alternation of sleeping sites by baboons (Hausfater & Meade, 1982) and slapping at insects (Dudley & Milton, 1990). Chimpanzees have been shown to use some plants for medicinal reasons (Huffman & Seifu, 1989; Rodriguez & Wrangham, 1993) and some plants, eaten primarily for their nutritional value, are likely to have secondary medicinal value, e.g., piths of Zingiberaceae herbs which have anti-helminth qualities. These behaviors limit infection with intestinal and ectoparasites, but do not protect against infectious diseases. More general mechanisms that will reduce the risk of infectious diseases spreading to other groups within a population, include: the general rarity both of close contact between members of different groups and of transfers of individuals between groups; and the tendency to avoid sick conspecifics and cadavers.

Susceptibility to disease varies between and within individuals and infants are particularly vulnerable. Disease, including parasites, is probably a major factor of natural selection, eliminating less fit individuals, and can also play an important role in regulating population density, e.g., for the howler monkeys on Barro Colorado Island (Milton, 1982).

In summary, it seems likely that in the past disease will have caused local epidemics, which, in some cases, could have exterminated entire groups, but devastating population loss through infectious diseases in the recent past seems unlikely.

NATURAL DISASTERS

Floods, droughts and severe storms will have negative impacts on primates: in the short term by causing direct mortality or destroying food; and in the longer term by changing demography and altering habitats. Such events are rare and their effects likely to be localized, but lasting effects on behavior have been documented. For example, following a cyclone that devastated the habitat of toque macaques in Sri Lanka, high mortality occurred in the population and fusion of groups was subsequently observed, showing unexpected flexibility of social organization (Dittus, 1986). Behavioral innovations that arise after such catastrophes might endure if selective advantages accrue.

While dramatic natural catastrophes are uncommon, periodic years of food shortage, representing "ecological catastrophes", probably occur more frequently. Minor variations in climate can lead to community-wide fruit crop failures that constitute severe bottlenecks of food availability. In extreme cases mortalities occur among frugivores as, for example, on Barro Colorado Island where five famines were recorded over a 40-year period (Foster, 1982a). Long-term phenology studies have documented large inter-annual variations in the availability of fruit, seeds and young leaves in tropical forest habitats (Foster, 1982b; van Schaik et al., 1993; Tutin & White, 1998) and it has been argued that supra-annual fruiting rhythms typical of forests in south-east Asia explain the low primate biomass of the region (Terborgh & van Schaik, 1987).

DISCUSSION

This overview suggests that the recent evolutionary past of primate communities in the tropics has been shaped by a number of historical factors. Of these, modification of vegetation by climate change and by human agricultural activities were probably the most important. In Madagascar, 14 species of lemur became extinct after humans reached the island, 1500 years ago. Hunting probably joined habitat destruction as a major cause, but disease may also have played a role. This underlines a very important point: that interactions between the four classes of possible impacts are likely. For example, fires of human origin will be more devastating during periods of climatic stress, or where forest structure has been altered by farming; and, as forest habitats shrank 3000–2000 years ago, food shortages and overcrowding will have increased susceptibility to disease.

A striking example of interactions between human activities and climate change comes from the history of forest habitats of the primate community at Lopé. Forest cover shrank under the changed climate 3000 years ago, but from 2000 B.P., the climate became favorable to forest expansion into savanna and remains so to the present day. Figure 13.1 shows the distribution of forests and savanna at Lopé now, and 1400 years ago. At present, forest colonization of savanna is blocked by fires that are regularly lit by humans, and the spectacular forest expansion of the past corresponds to the 600- to 700-year period, starting 1,400 B.P., when archeological research suggests that the Lopé area was void of humans (Oslisly, in press). Within this relatively short time period the members of the primate community experienced major upheavals as forests shrank and then expanded and humans came and went. While likely responses to forest habitat shrinkage can be inferred from the comparison of primate use of present-day savanna and forest, we do not know how rapidly each species is able to colonize expanding forests, nor exactly what botanical features define potential habitat for each species. It is possible that the low total biomass of the primate community at Lopé compared to other African sites reflects the long time needed for primate populations to reach carrying capacity in new areas of habitat (White, 1994a; Tutin et al., 1997b).

Few data exist on the speed at which primates can adjust to, or recover from, environmental changes. Given the slow reproductive rates of primates and the even slower rate of regeneration of forest habitats after felling, or passage of fire, major events of the past will certainly leave traces for many generations. Behavioral adjustments can be made more rapidly as shown by the modification of social structure by toque macaques following a demographic crash (Dittus, 1986). Another example comes from behavior shown by *Cercopithecus* monkeys in the Conkouati Reserve, Congo, where hunting with shotguns has been intense for about a decade: on detecting humans, these small, arboreal monkeys drop from the trees and flee along the ground to hide in dense thickets (C. E. G. Tutin, pers. obs.), in contrast to the normal pattern of flight through tree canopies. At Lopé, where hunting with guns has never been common, these monkeys only drop to the ground in response to crowned hawk-eagles (Gautier-Hion & Tutin, 1989). This shows rapid behavioral adaptations are sometimes possible in the face of intense pressure, but species will vary in their options, for example, larger arboreal primates would probably suffer significant mortality if they tried to escape predation by dropping tens of meters from trees.

There can be little doubt that some aspects of present-day community structure and of morphology and behavior of individual species may have evolved under changed conditions in the recent past that intensified particular pressures. Precise reconstruction of events of the past three millennia is impossible at present, but the exciting developments in zoological and botanical archeology (Hather, 1994; Stahl, 1995a) will lead to a better understanding of human history in the tropics during this time period, and will shed light directly and indirectly on contemporary primate communities. New techniques in molecular genetics can also be used to test hypotheses about, for example, the degree of isolation that fragmented habitats imposed on populations of various primates in the past. It is important to understand as much as possible about the impact of the historical events discussed in this chapter as history is repeating itself at gathering speed. Climate change, human activities, disease and natural disasters continue to exert pressure on primate communities and an understanding of the past will help predict and thus counter some of the present and future threats (see Struhsaker, chapter 17 and Wright & Jernvall, chapter 18, this volume).

More and better data on the responses of species to habitat alteration and forest fragmentation, their ability to adapt to change, and to recover from population crashes, are needed if thriving communities of primates are to continue to share the tropical forest ecosystem with humans into the next millennium.

ACKNOWLEDGMENTS

We gratefully acknowledge core funding from the Centre International de Recherches Médicales de Franceville, Gabon and from the Wildlife Conservation Society, New York. Additional funding came from the Leakey Foundation and invaluable logistical support from the ECOFAC programme funded by the European Union (DG VIII) and managed by AGRECO-GEIE: we extend special thanks to Michel Fernandez for all his assistance. Many discussions with colleagues interested in the recent past of African forests stimulated the integration of data and the development of our interpretations, and we thank Kate Abernethy, Mike Fay, Alphonse Mackanga-Missandzou, Jean Maley, John Oates, Anna Roberts, Colin Chapman, Daphne Onderdonk and particularly Richard Oslisly.

This paper grew from the Workshop on Primate Communities held at Madison in August 1996 and we thank John Fleagle and The Wenner Gren Foundation for making this possible.

REFERENCES

Ashford, R. W., Reid, G. D. F. & Butynski, T. M. (1990). The intestinal fauna of man and mountain gorillas in a shared habitat. *Annals of Tropical Medicine and Parasitology*, **84**, 337–40.

Bailey, R. C., Head, G., Jenike, M., Owen, B., Rechtman, R. & Zechenter, E. (1989). Hunting and gathering in tropical rain forest: Is it possible? *American Anthropologist*, **91**, 59–82.

Baldwin, P. J., McGrew, W. C. & Tutin, C. E. G. (1982). Wide-ranging chimpanzees at Mt. Assirik, Senegal. *International Journal of Primatology*, **3**, 367–85.

Blakeslee, J. R., McClure, H. M., Anderson, D. C., Bauer, R. M., Huff, L. Y. & Olsen R. G. (1987). Chronic fatal disease in gorillas seropositive for simian T-Lymphotrpic Virus I antibodies. *Cancer Letters*, **37**, 1–6.

Boesch, C. & Boesch, H. (1990). Tool use and tool making in wild chimpanzees. *Folia Primatologica*, **54**, 86–99.

Butynski, T. M. (1990). Comparative ecology of blue monkeys (*Cercopithecus mitis*) in high- and low-density subpopulations. *Ecological Monographs*, **60**,1–26.

Butzer, K. W. (1972). *Environment and Archeology: An Ecological Approach to Prehistory*. London: Methuen.

Cheney, D. L., Lee, P. C. & Seyfarth, R. M. (1981). Behavioral correlates of non-random mortality among free-ranging female vervet monkeys. *Behavioral Ecology and Sociobiology*, **9**, 153–61.

Clutton-Brock, J. (1996). The legacy of Iron Age dogs and livestock in southern Africa. In *The Growth of Farming Communities in Africa from the Equator Southwards*, ed. J. E. G. Sutton, pp. 161–67. London: The British Institute in Eastern Africa.

Colyn, M., Gautier-Hion, A. & Verheyen, W. (1991). A re-appraisal of palaeoenvironmental history in Central Africa: Evidence for a major fluvial refuge in the Zaire Basin. *Journal of Biogeography*, **18**, 403–7.

Cousins, D. (1984). Notes on the occurrence of skin infections in gorillas (*Gorilla gorilla*). *Zool. Garten N. F., Jena*, **54**, 333–8.

Davies, C. R., Ayers, J. M., Dye, C. & Deane, L. M. (1991). Malaria infection rate of Amazonian primates increases with body weight and group size. *Functional Ecology*, **5**, 655–662.

de Langhe, E., Swennen, R. & Vuylsteke, D. (1996). Plantain in the early Bantu world. In *The Growth of Farming Communities in Africa from the Equator Southwards*, ed. J. E. G. Sutton, pp. 147–60. London: British Institute in Eastern Africa.

de Maret, P. (1996). Pits, pots and the far-west streams. In *The Growth of Farming Communities in Africa from the Equator Southwards*, ed. J. E. G. Sutton, pp. 318–23. London: The British Institute in Eastern Africa.

de Maret, P., Clist, B. & van Neer, W. (1987). Résultats des premières fouilles dans un abri de Shum Lae et d'Abeke au nord-ouest Cameroun. *L'Anthropologie*, **91**, 559–84.

Dittus, W. P. J. (1986). Sex differences in fitness following a group take-over among toque macaques: Testing models of social evolution. *Behavioral Ecology and Sociobiology*, **19**, 257–66.

Dudley, R. & Milton, K. (1990). Parasitic deterence and the energetic costs of slapping in howler monkeys, *Alouatta palliata*. *Journal of Mammalogy*, **71**, 463–465.

Eberle, R. & Hilliard, J. K. (1989). Serological evidence for variation in the incidence of Herpesvirus infections in different species of apes. *Journal of Clinical Microbiology*, **27**, 1357–66.

Eggert, M. K. H. (1996). Pots, farming and analogy: Early ceramics in the equatorial rainforest. In *The Growth of Farming Communities in Africa from the Equator Southwards*, ed. J. E. G. Sutton, pp. 332–38. London: The British Institute in Eastern Africa.

Fay, J. M. (1997). *The Ecology, social organisation, habitat and history of the western lowland gorilla (Gorilla gorilla gorilla Savage and Wyman 1847)*. PhD Dissertation, Washington University, St. Louis.

Fimbel, C. (1994). The relative use of abandoned farm clearings and old forest habitats by primates and a forest antelope at Tiwai, Sierra Leone, West Africa. *Biological Conservation*, **70**, 277–86.

Foster, R. B. (1982a). Famine on Barro Colorado Island. In *The Ecology of a Tropical Forest: Seasonal Rhythms and Long-term Changes*, ed. E. G. Leigh, A. S. Rand & D. M. Windsor, pp. 201–212, Washington, DC: Smithsonian Institution Press.

Foster, R. B. (1982b). The seasonal rhythm of fruitfall on Barro Colorado Island. In *The Ecology of a Tropical Forest: Seasonal Rhythms and Long-term Changes*, ed. E. G. Leigh, A. S. Rand & D. M. Windsor, pp. 151–172. Washington, DC: Smithsonian Institution Press.

Gautier-Hion, A. & Tutin, C. E. G. (1989). Mutual attack by a polyspecific association of monkeys against a crowned hawk eagle. *Folia Primatologica*, **51**, 149–51.

Goodall, J. (1986). *The Chimpanzees of Gombe*. Cambridge, MA: The Belknap Press.

Hart, T. B. & Hart, J. A. (1986). The ecological basis of hunter-gatherer subsistence in African rainforests: The Mbuti of eastern Zaire. *Human Ecology*, **14**, 29–55.

Hart, T. B., Hart, J. A., Dechamps, R., Fournier, M. & Ataholo, M. (1996). Changes in forest composition over the last 4000 years in the Ituri basin, Zaire. In *The Biodiversity of African Plants*, ed. L. J. G. van der Maesen, X. M. van der Burgt & J. M. van Medenbach de Rooy, pp. 545–63. Dordrecht: Kluwer Academic Publishers.

Hather, J. G. (ed.) (1994). *Tropical Archaeobotany: One World Archaeology*. 22. London: Routledge.

Hausfater, G. & Meade, B. J. (1982). Alternation of sleeping groves by yellow baboons (*Papio cynocephalus*) as a strategy for parasite avoidance. *Primates*, **23**, 419–23.

Hladik, C. M., Hladik, A., Linares, O. F., Pagezy, H., Semple, A. *et al.* (eds.) (1993). *Tropical Forests, People and Food: Man and the Biosphere Series*, **13**. Paris: Unesco.

Hosaka, K. (1995). Epidemics and wild chimpanzee study groups. *Pan Africa News*, **2**, 1.

Houghton, R. A. (1995). Tropical forests and climate. *Ecology, Conservation, and Management of Southeast Asian Rainforests*, ed. R. B. Primack & T. E. Lovejoy, pp. 263–290, New Haven: Yale University Press.

Huffman, M. A. & Seifu, M. (1989). Observation on the illness and consumption of a possibly medicinal plant *Veronia amygdalina* (Del.), by a wild chimpanzee in the Mahale Mountains National Park, Tanzania. *Primates*, **30**, 51–63.

Izawa, K. (1970). Unit groups of chimpanzees and their nomadism in the savanna woodland. *Primates*, **11**, 1–45.

Johns, A. D. & Skorupa, J. P. (1987). Responses of rainforest primates to habitat disturbance: A review. *International Journal of Primatology*, **8**, 157–91.

Jones, E. W. (1956). Ecological studies on the rain forest of southern Nigeria. IV. The plateau forest of the Okoumu Forest Reserve, Part II, The reproduction and history of the forest. *Journal of Ecology*, **44**, 83–117.

Kano, T. (1972). The chimpanzees of Filibanga, Western Tanzania. *Primates*, **12**, 229–46.

Kershaw, A. P. (1992). The development of rainforest-savanna boundaries in tropical Australia. In *Nature and Dynamics of Forest-savanna Boundaries*, ed. P. A. Furley, J. Proctor & J. A. Ratter, pp. 255–271. London: Chapman & Hall.

Lahm, S. (1986). Diet and habitat preference of *Mandrillus sphinx* in Gabon: Implications of foraging strategy. *American Journal of Primatology*, **11**, 9–26.

Le Guenno, B., Formentry, P., Wyers, M., Gounon, P., Walker, F. &Boerch, C. (1995). Isolation and partial characterisation of a new strain of Ebola. *Lancet*, **345**, 1271–4.

Leininger, J. R., Donham, K. J. & Rubino, M. J. (1978). Leprosy in a chimpanzee. *Veterinary Pathology*, **15**, 339–46.

Livingstone, D. A. (1975). Late Quaternary climatic change in Africa. *Annual Review of Ecology and Systematics*, **6**, 249–80.

Mackie, C. (1984). The lessons behind East Kalimantan's forest fires. *Borneo Research Bulletin*, **16**, 63–74.

Maley, J. (1996). The African rain forest to main characteristics of changes in vegetation and climate from the Upper Cretaceous to the Quaternary. *Proceedings of the Royal Society in Edinburgh*, **104B**, 31–73.

Maley, J. (1997). Middle to late Holocene changes in tropical Africa and other continents: Paleomonsoon and sea surface temperature variations. In *Third Millennium BC Climate Change and Old World Collapse*, ed. H. N. Dalfes, G. Kukia & H. Weiss, pp. 611–40. Berlin, Springer-Verlag.

Maloney, B. K. (1994). The prospects and problems of using palynology to trace the origins of tropical agriculture: The case of Southeast Asia. In *Tropical Archaeobotany*, ed. J. G. Hather, pp. 139–71. London: Routledge.

McGrew, W. C. (1992). *Chimpanzee Material Culture*. Cambridge: Cambridge University Press.

McGrew, W. C., Ham, R. H., White, L. J. T., Tutin, C. E. G. & Fernandez, M. (1997). Why don't chimpanzees in Gabon crack nuts? *International Journal of Primatology*, **18**, 353–74.

Milton, K. (1982). Dietary quality and demographic regulation in a howler monkey population. In *The Ecology of a Tropical Forest*, ed. E. G. Leigh, A. S. Rand & D. M. Windsor, pp. 273–89. Washington: Smithsonian Institution Press.

Mittermeier, R. A. (1987). Effects of hunting on rain forest primates. In *Primate Conservation in the Tropical Rain Forest*, ed. C. W. Marsh & R. A. Mittermeier, pp. 109–46. New York: Alan R. Liss.

Mutombo, M. W., Arita, I. & Jezek, Z. (1983). Human monkeypox transmitted by a chimpanzee in a tropical rain-forest area of Zaire. *Lancet*, **(April 2)**, 735–7.

Nishida, T., Takasaki, H. & Takahata, Y. (1990). Demography and reproductive profiles. In *The*

Chimpanzees of the Mahale Mountains, ed. T. Nishida, pp. 63–97. Tokyo: University of Tokyo Press.

Oates, J. F., Whitesides, G. H., Davies, A. G., Waterman, P. G., Green, S. M., Dasilva, G. L. & Mole, S. (1990). Determinants of variation in tropical forest primate biomass: New evidence from West Africa. *Ecology*, **71**, 328–43.

Onderdonk, D. A. (1998). *Coping with Forest Fragmentation: The primates of Kibale National Park, Uganda*. Ph.D. thesis, University of Florida.

Oslisly, R. (1993). *Préhistoire de la moyenne vallée de l'Ogooué (Gabon)*. Paris: Editions de l'ORSTOM.

Oslisly, R. (1996). The middle Ogooué valley, Gabon: Cultural changes and palaeoclimatic implications of the last four millennia. In *The Growth of Farming Communities in Africa from the Equator Southwards*, ed. J. E. G. Sutton, pp. 324–331. London: The British Institute in Eastern Africa.

Oslisly, R. (in press). The history of human settlement in the Middle Ogooué valley (Gabon): Implications for the environment. In *African Rain Forest Ecology and Conservation*, ed. W. Weber, L. J. T. White, A. Vedder & H. Simons Morland. New Haven: Yale University Press.

Oslisly, R. & Peyrot, B. (1992). Un gisement du paléolithique inférieur: La haute terrasse d'Elarmekora (Moyenne vallée de l'Ogooué) Gabon. Problèmes chronologiques et paléogéographique. *Comptes Rendues de L'Academie de Science de Paris*, **314**, 309–12.

Oslisly, R., Peyrot, B., Abdessadok, S. & White, L. J. T. (1996). Le site de Lopé 2: Un indicateur de transition écosystémique ca 10 000 BP dans la moyenne vallée de l'Ogooué (Gabon). *Comptes Rendues de L'Academie de Science de Paris, série IIa*, **323**, 933–9.

Oslisly, R. & White, L. J. T. (1996). La rélation homme/milieu dans la Réserve de la Lopé (Gabon) au cours de l'Holocene: Les implications sur l'environnement. *Symposium ECOFIT*. Paris, ORSTOM.

Pearsall, D. M. (1994). Investigating New World tropical agriculture: Contributions from phytolith analysis. In *Tropical Archeobotany*, ed. J. G. Hather, pp. 115–38. London: Routledge.

Peeters, M., Honoré, C., Huet, T., Bedjabaga, L., Ossari, S., Bussi, P., Cooper, R. W. & Delaporte, E. (1989). Isolation and partial characterisation of an HIV-related virus occurring naturally in chimpanzees in Gabon. *Aids*, **3**, 625–30.

Plumptre, A. J. & Reynolds, V. (1994). The effects of selective logging on the primate populations in the Budongo Reserve, Uganda. *Journal of Applied Ecology*, **31**, 631–41.

Reynaud-Ferrera, I., Maley, J. & Wirrmann, D. (1996). Végétation et climat dans les forêts du sud-ouest Cameroun depuis 4770 ans BP: Analyse pollinique des sédiments du lac Ossa. *Comptes Rendues de L'Academie de Science de Paris*, **322**, 749–55.

Rietkerk, M., Ketner, P. & de Wilde, J. J. F. E. (1995). Caesalpinioideae and the study of forest refuges in Gabon: preliminary results. *Adansonia*, **1–2**, 95–105.

Richards, P. W. (1996). *The Tropical Rain Forest: An Ecological Study,* (Second Edition). Cambridge: Cambridge University Press.

Rodriguez, E. & Wrangham, R. W. (1993). Zoopharmacognosy: The use of medicinal plants by animals. In *Phytochemical Potential of Tropical Plants*, ed. K. R. Downum, pp. 89–105. New York: Plenum Press.

Rogers, M. E., Abernethy, K. A., Fontaine, B., Wickings, E. J., White, L. J. T. & Tutin C. E. G. (1996). Ten days in the life of a mandrill horde in the Lopé Reserve, Gabon. *American Journal of Primatology*, **40**, 297–313.

Sabater Pi, J. (1966). Gorilla attacks against humans in Rio Muni, west Africa. *Journal of Mammalogy*, **47**, 123–4.

Sanford, R. L., Saldarraga, J., Clark, K. E., Uhl, C. & Herrera, R. (1985). Amazon rain-forest fires. *Science*, **227**, 53–55.

Schwartz, D. (1992). Assèchement climatique vers 3,000 B. P. et expansion Bantu en Afrique centrale atlantique: quelques réflexions. *Bulletin de la Societé Géologique de France*, **163**, 353–361.

Shaw, T., Sinclair, P., Andah, B. & Okpoko, A. (ed.) (1993). *The Archaeology of Africa*. One World Archaeology. 20. London: Routledge.

Sheldon, B. C. & Read, A. F. (1997). Comparative biology and disease ecology. *TREE*, **12**, 43–4.

Stahl, P. W. (ed.) (1995a). *Archaeology in the Lowland American tropics*. Cambridge: Cambridge University Press.

Stahl, P. W. (1995b). Differential preservation histories affecting the mammalian zooarchaeological record from the forested Neotropical lowlands. In *Archaeology in the lowland American tropics*, ed. P. W. Stahl, pp. 154–80. Cambridge: Cambridge University Press.

Sugiyama, Y. & Koman, J. (1979). Tool-using and -making behavior in wild chimpanzees at Bossou, Guinea. *Primates*, **20**, 513–24.

Sussman, R. W., Richard, A. F. & Ravelojaona, G. (1985). Madagascar: Current projects and problems in conservation. *Primate Conservation*, **5**, 53–9.

Sutton, J. E. G. (ed.) (1996). *The Growth of Farming Communities in Africa from the Equator Southwards*. London: The British Institute in Eastern Africa.

Suzuki, A. (1969). An ecological study of chimpanzees in a savanna woodland. *Primates*, **10**, 103–48.

Swaine, M. D. (1992). Characteristics of dry forest in West Africa and the influence of fire. *Journal of Vegetation Science*, **3**, 365–74.

Terborgh, J. & van Schaik, C. P. (1987). Convergence vs. nonconvergence in primate communities. In *Organization of Communities: Past and Present*, ed. J. H. R. Gee & P. S. Giller, pp. 205–26. Oxford: Blackwell Scientific.

Tsujimoto, H., Cooper, R. W., Kodama, T., Fukasawa, M., Miura, T. *et al.* (1988). Isolation and characterization of Simian Immunodeficiency Virus from mandrills in Africa and its relationship to other Human and Simian Immunodeficiency Viruses. *Journal of Virology*, **62**, 4044–50.

Tutin, C. E. G. (1999). Fragmented living: Behavioural ecology of primates in a forest fragment in the Lopé Reserve, Gabon. *Primates*, **40**, 249–65.

Tutin, C. E. G. & Fernandez, M. (1993a). Relationships between minimum temperature and fruit production in some tropical forest trees in Gabon. *Journal of Tropical Ecology*, 9. 241–8.

Tutin, C. E. G. & Fernandez, M. (1993b). Composition of the diet of chimpanzees and comparisons with that of sympatric lowland gorillas in the Lopé Reserve, Gabon. *American Journal of Primatology*, **30**, 195–211.

Tutin, C. E. G. & Oslisly, R. (1995). *Homo, Pan* and *Gorilla*: Co-existence over 60 000 years at Lopé in central Gabon. *Journal of Human Evolution*, **28**, 597–602.

Tutin, C. E. G. & White, L. J. T. (1998). Primates, phenology and frugivory: Present, past and future patterns in the Lopé Reserve, Gabon. In *Dynamics of Populations and Communities in the Tropics*, BES Symposium 37, ed. D. M. Newbery, H. H. T. Prins & N. Brown, pp. 309–37. Oxford: Blackwell Science.

Tutin, C. E. G., Fernandez, M., Rogers, M. E., Williamson, E. A. & McGrew, W. C. (1991). Foraging profiles of sympatric lowland gorillas and chimpanzees in the Lopé Reserve, Gabon. *Philosophical Transactions of the Royal Society. London Series B*, **334**, 179–86.

Tutin, C. E. G., White, L. J. T. & Mackanga-Missandzou, A. (1996). Lightning strike burns large forest tree in the Lopé Reserve, Gabon. *Global Ecology and Biogeography Letters*, **5**, 36–41.

Tutin, C. E. G., White, L. J. T. & Mackanga-Missandzou, A. (1997a). The use by rain forest mammals of natural forest fragments in an equatorial African savanna. *Conservation Biology*, **11**, 1190–203.

Tutin, C. E. G., Ham, R. M., White, L. J. T. & Harrison, M. J. S. (1997b). The primate community of the Lopé Reserve, Gabon: Diets, responses to fruit scarcity and effects on biomass. *American Journal of Primatology*, **42**, 1–24.

van Schaik, C. P., Terborgh, J. W. & Wright, S. J. (1993). The phenology of tropical forests: Adaptive significance and consequences for primary consumers. *Annual Review of Ecology and Systematics*, **24**, 353–77.

Vansina, J. (1990). *Paths in the Rainforests: Toward a History of Political Tradition in Equatorial Africa*. London: James Currey.

White, L. J. T. (1994a). Biomass of rain forest mammals in the Lopé Reserve, Gabon. *Journal of Animal Ecology*, **63**, 499–512.

White, L. J. T. (1994b). Patterns of fruit-fall phenology in the Lopé Reserve, Gabon. *Journal of Tropical Ecology*, **10**, 289–312.

White, L. J. T. & Abernethy, K. (1997). *Guide to the Végétation of the Réserve de la Lopé, Gabon*. Libreville, Multipress/WCS.

White, L. J. T. & Tutin, C. E. G. (in press). Why chimpanzees and gorillas respond differently to logging: A cautionary tale from Gabon. In *African Rain Forest Ecology and Conservation*, ed. B. Weber, L. J. T. White, A. Vedder & H. Morland Simons, New Haven: Yale University Press.

Whitmore, T. C. & Prance, G. T. (ed.) (1987). *Biogeography and Quaternary History in Tropical America*. Oxford Monographs in Biogeography. 3. Oxford: Clarendon Press.

Wilkie, D. S. & Curran, B. (1991). Why do Mbuti hunters use nets? Ungulate hunting efficiency of bows and nets in the Ituri rain forest. *American Anthropologist*, **93**, 680–9.

Zeidler, J. A. (1995). Archaeological survey and site discovery in the forested neotropics. *Archaeology in the Lowland American Tropics*, ed. P. W. Stahl, pp. 7–41. Cambridge: Cambridge University Press.

14 • Resources and primate community structure

CHARLES H. JANSON AND COLIN A. CHAPMAN

INTRODUCTION

The concept of carrying capacity is fundamental in determining a species' density and biomass, which enter into the equations used in ecological theory to predict species diversity and community structure (MacArthur, 1972). Therefore, you might think that a great deal was known about how to determine the carrying capacity of species in nature; however, this is not the case. In the vast majority of species, if the carrying capacity is measured at all, it is estimated as the population density of individuals associated with a zero growth rate (Dennis & Taper, 1994). The more fundamental question of what resources determine that population level is often left unanswered.

Debate has raged over the past two decades as to the extent or even existence of resource limitation in primate populations. Early observations (Struhsaker, 1969) suggested that many fruit-eating primates faced an over-abundance of resources in the trees that they used. Later study quantitatively confirmed that some trees produced more fruit than a group of monkeys could consume in one sitting (Janson, 1988b), but also showed that many trees did not produce enough fruit to satiate even a single animal. In one of the few attempts to relate primate population densities to resources directly, Coehlo et al. (1976) compared the biomass and consumption rates of howler and spider monkeys in Guatemala to the availability of the major food resources that they used. They found that over an annual cycle total food production vastly exceeded the requirements of the population density in that site. However, in Peru, Janson (1984) found that the productivity per unit area of food species used by capuchin monkeys varied enough so that in the dry season insufficient fruit was produced to sustain the population. At that time, these

fruit-eating monkeys were seen to switch to alternate plant products. In the same study site, Goldizen et al. (1988) showed that fruit-eating tamarins of both sexes lost weight during the dry season. The general scarcity of fruit during the dry season in Peru and the convergence of many primate species on a few resources then (Terborgh, 1983) led to the concept of keystone resources: a relatively few plant species that are predictably available during periods of food scarcity and may serve to sustain primate populations at these critical times (Terborgh, 1986).

At the level of the entire community, Janson & Emmons (1990) found that the estimated metabolic requirements of frugivorous mammals at Cocha Cashu, Peru, was only 20% less than the total annual production of fleshy fruits. At least for this system, it appears reasonable to argue that the major resources consumed by fruit-eating primates indeed limit their population density. However, this argument does not work either for seed consuming or leaf-consuming primates. The same study showed that the production of seeds in all fruit types exceeded by a factor of 20 the metabolic needs of seed consuming mammals in the community. Similarly, it is clear that in no forest study site do folivorous mammals consume more than a small fraction of the biomass of leaves available on a steady basis (Hairston et al., 1960). Either the density of leaf-eaters is controlled by factors other than food, or they must be exceedingly selective about what they consider to be "food". Such selectivity can also occur among 'omnivorous' primates (Altmann, 1998), and it aggravates the problem of estimating the potential carrying capacity of a habitat for any given primate species (see Appendix 14.1).

Given the difficulty of estimating the limiting resources for individual primate species or ensembles at a single site, it is hardly surprising that few analyses have focused on how resources limit primate biomass. Most of the broad comparative surveys of primate community structure have dealt with species diversity (Bourlière, 1985; Reed & Fleagle, 1995; Kay *et al.*, 1997; Peres & Janson, chapter 3, this volume). Although it is well documented that the population density of a primate species varies inversely with its body mass (Robinson & Ramirez, 1982; Fa & Purvis, 1997), the resource basis for these correlations is not understood beyond the speculative inference that energy flow is independent of body mass in mammals (Brown, 1995). Only for folivores do there exist a few studies that have analyzed ecological factors correlated with biomass (Oates *et al.*, 1990; Ganzhorn, 1992; Peres, 1997*b*). The purpose of this chapter is to review what is known about resources used by primates, discuss methods to measure the abundance or quality of these resources, and describe the relationships between primate diversity or biomass and resources across study sites within and between continents. We do not tackle more detailed questions of primate community structure such as the mechanisms of niche separation among primate species (Ganzhorn, 1988) or between primate species and other animals (see Emmons, chapter 10, this volume).

PRIMATE DIETS AND THEIR NUTRITIONAL CHARACTERISTICS

Primate diets typically are divided into three major resource categories: insects, fruit, and leaves (Kay, 1984). However, these categories do not describe the full range of foods commonly used by primates. For instance, many primate species use flowers extensively for a short period each year (Janson *et al.*, 1981), although flowers do not comprise the majority of the diet of any known primate. A further problem is that the three major food categories often contain several quite distinct types of foods. For instance, species that feed on insects will also consume small vertebrates, so this category is more accurately labeled faunivores. Similarly, primates that eat fruit pulp or seeds are both called frugivorous, but the nutritional content, density, and temporal availability of these two components of fruits are very distinct (van Roosmalen, 1984). Finally, there are at least two distinct leaf digestion strategies shown by mammals: slow but relatively complete fermentation of foods in the forestomach vs. relatively rapid extraction of soluble nutrients with minimal

fermentation in the mid- or hind-gut (e.g., cows vs. horses, colobines vs. lemurs or howler monkeys). Even within a single community of primates, there can be many distinct constraints on eating leaves with major implications for resource availability and foraging strategies (Ganzhorn, 1989). We shall now discuss these dietary categories in more detail.

Insects and other animals

Eating animal prey generally requires little in the way of digestive specializations, as the body composition of most animals is fairly similar and the consumer can obtain nearly all essential nutrients from its prey (Moir, 1994). Thus the digestive systems of faunivores tend to be rather simple and comparable between species (Chivers & Hladik, 1980). Nevertheless, there are a few major differences between insects and other animals as food items. First, there is the chitinous exoskeleton of adult insects and other arthropods. It is hard to chew into pieces, compared to the soft muscle tissues of vertebrate prey. Primarily insectivorous primates fracture the chitin with very high-crowned molars (Strait, 1997). Chitin also is hard to digest. Many frugivore–insectivore primates appear to avoid ingesting chitin (Janson & Boinksi, 1992) and the chitinous parts of their insect diet pass nearly unaltered through their digestive systems. However, more dedicated insectivores do appear to break down some of the chitin they ingest (Kay & Sheine, 1979). Second, many insects (but few warm-blooded vertebrates, see Dumbacher *et al.*, 1992) are toxic or distasteful. Some nocturnal prosimian primates appear to specialize on such noxious prey (Charles-Dominique, 1977), whereas diurnal monkeys have developed a variety of behavioral mechanisms to get around stinging hairs and other defenses of insect prey (Janson & Boinski, 1992). Third, there may be nutritional differences between insects and vertebrates, although existing composition data show no consistent contrasts (Leung, 1968). Many primate species that depend on insects for protein seem eager to consume vertebrate prey on occasion, and some even focus their foraging efforts largely on vertebrate prey seasonally (Terborgh, 1983; Fedigan, 1990).

Absolute insect availability may be similar for both small and large primates within a given site. The size of insects ingested does not appear to be limited by primate body size, at least within the range of 300–3000 g in the

New World (Terborgh, 1983). Thus, larger primates should not have a markedly greater pool of insects available to them than do smaller species (one possible exception being the nests of social insects that are too tough for small primates to tear apart). Yet, larger primate species must satisfy a larger metabolic demand. Thus, as primate species become larger, they should and do spend more time foraging for insects (Charles-Dominique, 1977; Terborgh, 1983), and the fraction of the diet that is composed of insects declines (Hladik & Hladik, 1969).

Fruits

Because fruit pulp serves the apparent evolutionary purpose of attracting seed dispersers, it generally presents far fewer problems for primates to find and catch than do insects, and pulp is easier to ingest and digest than are leaves. However, plants also have been selected to provide fruits that cause dispersers to leave the tree; otherwise dispersal is not effected (Herrera, 1982). One mechanism to guarantee that dispersers eventually leave a tree crown is to make fruit pulp an incomplete diet (Janzen, 1983). Most fruit pulps are high in sugars but low in fats and proteins. Although some fruit pulps are high in particular minerals (such as potassium in bananas), it seems unlikely that primates would actively seek out such fruits as there is little evidence for taste receptors for minerals other than sodium (Hladik & Simmen, 1997; but see O'Brien et al., 1998). Because of the nutritional deficiencies of fruit as a diet, every predominantly fruit-eating primate species must complement its diet with either insects or leaves or both. The constraints on ranging behavior set by these alternative resources may have important effects on primate social ecology (Janson, 1988a).

Fruit-eating primates have to cope with other problems as well. First, fruits are often chemically defended against insect or mammalian herbivores before the pulp and seed mature, and some continue to be defended even when the pulp is ripe (Cipollini & Levey, 1997). Second, plants have evolved a variety of ways to restrict dispersal of their fruits to a fraction of all of the potential fruit-eating animals in the forest: particular fruit presentations (Denslow & Moermond, 1982), morphologies (Janson, 1983), ripening schedules (McKey, 1975), and taste or defensive chemicals (Janzen, 1983). For instance, the chemicals that make red peppers spicy to humans and other mammals are apparently not perceived by birds; conversely at least one taste compound in fruits, methyl anthranilate, is palatable to mammals but so noxious to birds that it is used as a commercial bird repellent on fruits, grains, and lawns (Mason et al., 1989). Because of some combination of fruit size, protection, taste, toxicity, inaccessibility, or slow ripening rate, primates will use only a fraction of the many hundreds of fruit species in a forest.

Fruit-eating primates have to solve the challenge of locating ripe fruit crops that are often sparsely distributed in tropical forests. Searching for rare fruit trees is likely to be inefficient because detection distances for fruit crops are probably short (Janson & Di Bitetti, 1997). Instead, many primates appear to remember the locations of fruit crops over periods of days or weeks, returning at relatively predictable intervals to the same tree crown and moving in relatively straight lines from one resource to the next (Garber, 1989; Janson, 1998). Spatial memory can increase foraging efficiency up to 300% relative to random searching (Janson, 1998). Thus, the use of spatial memory may substantially increase primate biomass in a forest.

Seeds

Although often lumped under the category of fruit, the seeds consumed by primates should be considered a distinct category for several reasons. First, unlike fruit pulp, the seed must contain complete nutrition for the growing seedlings. Thus, seeds typically are higher in fats, proteins, and minerals (especially phosphorus, a potentially limiting nutrient for vegetarian animals, see below) compared to fruit pulp. Although the nutritional needs of growing plants and primates are not likely to be identical, some primates appear capable of surviving on seeds alone for extended periods (Ayres, 1989; Norconk et al., 1997), whereas no primate species is known to feed strictly on fruit pulp. Second, plants typically defend their seeds either with hard shells or noxious chemicals, making it time-consuming or dangerous for a primate to eat the seeds. However, there may be limits to the kinds and concentration of toxic chemicals that a plant can put into its seeds (Orians & Janzen, 1974). Third, seeds are often available over much longer time spans than is ripe fruit pulp in the same tree crowns. Not only do seeds typically mature somewhat earlier than fruit pulp ripens within a fruit, but heavily defended seeds may remain available on the forest floor long after they are dispersed and serve as primate food sources for an extended period (e.g., palm seeds: Terborgh, 1983). Altogether, seeds present characteristics somewhat intermediate between those of fruit pulp and leaves. Like fruits, seeds are produced in large clumps and have relatively high energy concentrations, but like leaves, seeds often contain toxins as well as scarce mineral nutrients, and they may be available over longer time spans than ripe fruit pulp.

Flowers

Little attention has been paid to flowers as a primate resource (but see Terborgh & Stern, 1987). Generally, primates eat only a few flower species out of hundreds or thousands of plant species in a community (Janson et al., 1981; Overdorff, 1992). Although flowers rarely comprise more than 10% of a primate's annual diet, they can provide over 70% of the feeding time in a given month (Janson et al., 1981). Flowers appear to be fall-back resources for primates, used when little else is available, because nectar is usually dilute compared to fruit pulp and produced in minute quantities per flower per day (Terborgh and Stern, 1987). Because flowers are generally frail compared to fruits, primates need no specializations to eat them. At least in Peru, flower-feeding is seen in primate species from the smallest to largest in a given community, including gummivores, frugivore–insectivores, frugivore–folivores, and folivores (Janson et al., 1981 and unpubl. data). However, even primate species of similar size may differ substantially in how they feed on flowers of a given species and hence their effects on floral survival and pollination (Overdorff, 1992).

Leaves

Eating and digesting leaves undoubtedly present the biggest dietary challenges to primates. It is not that leaves are a 'low-quality' resource in absolute terms. On a per-weight basis, leaves are often richer in protein than other primate vegetal resources (Milton, 1980; Waterman, 1984; Barton et al., 1993), and the potential energy density is very high as the major component of leaves, cellulose, is nothing more than a polymer of sugar molecules. However, plants have evolved a dazzling array of defensive chemicals to protect their leaves from herbivores. Cellulose is the first line of defense. Despite its similarity to starch, this ubiquitous structural compound of plants is configured in such a way that no vertebrate can digest it without the help of protozoans or bacterial symbionts. How then do primates choose and cope with ingesting and digesting leaves? I will present here only a brief outline of this issue, which is covered in much greater detail in a recent review by Lambert (1998).

The simplest strategy to use leaves is to avoid the problem of cellulose digestion altogether. Many fruit-eating primates obtain protein from leaves by selecting low-fiber parts of plants, which are processed in the gut relatively quickly to extract soluble carbohydrates and proteins; little emphasis is placed on digestion of cellulose. Although this method of eating leaves seems to provide only slightly more digestive challenge than eating fruit, two difficulties remain. First, finding these leaves may require almost as much searching as looking for sparsely distributed insects because many tropical evergreen trees produce tender young leaves at low, relatively even rates (Sterck, 1995). In seasonally dry areas of the tropics, trees may put out young leaves synchronously in a single seasonal flush that may attract primates that otherwise eat few leaves (Terborgh & van Schaik, 1987), but this food source is available only briefly. Second, even young growing leaves are chemically defended, usually with small, often highly toxic compounds such as cyanide, alkaloids, and terpenes (Coley & Barone, 1996). Either the primate must restrict its leaf intake to poorly defended species or parts (such as petioles), or it must possess the ability to detoxify these poisons. In sum, high-quality vegetation presents a more sparsely distributed and chemically better-defended resource than do fruits, although certainly more abundant and easier to capture than insects. For fruit-eating primates that depend on leaves for protein, fitness and population density may be limited by ability to find and extract protein rather than energy (Whiten et al., 1991).

There are two major methods of microbial digestion of cellulose in vertebrates, forestomach and caeco-colic fermentation (Lambert, 1998). In either type, fiber is exposed to complex communities of microbes in the primate's gut for relatively long periods of 8–24 hours. Forestomach (often known as "foregut") fermentation requires anatomical specializations of the stomach and complex relationships between the microbes and their host, which appear to arise relatively rarely in evolution. In extant primates, stomach-based fermentation has arisen only once, in the ancestors of the colobines.

Forestomach fermentation has advantages and costs. The advantages are (1) that any microbes that wash out of the stomach are digested in the intestines, thus reclaiming their precious proteins, and (2) it may help to detoxify plant defensive chemicals (Parra, 1978; Janzen, 1979; Langer, 1986), as well as break apart phytin. Phytin is one of the major chemical 'storage' compounds for phosphorous, a mineral that is often limiting to large herbivores (T. Pope, pers. comm.). The major costs of forestomach fermentation include (1) holding the fiber for long periods in the uppermost part of the digestive tract, thereby hindering further ingestion while it is fermenting, (2) the loss of the easily absorbed parts of leaf cell contents, which are taken up and used by the microbes in the stomach rather than being absorbed by the primate host directly through its intestine, and (3) the need to regulate stomach acidity, as excess acidity can disrupt the microbial community and even lead to host death. To reduce the cost of fiber 'storage' in the stomach, primate forestomach fermenters that use both low- and high-fiber parts in the diet may have the capacity to 'shunt' non-fibrous foods past the fermenting chambers of the stomach to the acid-enzymatic hindstomach, where digestion is faster (Cork, 1994). To prevent excess acid accumulation, foregut fermenters must avoid ingesting large amounts of acid fruit pulp or a diet too high in fiber (Lambert, 1998). Because of these constraints forestomach fermenters should choose relatively high-nitrogen plant matter of low or moderate fiber content with near-neutral pH. Thus, primate forestomach fermenters are selective browsers on leaves and seeds.

Fermentation in the mid-gut (caecum and right colon) or hind-gut (descending colon) is much more widespread than forestomach fermentation. Indeed, some fermentation in the caecum or colon probably occurs inevitably even in omnivorous species that ingest leaves as a source of protein. Fermentation can be enhanced by enlarging the caecum or colon, because cellulose-rich material remains exposed longer to microbial action. Such caeco-colic (previously referred to simply as "hindgut") fermentation has evolved at least four times in the primates, once in each of the major primate radiations (Lambert, 1998).

Caeco-colic fermentation of cellulose has its own distinct advantages and costs. Its advantages include (1) that it does not hinder further ingestion of food if the upper parts of the digestive tract are empty, and (2) any easily-absorbed cell contents are taken up directly by the host before being exposed to microbial symbionts. Thus, when foliage is low in nitrogen and fiber-rich (dead grass, for instance), a caeco-colic fermenter can survive by pushing large amounts of material through the gut and extracting only the easily-obtained fraction while ignoring fermentation as a major option; this is the digestive strategy of horses. The costs of caeco-colic fermentation include (1) the loss in feces of the symbiotic microbes (and their proteins), (2) exposure to plant defensive chemicals in the gut before the microbes have a chance to detoxify them (Waterman et al., 1988), and (3) little breakdown of phosphorous-rich phytin, thus limiting the absorption of phosphorous from foliage to between 44–65% of that achievable by forestomach fermenters (T. Pope, pers. comm.). To reduce the cost of lost microbial proteins, caeco-colic fermenters can (and do) ingest their feces and have evolved intra-colonic separation mechanisms to retain the microbes and shunt them back to the fermentation chamber (Björnhag, 1994). Their inability to avoid exposure to plant toxins in the fore- and mid-gut may account for the often extreme food selectivity of some caeco-colic fermenting primates (Glander, 1982). However, it is not clear whether forestomach and caeco-colic fermenters differ systematically in their ability to detoxify plant defensive chemistry (Foley & McArthur, 1994).

For either strategy of microbial fermentation to be effective, two requirements must be met. First, leaves must be shredded into fine particles to expose as much surface area of cellulose to microbial action as possible (Kay & Hylander, 1978). Second, fermentation takes time, so that gut passage must be relatively slow compared to insect- or fruit-eaters. A comparison of leaf- versus fruit-eating primates shows that the passage time in folivores is much slower than in fruit-eating species of comparable body size (Milton, 1984; but see Lambert, 1998). Longer gut retention times effectively limit the maximum daily rates of nutrient extraction by leaf-eating primates, which in turn affects their energetics and social behavior (e.g., Janson & Goldsmith, 1995). The lower the metabolic needs of a folivore relative to its gut capacity, the lower the quality of leaves it can afford to eat (Cork, 1994). Because mass-specific metabolic rates tend to decline with larger body mass, whereas relative gut capacity stays roughly constant, larger herbivores usually can survive on lower-quality foliage (Cork, 1994). All primate folivores of less than 15 kg ingest more easily-digested non-foliage items (immature fruits, seeds, gums) as well as leaves, perhaps because they are too small to rely completely on a leaf diet regardless of digestive strategy (Cork, 1994). Primate species that succeed in fermenting cellulose can achieve very high densities and biomass compared to non-fermenting fruit- or insect-eating species (see below).

How do the two distinct fermenting strategies compare? The answer may depend on body size. Theoretical studies suggest that forestomach fermentation is preferable in larger herbivores, while caeco-colic fermentation is best for small species, but neither strategy may be clearly superior within the 5–15 kg mass range of most arboreal folivores (Parra, 1978; C. M. Hladik, 1978; Illius & Gordon, 1992). Caeco-colic fermentation may be more flexible, with the primate having considerable control over the length of time material ferments in the gut, as well as over the fraction of material that is shunted past the fermenting chamber (Lambert, 1998). Published data on digestive efficiency suggest that fore-stomach fermenting colobine monkeys are nearly twice as efficient at digesting cell-wall contents as are the caeco-colic fermenting howler monkey (Kay & Davies, 1994; Milton, 1998); however, the diets used in different studies were not standardized, and more controlled comparisons are desirable.

Gums

This category usually combines both true gums, which consist of branched carbohydrate polymers (as well as soluble sugars; Hladik & Chivers, 1994), and sap, which may be little more than a solution of soluble sugars and proteins being transported by the plant from roots to leaves or vice versa. These two classes of foods should be distinguished, however. Despite the lack of cellulose and cell walls, true gums may not be much easier to digest than are leaves (Hladik & Chivers, 1994). Recent studies have shown that different gum-eating prosimians and callitrichines differ in their ability to digest the complex carbohydrates in gums (Nash, 1986; Power & Oftedal, 1996), at least in part through caecal fermentation. In contrast, sap-eating primates are quite small (including the smallest anthropoid, the pygmy marmoset) and most necessarily have short gut passage times (but see Power, 1996). Such a digestive strategy works well for dissolved nutrients such as occur in plant phloem contents.

Why are primates such generalized feeders?

Dietary specialists are uncommon in primates. Because of the dearth of protein in fruit pulp, all fruit-eating primate species complement their diet with proteins (and probably minerals) from either animals or leaves (Kay, 1984). Depending on their body size, primates species vary in the degree to which energy vs. proteins limit their foraging effort. For instance, Terborgh (1983) showed that the percentage of the activity budget devoted to obtaining fruit was essentially constant across a tenfold range of primate body masses in Peru, whereas the time devoted to searching for insects increased markedly with increasing body size. For the largest study species, the brown capuchin monkey (*Cebus apella*), foraging for dispersed insects occupied two to three times as many hours per day as did feeding on fruit. In this species, it can be shown that energetic returns from insect searching are negative – that is to say a brown capuchin would starve to death if all it could eat was insects (Janson, 1985). The relative superabundance of fruit at most times compared to the scarcity of insects implies that, at least in this case, fruit feeding usually subsidizes energetically the search for protein. For fruit-eating species larger than brown capuchin monkeys, it is difficult to meet protein needs by using insects alone, so these larger species include or depend entirely upon leaves as a protein source (see also Kay, 1984). Because they do not require the energy from cellulose in the leaves, these omnivores can afford to choose only high-protein, low-fiber leaves, thus minimizing problems of digestion. Even so, however, these frugivore-folivore species typically ingest leaves at the end of their foraging day, a time when they can fill up their gut with relatively indigestible material without inhibiting the digestion or intake of less fibrous fruit or insect material (Chapman & Chapman, 1991; but see Ganzhorn & Wright, 1994). Many Old World monkeys (*Cercopithecus*) appear to use slow passage time and fermentation of fiber as a digestive strategy despite little obvious enlargement of the mid- or hind-gut (Lambert, 1998).

Primate faunivores and folivores can be dietary specialists. Nearly all specialized faunivorous primates are small-bodied. Their short digestion times force them to use easily absorbed foods such as insects or nectar, while their low total metabolic demands allow them to specialize on these scarce high-quality foods. By contrast, folivores tend to be large bodied (Kay, 1984). Their large body size and long gut-passage times permit them to ferment relatively common but complex vegetation, while their high absolute metabolic demands preclude their being specialized on relatively scarce insect resources. Even if they eat only leaves, however, leaf-eating primates typically use vegetative parts of many dozens to hundreds of plant species in a single area. Given that most tropical trees are evergreen, why do leaf eating primates not specialize on just the few types of leaves that provide the highest protein–fiber ratio? Because most tropical trees species occur at low densities, a primate that favored only a few tree species in its diet might be forced to travel long distances to obtain a stomach full of food. In addition, for caeco-colic fermenters, there is no easy way to avoid the toxins in leaves, so that excessive consumption of one or a few plant species might overload the capacity of the digestive system and liver to detoxify the plant defensive chemicals encountered (Milton, 1979; Glander, 1982). Forestomach fermenters may be less hindered in this regard, and at least some colobine monkeys are quite specialized in their choice of leaves (*Colobus guereza*: Oates, 1977). Studies that contrast folivores that vary in their degree of specialization can reveal interesting dietary strategies (Struhsaker & Oates, 1975).

Summary: primate dietary niches

Primates have diversified into a broad array of dietary types that span most of the dietary specializations among mammals in general (Chivers *et al.*, 1984). Body size is a major correlate of primate diets (Kay, 1984). The smallest species (50–500 g) tend to be restricted to foods of relatively high digestibility – insects, small vertebrates, saps and gums. Species of intermediate size (500–5000 g) tend to rely largely on fruit or seeds, but obtain their protein needs from insects (species up to ca. 3000 g) or leaves (species above 3000 g). The largest species (>5000 g) are capable of fermenting sufficient fiber to obtain at least part of their energy requirements from leaves in addition to fruits, or they may rely entirely on leaves for brief periods (species up to ca. 10 000 g) or prolonged periods (species above 10 000 g). These generalizations are by no means absolute, as there are relatively small-bodied species that are entirely folivorous (*Lepilemur mustelinus* at 650 g) and

large species that are extremely frugivorous (chimpanzees, *Pan troglodytes*, at 45 000 g).

Within dietary types, there can be finer-scale differences as well. The major ecological differences that distinguish primate faunivores are the methods used to find prey or to hunt them. Such differentiation is well illustrated by Terborgh's (1983) study of five insectivorous-frugivorous primates in Manu National Park, Peru or Charles-Dominique's (1977) study of prosimians in Africa. In both studies, species differed in the height in the forest and in the size of substrates they moved on while searching for insects. In Peru, the major difference among primate species was in the toughness of substrates manipulated to extract insect prey, from open-leaf gleaning by smaller species to ripping apart palm fronds and tree trunks by the largest (Terborgh, 1983). However, some species have developed behaviors to cope with the defenses of caterpillars (squirrel monkeys: Janson & Boinski, 1992), and others even appear to specialize on slow-moving distasteful insects (slender lorises: Charles-Dominique, 1977).

Niche differences among fruit-eating species are less obvious. There is a positive correlation between body size and mean fruit size used among New World monkeys (Janson, unpubl. data), but even very small species such as tamarins may ingest the pulp of some of the larger fruits (Terborgh, 1983). In Africa and Asia, where most of the frugivorous primates belong to a single genus (either *Cercopithecus* or *Macaca*), body-size-related differences in fruit use are much less pronounced. Many of the differences in fruit use among primate species in both areas show up as preferences for species with distinct phenological patterns – sudden-ripening, large-crowned figs vs. slowly-ripening, small-crowned understorey plants (e.g., Terborgh, 1983). Chemical differences among preferred fruits may exist, but have not been studied systematically (see Appendix 14.1). Oddly, differences in overall fruit species use were far less than in insect-foraging techniques among the Peruvian primates in Terborgh's (1983) study, even though fruit comprised a majority of the energy intake for all of them. However, recent ecological theory argues that species may overlap broadly on common resources but diverge in their use of rare but fitness-enhancing ones (Robinson & Wilson, 1998).

What major differences in leaf choice are predicted from the three types of digestive strategies (non-fermenting, forestomach, and caeco-colic fermentation)? Because fiber (cellulose) always presents a cost of retaining relatively slowly-digestible bulk relative to rapidly-absorbed soluble proteins and carbohydrates, nearly all primates that eat leaves strongly prefer leaves with high protein-to-fiber ratios (Cork, 1994). Non-fermenting primates should be restricted to leaves with low absolute fiber levels. Small-bodied caeco-colic fermenters are likely to be relatively specialized feeders on relatively rare high-protein, low-fiber plant parts. They may evolve to deal physiologically with large amounts of particular plant toxins (Glander *et al.*, 1989), with different primate species specializing on different kinds of plant defensive chemicals (Ganzhorn, 1989). Most large-bodied caeco-colic primates are capable of surviving on leaves of a broad range of protein/fiber ratios, but may need to avoid ingesting large amounts of the secondary chemicals of any one plant species (Milton, 1979; Glander, 1982). They can add fruit to the diet opportunistically as availability dictates. Forestomach fermenters should generally browse only relatively high protein, low- to medium-fiber leaves and seeds, while avoiding most ripe fruit pulp to prevent acidosis (Kay & Davies, 1994). They can become fairly specialized (Oates, 1977) because of the ability of the forestomach microbes to detoxify plant defensive chemicals at the first stages of digestion. When two or more species of colobines coexist, their diets may differ in leaf maturity and degree of specialization on few vs. many food species (Struhsaker & Oates, 1975).

PATTERNS OF RESOURCE AVAILABILITY IN PRIMATE HABITATS

A brief sketch of tropical forests

There is no such thing as a 'typical' tropical forest, which can include relatively short-stature deciduous woodlands with only 500 mm of rainfall per year and evergreen forests with 60 meter tall emergent trees in areas of more than 2500 mm rainfall. Although the bewildering diversity of plants associated with tropical forests is typical of unflooded forests, in some areas (permanently-flooded oxbow lakes, swamps, tidal river edges), species diversity is typically low, and virtually mono-specific stands of palms or fig trees occur both in New and Old World tropical sites. Even areas of relatively similar physical and climatic characteristics can produce forests dominated by

quite distinct plant families with unique biological traits, depending on their evolutionary and biogeographic history. The myth that tropical forest is a single biome is only one of many; we shall briefly review some others.

Of all the misconceptions common about tropical forests, the most enduring is that they lack definite seasons. It is true that average temperatures are high (above 20 °C) and seasonal temperature variation is slight in the tropics. However, most tropical forests on continents have distinct rainy and dry periods severe enough to limit plant production, and predictable enough to produce repeatable seasonal patterns of plant reproduction and leaf loss (e.g., Fig. 14.1). Rainfall patterns are driven by total solar radiation received per unit land area, and are therefore affected both by latitude and the inclination of the earth's axis relative to the plane of the earth's orbit (MacArthur, 1972). Areas more than 5 degrees from the equator typically possess a single dry season of 1–3 months alternating with a longer rainy season per year (Terborgh, 1992). However, this pattern also occurs closer to the equator near the east-

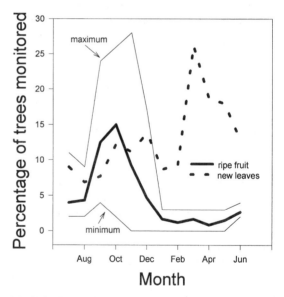

Fig. 14.1. Curves of monthly phenological changes in new leaves and ripe fruit in a sample of 58 Malaysian rainforest trees (data from Whitmore, 1975). Data are averaged across six years for both categories of foods, and the range of values per month is shown for ripe fruits. Although the seasonal pattern of fruit and leaf production is quite predictable, there are notably 'good' and 'bad' years for total amount of fruit produced.

ern edges of Africa and South America (Whitmore, 1975). Within 5 degrees of the equator near continental edges, one can find two brief dry seasons alternating with two longer rainy seasons per year (Martin, 1991, pp. 52–53). Alternatively, rainfall patterns are virtually aseasonal in a few equatorial sites far from the ocean in Africa and South America, and, oddly, also in the Indonesian islands (Whitmore, 1990, p. 11; Terborgh, 1992, p. 180). Such even rainfall can be associated with highly unpredictable fruit production (e.g., at Lomako in Democratic Republic of Congo: Malenky, 1990). However, the overall prevalence of seasonal food production in tropical forests means that most primate species need to be flexible enough to shift dietary preferences rapidly as preferred foods dwindle or increase.

A more subtle aspect of seasonality in tropical areas is variation between years. Some plants fruit only every other year. Others store nutrients for reproduction over several years before producing a large burst of flowering and fruiting, known as mast crops. Yet even among those species that attempt to reproduce every year, fruiting failure is common in many years due to between-year climatic variation. Sometimes this variation is global in extent (such as in El Niño years), albeit with different effects on different parts of the world. Such large-scale patterns can produce reproductive failure in a large percentage of the plant species in a tropical site, even leading to starvation among the fruit- and seed-eating animals (Foster, 1982). Even at a more local scale, however, the reproductive output of a given plant species can vary many-fold from year to year (Chapman et al., in review), so that the composition of available fruit resources is never the same two years in a row. Far from being a constant aseasonal source of food, tropical forests vary markedly in food production both seasonally within years and on average across years. Again, primates need to be flexible in their foraging to cope with such variation.

Another prevailing myth about tropical forests, spawned by their lushness and species diversity, is that they are areas of extraordinary fertility. Soil fertility varies a great deal, from high-fertility sites near volcanoes or young mountain ranges like the Andes to old, weathered soils dating back to the pre-Cambrian era, 600 million years ago. Often the fertility of a tropical area can be surmised by looking at the color of the rivers that lead from it (Janzen, 1974). Fertile soils produce muddy runoff leading to so-called 'white-water' rivers. By contrast, the low productivity on infertile soils leads to clear runoff that is often tinged dark brown by leaching from tannin-rich leaves, thus producing 'black-water' rivers. What nutrients are present in most tropical sites are not stored in the soil (unlike many temperate areas), but are sequestered in the living mass of plant tissue. Thus, when tropical forests are cleared, the nutrients are released rapidly to the soil and wash away, producing sites that can sustain agricultural crops for a few years at most. The soils under some diverse forests in Brazil are so nutrient-poor that they cannot sustain a single crop of corn after clearing (Lathwell & Grove, 1986). Although poorly documented, soil fertility is thought to be an important determinant of resource availability for primates (Peres, 1997b).

A final widespread view of the tropics, even among biologists, is that they are more 'stable' in evolutionary time than temperate forests. This perspective is not entirely false – something like tropical forest has been present for over 80 million years in areas we now consider tropical. What is now apparent, however, is that many areas in today's temperate zone were once at least subtropical when world climate was much warmer than it is today, and that much of the expanse of rainforest we know today (at least before recent human destruction) was much drier woodland or savanna during recent glacial maxima (Whitmore & Prance, 1987). Thus, the extent and continuity of tropical forests have changed markedly over the past 20 million years. More recent changes brought about by human occupation, including the frequent use of fire, may have radically altered the balance of continuous forest vs. savanna woodland in tropical Africa (see Tutin & White, chapter 13, this volume). One consequence of long-term changes is that the potential pool of primate species available to colonize a given forest is not the same between different regions even on the same continent, with obvious effects on primate community diversity, species composition and perhaps biomass (see Peres & Janson, chapter 13 and Struhsaker, chapter 17, this volume).

Because of variation in total amount and seasonality of rainfall, soil types, and history, tropical forests vary considerably across space and time in resource availability for primates. We shall now explore these availability patterns in more detail.

Insects

In virtually all tropical sites sampled, insects show strong seasonal variation in apparent abundance (Smythe, 1982; Kato et al., 1995). Because many insect sampling methods require movement of the insects, some of the seasonal pattern may be due to inactivity as insects respond to drought or cold (however, see Wolda & Wright, 1992). Peak abundance is usually correlated with periods of new leaf flush at the end of the dry season in seasonally dry forests, or the onset of warmer weather in the subtropics (Coley & Barone, 1996). Because most methods of insect sampling have little control over the area sampled, it is not usually possible to compare the abundance of insects absolutely across different study sites. Anecdotal data suggest that insects are scarce in low-productivity areas of highly-weathered soils (Coley & Barone, 1996).

Fruits

Differences in fruit production between distinct communities are not well-documented, but some broad patterns have been suggested. Janzen's classic article (1974) noted that white-water river drainages were characterized by high fleshy fruit production, high animal densities, lots of mosquitoes, and relatively thin, short-lived leaves, with black water rivers typically being the opposite. His interpretation of this pattern was that in black-water rivers, low nutrient availability favored slow growing, chemically heavily defended plants, thus compounding the effect of low productivity in limiting the abundance of animals. As part of the same pattern, plants in black-water rivers should tend to produce fewer fleshy fruits than in white-water systems, thereby favoring seed predators over seed dispersers.

Plants characteristic of different successional stages may also produce different amounts of fruit. Early successional plants typically produce large quantities of fruit relative to their size, probably because they are selected for rapid growth, early maturation and prolific reproduction in a relatively energy rich environment (Levey, 1990). In later stages of succession, plants suffer more competition for light and so must produce fewer and larger-seeded fruits (Janzen, 1969; Foster & Janson, 1985).

Because fruit may be a major energy source sustaining primate populations (Janson & Emmons, 1990), the density

of fruit eating primates may be limited by the lowest seasonal level of fruit availability (Terborgh, 1983; Janson, 1984). In nearly all sites known, fruit production changes seasonally, although much between-year variation of phenological patterns is noted in virtually all long-term studies (Fig. 14.1; Whitmore, 1975; Terborgh et al., 1986; Tutin & Fernandez, 1993; van Schaik, 1986; Chapman et al., in review). Typically, fruit production peaks in the late dry season or early rainy season. This pattern has been explained by several hypotheses based on optimal reproductive patterns in plants (review in van Schaik et al., 1993).

Leaf quantity

As noted above, a leaf is not a leaf is not a leaf. Thus, measuring differences in leaf production may be meaningless in comparison to differences in leaf quality. Therefore we shall review as distinct topics the ecological factors that affect total leaf production vs. leaf quality. Often data are reported only on net primary production, the sum of all new plant parts produced per area, but leaves typically form the major part of the total. It is generally acknowledged that the high productivity and diversity of tropical forests depends on high levels of annual rainfall (Richards, 1996). However, recent analyses suggest that primary productivity actually declines above 2500 mm of rain per year (Fig. 14.2, redrawn after Kay et al., 1997), either because of nutrient leaching or low effective levels of solar irradiance due to dense cloud cover (cf. van Schaik et al., 1993). Maximum primary productivity probably occurs in areas that receive between 2000 and 2500 mm of rainfall, dropping off more than 50% below 2000 mm and less so (by ca. 15%) above 2500 mm. The vast majority of primate study sites occur in areas with 1000–3000 mm of rainfall (see Table 14.1 and other chapters in this volume).

Because of the relatively slight effect of rainfall on average productivity within the moist tropics, much site-to-site variation in leaf production may be due to other causes. In fact, traditional analyses of tropical forest productivity do not even mention rainfall, but instead emphasize the importance of soil type (Lathwell & Grove, 1986; Vitousek & Sanford, 1986). Weathered, old soils do not readily retain cations (calcium, potassium, magnesium) and most of the phosphorus has been lost to leaching or is chemically bound and unavailable to plants (see Schlesinger,

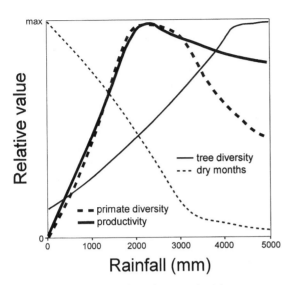

Fig. 14.2. Curves relating plant primary productivity, tree species diversity, number of dry months and primate species diversity to total annual rainfall in neotropical forests (based on Kay *et al.*, 1997). The relative value axis ranges from 0 to the maximum value appropriate to each curve; see Kay *et al.* (1997) for methods and definitions.

1997). Such low–nutrient soils cover over 70% of the tropics (Vitousek & Sanford, 1986) and have relatively low annual carbon fixation rates (Lathwell & Grove, 1986). Although soil weathering is affected by rainfall and temperature (Drever, 1994), it also depends on recency of origin, as weathering is a time-dependent process (Schlesinger, 1997). Geologically active areas in the tropics (Central America, areas close to Andes mountains, east Africa, Indonesia) have young soils that are usually productive despite high rainfall (Whitmore, 1990, pp. 138). Other sites (Guianan Shield) have old weathered soils despite average rainfall. Some south-east Asian sites are in regions of high volcanic activity and thus should have relatively young and productive soils (Burnham, 1975), but they receive so much rainfall that productivity may be low because of low solar irradiance.

Leaf quality

All mammalian folivores have to solve the problem of digesting fiber, and they do so in the same essential way, by fermentation (Cork, 1994). Thus, it is fairly easy to develop criteria of leaf digestibility that apply to folivores anywhere (Milton, 1979; Oates *et al.*, 1990; Ganzhorn, 1992; Barton *et al.*, 1993). In contrast, how folivores deal with plant defensive compounds can be quite varied (Ganzhorn, 1989). One species, the golden bamboo lemur (*Hapalemur aureus*), daily ingests enough cyanide to kill an adult human (Glander *et al.*, 1989), while other lemur species in the same forest specialize on leaves rich in alkaloids or tannins (Ganzhorn, 1989). If each folivorous primate species were restricted to using a distinctive set of plant species because of such chemical specialization, estimating food abundance (and hence population density) for a species would require a detailed understanding of both the consumer's detoxification abilities and the chemical makeup of all plant species in a given site. However, at present there is little direct evidence that secondary compounds limit leaf choice among primate folivores in general (Milton, 1998). Furthermore, at least in Africa and Madagascar, it appears that site-to-site differences in leaf digestibility (protein to fiber ratios) are adequate to predict differences in folivore biomass (Oates *et al.*, 1990; Ganzhorn, 1992), suggesting that chemical specialization does not override the digestibility problem in setting limits to total folivore densities. This significant effect between sites of leaf digestibility on folivore biomass does not preclude the possibility that leaf-eating primate species diverge ecologically within sites based on detoxification ability, because the digestibility of plant species with distinct chemical profiles may vary in parallel between sites.

Are there habitat predictors of community-wide leaf digestibility? Allocation theory in the literature on plant defensive chemistry (Coley *et al.*, 1985) suggests that when plants can easily replace leafy tissue by growth, they should invest relatively little in chemical defense because toxins are often costly to produce (Zangerl *et al.*, 1997). Thus, areas of high nutrient soils should produce both relatively large amounts of leaves per year and leaves of relatively high quality. In fact, the availability of mineral nutrients in tropical soils generally correlates well with nutrient availability in the plants growing on them (Vitousek & Sanford, 1986). Peres (1997*b*) found that the densities of howler monkeys were much higher in floodplain soils than unflooded upland areas, and the former are known to contain higher concentrations of nutrients than the latter in Amazonian white-water river systems (but not

Table 14.1. *Estimates of primate biomass per km², primate community richness and rainfall at selected sites*

Site	Total biomass	Frugivore biomass	Folivore biomass	No. of species of frugivores	No. of species of folivores	Rainfall	Source
Asia							
Kutai Nature Reserve, Borneo	335	273	82	7	3	2177	1
Kuala Lompat, Malaysia	933	596	337	6	2	2120	2
Ketambe, Sumatra	837	702	135	6	1	3000	3
Polonnaruwa, Sri Lanka	2480	300	2180	3	2	1670	4
Sepilok, Sabah	268	184	82.5	6	2	3000	5
Africa							
Kibale National Park, Uganda	2759	682	2077	11	2	1570	6
Kibale National Park, Uganda	1954	754	1200	11	2	1490	7
Tai, Cote d'Ivoire	802	244	558	9	3	1800	8
Tiwai Island, Sierra Leone	1379	785.5	599.5	11	3	2708	9
Douala-Edea, Cameroon	409	198	217	7	1	4000	10
Lopé Reserve, Gabon	318.6	227.9	90.7	8	1	1506	11
Budongo Forest Reserve, Uganda	545	261.5	283.5	5	1	1495	12
Ituri Forest, Zaire	709.6	401.9	307.7	12	3	1802	13
South/Central America							
Manu National Park, Peru	636	456	180	11	1	2080	14
Urucu River, Amazonas, Brazil	381	344	37	13	1	3256	15
Barro Colorado Island, Panama	445	29	416	5	1	2730	16
Raleighvallen, Suriname	251	157	94	8	1	3000	17
La Macarena, Columbia	497.55	411.3	86.25	7	1	2600	18
Santa Rosa, Costa Rica	327	180.7	145.8	3	1	1527	19
Los Tuxtlas, Mexico	171.3	10.3	161	2	1	4900	20
Madagascar							
Morondava N5, Madagascar	685	247	438	7	2	950	21
Morondava CS7, Madagascar	583	357	226	7	2	950	21
Ampijoroa, Madagascar	771	444	327	7	3	1250	21
Ankarana, Madagascar	346	165	181	10	1	1900	21
Analamazaotra, Madagascar	375	208	167	11	3	1700	21
Ranomafana, Madagascar	290	249	41	12	3	2600	21

Note: 1. Rodman, 1978; Rodman pers. comm.; Waser, 1987.

2. Terborgh & van Schaik, 1987; Raemaekers & Chivers, 1980.

3. Terborgh & van Schaik, 1987.

4. Eisenberg *et al.* 1972, Dittus, 1975, Oates *et al.*, 1990.

5. Oates *et al.*, 1990, Davies *et al.*, 1988.

6. Struhsaker, 1975, 1978, 1981, Struhsaker & Leland 1979; Weisenseel *et al.*, 1993, Chapman & Wrangham, 1993, Chapman, unpublished data.

7. Butynski, 1990 (Ngogo only).

8. Terborgh & van Schaik, 1987; Galat & Galat-Luong, 1985, with *Pan* added from Boulière, 1985.

9. Oates *et al.*, 1990 (mid-point in range used).

10. Oates *et al.*, 1990.

in black-water ones: Furch & Klinge, 1989). However, Oates *et al.* (1990) found leaves of high protein/fiber ratios and a high biomass of folivorous primate on very infertile soils in Sierra Leone, apparently because of the prevalence of nitrogen-fixing leguminous trees in the forest (see also Harrison, 1986).

The effect of rainfall on leaf digestibility may be complex. If leaf digestibility parallels total primary productivity, both should reach a maximum between 2000 mm and 2500 mm of rainfall (Fig. 14.2, Kay *et al.*, 1997). However, lower amounts of rainfall are almost always associated with stronger patterns of seasonality in rainfall as well, with areas of less than 2000 mm of rain often having a distinct dry season of three or more months, leading to at least some canopy trees shedding their leaves (Richards, 1996, pp. 167–8). Because deciduousness necessarily limits leaf life span, plants in seasonally dry forests invest less in defensive chemicals for their short-lived leaves than do plants in ever-wet areas with long leaf life spans (up to 14 years: Coley & Barone, 1996). Thus, the effect of seasonality may mean that leaf digestibility increases, rather than decreases, as rainfall drops below 2000 mm.

Leaf quality may also vary over very short spatial and temporal scales. For instance, leaves that receive more light may have substantially higher levels of digestible carbohydrates, proteins, and tannins than shaded leaves on the same tree or species (Ganzhorn, 1995). Likewise, leaves accumulate carbohydrate during the day when they photosynthesize, so that leaves sampled at the end of the day are of significantly higher quality than those sampled in the early morning (Ganzhorn & Wright, 1994). Unless leaves are sampled at a standard time of day and tree position, measures of leaf quality may be unreliable and difficult to compare across species or sites.

Leaf quality may also vary between plant families (see Gupta & Chivers, chapter 2, this volume). In particular, legumes (the bean family) frequently have nitrogen-fixing bacteria associated with their roots, thereby allowing them to produce foliage of relatively high nitrogen content (as well as a high concentration of nitrogen-containing defensive compounds, such as alkaloids). Although nitrogen does not appear to be limiting to productivity in most tropical soils (Vitousek, 1984), the relative abundance of legumes in a forest may determine the productivity for leaf-eating primates (Davies, 1994). In addition to leaf quality, different plant families are distinguished by particular sets of defensive chemicals that may favor particular consumers adapted to cope with these defenses (Ehrlich & Raven, 1964; Ganzhorn, 1989; Jaenike, 1990). Folivore species compositions may respond to dominance by different plant families in different regions.

Finally, leaf quality is expected to decrease in older successional stages of a forest (Coley & Barone, 1996). Because leaf life spans are short (average of 6 months) in tropical trees of early successional stages, they should be less defended chemically than those of mature forest species. In fact, leaves of early successional tree species sustain over four times more damage per leaf than do leaves of mature forest species (48 vs. 11%: Coley & Barone, 1996). Thus, areas that differ in disturbance history may well differ in mean leaf digestibility despite similar rainfall and seasonality (Ganzhorn, 1995).

Altogether, leaf quality for primates appears to be indexed well by the ratio of protein/fiber or protein/(fiber

Notes to Table 14.1 (*cont.*)

11. White, 1994 (mean of 5 neighboring sites).
12. Plumptre *et al.*, 1994; Plumptre & Reynolds, 1995 (density – unlogged forest only); Harvey *et al.*, 1987 (weights assuming 50/50 sex ratio and 1/2 group immature weighing 1/2 adult); Eggling, 1947 (rainfall).
13. Thomas, 1991 (density); Hart, 1985 (rainfall).
14. Terborgh, 1983 and Symington, 1988 for *Ateles belzebuth*; Harvey *et al.* 1987 (weights).
15. Peres, 1993.
16. Glanz, 1982; Eisenberg & Thorington, 1973.
17. Mittermeier & van Roosmalen, 1981 (body weights adjusted).
18. Stevenson, 1996 (assuming 1/2 group immature weighing 1/2 adult).
19. Chapman, 1988; Chapman, 1990*b*, Glander *et al.*, 1991; Chapman, unpublished.
20. Estrada & Coates-Estrada 1985.
21. Ganzhorn, 1992.

+ tannins) despite some specializations among folivorous species for particular classes of plant defensive compounds. These simple indexes of leaf quality increase in more fertile soils, younger successional stages of forest, more seasonally deciduous forests, sunnier locations within a tree crown, at the end of the day, and in selected plant families such as the legumes. These diverse factors make it difficult to predict accurately the leaf quality of a given area from general climatic or latitudinal measures, or how a given primate species will choose leaves within a site.

Covariation in productivity among dietary items

Should the productivity of different diet categories covary among communities? Janzen's (1974) and Coley et al.'s (1985) arguments suggest the answer should be: yes: high-nutrient soils should favor plants that have relatively high growth rates, which should lead to higher leaf turnover (i.e., more flush leaves), and less heavily-defended leaves. All these characters should increase leaf quality, thereby benefitting folivorous monkeys. Insect production should also flourish, which will favor insectivorous primates. Higher fertility should also allow more fruit production, which will support more dedicated frugivores and may encourage greater investment in fleshy fruits as opposed to the mechanically well-defended, non-fleshy species such as Lecithydaceae and Dipterocarpaceae typically found in low-nutrient, low-production areas of the tropics (Janzen, 1974; see also Ganzhorn et al., chapter 4, this volume). Leaf quality and fruit production may also parallel each other across stages of succession, both being highest in early succession, albeit for different reasons (see above). Even though rainfall and primary productivity are broadly correlated (except in the wettest sites: Kay et al., 1997), rainfall may affect fruit and leaf availability to primates in different ways. Fruit-eating species should benefit from higher overall productivity as rainfall increases, but folivore biomass may peak in seasonally dry areas with modest total productivity because these would contain short-lived leaves of high digestibility. Existing data are not sufficient to test whether some resource categories increase more than others as a function of site-related differences in total productivity, given that insect abundance is rarely measured on a per-area basis, and few studies break down primary productivity into its components (Kay et al., 1997). To facilitate future comparisons of resource production among primate communities, we briefly review below (Appendix 14.1) methods for sampling distinct resource types, along with some suggestions for improving comparability of results across studies.

Distinct resources often vary in parallel across the tropical seasons. Many primate ecological studies have measured the availability over an annual cycle of at least two dietary components (insects and fruits, fruits and flush leaves). Peaks of fruit and insect abundance often overlap or coincide markedly (Terborgh et al,. 1986), and the peak of flush leaf production usually occurs only shortly before the peak of ripe fruit production in seasonally dry forests (Terborgh & van Schaik, 1987). However, in the moist and wet forests which comprise most of the tropics, there is little or no correlation between fruit and new leaf production (Terborgh & van Schaik, 1987). Initial comparisons suggested that South America might differ from Old World sites in having greater synchrony between fruit and new leaf production patterns, so that frugivore–folivores might suffer a real dearth of resources in the early dry season (Terborgh & van Schaik, 1987). However, this conclusion was based on data from only one Neotropical study site, and more recent data on other study areas in South America show no or even a negative correlation between the availability of fruit and new leaves (Peres, 1993; Chapman, unpubl. data).

WHAT RESOURCES LIMIT PRIMATE DENSITIES?

There are three major candidate explanations for what determines differences in primate densities across undisturbed sites: (1) total or seasonally low plant productivity; (2) resource diversity; (3) mineral availability. We will deal with these in reverse order, as the evidence for the latter two is limited at present.

Minerals

The evidence that any mineral nutrient may limit primate densities is largely anecdotal. Primates occasionally ingest substances not generally considered as food, such as dirt, and seem to go out of their way to obtain certain foods that would seem to offer few advantages over other dietary items in terms of either energy or protein yield. Leaf eating colobine monkeys have been observed to prefer to eat water plants that contain relatively high levels of sodium (Oates, 1977). Many primates, as well as other herbivorous mammals, congregate at "salt licks" in the New World, where they eat layers of dirt rich in sodium (Emmons & Stark, 1979). However, many primates and other herbivores consume clay-like soils even when these do not contain high concentrations of sodium or other ions (Mahaney et al., 1995). It has been suggested that soil consumption may aid in digesting leaves with high tannin content by precipitating the tannins (Hladik & Chivers, 1994; Struhsaker et al., 1997) or that the clay content in the soil may help to stop diarrhea (Mahaney et al., 1995).

For many large herbivores, phosphorus is a limiting nutrient (Freeland & Choquenot, 1990) because bones require a ratio of calcium to phosphorus between 1:1 and 1:2, whereas most leaf material contains ratios of 5:1 or greater, both for crop plants (Lloyd et al., 1978) and tropical forest species (Vitousek & Sanford, 1986; Theresa Pope, pers. comm.). Furthermore, on highly weathered or acidic tropical soils, absolute phosphorus availability is extremely low (Vitousek & Sanford, 1986; Schlesinger, 1997). High ratios of calcium to phosphorus are detrimental even if phosphorus levels are absolutely high because excess calcium prevents the absorption of phosphorus. Phosphorus limitation might help to explain the striking differences in folivore biomass between African and New World rainforests (cf. Terborgh & van Schaik, 1987), as forestomach fermenters can access from 50–110% more phosphorus in leaf tissue than can caeco-colic fermenters (see above). At least one colobine primate, the proboscis monkey, prefers leaves with higher phosphorus concentrations (Yeager et al., 1997).

Fruit-eating animals may be limited by other minerals, most of which are scarce in fruits. The nearly universal popularity of figs as a major fruit resource in the tropics might be linked to the fact that they have higher absolute concentrations of calcium than do other types of fruits in the same forests (O'Brien et al., 1998). However, fig trees are also often immense and tremendously productive trees, so many animals may prefer them simply because they offer a vast abundance of food in one place (Terborgh, 1986). Calcium availability clearly does not explain the apparent love of anthropoid primates for domesticated bananas, as they contain less than one quarter of the calcium of figs and have the lowest levels of calcium of any domesticated fruit (from data in Watt & Merrill, 1975).

Food species diversity

Ecological theory has predicted that the diversity of consumers should relate to the diversity of resources they can use. As most primates depend largely on plants, plant species diversity should be a simple predictor of primate diversity and perhaps biomass (if only because the more consumer species can coexist in an area, the greater the total density of animals if they do not compete strongly for resources). The relationship of food plant diversity to primate species diversity, however, is complex (Ganzhorn et al., 1997; Kay et al., 1997). Although both increase in parallel with rainfall up to 2500 mm, plant species diversity increases slowly or not at all above this level, whereas primate species diversity actually declines (Fig. 14.2). Ganzhorn et al. (1997) attribute this decline to the rarefaction problem – as plant species diversity increases, the density of each species necessarily declines if total stem density is roughly constant. Thus, at very high plant diversities, consumers may have difficulty locating or foraging efficiently on the plants they need to survive. In this explanation, it is not clear why each consumer species does not simply expand its diet, as expected from foraging theory when preferred resource density is low (Stephens & Krebs, 1986).

Quantitative field comparisons of primate biomass and the biomass of their foods are rare. Gautier-Hion (1983) proposed that the higher consumption of fruit by Lophocebus albigena and Colobus guereza in Makokou relative to Kibale, and the generally more frugivorous diet of the entire community at Makokou, were a result of the fact that fruit is available in greater quantities and there are more fruiting species in central Africa than east Africa. At Makokou, 95 species (> 5 cm DBH) were identified in a 0.4 ha plot (A. Hladik, 1978), while at Kibale, Struhsaker (1975) describes only approximately 34 species in a 1 ha sampling area. Yet, the primate community richness is similar at these two sites and the biomass of primates is higher at Kibale than Makokou (Waser, 1987). In a 3-year study of red colobus monkeys (Procolobus badius), Chapman & Chapman (in review) assessed red colobus density and diet and quantified the diversity of their food trees at six sites in Kibale National Park, Uganda and found that higher-density populations tended to be those with richer diets.

Productivity

At the level of single species, authors have sometimes been successful in explaining variation in primate population density using ecological measures. For instance, Mather (1992) found a nearly perfect ($r = 0.99$) correlation across sites in south-east Asia between the biomass of gibbons (including siamangs) and the proportion of trees that were gibbon food trees. Using ecological variables related to food production, Dunbar (1992) was able to explain the limits to the distribution of gelada baboons, in effect predicting where gelada density reached zero. Such single-species analyses are worth attempting for other primates.

At the level of entire communities, direct evidence relating productivity to primate community structure is scarce because most methods of measuring food production only estimate relative, not absolute, abundances (see above). Comparisons between sites, even when the method used is similar (phenological surveys, plant structure indices, or fruit traps), show little relationship between primate biomass or species diversity and measured food abundance (Chapman & Onderdonk, unpubl. data). Fruit traps could provide comparable absolute measures of fruit production between sites, but the criteria for what is potential primate food and the methods used to interpret the fruit biomass captured have varied between studies (Terborgh, 1983; Janson, 1984; Chapman et al., 1994; see Appendix 14.1). Therefore, researchers have often looked for simple surrogates for resource productivity that can be used in predictive models of primate species richness or biomass (Coe, 1984; Oates et al., 1990; Davies, 1994; Reed & Fleagle, 1995).

The most widespread such substitute for resource productivity is rainfall. At some level there is an obvious relationship between the amount of rain an area receives and primate community structure (e.g., desert primate communities are impoverished and have a relatively low biomass when compared to rainforest communities). Positive correlations between rainfall and primate species diversity are strong within each major primate radiation except Asia (Reed & Fleagle, 1995; but see Gupta & Chivers, chapter 2, this volume). Nonetheless, plant productivity does not increase indefinitely with rainfall (Fig. 14.2), reaching a peak in the New World tropics between 2000 and 2500 mm. Above this level, both plant productivity and primate species diversity in the New World decline

(Kay et al., 1997; see also Peres & Janson, chapter 3, this volume). Similar declines seem to be present in Madagascar and south-east Asia, and may hold true in Africa (Kay et al., 1997). A possible reason for the relatively low primate diversity found in south-east Asia compared to other areas could be the very high levels of rainfall (see Gupta & Chivers, this volume).

Further support for the effect of productivity on primate species diversity comes from comparison of communities on distinct soil types in Amazonia. The impoverished soils of upland terra firme forest support fewer primate species than those of more productive terra firme forests near rivers (see Peres, chapter 15, this volume). Although the most productive chronically-flooded alluvial forests have fewer species than either type of terra firme forest, this low diversity may be caused simply by the direct effect of flooding on insect availability, as most of the species that are absent from flooded forest are small-bodied forms that rely on insects for protein (Peres, 1997a). Thus, primate species diversity in general appears to track total forest productivity. Does the same trend hold for primate biomass?

Because so few studies of primate communities have measured local primary production, most studies on primate biomass use correlates of productivity such as rainfall. Attempts to relate rainfall to primate biomass have revealed somewhat surprising results. Contrary to the common view that primate biomass should increase with total productivity and hence with annual rainfall, Peres (1997b) discovered a negative relationship among sites between the biomass of Alouatta populations and rainfall. Similarly, Ganzhorn (1992) documented that the biomass of folivorous lemur species was negatively related to the annual rainfall of the area, and Gupta & Chivers (this volume) find a similar negative trend of total primate biomass with rainfall in south-east Asia. In an effort to test the generality of these patterns, we present here an analysis of primate community biomass and rainfall from well-studied primate communities across the tropics (Chapman & Onderdonk, unpublished). Our first clear result, consistent with the observations of previous studies (Terborgh & van Schaik, 1987) is that present-day primate biomass is significantly greater in regions with forestomach fermenters (Africa and Asia) than where only caeco-colic fermenters occur (New World and Madagascar). These regional differences parallel the difference in cell-wall

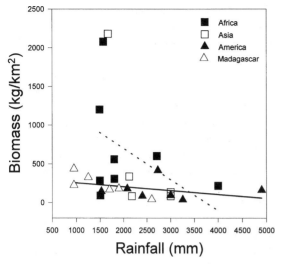 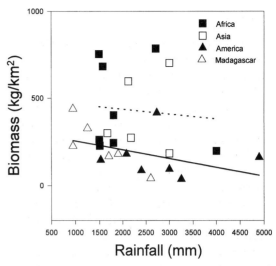

Fig. 14.3. Regression of folivore biomass on rainfall across study sites (data from Table 14.1). Regressions are separated by forestomach (Africa and Asia: dotted line) vs. caeco-colic (South America and Madagascar: solid line) fermenters because of the large mean difference in absolute biomass maintained by folivores of these two types. In an analysis of covariance, forestomach fermenters have higher mean biomass than caeco-colic fermenters ($F(1,22) = 4.95$, $P = 0.039$), and the biomass of both types decreases with increasing rainfall ($F(1,22) = 3.97$, one-tailed $P = 0.030$). The difference in the slopes between digestive types is not significant ($F(1,22) = 2.44$, $P = 0.132$).

Fig. 14.4. Regression of frugivore biomass on rainfall across study sites. Data are separated by region as in Fig. 14.1. In an analysis of covariance, the effect of region is significant, but that of rainfall is not (respectively, $F(1,22) = 7.25$, $P = 0.013$; $F(1,22) = 1.84$, one-tailed $P = 0.094$). Although the difference in the slopes of the rainfall effect between areas is not significant ($P = 0.64$), the regression for America and Madagascar alone is significant ($F(1,11) = 4.3453$, $P = 0.03$, one-tailed), whereas that for Africa and Asia is not ($F(1,11) = 0.09$, $P = 0.38$, one-tailed).

digestive efficiency between howlers and colobine monkeys (Milton, 1998). Once these large-scale differences are accounted for, the biomass of folivores decreases with increasing rainfall (Fig. 14.3) and frugivores show similar, but not statistically significant, trends (Fig. 14.4). The parallel decline of primate biomass (Fig. 14.3) and primary production with rainfall above 2500 mm (Fig. 14.2) might suggest that primate biomass is broadly limited by productivity.

The inferential link between primate biomass and productivity is at best weak, however. First, rainfall explains only a small part of the variation in primate biomass in our regressions (0.1–23%) and in primary production in moist New World sites (< 10%: Kay *et al.*, 1997). Second, nearby sites of similar rainfall and seasonality can differ markedly in primate biomass, such as the neighboring forest communities of Kanyawara and Ngogo (2759 kg/km² vs. 1954 kg/km², 1660 mm vs. 1490 mm rainfall) in Kibale

National Park, Uganda (Chapman, unpubl. data). Almost surely, some of the remaining variation in both primary productivity and primate biomass among sites is related to soil fertility. In fact, Peres (chapter 15, this volume) found that primate biomass increases markedly with expected soil fertility among sites of similar rainfall in Amazonia. How primate biomass relates to productivity will remain a puzzle until we possess much better measures of primary production for many sites with known primate densities.

Food quality

Food quality is an important determinant of food choice for dedicated as well as opportunistic folivores (Milton, 1979; Barton & Whiten, 1994). Do differences in leaf digestibility between communities contribute substantially to explaining variation in primate biomass? Researchers have met with considerable success predicting leaf-eating primate biomass using indices of leaf quality alone. Studies of colobines (Waterman *et al.*, 1988; Oates *et al.*, 1990) and folivorous lemurs (Ganzhorn, 1992) have found a significant positive correlation between the biomass of folivorous primates and the simplest index of leaf digestibility, the ratio of protein/fiber or protein/(fiber + tannins). Leaf quality differences explained 90% of the variation in colobine biomass and 76% of the variation in folivorous lemur biomass. Such strong relationships are surprising, given the simplicity of this leaf quality index, the difficulty of determining primate biomass, and the number of alternative explanations available to explain primate biomass (Freeland, 1977; Cant, 1980; Terborgh & van Schaik, 1987). Less clear results are found in southern Asia, where the protein/fiber ratio of dominant plant families varies nearly 20-fold among sites, but does not correlate well with primate biomass for the region as a whole (Gupta & Chivers, chapter 2, this volume).

Another way to test the importance of leaf digestibility to primate biomass uses the observation that plants invest less in structural toughness and chemical defense of short-lived leaves in deciduous forests than in leaves that will last for several years as in everwet forests (Janzen, 1975; Coley & Barone, 1996). Thus, leaves in more seasonal environments should have higher digestibility. This relationship allows a direct test of the importance of leaf quality vs. leaf production in determining folivore biomass. The total primary production of more deciduous tropical forests (at 1000–2000 mm of rainfall) is generally less than those of moist or wet forests with more than 2000 mm of rain (Fig. 14.2), but the leaf quality in deciduous forests is expected to be greater. If folivore biomass depends primarily on total primary production, it should be lower in more deciduous forests than in moister forests; the opposite pattern should hold if leaf quality is more important to folivore biomass.

Existing data support the importance of leaf quality. In Madagascar, the biomass of folivorous lemurs (Ganzhorn, 1992) increases with increasing seasonality, as does that of howler monkeys in the New World (Peres, 1997b). Our regressions show the same trend. We separated sites with less than 2000 mm as seasonal/more deciduous from those with more than 2000 mm of rainfall. As expected from the leaf quality hypothesis, folivore biomasses tended to be higher in the former than in the latter ($t = -1.86$, $P = 0.077$), after accounting for digestive strategy. No such difference is apparent for frugivorous species in a comparable analysis ($t = -0.01$, $P = 0.99$), which reinforces the

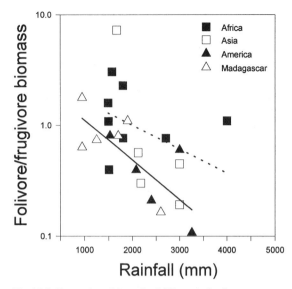

Fig. 14.5. Regression of the ratio of folivore to frugivore biomass on rainfall across study sites. Data are separated as in previous figures. Data for Barro Colorado Island (Panama) and Los Tuxtlas (Mexico) are removed because these had strongly deviant values of this ratio within the America/Madagascar data set (the two sites were, respectively, 2.1 and 2.4 standard deviations from the overall mean). Both sites are actual or habitat islands and so may not represent the equilibrium tendency among tropical forests in general (see also Kay et al., 1997). The effect of rainfall on log(folivore/frugivore biomass) is highly significant ($F(1,22) = 5.02$, $P = 0.018$, one-tailed), but neither the slope nor the intercept of the regression differs between areas.

inference that the result for folivores is not due to overall productivity. Furthermore, the fraction of total primate biomass that consists of folivores declines linearly from the driest sites in our sample toward the wettest (Fig. 14.5). It appears that leaf quality really is more important than total leaf production in explaining folivore biomass, although presumably the critical variable is really the total production of high-quality leaves.

CONCLUSIONS

Despite the initial appearance of superabundant food for primates in tropical forest, many studies support the idea that primates do face food shortages, either seasonally in the case of frugivores, or in terms of sparse or ephemeral high-quality leaves for folivores. Although food abundance has been related directly to the population density of few primate species (Janson, 1984), general measures or correlates of habitat productivity have been successful in explaining some of the between-site variation in primate species richness and biomass. Because folivorous primates often comprise a major fraction of the total biomass of a primate community (Fig. 14.5), nearly three-quarters of the variation in total primate biomass derives from variation in folivore biomass in our data. Therefore, explaining variation in total community biomass of primates largely is an exercise in explaining variation in folivore biomass. The somewhat surprising conclusion from our review of the literature and our own data is that folivore biomass actually declines with increasing rainfall, and may decline even in areas of greater primary productivity. The success of a simple index of average leaf quality in predicting between-site variation in folivore biomass suggests either that leaf quality is more important than leaf quantity, or that this index of leaf quality is a more precise reflection of habitat productivity than is rainfall. Perhaps the constraints of being arboreal folivores of modest body size place leaf quality above leaf quantity in affecting their population density.

To distinguish among these hypotheses will require far more complete data on various measures of food production than are currently available. We emphasize in Appendix 14.1 that the major common methods of estimating fruit and insect availability do not currently allow consistent comparisons of absolute production between study sites. Phenological methods should ideally be complemented with litterfall (fruit, seed, leaf) traps and vice versa to provide more informed and objective absolute measures of fruit and leaf production. Insect samples based on known areas or volumes of vegetation, although tedious, are the only method likely to provide absolute density data that can be compared between sites. Leaf quality may vary across short distances and time spans, so care must be taken to sample leaves in the same location in the tree and time of day that the primates actually ingest them.

Many interesting questions about the determinants of primate diversity and biomass remain to be investigated. First, are these aspects of community structure determined more by the total annual or the minimum seasonal availability of foods. Resolving this question will have to await better data on absolute food production for a variety of sites. Second, is the niche partitioning of leaves according to toxin chemistry, as found in lemurs, more general to folivores, at least where more than one species coexist? Although comparisons of general foraging ecology of coexisitng colobines do exist, focused studies of comparative leaf chemistry of their food choices appear still to be lacking. Third, do forestomach fermenters have any real advantage over caeco-colic fermenters in terms of energy extraction per unit of time or ability to handle plant defensive compounds? What really is the basis for the more than two-fold difference in folivore biomass between areas with forestomach- and those with only caeco-colic-fermenting primate species (Fig. 14.3)? Why does this difference in biomass appear both in folivorous species and in frugivore-folivores in the same regions (Fig. 14.4)? Fourth, does disturbance history affect the quality of an existing habitat differently for folivores vs. frugivores (see Ganzhorn, 1995)? This last question is of special concern as more and more primate habitats become disturbed by human activity.

APPENDIX 14.1: METHODS AND CAUTIONS FOR SAMPLING THE ABUNDANCE OF PRIMATE DIETARY RESOURCES

Few quantitative data exist to support the claim that primate populations are limited by the availability of food resources. Few studies provide data on the densities of every primate species in a community and have simultaneously collected data on the food available to those species. The study of primates resources is greatly hindered by methodological difficulties that reduce the reliability and comparability of results across study sites.

The first methodological difficulty is answering the seemingly simple question of: "what is primate food"? Tropical rainforests are famously species-rich in plants, with up to 200 plant species in a 25 by 25 m plot in Africa (Hall, 1977; Hall & Swaine, 1981), and over 1000 woody species in an area of a few square kilometers (Croat, 1978). Only a fraction of this diversity is consumed by any one primate species, but knowing exactly which plants are potential foods is difficult. Typical lists of plant foods for a single primate species vary from 60 to over 200 species, and often these lists grow even after several years of study.

Some food items may be so low on a primate species' preference list that they would be eaten only during periods of severe food scarcity (cf. Stephens & Krebs, 1986). During five years of observation of a spider monkey (*Ateles geoffroyi*) community in Santa Rosa National Park, Costa Rica (Chapman, 1990*a*), animals were rarely seen to eat the fruits of *Bursera simaruba*; however in the sixth year, the community fed on this fruit extensively during a month when little else was available. Furthermore, even the fraction of the diet that is fruit vs. other components can vary greatly over time (Terborgh, 1983; Chapman & Chapman, 1990), making it difficult to assess the potential diet of a species. For example, MacKinnon (1974, 1977) studied the diet of orangutans (*Pongo pygmaeus*) in Sumatra. In one month the orangutans were primarily frugivorous, spending 90% of their feeding time eating fruit and only 5% eating leaves, and 5% eating insects. In another month this population was primarily folivorous, spending 75% of their time eating leaves, 15% eating bark, and only 10% eating fruit. Similar variation can be found in spider monkeys, that are typically considered ripe fruit specialists (Klein & Klein, 1977). Chapman (1988) described that in one month the spider monkey community in Santa Rosa National Park, Costa Rica, ate only fruit (100% of their feeding time); however, in another month they ate primarily leaves (86%) and little fruit (14%), and in yet another month, insects were a common component of their diet (30%). Only long-term study can reveal a reasonably complete profile of the foods used by one or more primate species.

Without complete food lists, methods of estimating food availability must make some assumptions about what to consider food. At one extreme, some studies assume that any fleshy fruit is a potential food for a fruit-eating primate, even when the study species has not been observed to eat that fruit (Terborgh, 1983). This assumption is partly justified by the fact that there is much overlap in the fruit part of the diets of co-existing primate species (Terborgh, 1983), and the fact that primates eat many species of fruit that apparently have evolved to be dispersed by other animals (Janson, 1983). However, this premise is clearly only an approximation; it is often made for convenience when fruit traps are used to measure fruit production, as it is much easier to assign the half-decomposed remains of a fruit to broad categories (fleshy, dry, winged) than to particular species (but see below). At the other extreme, many observational studies consider as food only the species which the study animals have been seen to eat in the study site. Because behavioral and ecological data are often collected at the same time, it usually happens that new food species are being added to the diet during a study; the availability of these resources can only be monitored after they are known to be eaten. An intermediate solution for a primate species that has been studied in several sites is to include as possible foods all the species known to be consumed by that primate in any area; any of these foods that occur in a given study site should then be monitored and measured. Even if you know which foods you need to measure, assessing their production for an entire primate community requires many distinct methods, as most primate communities are diverse, including species that use very different types of resources, as well as many that eat a mixed diet of fruit and either insects or leaves. We now review what is known about variation in the different major classes of primate resources and how they affect primate community structure.

Insects

Insects make up a large proportion of animal biomass in tropical habitats (Fittkau & Kline, 1973) and are very diverse even in single localities. For example, Stork (1991) used insecticide fogging to sample the arboreal arthropod fauna in 10 individual Bornean rainforest trees and altogether found 23 874 individuals of at least 3000 species. Despite the abundance and richness of arboreal insects, rarely has insect availability been estimated in conjunction with studies of primates. The likely reason is that the great diversity of insects, their various specialized microhabitats and seasonal changes in abundances make rigorously quantifying their abundance a difficult task. The problem would be less daunting for a specialized insect-eater, but primates often search for insects in many distinct places and ways (see above). Thus, they almost certainly consume many dozens to hundreds of insect species, the availability of which to the primate may depend on more than just the absolute abundance of the insect. Humans might mimic the foraging style of primates to estimate their foraging success, but most primates forage in the forest canopy, a difficult task for a researcher to replicate.

Several means have been employed to quantify insect abundance ranging from traditional sweep samples, to sticky boards, to fogging of entire trees (Southwood,

1978). Hladik *et al.* (1980) used a combination of methods to quantify seasonal changes in the insects available to the prosimian community of Morondava, Madagascar. An ultraviolet lamp was used to attract flying insects, and the biomass of insects coming to the light during two hours at the beginning of each night was measured. To estimate the seasonal abundance of caterpillars, feces were collected from litter traps. Sampling regimes must be designed to accommodate the fact that invertebrate distributions are not homogenous. For example, Hladik *et al.* (1980) found that invertebrate abundance was twice as great in dense vegetation neighboring a temporary pond than in dryer areas. In any case, such indirect methods can only index relative abundance of insects. Such an index may be adequate for estimating seasonal trends in insect availability (Janzen, 1973), but they are not likely to be useful to compare absolute abundances between different areas even within the same general study site. Even as a measure of seasonal availability, indirect methods may be biased toward a few insect orders (e.g., nocturnal lepidoptera for ultraviolet lamps). These may not be the taxa used by primates nor is their abundance necessarily correlated with the kinds found in primate diets. To increase comparability of insect abundance across study sites, effort should be directed toward direct censuses of the actual prey used by primates whenever this technique is feasible (e.g. searching a known number of leaves by hand or visually: Terborgh *et al.*, 1986).

Fruit

Fruit resources have primarily been quantified using either phenology methods or fruit traps (Blake *et al.*, 1990; Hutto, 1990; Janson & Emmons, 1990; Chapman *et al.*, 1994). Phenology transects (or quadrats) involve the establishment of areas in which trees are routinely monitored for the presence of fruit. Typically a subset of all trees within the sampling area is selected for monitoring; a standard that is often used is to monitor all trees >10 cm DBH (Diameter at Breast Height) on a monthly basis. By selecting trees above a specific size, one is making the assumptions that trees smaller than the size criteria used are incapable of producing fruit that are useful to primates; this assumption may often be correct for large-bodied canopy species, but is clearly false for small-bodied understory species (Terborgh, 1983). Instead of using a criterion that is based on size, researchers can monitor only those species

known to be eaten by particular primates. The problem with this alternative is that new food species often appear as a study progresses.

Various means are used to assess the size of fruit crops on individual trees. However, there are few guidelines available to indicate which procedures are most appropriate in different situations (Peters *et al.*, 1988; Chapman *et al.*, 1992). In some instances no estimate of crop size is made and ripe fruit availability is indexed simply as the number of trees bearing fruit during a given sampling period. This method suffers in that it ignores large differences in production of fruit between individual large and small trees. For many species, determining if the fruit on the tree is ripe is not difficult, because ripening is often associated with a color change, fruits ripen over a short period, and ripe fruits often fall to the forest floor soon after the time when frugivores start eating them (but they may be delayed longer in falling, see Zhang & Wang, 1995). However, for some species, it is difficult or impossible to distinguish ripe from unripe fruit based on a visual assessment through binoculars. For example, *Monodora myristica* bears large (16 cm diameter) green fruit that exhibit no visual signs of ripening. Although its fruits may remain on the tree at full size for up to 3 months, when chimpanzees (*Pan troglodytes*) and baboons (*Papio anubis*) view these fruits as being ripe, they often remove an entire fruit crop in a few days (Balcomb, pers. comm).

Other researchers have assumed that larger trees will produce more fruit than smaller trees and have used a variety of indices of tree size to weight the value of a fruiting tree (e.g., DBH – Leighton & Leighton, 1982; Peters *et al.*, 1988; Chapman, 1990*b*; crown volume – Symington, 1988). Using such indices assumes that the relationship between tree size and fruit production is the same for a variety of species, which is likely inappropriate. Other researchers have attempted to visually assess fruit crop size and assign a relative rank to crop size. Chapman *et al.* (1992) compared the accuracy and precision of three methods of estimating fruit abundance on tropical trees: tree diameter (DBH), crown volume, and visual estimation. DBH was the most consistently accurate method and exhibited low levels of inter-observer variability. Generally, crown volume was neither precise nor accurate. The visual estimation method was accurate for trees with very large fruits, but inter-observer variability was high and this method required a large time investment per tree.

Although DBH is simple and accurate to measure, the relationship between DBH and fruit production is not known either within or between species. It is very unlikely to be perfectly proportional even within species. DBH is a linear function of tree size, whereas fruit production within a species should scale with the area or volume of the tree crown, which are respectively quadratic or cubic measures of tree size. For instance, a tree of 100 cm DBH is only twice as big in linear dimension as a tree of 50 cm DBH, but the difference in fruit production between them is likely to be in the range of fourfold to eightfold. Between plant species, additional problems creep in. Different life-history adaptations or habitat light regimes may lead to very different patterns of allocation to fruit production among species of similar DBH. It appears that sunlit canopy and emergent trees are much more productive per unit of leaf area than are severely shaded understory trees and shrubs (cf. Levey, 1990; Ganzhorn, 1995). Because of these problems, summing or averaging DBHs is unlikely to be a reliable way of gauging mean production. If the regression between log(fruit production) and log(DBH) is linear, then the most reliable gauge of mean fruit production is first to calculate the mean of the log(DBH) of the sampled trees, and then use the regression equation to estimate the mean log(fruit production).

Counting trees is relatively easy, detecting ripe fruit is generally not too difficult, but quantifying biomass is difficult. Although it might be possible to count all the fruit in a tree, doing so would be tedious and prone to many inaccuracies. Therefore, nearly all phenology methods use an index of fruit biomass which depends on a number of assumptions. Typically, the fruit in a tree will be scored on a relative scale from 0 (no fruit) to a small integer (often 3 or 4) which represents 'many' (e.g., Koenig et al., 1997). Usually, these scores are species-specific, so a 'many' score means different amounts of fruit for different tree species. The scaling of fruit abundance between 0 and 'many' is not necessarily linear and is usually not specified in detail. An alternative semi-quantitative index is to note the $\log_{10}$ of the estimated number of fruits in the tree: 0 for 1–9, 1 for 10–99, 2 for 100–999, etc. Although estimating the order of magnitude of number of fruits is more time-consuming than a 0-to-3 scale, it is not very much more so, and at least provides a rough idea of absolute fruit numbers. Pulp mass per fruit can be determined by dissection of individual fruits (Janson, 1985).

In contrast to phenology transects, fruit traps are used to estimate fruit production directly. Fruit traps are simply baskets that are placed, usually at regular intervals, somewhat away from existing trails to catch falling fruit. Because one can weigh the fruits that fall into traps, this method has the benefit of providing information directly related to biomass. However, fruit traps measure fruit fall, not fruit production. Terborgh (1983) points out a number of biases inherent in the fruit trap method. Fruit fall is potentially underestimated during periods of fruit scarcity when frugivores consume a greater proportion of the available fruit than during period of fruit abundance. Fruit traps are biased against catching more preferred fruits that are removed by frugivores and do not fall into the traps in the same proportion as less preferred fruits. Fruit traps also may contain a large number of fruits that are aborted by their parent trees. Finally, this method is likely biased against plant species that produce fruits that ripen slowly over a long period, since it is probable that a greater proportion of the fruits of slow ripening species are eaten and therefore do not fall into the traps. All these biases can be markedly reduced by counting not fruits that fall into traps, but rather mature seeds, and then estimating the equivalent biomass of pulp produced by the observed catch of seeds per species, based on detailed measurement of the components of freshly-collected ripe fruits (Janson, 1984; Janson & Emmons, 1990). With this procedure, the removal of fruit by frugivores does not affect the estimated production as long as seeds are not destroyed in the animals' digestive systems – a trap will catch the seed either below the tree that produced it or after it is deposited by the frugivore. Likewise, aborted fruits are not counted as they do not produce mature seeds. There is still the problem that seeds typically fall into traps up to a month after the fruits are available on the tree (Zhang & Wang, 1995), so that ideally production of pulp for each species should be allocated seasonally according to independent phenological estimates of relative ripe fruit availability for that species (see Janson, 1984).

In a review of studies using fruit traps to quantify habitat-wide fruit availability, Chapman et al. (1994) point out that fruit traps typically sample only a very small proportion of a focal species home range (on average 0.004% of the area used by the study animal). The fact that fruit traps cover only a small proportion of the total area of interest can lead to biases. If traps are placed under rare tree species, which produce many fruits (e.g., *Ficus* sp.) or large fruits (e.g., *Monodora myristica*), fruit traps may overestimate habitat-wide fruit abundance. Such biases are exaggerated by non-random fruit trap placement, so traps should be located either randomly or at fixed intervals with a random starting point to reduce the possibility of sampling only more productive trees. Another potential criticism of fruit traps is that rodents or other frugivores could systematically remove fruit or seeds from the traps. However, data collected by Goldizen *et al.* (1988) and Chapman *et al.* (1994), suggest that this is not typically a large problem.

Although sympatric primate species overlap broadly in their use of fruit (Terborgh, 1983), not all fruits are of similar quality and not all primates have similar abilities to handle the various secondary compounds found in fruits. For example, whereas tannins are avoided by many primates (Oates *et al.*, 1977, 1980; Waterman *et al.*, 1980; Glander, 1982; Leighton, 1993) this does not appear to be the case with gorillas (*Gorilla gorilla*, Calvert, 1985; Rogers *et al.*, 1990) or spider monkeys (Howe & Vande Kerkhove, 1981). Similarly, the fruits of *Strychnos mitis* are readily eaten by redtail monkeys (*Cercopithecus ascanius*, Lambert, 1997) and blue monkeys (*C. mitis*, Rudran, 1978), while they are steadfastly ignored by chimpanzees (Lambert, 1997). This particular genus of tropical tree is laden with compounds that are toxic to mammals: the fruit pulp, root, bark and leaves of this genus are reported to contain high levels of phenolics, terpenes, flavonol glycosides, and various alkaloids (Thepenier *et al.*, 1990). Thus, an accurate assessment of fruit availability should take into account any specialization in fruit consumption for each primate species. Fruit or seed traps allow the researcher to assess production of all fruiting species (identified using seed reference collections), subsets of which can be used afterwards to produce quantitative measures of resource availability tailored for a particular study species (Janson, 1984).

Leaves

When quantifying the availability of leaf resources, one faces many of the same issues considered when dealing with fruit. Correspondingly, researchers have used the same sorts of methods to assess leaf availability: transects (Oates, 1977) or leaf-fall traps (Proctor, 1980), with the same range of problems and biases as for fruit. In addition, however, the quality of leaves is probably far more important a confounding issue than it is for fruits (see above). If the fraction of leaf material in a forest that is edible varies considerably from site to site or season to season, then measures of gross production may have little value.

Although the variety of possible plant defensive chemicals is vast, as is the list of chemical techniques needed to detect them, some fairly simple measures of leaf chemistry appear to be important predictors of food quality across primate species (Barton *et al.*, 1993). The ratio of protein to fiber or to fiber-plus-tannin has been shown to predict diet choice in both folivores (Milton, 1979; Oates *et al.*, 1990) and in frugivore–folivores (Whiten *et al.*, 1990). It is odd that this index works at all given the diversity and toxicity of many plant defensive chemicals, but several factors may help to explain this seeming paradox. First, generalized mammals may have the ability to detoxify at least small quantities of many kinds of plant defensive chemicals, so that these do not represent an absolute barrier to ingestion. Second, the cost of detoxifying a given plant defensive chemical may be offset by high protein content, so that primates may eat a nutritious plant leaf despite high defensive chemical content. Third, primates prefer to eat expanding leaves that typically possess low fiber content, and may also be limited in the concentration or kinds of secondary chemicals they possess (Orians & Janzen, 1974; Coley and Barone, 1996). Although different leaf-eating lemurs in Madagascar may specialize on the use of distinct plant defensive chemicals (Ganzhorn, 1988), it is not known to what extent this is true of other folivorous primates.

Despite the seeming simplicity of the protein/fiber ratio as a measure of leaf quality, some possible problems must be kept in mind. As noted above, leaf chemistry can vary markedly in a single tree across space (sunlit vs. shaded branches, Ganzhorn, 1995) and time (morning vs. evening, Ganzhorn & Wright, 1994). Greater comparability of leaf quality may be achieved in future surveys across study sites by standardizing the place and time of leaf collection (for instance, always collect leaves at midday from branches exposed to sunlight). Researchers interested in understanding leaf selection by individual primates should take care to sample the leaves at the same location and time of day that the animals actually ate them.

REFERENCES

Altmann, S. A. (1998). *Foraging for Survival. Yearling Baboons in Africa*. Chicago: University of Chicago Press.

Ayres, J. M. (1989). Comparative feeding ecology of the uakari and bearded saki *Cacajao* and *Chiropotes*. *Journal of Human Evolution*, 18, 697–716.

Barton, R. A., Whiten, A. Byrne, R. W. & English, M. (1993). Chemical composition of baboon plant foods: implications for the interpretation of intra- and interspecific differences in diet. *Folia primatologica*, 61, 1–20.

Barton, R. A. & Whiten, A. (1994). Reducing complex diets to simple rules: food selection by olive baboons. *Behavioural Ecology and Sociobiology*, 35, 283–93.

Björnhag, G. (1994). Adaptations in the large intestine allowing small animals to eat fibrous foods. In *The Digestive System in Mammals: Food, Form, and Function*, ed. D. J. Chivers & P. L. Langer, pp. 287–309. Cambridge: Cambridge University Press.

Blake, J. G., Loiselle, B. A., Moermond, T. C., Levey, D. J. & Denslow, J. S. (1990). Quantifying abundance of fruits for birds in tropical habitats. *Studies in Avian Biology*, 13, 73–9.

Bourlière, F. (1985). Primate communities: their structure and role in tropical ecosystems. *International Journal of Primatology*, 6, 1–26.

Brown, J. H. (1995). *Macroecology*. Chicago: University of Chicago Press.

Burnham, C. P. (1975). The forest environment: soils. In *Tropical Rain Forests of the Far East*, ed. T. C. Whitmore, pp. 103–20. Oxford: Clarendon Press.

Butynski, T. M. (1990). Comparative ecology of blue monkeys (*Cercopithecus mitis*) in high- and low-density subpopulations. *Ecological Monographs*, 60, 1–26.

Calvert, J. J. (1985). Food selection by western gorilla (*G. g. gorilla*) in relation to food chemistry. *Oecologia*, 65, 236–46.

Cant, J. G. H. (1980). What limits primates? *Primates*, 21, 538–44.

Chapman, C. A. (1988). Patterns of foraging and range use by three species of neotropical primates. *Primates*, 29, 177–94.

Chapman, C.A. (1990*a*). Association patterns of spider monkeys: The influence of ecology and sex on social organization. *Behavioral Ecology and Sociobiology*, 26, 409–14.

Chapman, C. A. (1990*b*). Ecological constraints on group size in three species of neotropical primates. *Folia Primatologica*, 55, 1–9.

Chapman, C. A. & Chapman, L. J. (1990). Dietary variability in primate populations. *Primates*, 31, 121–8.

Chapman, C. A. & Chapman, L. J. (1991). The foraging itinerary of spider monkeys: When to eat leaves? *Folia Primatologica*, 56, 162–6.

Chapman, C. A., Chapman, L. J., Wrangham, R. W., Hunt, K., Gebo, D. & Gardner, L. (1992). Estimators of fruit abundance of tropical trees. *Biotropica*, 24, 527–31.

Chapman, C. A. & Wrangham, R. W. (1993). Range use of the forest chimpanzees of Kibale: implications for the evolution of chimpanzee social organization. *American Journal of Primatology*, 31, 263–73.

Chapman, C. A., Wrangham, R. W. & Chapman, L. J. (1994). Indices of habitat-wide fruit abundance in tropical forests. *Biotropica*, 26, 160–71.

Chapman, C. A., Wrangham, R. W., Chapman, L. J., Kennard, D. K. & Kaplan, A. E. (in review). Fruit and flower phenology at two sites in Kibale National Park, Uganda.

Charles-Dominique, P. (1977). *Ecology and Behaviour of Nocturnal Primates. Prosimians of Equatorial West Africa*. New York, NY: Columbia University Press.

Chivers, D. J., Andrews, P., Preuschoft, H., Bilsborough, A. & Wood, B. A. (1984). Food acquisition and processing in primates: concluding discussion. In *Food Acquisition and Processing in Primates*, ed. D. J. Chivers, B. A. Wood & A. Bilsborough, pp. 545–56. New York: Plenum Press.

Chivers, D. J. & Hladik, C. M. (1980). Morphology of the gastro-intestinal tract in primates: comparisons with other mammals in relation to diet. *Journal of Morphology*, 166, 337–86.

Cipollini, M. L. & Levey, D. J. (1997). Secondary metabolites of fleshy vertebrate-dispersed fruits: Adaptive hypotheses and implications for seed dispersal. *American Naturalist*, 150, 346–72.

Coe, M. (1984). Primates: Their niche structure and habitats. In *Food Acquisition and Processing in Primates*, ed. D. J. Chivers, B. A. Wood & A. Bilsborough, pp. 1–32. New York: Plenum Press.

Coehlo, A. M., Bramblett, C. A., Quick, L. B. & Bramblett, S. S. (1976). Resource availability and population density in primates: a sociobioenergetic analysis of the energy budgets of Guatemalan howler and spider monkeys. *Primates*, 17, 63–80.

Coley, P. D. & Barone, J. A. (1996). Herbivory and plant defenses in tropical forests. *Annual Review of Ecology and Systematics*, 27, 305–35.

Coley, P. D., Bryant, J. P. & Chapin, F. S. III. (1985).

Resource availability and plant anti-herbivore defense. *Science*, **230**, 895–99.

Cork, S. J. (1994). Digestive constraints on dietary scope in small and moderately-sized mammals: how much do we really understand? In *The Digestive System in Mammals: Food, Form, and Function*, ed. D. J. Chivers & P. L. Langer, pp. 337–91. Cambridge: Cambridge University Press.

Croat, T. B. (1978). *Flora of Barro Colorado Island*. Stanford, CA: Stanford University Press.

Davies, A. G. (1994). Colobine populations. In *Colobine Monkeys: Their Ecology, Behaviour and Evolution*, ed. A. G. Davies & J. F. Oates, pp. 285–310. Cambridge: Cambridge University Press.

Davies, A. G., Bennett, E. L. & Waterman, P. G. (1988). Food selection by two southeast Asian colobine monkeys (*Presbytis rubicunda* and *Presbytis melalophos*) in relation to plant chemistry. *Biological Journal of the Linnean Society*, **34**, 33–56.

Dennis B. & Taper M L. (1994). Density dependence in time series observations of natural populations: Estimation and testing. *Ecological Monographs*, **64**, 205–24.

Denslow, J. S. & Moermond, T. C. (1982). The effect of accessibility on rates of fruit removal from neotropical shrubs: an experimental study. *Oecologia*, **54**, 170–6.

Dittus, W. P. J. (1975). Population dynamics of the toque monkey, *Macaca sinica*. In *Socioecology and Psychology of Primates*, ed. R. H. Tuttle, pp. 125–151. The Hague, Netherlands: Mouton.

Drever, J. I. (1994). The effect of land plants on weathering rates of silicate minerals. *Geochemica et Cosmochimica Acta*, **58**, 2325–32.

Dumbacher, J. P., Beehler, B. M., Spande, T. F., Garraffo, H. M. & Daly J. W. (1992). Homobatrachotoxin in the genus *Pitohui*: chemical defense in birds? *Science*, **258**, 799–801.

Dunbar, R. I. M. (1992). A model of the gelada socioecological system. *Primates*, **33**, 69–83.

Eggling, W. (1947). Observations on the ecology of the Budongo rainforest, Uganda. *Journal of Ecology*, **34**, 20–87.

Ehrlich, P. R. & Raven, P. H. (1964). Butterflies and plants: a study in coevolution. *Evolution*, **18**, 586–608.

Eisenberg, J. F., Muckenhirn, N. A. & Rudran, R. (1972). The relation between ecology and social structure in primates. *Science*, **176**, 863–74.

Eisenberg, J. F. & Thorington, jr., R. W. (1973). A preliminary analysis of a Neotropical mammal fauna. *Biotropica*, **5**, 150–161.

Emmons, L. H. & Stark, N. M. (1979). Elemental composition of a natural mineral lick in Amazonia. *Biotropica*, **11**, 311–13.

Estrada, A. & Coates-Estrada, R. (1985). A preliminary study of resource overlap between howling monkeys (*Alouatta palliata*) and other arboreal mammals in the tropical rain forest of Los Tuxtlas, Mexico. *American Journal of Primatology*, **9**, 27–37.

Fa, J. E. & Purvis, A. (1997). Body size, diet and population density in afrotropical forest mammals: A comparison with neotropical species. *Journal of Animal Ecology*, **66**, 98–112.

Fedigan, L. M. (1990). Vertebrate predation in *Cebus capucinus*: meat eating in a neotropical monkey. *Folia Primatologica*, **54**, 196–205.

Fittkau, E. J. & Kline, H. (1973). On biomass and trophic structure of the central Amazonian rain forest ecosystem. *Biotropica*, **5**, 2–14.

Foley, W. J. & McArthur, C. (1994). The effects and costs of allelochemicals for mammalian herbivores: an ecological perspective. In *The Digestive System in Mammals: Food, Form, and Function*, ed. D. J. Chivers & P. L. Langer, pp. 370–91. Cambridge: Cambridge University Press.

Foster, R. B. (1982). Famine on Barro Colorado Island. In *The Ecology of a Tropical Forest: Seasonal Rhythms and Long-term Changes*, ed. E. G. Leigh, A. S. Rand & D. M. Windsor, pp. 201–12. Washington, DC: Smithsonian Instituion Press.

Foster, S. A. & Janson, C. H. (1985). The relationship between seed size, gap dependence, and successional status of tropical rainforest woody species. *Ecology*, **66**, 773–80.

Freeland. W. (1977). *The Dynamics of Primate Parasites*. Ph.D. Thesis. University of Michigan, Ann Arbor.

Freeland, W. J. & Choquenot, D. (1990). Determinants of herbivore carrying capacity: plants, nutrients and *Equus asinus* in northern Australia. *Ecology*, **71**, 589–97.

Furch, K. & Klinge, H. (1989). Chemical relationships between vegetation, soil and water in contrasting inundation areas of Amazonia. In *Mineral Nutrients in Tropical Forest and Savanna Ecosystems*, ed. J. Proctor, pp. 189–204. Oxford: Blackwell Scientific.

Galat, G. & Galat-Luong, A. (1985). La communauté de primates diurnes de la forêt de Taï, Cote d'Ivoire. *Revue d'Ecologie (Terre et Vie)*, **40**, 3–32.

Ganzhorn, J. U. (1988). Food partitioning among Malagasy primates. *Oecologia*, **75**, 436–50.

Ganzhorn, J. U. (1989). Niche separation of the seven lemur

species in the eastern rainforest of Madagascar. *Oecologia*, **79**, 279–86.

Ganzhorn, J. U. (1992). Leaf chemistry and the biomass of folivorous primates in tropical forests: test of a hypothesis. *Oecologia*, **91**, 540–7.

Ganzhorn, J. U. (1995). Low level forest disturbance effects on primary production, leaf chemistry, and lemur population. *Ecology*, **76**, 2084–96.

Ganzhorn, J. U., Malcomber, S., Andrianantoanina, O. & Goodman, S. M. (1997). Habitat characteristics and lemur species richness in Madagascar. *Biotropica*, **29**, 331–43.

Ganzhorn J. U. & Wright P. C. (1994). Temporal patterns in primate leaf eating: The possible role of leaf chemistry. *Folia Primatologica*, **63**, 203–8.

Garber, P. A. (1989). Role of spatial memory in primate foraging patterns: *Saguinus mystax* and *Saguinus fuscicollis*. *American Journal of Primatology*, **19**, 203–16

Gautier-Hion, A. (1983). Leaf consumption by monkeys in western and eastern Africa: A comparison. *African Journal of Ecology*, **21**, 107–13.

Glander, K. E., (1982). The impact of plant secondary compounds on primate feeding behavior. *Yearbook of Physical Anthropology*, **25**, 1–18.

Glander, K., Fedigan, L. M., Fedigan, L. & Chapman, C. A. (1991). Field methods for capture and measurement of three monkey species in Costa Rica. *Folia Primatologica*, **57**, 70–82.

Glander, K. E., Wright, P. C., Seigler, D. S., Randrianasolo, V. & Randrianasolo, B. (1989). Consumption of cyanogenic bamboo by a newly discovered species of bamboo lemur. *American Journal of Primatology*, **19**, 119–24.

Glanz, W. E. (1982). The terrestrial mammal fauna of Barro Colorado Island: censuses and long-term changes. In *The Ecology of a Tropical Forest: Seasonal Rhythms and Long-term Changes*, ed. E. G. Leigh, Jr., A. S. Rand & D. M. Windsor, pp. 455–68. Washington, DC: Smithsonian Institution Press.

Goldizen, A. W., Terborgh, J., Cornejo, F., Porras, D. T. & Evans, R. (1988). Seasonal food shortages, weight loss, and the timing of births in saddle-backed tamarins (*Saguinus fuscicollis*). *Journal of Animal Ecology*, **57**, 893–902.

Hairston, N. G., Smith, F. E. & Slobodkin, L. B. (1960). Community structure, population control, and competition. *American Naturalist*, **94**, 421–5.

Hall, J. B. (1977). Forest types in Nigeria: an analysis of pre-exploitation forest enumeration. *Journal of Ecology*, **65**, 187–99.

Hall, J. B. & Swaine, M. D. (1981). *Distribution and Ecology of Vascular Plants in a Tropical Rain Forest*. The Hague, Netherlands: Dr. W. Junk Publ.

Harrison, M. J. S. (1986). Feeding ecology of black colobus, Colobus satanas, in central Gabon. In *Primate Ecology and conservation*, ed. J. G. Else & P. C. Lee, pp. 31–7. Cambridge: Cambridge University Press.

Hart, J. (1985). *Comparative Dietary Ecology of a Community of Frugivorous Forest Ungulates in Zaire*. Ph.D. Thesis. Michigan State University, Lansing, Michigan.

Harvey, P. H., Martin, R. D. & Clutton-Brock, T. H. (1987). Life histories in comparative perspective. In *Primate Societies*, ed. B. B. Smuts, D. L. Cheney, R. M. Seyfarth, R. W. Wrangham & T. T. Struhsaker, pp. 181–96. Chicago: Chicago University Press.

Herrera, C. M. (1982). Defense of ripe fruit from pests: its significance in relation to plant-disperser interactions. *American Naturalist*, **120**, 218–41.

Hladik, A. (1978). Phenology of leaf production in rain forest of Gabon: distribution and composition of food for folivores. In *The Ecology of Arboreal Folivores*, ed. G. G. Montgomery, pp. 51–71. Washington, DC: Smithsonian Institution Press.

Hladik, A. & Hladik, C. M. (1969). Rapports trophiques entre végétation et Primates dans la forêt de Barro Colorado (Panama). *Terre et Vie*, **23**, 25–117.

Hladik, C. M. (1978). Adaptive strategies of primates in relation to leaf-eating. In *The Ecology of Folivores*, ed. G. G. Montgomery, pp. 373–95. Washington, DC: Smithsonian Institution Press.

Hladik, C. M., Charles-Dominique, P. & Peter, J. J. (1980). Feeding strategies of five nocturnal prosimians in the dry forest of the west coast of Madagascar. In *Nocturnal Malagasy Primates: Ecology, Physiology, and Behavior*, ed. P. Charles-Dominique, H. M. Cooper, A. Hladik, C. M. Hladik, E. Pages, G. F. Pariente, A. Petter-Rousseaux, A. Schilling & J. J. Peter, pp. 41–73. New York: Academic Press.

Hladik, C. M. & Chivers, D. J. (1994). Foods and the digestive system. In *The Digestive System in Mammals: Food, Form, and Function*, ed. D. J. Chivers & P. L. Langer, pp. 65–73. Cambridge: Cambridge University Press.

Hladik, C. M. & Simmen, B. (1997). Taste perception and feeding behavior in non-human primates and human populations. *Evolutionary Anthropology*, **5**, 58–71.

Howe, H. F. & Vande. Kerkhove, G. A. (1981). Removal of wild nutmeg (*Virola surinamensis*) crops by birds. *Ecology*, **62**, 1093–106.

Hutto, R. L. (1990). Measuring the availability of food resources. *Studies in Avian Biology*, **13**, 20–28.

Illius, A. W. & Gordon, I. J. (1992). Modeling the nutritional ecology of ungulate herbivores – evolution of body size and competitive interactions. *Oecologia*, **89**, 428–34.

Jaenike, J. (1990). Host specialization in phytophagous insects. *Annual Review of Ecology and Systematics*, **21**, 243–73.

Janson, C. H. (1983). Adaptation of fruit morphology to dispersal agents in a neotropical rainforest. *Science*, **219**, 187–9.

Janson, C. H. (1984). Female choice and mating system of the brown capuchin monkey *Cebus apella* (Primates: Cebidae). *Zeitschrift für Tierpsychologie*, **65**, 177–200.

Janson, C. H. (1985). Aggressive competition and individual food intake in wild brown capuchin monkeys. *Behavioural Ecology and Sociobiology*, **18**, 125–38.

Janson, C. H. (1988a). Intra-specific food competition and primate social structure: a synthesis. *Behaviour*, **105**, 1–17.

Janson, C. H. (1988b). Food competition in brown capuchin monkeys (*Cebus apella*): quantitative effects of group size and tree productivity. *Behaviour*, **105**, 53–76.

Janson, C. H. (1998). Experimental evidence for spatial memory in wild brown capuchin monkeys (*Cebus apella*). *Animal Behaviour*, **55**, 1129–43.

Janson, C. H. & Boinski, S. (1992). Morphological and behavioral adaptations for foraging in generalist primates: the case of the cebines. *American Journal of Physical Anthropology*, **88**, 483–98.

Janson, C. H. & Di Bitetti, M. S. (1997). Experimental analysis of food detection in capuchin monkeys: effects of distance, travel speed, and resource size. *Behavioral Ecology and Sociobiology*, **41**, 17–24.

Janson, C. H. & Emmons, L. E. (1990). Ecological structure of the non-flying mammal community at the Cocha Cashu Biological Station, Manu National Park, Peru. In *Four Neotropical Rain Forests*, ed. A. Gentry, pp. 314–38. New Haven: Yale University Press.

Janson, C. H. & Goldsmith, M. (1995). Predicting group size in primates: foraging costs and predation risks. *Behavioural Ecology*, **6**, 326–36.

Janson, C. H., Terborgh, J. W. & Emmons, L. E. (1981). Non-flying mammals as pollinating agents in the Amazonian forest. *Biotropica*, **13**, 1–6.

Janzen, D. H. (1969). Seed-eaters versus seed size, number, toxicity, and dispersal. *Evolution*, **23**, 1–27.

Janzen, D. H. (1973). Sweep samples of tropical foliage insects: Effects of seasons, vegetation types, elevation, time of day and insularity. *Ecology*, **54**, 687–701.

Janzen, D. H. (1974). Tropical blackwater rivers, animals,

and mast fruiting by the Dipterocarpaceae. *Biotropica*, **6**, 69–103.

Janzen, D. H. (1975). *Ecology of Plants in the Tropics*. London: Edward Arnold.

Janzen, D. H. (1979). New horizons in the biology of plant defenses. In *Herbivores: Their Interactions with Secondary Plant Metabolites*, ed. G. Rosenthal & D. H. Janzen, pp. 331–51. New York, NY: Academic Press.

Janzen, D. H. (1983). Dispersal of seeds by vertebrate guts. In *Coevolution*, ed. D. J. Futuyma & M. Slatkin, pp. 232–62. Sunderland, Mass: Sinauer.

Kato, M., Inoue, T., Hamid, A. A., Nagamitsu, T., Ben Merdek, M., Nona,, A. R., Itino, T., Yamane, S. & Yumoto, T. (1995). Seasonality and vertical structure of light-attracted insect communities in a dipterocarp forest in Sarawak. *Researches on Population Ecology (Kyoto)*, **37**, 59–79.

Kay, R. F. (1984). On the use of anatomical features to infer foraging behavior in extinct primates. In *Adaptations for Foraging in Nonhuman Primates: Contributions to an Organismal Biology of Prosimians, Monkeys, and Apes*, ed. P. S. Rodman & J. G. H. Cant, pp. 21–53. New York, NY: Columbia University Press.

Kay, R. F. & Hylander, W. L. (1978). The dental structure of mammalian folivores with special reference to primates and phalangeroidea (Marsupialia). In *The Ecology of Arboreal Folivores*, ed. G. G. Montgomery, pp. 173–91. Washington, DC: Smithsonian Institution Press.

Kay, R. F., Madden, R. H., van Schaik, C. P. & Higdon, D. (1997). Primate species richness is determined by plant productivity: implications for conservation. *PNAS*, **94**, 13023–7.

Kay, R. F. & Sheine, W. S. (1979). On the relationship between particle size and digestibility in the primate *Galago senegalensis*. *American Journal of Physical Anthropology*, **50**, 301–8.

Kay, R. N. B. & Davies, A. G. (1994). Digestive physiology. In *Colobine Monkeys: Their Ecology, Behaviour and Evolution*, ed. A. G. Davies & J. F. Oates, pp. 229–59. Cambridge: Cambridge University Press.

Klein, L. L. & Klein, D. (1977). Feeding behavior of the Columbian spider monkey. In *Primate Ecology*, ed. T. H. Clutton-Brock, pp. 153–81. London: Academic Press.

Koenig A., Borries C., Chalise M. K. & Winkler P. (1997). Ecology, nutrition, and timing of reproductive events in an Asian primate, the Hanuman langur (*Presbytis entellus*). *Journal of Zoology*, **243**, 215–35.

Lambert, J. (1997). *Fruit processing and seed dispersal by* *chimpanzees (*Pan troglodytes schweinfurthii*) and redtail monkeys (*Cercopithecus ascanius schmidti*) in the Kibale National Park, Uganda*. Ph.D. Thesis. University of Illinois, Urbana.

Lambert, J. (1998). Primate digestion: interactions between anatomy, physiology, and feeding ecology. *Evolutionary Anthropology*, **7**, 8–20.

Langer, P. (1986). Large mammalian herbivores in tropical forests with either hindgut or forestomach fermentation. *Zeitschrift für Säugetierkunde*, **51**, 173–87.

Lathwell, D. J. & Grove, T. L. (1986). Soil-plant relationships in the tropics. *Annual Review of Ecology and Systematics*, **17**, 1–16.

Leighton, M. (1993). Modeling dietary selectivity by Bornean orangutans: evidence for integration of multiple criteria in fruit selection. *International Journal of Primatology*, **14**, 257–314.

Leighton, M. & Leighton, D. R. (1982). The relationship of size of feeding aggregate to size of food patch: Howler monkeys (*Alouatta palliata*) feeding in *Trichilia cipo* fruit trees on Barro Colorado Island. *Biotropica*, **14**, 81–90.

Leung, W. W. (1968). *Food Composition Table for Use in Africa*. Bethesda, MD: U.S. Dept. of Health, Education, and Welfare.

Levey D. J. (1990). Habitat-dependent fruiting behavior of an understory tree *Miconia centrodesma* and tropical treefall gaps as keystone habitats for frugivores in Costa Rica. *Journal of Tropical Ecology*, **6**, 409–20.

Lloyd, L. E., McDonald, B. E. & Crampton, E. W. (1978). *Fundamentals of Nutrition*. San Francisco: W. H. Freeman.

MacArthur, R. H. (1972). *Geographical Ecology*. New York, NY: Harper & Row.

MacKinnon, J. R. (1974). The behavior and ecology of wild orangutans (*Pongo pygmaeus*). *Animal Behaviour*, **22**, 3–74.

MacKinnon, J. R. (1977). A comparative ecology of Asian apes. *Primates*, **18**, 747–72.

Mahaney, W. C., Stambolic, A., Knezevich, M., Hancock, R. G. V., Aufreiter, S., Sanmugadas, K., Kessler, M. J. & Grynpas, M. D. (1995). Geophagy amongst rhesus macaques on Cayo Santiago, Puerto Rico. *Primates*, **36**, 323–33.

Malenky, R. K. (1990). *Ecological Factors Affecting Food Choice and Social Organization in Pan paniscus*. Ph.D. Thesis, State Univ. of New York at Stony Brook, Stony Brook, NY.

Martin, C. (1991). *The Rainforests of West Africa*. Basel: Birkhäuser Verlag.

Mather, R. (1992). A Field Study of Hybrid Gibbons in Central Kalimantan, Indonesia. Ph.D. Thesis, Cambridge University, Cambridge, UK.

Mason, J. R., Adams, M. A. & Clark, L. (1989). Anthranilate repellency to starlings: chemical correlates and sensory perception. *Journal of Wildlife Management*, **53**, 55–64.

McKey, D. (1975). The ecology of coevolved seed dispersal systems. In *Coevolution of Animals and Plants*, ed. L. E. Gilbert & P. H. Raven, pp. 159–91. Austin: University of Texas Press.

Milton, K. (1979). Factors influencing leaf choice by howler monkeys: a test of some hypotheses of food selection by generalist herbivores. *American Naturalist*, **114**, 362–78.

Milton, K. (1980). *The Foraging Strategy of Howler Monkeys*. New York, NY: Columbia University Press.

Milton, K. (1984). The role of food-processing factors in primate food choice. In *Adaptations for Foraging in Nonhuman Primates: Contributions to an Organismal Biology of Prosimians, Monkeys, and Apes*, ed. P. S. Rodman & J. G. H. Cant, pp. 249–79. New York, NY: Columbia University Press.

Milton, K. (1998). Physiological ecology of Howlers (*Alouatta*): energetic and digestive considerations and comparison with the Colobinae. *International Journal of Primatology*, **19**, 513–48.

Mittermeier, R. A. & van Roosmalen, M. G. M. (1981). Preliminary observations on habitat utilization and diet in eight Surinam monkeys. *Folia Primatologica*, **36**, 1–39.

Moir, R. J. (1994). The 'carnivorous' herbivores. In *The Digestive System in Mammals: Food, Form, and Function*, ed. D. J. Chivers & P. Langer. pp. 87–102. Cambridge: Cambridge University Press.

Nash, L. T. (1986). Dietary, behavioral, and morphological aspects of gummivory in primates. *Yearbook of Physical Anthropology*, **29**, 113–37.

Norconk, M A., Wertis, C. & Kinzey, W. G. (1997). Seed predation by monkeys and macaws in Eastern Venezuela: Preliminary findings. *Primates*, **38**, 177–84.

Oates, J. F. (1977). The guereza and its foods. In *Primate Ecology*, ed. T. H. Clutton-Brock, pp. 275–321. London: Academic Press.

Oates, J. F., Swain, T. & Zantovska, J. (1977). Secondary compounds and food selection by colobus monkeys. *Biochemical Systematics and Ecology*, **5**, 317–21.

Oates, J. F., Waterman, P. G. & Choo, G. M. (1980). Food selection by the south Indian leaf-monkey *Presbytis johnii*, in relation to leaf chemistry. *Oecologia*, **45**, 45–56.

Oates, J. F., Whitesides, G. H., Davies, A. G., Waterman,

P. G., Green, S. M., Dasilva, G. L. & Mole, S. (1990). Determinants of variation in tropical forest primate biomass: new evidence from West Africa. *Ecology*, **71**, 328–43.

O'Brien, T. G., Kinnaird, M. F., Dierenfeld, E. S., Conklin-Brittain, N. L., Wrangham, R. W. & Silver, S. C. (1998). What's so special about figs? *Nature*, **392**, 668.

Orians, G. H. & Janzen, D. H. (1974). Why are embryos so tasty? *American Naturalist*, **108**, 581–92.

Overdorff, D. J. (1992). Differential patterns in flower feeding by *Eulemur fulvus rufus* and *Eulemur rubriventer* in Madagascar. *American Journal of Primatology*, **28**, 191–203.

Parra, R. (1978). Comparison of foregut and hindgut fermentation in herbivores. In *The Ecology of Arboreal Folivores*, ed. G. G. Montgomery, pp. 205–29. Washington, DC: Smithsonian Institution Press.

Peres, C. A. (1993). Structure and spatial organization of an Amazonian terre firme primate community. *Journal of Tropical Ecology*, **9**, 259–76.

Peres, C. A. (1997a) Primate community structure at twenty western Amazonian flooded and unflooded forests. *Journal of Tropical Ecology*, **13**, 381–405.

Peres, C. A. (1997b). Effects of habitat quality and hunting pressure on arboreal folivore density in Neotropical forests: A case study of howler monkeys (*Alouatta* spp.) *Folia Primatologica*, **68**, 199–222.

Peters, R., Cloutier, S., Dube, D., Evans, A., Hastings, P., Kohn, D. & Sawer-Foner, B. (1988). The ecology of the weight of fruit on trees and shrubs in Barbados. *Oecologia*, **74**, 612–16.

Plumptre, A. J. & Reynolds, V. (1995). The effect of selective logging on the primate populations in the Budongo Forest Reserve, Uganda. *Journal of Applied Ecology*, **31**, 631–41.

Plumptre, A. J., Reynolds, V. & Bakuneeta, C. (1994). The contribution of fruit eating primates to seed dispersal and natural regeneration after selective logging. ODA Report.

Power, M. L. (1996). The other side of callitrichine gummivory. In *Adaptive Radiations of Neotropical Primates*, ed. M. A. Norconk, A. L. Rosenberger & P. A. Garber, pp. 97–110. New York, NY: Plenum Press.

Power, M. L. & Oftedal, O. T. (1996). Differences among captive callitrichids in the digestive responses to dietary gum. *American Journal of Primatology*, **40**, 131–44.

Proctor, J. (1980). Tropical forest litter fall. I. Problems of data comparisons. In *Tropical Rainforest: Ecology and Management*, ed. S. L. Sutton, T. C. Whitmore & A. C. Chadwick, pp. 267–73. Oxford: Blackwell Scientific.

Raemaekers, J. & Chivers, D. J. (1980). Socio-ecology of Malayan forest primates. In *Malayan Forest Primates: Ten Years Study in Tropical Rain Forest*, ed. D. J. Chivers, pp. 29–61. New York, NY: Plenum Press.

Reed, K. E. & Fleagle, J. G. (1995). Geographic and climatic control of primate diversity. *PNAS*, **92**, 7874–6.

Richards, P. W. (1996). *The Tropical Rainforest* (2nd edition). Cambridge: Cambridge University Press.

Robinson, B. W. & Wilson, D. S. (1998). Optimal foraging, specialization, and a solution to Liem's paradox. *American Naturalist*, **151**, 223–35.

Robinson, J. G. & Ramirez, J. (1982). Conservation biology of Neotropical primates. In *Mammalian Biology in South America*, ed. M. A. Mares & H. Genoways, pp. 329–44. Pittsburgh, PA: University of Pittsburgh Press.

Rodman, P. S. (1978). Diets, densities, and distributions of Bornean primates. In *The Ecology of Arboreal Folivores*, ed. G. G. Montgomery, pp. 465–78. Washington, DC: Smithsonian Institution Press.

Rogers, M. E., Maisels, F., Williamson, E. A., Fernandez, M. & Tutin, C. E. G. (1990). Gorilla diet in the Lopé Reserve, Gabon: A nutritional analysis. *Oecologia*, **84**, 326–39.

Roosmalen, M. G. M. van. (1984). Subcategorizing food in primates. In *Food Acquisition and Processing in Primates*, ed. D. J. Chivers, B. A. Wood & A. Bilsborough, pp. 167–75. New York, NY: Plenum Press.

Rudran, R. (1978). Socioecology of the blue monkey (*Cercopithecus mitis stuhlmanni*) of the Kibale Forest, Uganda. *Smithsonian Contribution to Zoology* 249.

Schlesinger, W. H. (1997). *Biogeochemistry: an analysis of global change*. 2nd edn. Academic Press: San Diego.

Smythe, N. (1982). The seasonal abundance of night-flying insects in a Neotropical forest. In *The Ecology of a Tropical Forest: Seasonal Rhythms and Long-term Changes*, ed. E. G. Leigh, A. S. Rand & D. M. Windsor, pp 309–18. Washington, DC: Smithsonian Institution Press.

Southwood, T. R. E. (1978). *Ecological Methods, With Particular Reference to the Study of Insect Populations*. London: Chapman & Hall.

Sterck, L. (1995). *Females, Foods, and Fights*. Ph.D. thesis, Utrecht University, Utrecht, the Netherlands.

Stephens, D. W. & Krebs, J. R. (1986). *Foraging Theory*. Princeton: Princeton University Press.

Stevenson, P. R. (1996). Censos diurnos de mamíferos y aves de gran tamaño en el Parque Nacional Tinigua, Colombia. *Universitas Scientiarum (Bogota)*, **3**, 67–81.

Stork, N. E. (1991). The composition of the arthropod fauna

of Bornean lowland rain forest trees. *Journal of Tropical Ecology*, **7**, 161–80.

Strait, S. G. (1997). Tooth use and the physical properties of food. *Evolutionary Anthropology*, **5**, 199–211.

Struhsaker T. T. (1969). Correlates of ecology and social organization among African cercopithecines. *Folia primatologica*, **11**, 80–118.

Struhsaker T. T. (1975). *The Red Colobus Monkey*. Chicago: University of Chicago Press.

Struhsaker, T. T. (1978). Interrelations of red colobus monkeys and rain-forest trees in the Kibale Forest, Uganda. In *The Ecology of Arboreal Folivores*, ed. G. G. Montgomery, pp. 397–422. Washington, DC: Smithsonian Institution Press.

Struhsaker, T. T. (1981). Polyspecific associations among tropical rain-forest primates. *Zeitschrift fur Tierpsychologie*, **57**, 268–304.

Struhsaker, T. T., Cooney, D. O. & Siex, K. S. (1997). Charcoal consumption by Zanzibar red colobus monkeys: its function and its ecological and demographic consequences. *International Journal of Primatology*, **18**, 61–72.

Struhsaker, T. T. & Leland, L. (1979). Socioecology of five sympatric monkey species in the Kibale Forest, Uganda. *Advances in the Study of Behavior*, **9**, 159–228.

Struhsaker, T. T. & Oates, J. F. (1975). Comparison of the behavior and ecology of red colobus and black-and-white colobus monkeys in Uganda: A summary. In *Socio-Ecology and Psychology of Primates*, ed. R. H. Tuttle, pp. 103–123. The Hague: Mouton.

Symington, M. M. (1988). Food competition and foraging party size in the black spider monkey (*Ateles paniscus chamek*). *Behaviour*, **105**, 117–34.

Terborgh, J. W. (1983). *Five New World Primates*. Princeton: Princeton University Press.

Terborgh, J. W. (1986). Community aspects of frugivory in tropical forests. In *Frugivores and Seed Dispersal*, ed. A. Estrada & T. H. Fleming, pp. 371–84. Dordrecht: D. W. Junk Publ.

Terborgh, J. W. (1992). *Diversity and the Rain Forest*. New York: Scientific American Library.

Terborgh, J. W., Janson, C. H. & Munn, M. B. (1986). Cocha Cashu: su vegetacion, clima y recursos. In *Reporte Manu*, ed. M. A. Rios, pp. 1–18. Lima, Peru: Centro Datos para la conservacion.

Terborgh, J. W. & Stern, M. (1987). The surreptitious life of the saddle-backed tamarin. *American Scientist*, **75**, 260–9.

Terborgh, J. W. & van Schaik, C. P. (1987). Convergence vs. nonconvergence in primate communities. In

Organization of Communities, Past and Present, ed. J. H. R. Gee & P. S. Giller, pp. 205–26. Oxford: Blackwell Scientific.

Thepenier, P., Jacquier, M. J., Massiot, G., Men-Olivier, L. L. & Delaude, C. (1990). Alkaloids from seeds of *Strychnos variabilis* and *S. longicaudata*. *Phytochemistry*, **29**, 686–7.

Thomas, S. C. (1991). Population densities and patterns of habitat use among anthropoid primates of the Ituri Forest, Zaire. *Biotropica*, **23**, 68–83.

Tutin, C. E. G. & Fernandez, M. (1993). Relationships between minimum temperature and fruit production in some tropical forest trees in Gabon. *Journal of Tropical Ecology*, **9**, 241–8.

van Schaik, C. P. (1986). Phenological changes in a Sumatran rain forest. *Journal of Tropical Ecology*, **2**, 327–47.

van Schaik, C. P., Terborgh, J. W. & Wright, S. J. (1993). The phenology of tropical forests: adaptive significance and consequences for primary consumers. *Annual Review of Ecology and Systematics*, **24**, 353–77.

Vitousek, P. M. (1984). Litterfall, nutrient cycling and nutrient limitation in tropical forests. *Ecology*, **65**, 285–98.

Vitousek, P. M. & Sanford, R. L. (1986). Nutrient cycling in moist tropical forest. *Annual Review of Ecology and Systematics*, **17**, 137–67.

Waser, P. M. (1987). Interactions among primate species. In *Primate Societies*, ed. B. B. Smuts, D. L. Cheney, R. M. Seyfarth, R. W. Wrangham & T. T. Struhsaker, pp. 210–26. Chicago: Chicago University Press.

Waterman, P. G. (1984). Food acquisition and processing as a function of plant chemistry. In *Food Acquisition and Processing in Primates*, ed. D. J. Chivers, B. A. Wood & A. Bilsborough, pp. 177–211. New York, NY: Plenum Press.

Waterman, P. G., Mbi, C. N., McKey, D. B. & Gartlan, J. S. (1980). African rainforest vegetation and rumen microbes: Phenolic compounds and nutrients as correlates of digestibility. *Oecologia*, **47**, 22–33.

Waterman, P. G., Ross, J. A. M., Bennett, E. L. & Davies, A. G. (1988). A comparison of the floristics and leaf chemistry of the tree flora in two Malaysian rain forests and the influence of leaf chemistry on population of colobine monkeys in the old World. *Biological Journal of the Linnean Society*, **34**, 1–16.

Watt, B. K. & Merrill, A. L. (1975). *Handbook of the Nutritional Contents of Foods*. New York, NY: Dover Publ.

Weisenseel, K., Chapman, C. A. & Chapman, L. J. (1993).

Nocturnal primates of Kibale Forest: effects of selective logging on prosimian densities. *Primates*, **34**, 445–50.

White, L. J. T. (1994). Biomass of rain forest mammals in the Lope Reserve, Gabon. *Journal of Animal Ecology*, **63**, 499–512.

Whiten, A. Byrne, R. W., Barton, R. A. Waterman, P. E. & Henzi, S. P. (1991). Dietary and foraging strategies of baboons. *Philosophical Transactions of the Royal Society, London, Series B.*, **334**, 187–97.

Whiten, A. Byrne, R. W. Waterman, P. G., Henzi, S. P. & McCulloch, F. M. (1990). Specifying the rules underlying selective foraging in wild mountain baboons, *P. ursinus*. In *Baboons: Behaviour and Ecology, Use and Care*, ed. M. T. de Mello, A. Whiten & R. W. Byrne, pp. 5–22. Brasilia: University of Brasilia Press.

Whitmore, T. C. (1975). *Tropical Rain Forests of the Far East*. Oxford: Clarendon Press.

Whitmore, T. C. (1990). *An Introduction to Tropical Rain Forests*. Oxford: Clarendon Press.

Whitmore, T. C. & Prance, G. T. (eds) (1987). *Biogeography and Quaternary History in Tropical America*. Oxford: Clarendon Press.

Wolda, H. & Wright, S. J. (1992). Artificial dry season rain and its effects on tropical insect abundance and seasonality. *Proc. Koninklijke Nederlandse Akademie van Wetenschappen – Biological Chemical Geological Physical and Medical Sciences*, **95**, 535–48.

Yeager, C. P., Silver, S. C. & Dierenfeld, E. S. (1997). Mineral and phytochemical influences on foliage selection by the proboscis monkey (*Nasalis larvatus*). *American Journal of Primatology*, **41**, 117–28.

Zangerl, A R., Arntz, A M. & Berenbaum, M R. (1997). Physiological price of an induced chemical defense: Photosynthesis, respiration, biosynthesis, and growth. *Oecologia*, **109**, 433–41.

Zhang, S. & Wang, L. (1995). Comparison of three fruit census methods in French Guiana. *Journal of Tropical Ecology*, **11**, 281–94.

15 · Effects of subsistence hunting and forest types on the structure of Amazonian primate communities

CARLOS A. PERES

INTRODUCTION

Diurnal primates contribute a conspicuous component of the tropical forest vertebrate fauna (Terborgh, 1983; Bourlière, 1985), and are considered to be fair game species by subsistence and market hunters in most African (Fa *et al.*, 1995; Oates, 1996; McRae, 1997), southeast Asian (Bennett *et al.*, in press), and neotropical forests (Redford & Robinson, 1987; Peres, 1990). This is particularly the case for large-bodied species that are preferred by hunters because they yield the most amount of meat per unit of ammunition cost (Peres, 1990; Bodmer, 1995).

As sources of protein, however, primates are largely incompatible with a sustainable harvest regime because of their small litter sizes (typically one), long interbirth intervals, delayed age of first reproduction, heavy burden of prolonged parental investment, and in most species, severe competition for reproductive opportunities among the breeding members of the population (Smuts *et al.*, 1987). These life-history and socioecological characteristics usually result in extremely low per capita reproductive rates which largely explain why primates and other long-lived mammals sharing a similar reproductive biology are so prone to severe reductions in numbers, if not local extinctions, when exposed to even relatively light hunting regimes. Yet overhunting of tropical forest vertebrates is now a rampant and nearly universal phenomenon (Robinson & Redford, 1991; Redford, 1992; Robinson & Bennett, in press), rendering primates particularly susceptible to widespread and profound shifts in population and community structure. Ultimately, the most vulnerable target species can succumb to hunting-mediated extinctions, ranging in extent from a local (Peres, 1991) to a global scale (Dewar, 1984; Culotta, 1995), that

can result in irreversible changes in community composition.

This chapter examines the impact of subsistence hunting on the structure of primate communities in neotropical forests of northern South America. I present data based on a comprehensive review of Amazonian forest sites from which reasonably reliable density estimates are available for all sympatric diurnal primate species. I focus primarily on the evidence from the lowland Amazon basin (hereafter including the Guianan Shield) because this vast region of largely continuous forest (1) encompasses the most number of primate surveys conducted to date in the neotropics (Muckenhirn *et al.*, 1975; Freese *et al.*, 1982; Peres, 1990; 1997*a*; Peres & Janson, chapter 3, this volume); (2) sustains many tribal and non-tribal groups of forest dwellers who still rely heavily on meat of forest vertebrates as a prime source of protein and energy (Peres, in press *a*); and (3) this minimizes potential biogeographic problems involved in comparative analyses addressing appreciably different source biotas. In particular, I consider the synergistic effects of forest nutrient status and hunting pressure on primate abundance and guild structure, once differences in species packing are taken into account. Recent cross-site comparisons of the residual density and biomass of large-bodied birds and mammals in Amazonian forests suggest that the impact of subsistence hunting is strongly mediated by differences in soil fertility and forest types (Eisenberg, 1979; Emmons, 1984; Peres, in press, *a,b,c*). Community-wide responses to differential hunting pressure should then be particularly detectable in mammalian game taxa exhibiting slow rates of recovery such as primates, although shifts in abundance are expected to vary across different species occurring in different habitats.

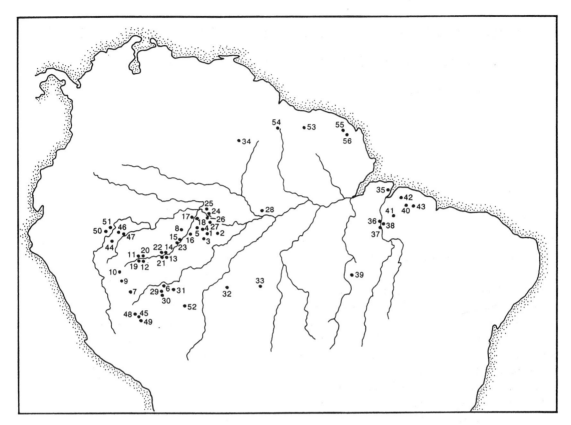

Fig. 15.1. Location of the 56 Amazonian forest sites from which community-wide density estimates are available for diurnal primate species. Site numbers correspond to those listed in Table 15.1.

METHODS AND SAMPLING SITES

This analysis is based on a comparison of primate communities occurring at 56 forest sites in six northern South American countries from which density estimates are available for all diurnal species (Fig. 15.1). These include 41 sites from central-western Amazonia, nine sites from eastern Amazonia, and six sites from the Guianan Shield. Primate density estimates were obtained using comparable line-transect censuses conducted during the day, which exclude only one species at any given site – the owl monkey (*Aotus* spp.) – the densities of which are available for only a few sites. Nearly half of all sites (25 of 56) were surveyed during an ongoing standardized series of primate censuses, which I began in 1984 (Peres, 1989, 1990, 1997a). Survey methodology described here applies to those sites only. All additional data were compiled from either similar line-transect censuses sharing a reasonably robust sampling effort, or more intensive synecological studies reporting local densities for all sympatric primate species (see Table 15.1). I therefore excluded from this analysis several primate community surveys available in the literature (Bernstein *et al.*, 1976; Branch, 1983; most sites in Freese *et al*, 1982), because they consisted of relatively low sampling efforts, provided density estimates which were thought to be mathematically unreliable, or did not cover the full diurnal complement of species occurring at a given site.

Primate species were grouped into three size classes scaled to take into account the skewed distribution of body size towards small-bodied species (<1.5 kg), which included all callitrichines (*Cebuella pygmaea*, *Callithrix* spp., *Saguinus* spp., and *Callimico goeldii*), squirrel monkeys

Table 15.1. *Profile of the 56 Amazonian primate communities considered in this analysis. Site numbers correspond to those shown on Figure 15.1*

Forest site	Location	Forest[a] type	Hunting[b] pressure	No. of[c] species	Density (ind/km^2)	Biomass (kg/km^2)	Source[d]
1. Urucu	Amazonas, Brazil	OTF	N	11	137	324	1
2. Igarapé Açú	Amazonas, Brazil	OTF	N	10	106	313	2
3. SUC-1	Amazonas, Brazil	OTF	N	10	77	217	2
4. Oleoduto	Amazonas, Brazil	OTF	N	9	103	253	3
5. Curimatá	Amazonas, Brazil	OTF	N	12	79	222	3
6. São Domingos	Acre, Brazil	OTF	H	8	175	138	2
7. Kaxinawá Area	Acre, Brazil	MTF	H	8	145	173	4
8. Riozinho	Amazonas, Brazil	OTF	M	12	157	226	4
9. Porongaba	Amazonas, Brazil	MTF	H	11	210	163	4
10. Sobral	Amazonas, Brazil	MTF	H	9	175	122	4
11. Condor	Amazonas, Brazil	MTF	L	9	195	279	4
12. Penedo	Amazonas, Brazil	MTF	H	7	129	118	4
13. Altamira	Amazonas, Brazil	MTF	L	11	227	463	4
14. Barro Vermelho	Amazonas, Brazil	MTF	M	12	165	261	4
15. Fortuna	Amazonas, Brazil	MTF	M	11	216	297	4
16. Igarapé Jaraquí	Amazonas, Brazil	OTF	M	9	137	131	4
17. Vira Volta	Amazonas, Brazil	MTF	L	12	182	282	4
18. Vai Quem Quer	Amazonas, Brazil	MTF	M	10	124	176	4
19. Sacado	Amazonas, Brazil	VZF	M	4	126	245	4
20. Nova Empresa	Amazonas, Brazil	VZF	M	5	185	410	4
21. Boa Esperança	Amazonas, Brazil	VZF	L	5	355	953	4
22. Barro Vermelho II	Amazonas, Brazil	VZF	M	7	213	361	4
23. Lago Fortuna	Amazonas, Brazil	VZF	M	6	358	627	4
24. Lago Teiú	Amazonas, Brazil	VZF	L	4	181	377	5
25. Lago Mamirauá	Amazonas, Brazil	VZF	L	3	266	448	5
26. Ponta da Castanha	Amazonas, Brazil	OTF	M	10	156	188	6
27. Açaituba	Amazonas, Brazil	OTF	L	11	68	244	6
28. MCSE Reserve	Amazonas, Brazil	OTF	L[e]	6	24	81	7
29. Rio Iaco	Acre, Brazil	MTF	H	8	56	53	8
30. Fazenda União	Acre, Brazil	MTF	M	9	61	50	8
31. Antimarí	Acre, Brazil	MTF	H	9	40	33	9
32. Samuel	Rondônia, Brazil	MTF	L	8	77	175	10
33. Aripuanã	Mato Grosso, Brazil	MTF	L	10	70	197	11
34. Maracá Island	Roraima, Brazil	MTF	N	5	42	105	12
35. Marajó Island	Pará, Brazil	ALL[f]	L	2	71	128	13
36. Pucuruí	Pará, Brazil	MTF	L	5	81	157	6
37. Remansinho	Pará, Brazil	MTF	L	5	72	147	6
38. Vila Braba	Pará, Brazil	MTF	H	5	75	112	6
39. Kayapó Area	Pará, Brazil	MTF	L	5	68	177	3
40. Tailândia	Pará, Brazil	OTF	H	4	22	30	14
41. Rio Capim	Pará, Brazil	OTF	H	4	30	27	14
42. Irituia	Pará, Brazil	OTF	M	4	36	52	14
43. Gurupí	Pará, Brazil	OTF	L	5	35	70	14
44. Cahuana Island	Eastern Peru	ALL[g]	L	6	120	328	15

Table 15.1. (*cont.*)

Forest site	Location	Forest[a] type	Hunting[b] pressure	No. of[c] species	Density (ind/km^2)	Biomass (kg/km^2)	Source[d]
45. Cocha Cashu	Southern Peru	ALL	N	11	249	590	16
46. Tahuayo–Blanco	Eastern Peru	MTF	H	12	90	125	17
47. Yavari-Miri	Eastern Peru	MTF	L	10	158	310	17
48. Yomiwato	Southern Peru	MTF	M	9	75	191	18
49. Diamante	Southern Peru	ALL	M	8	150	287	18
50. Samiria	Eastern Peru	ALL	L	7	161	345	19
51. Upper Nanay	Eastern Peru	OTF[h]	H	6	40	51	19
52. Tahuamanu	Northern Bolivia	MTF	H	9	73	51	20
53. Raleighvallen	Central Surinam	OTF	N	7	103	244	21
54. Pakani Area	Central Guyana	OTF	N	6	135	242	22
55. Petit Saut	French Guiana	OTF	L	5	40	73	23
56. Saut Pararé	French Guiana	OTF	N	6	103	289	24

Note: [a]Forest type: (OTF) oligotrophic terra firme forests of remote interfluvial areas; (MTF) eutrophic terra firme forests, usually adjacent to fluvial sources of alluvial sediments; (ALL) mature alluvial floodplain forests; (VZF) seasonally flooded várzea forests.

[b]Level of hunting pressure: (N) none; (L) light; (M) moderate; (H) heavy.

[c]Number of sympatric diurnal primate species detected at each survey site.

[d]Sources of data: (1) Peres, 1993*b*; (2) Peres, 1988; (3) C. Peres & H. Nascimento, unpublished data; (4) Peres, 1997*a*; (5) Ayres, 1986; (6) Johns, unpublished data; (7) Rylands & Keuroghlian, 1988, L. Emmons, pers. comm.; (8) Martins, 1992; (9) Calouro, 1995; (10) Lemos de Sá, 1995; (11) Rylands, 1983; (12) Mendes-Pontes, 1994; (13) Peres, 1989; (14) Ferrari *et al.*, 1996; (15) Soini, 1986; (16) Terborgh, 1983, Janson & Emmons, 1990; (17) Puertas & Bodmer, 1993; (18) Mitchell & Raez Luna, 1993; M. Alvard, pers. comm.; (19) Freese *et al.*, 1982; (20) Cameron *et al.*, 1989; (21) Mittermeier, 1977; Mittermeier & Roosmalen, 1981; (22) Muckenhirn *et al.*, 1975; (23) Pack, 1994; (24) Guillotin *et al.*, 1994.

[e]Recorded as lightly hunted (L), as some hunting may have taken place prior to surveys (L. Emmons, in litt.)

[f]Recorded as ALL but best described as a brackish-water tidal gallery forest in the Amazon estuary (Peres, 1989).

[g]Recorded as ALL, but may be considered as intermediate towards várzea (Soini, 1986).

[h]Primarily a floodplain forest supplied by a black-water river, but in terms of geochemistry and productivity more similar to OTF (Freese *et al.*, 1982).

(*Saimiri* spp.), and titi monkeys (*Callicebus* spp.); medium-bodied species (1.5–5 kg), including the capuchin monkeys (*Cebus* spp.), and all pitheciines (*Pithecia* spp., *Chiropotes* spp., *Cacajao calvus*); and large-bodied species (5–10 kg), including howler monkeys (*Alouatta* spp.), woolly monkeys (*Lagothrix lagotricha*), and spider monkeys (*Ateles paniscus*).

Unhunted sites are defined as those entirely uninhabited by forest dwellers (i.e. native Amerindians, detribalized Amazonians, and rubber-tappers of European descent) and showing no enduring evidence of hunting activity, including axe marks on core hardwoods which can deter canoe

traffic along perennial forest streams, and old scars on the bark of large latex trees (e.g. *Couma macrocarpa, Brosimum* spp.) harvested by seasonal tappers. These sites were not accessible to hunters, and census work was only possible via helicopters made available by the Brazilian Oil Company (Petrobrás, S.A.). The term 'unhunted' as used here is thus restricted to entirely pristine forests, usually of remote interfluvial basins, rather than sites rarely visited by hunters.

Because reliable data on the local primate harvest were generally unavailable for hunted sites, it was difficult to accurately reconstruct their history of hunting over the last several decades. I thus simply assigned each site to one of three broad categories of hunting pressure (light, moderate, or heavy) on the basis of (1) semi-structured interviews with hunters who had lived at a given site for at least 2 years prior to the censuses, (2) present and past human population densities quantified on the basis of the number of households in each area, as revealed by high-resolution (1:250 000) maps of each survey area (RADAM 1973–1981), and (3) the number of shotgun blows heard during each line-transect census. Interviews with hunters were unbiased with respect to fear of disclosing illegal hunting activities, since interviewees in such remote areas appeared to be genuinely unaware of legality issues concerning primate hunting.

With one exception, hunting in all hunted sites was practiced purely for subsistence. At Igarapé Curimatá (site 5 in Table 15.1) game meat was harvested for both local consumption and sale, but the commercial motivation behind the latter was unimportant. Hunting both within and outside Indian Areas (Kaxinawá Reserve, Kayapó Reserve, Penedo) was carried out with 12-, 16-, 20-, and 36-gauge shotguns, as the rapid transition from aboriginal weapons (e.g. bow-and-arrow, blowguns) to firearms has now reached even the most remote parts of Amazonia. The abundance of species thought to be most susceptible to hunting was not used to infer the previous hunting pressure at a given site, because rates of game recovery may also be affected by habitat productivity (Peres, in press *b*). I therefore assume that the crude three-point scale of hunting pressure as used here was the most refined this methodology could reasonably afford in the absence of more accurate information on primate harvest by present or past human populations. Ranking of hunting pressure for all sites surveyed by other investigators was based on information explicitly provided with each published or unpublished source of data, or through personal communications to the author.

Sampling sites were also assigned to one of three broad categories of forest type according to large-scale differences in soil types and nutrient input. In lowland Amazonia these are intimately related to differences in seasonal influx of alluvial sediments which represent the only overflow sources of macronutrients of critical importance to plants including carbon, nitrogen, potassium, calcium, magnesium, and phosphorus (Irion, 1978). On the basis of the description of each study area I thus classified each site into (1) nutrient-poor (oligotrophic) terra firme forests,

which are usually located in remote interfluvial regions; (2) nutrient-rich (eutrophic) terra firme forests, which usually lie adjacent to major white-water rivers; or (3) floodplain forests, including sites which are flooded on a seasonal (várzea forests), supra-annual (alluvial forests), or daily basis (tidal forests of the Amazon estuary). A more detailed description of sampling sites can be found elsewhere (see Table 15.1 and references therein).

Line-transect censuses

Each survey consisted of an average cumulative census distance of 110 ± 81 km (range: 37–359 km), giving a total of 2761 km of census walks at the 25 sites. Line-transect censuses were conducted from early morning to mid-day (0600–1130 h) by independent observers using between two and four forest transects of 4–5 km in length, which were marked every 50 m. Transects were left to "rest" for at least one day after they had been cut before census walks could be initiated. At previously hunted sites, we avoided using 'hunter paths' and rubber-tapper trails ('estradas') regardless of their linearity because that could potentially introduce detection biases resulting in density underestimates. Censuses were conducted on clear or overcast (but not on rainy) days, at walking velocities of ca. 1.25 km/h by one observer per transect, and were usually completed within a 30-day period.

Population densities were calculated on the basis of group densities and mean group sizes at each site from the fraction of total group counts considered to be accurate. In order to calculate crude population biomass, I used the mean weight of an individual of a given species, defined as 80% of the average body mass of adult males and females in different Amazonian populations (data from Ayres, 1986; Janson & Emmons, 1990; Martins, 1992; Calouro, 1995; Peres, 1993*a*; C. Peres & H. Nascimento, unpublished data). Further details on the derivation of primate densities for the forest sites I censused can be found elsewhere (Peres, 1997*a*; in press *a,b*). A number of methods were used to obtain estimated primate densities from the line-transect census data obtained at other sites, including King's method, Fourier series analysis, and nonparametric estimators based on shape restrictions (see Burnham *et al.*, 1980; Johnson & Routledge, 1985; Buckland *et al.*, 1993 and associated software programs). Given the absence of raw sighting data obtained by other investigators, it is impossible to evaluate density discrepancies resulting from different methods of analysis. However, density estimates become more comparable and less sensitive to statistical variance, as the total number of independent detection events increases as a function of either greater sampling efforts or higher population densities. This is the main reason why several primate surveys based on relatively short cumulative distances, or deriving low group densities, were excluded from this analysis from the outset, with an aim of making the entire data set more comparable.

STATISTICAL ANALYSIS

Analyses of variance (ANOVAS) and covariance (ANCOVAS) presented here were always preceded by homogeneity of variance (Levene) tests. In most cases it thus became necessary to $\log^{10}$-transform the abundance (population density and biomass) data in order to prevent violation of these tests. In all ANCOVAS, the number of diurnal primate species occurring at any given site was entered as a covariate of total community biomass when the effects of hunting pressure, forest type, or both, were considered. This was thought to be appropriate because local species packing across primate communities in different parts of Amazonia can be highly variable for both historical (Ayres & Clutton-Brock, 1992) and ecological reasons (Peres, 1997a). It would, therefore, be unreasonable to expect resource partitioning in a species-poor community to be as refined as that in a species-rich community, regardless of whether interspecific competition actually takes place.

Detrended correspondence analysis (DCA: ter Braak, 1988), which reduces the dimensionality of n species across m forest sites to a few ordination axes, proved to be the most appropriate ordination technique for this matrix. The performance of other techniques applied to the same matrix was consistently poor and showed systematic distortions of the first axis (arch effect). Species occurrences were weighted according to their respective untransformed density estimates (ind./km) at each site. Forest sites are thus positioned on DCA plots according to their species composition and abundance, which in turn can be determined by different environmental gradients. Primate species were entered into the matrix as discrete functional groups (hereafter, ecospecies) based on extensive information on the behavioral ecology, patterns of resource use,

and distribution of Amazonian primates which are now reasonably well documented (for ecospecies classification see Peres, 1997a and references therein). Primate ecospecies corresponded to a single species or subspecies, or a few ecologically equivalent (and mutually exclusive) congeners, usually representing parapatric replacements across sharp geographic range boundaries. The only exception was the midas tamarin (*Saguinus midas midas* and *S. m. niger*) which was grouped within the moustached tamarin species group (*S. mystax*, *S. labiatus* and *S. imperator*) because of clear similarities in their foraging ecology (Peres, 1992). Several polytypic genera (e.g. *Callithrix*, *Pithecia*, *Chiropotes*, *Cacajao*, *Alouatta*, *Ateles*) were assigned to a single ecospecies group because their congeners never co-occur at any one site, and different congeners occurring at different sites were considered to fulfill similar ecological roles.

RESULTS

Primate abundance, hunting pressure and forest types

Although the density of all primate species combined was widely variable across all sites (mean = 125 ± 76 ind. km, range = 22–358 ind./km, $n=56$), there were no consistent changes in overall primate density in relation to different levels of hunting pressure (Table 15.2). Hunting regime thus failed to explain a significant component of the variation in the number of animals at each site, whether hunting was considered alone (one-way ANOVA, $F_{3,52}= 1.8$, $P = 0.16$) or in combination with forest type (Table 15.3). In other words, whatever impact hunting may have had on densities of large-bodied species must have been canceled out by a positive numeric response in the abundance of small- or medium-bodied species. As we shall return to later, this should not necessarily be interpreted as evidence for density compensation by species unaffected by hunting which may have resulted from the selective removal of large-bodied species.

The total biomass of diurnal primates across all 56 sites was widely variable, ranging from 27 to 953 kg/km (Tables 15.1 and 15.2). Total primate biomass, however, did not necessarily decline with increasing levels of hunting pressure. Indeed, the rank order of total primate biomass across all forest sites could not be easily predicted by

Table 15.2. *Average primate density, biomass, body mass, and biomass contribution of large-bodied species to the total primate community at 56 northern neotropical sites within different forest types and under varying degrees of hunting pressure*

	Primate density (ind./km^2)		Primate biomass (kg/km^2)		Mean primate body mass(kg)		% biomass of large-bodied taxa		
	Mean	SD	Mean	SD	Mean	SD	Mean	SD	n
Hunting pressure									
None	114.4	58.0	283.8	131.8	2.55	0.36	59.9	10.5	9
Light	127.6	86.5	273.8	197.9	2.25	0.58	56.8	12.7	20
Moderate	154.1	78.8	250.1	149.5	1.59	0.46	34.6	16.4	14
Heavy	96.8	63.1	91.8	52.5	1.02	0.28	15.2	12.1	13
Forest type									
Oligotrophic terra firme	87.9	50.0	170.7	99.3	2.03	0.86	43.6	26.7	20
Mesotrophic terra firme	116.8	60.1	175.5	100.3	1.57	0.65	34.6	18.3	24
Alluvial/várzea	202.9	89.9	424.9	215.1	2.09	0.36	54.5	13.7	12
All forest sites	125.0	76.4	227.2	166.5	1.85	0.72	42.1	21.9	56

Table 15.3. *Analysis of variance examining the simultaneous effect of hunting pressure and forest nutrient status on the total primate density and biomass[a], once the number of sympatric species co-occurring at any one site was entered as a covariate*

Source of variation	SS	df	MS	F-value	P
Primate density					
Forest type	1.081	2	0.541	12.77	0.000
Hunting pressure	0.059	3	0.020	0.47	0.708
Forest type × hunting pressure	0.268	5	0.054	1.27	0.295
Primate richness (covariate)	0.696	1	0.696	16.45	0.000
Explained	2.739[b]	11	0.249	5.88	0.000
Residual	1.863	44	0.042		
Total	6.706	66	1.602		
Primate biomass					
Forest type	1.200	2	0.600	14.56	0.000
Hunting pressure	1.003	3	0.334	8.118	0.000
Forest type × hunting pressure	0.194	5	0.039	0.94	0.462
Primate richness (covariate)	0.692	1	0.692	16.78	0.000
Explained	4.595[c]	11	0.418	10.14	0.000
Residual	1.813	44	0.041		
Total	9.497	66	2.124		

Note: [a]Primate density and biomass values were $\log_{10}$-transformed, and unhunted and lightly hunted sites were pooled together, in order to prevent violation of homogeneity of variance tests.
[b]$r^2 = 0.595$; adjusted $r^2 = 0.494$.
[c]$r^2 = 0.717$; adjusted $r^2 = 0.646$.
SS, sum of squares; MS, mean squares.

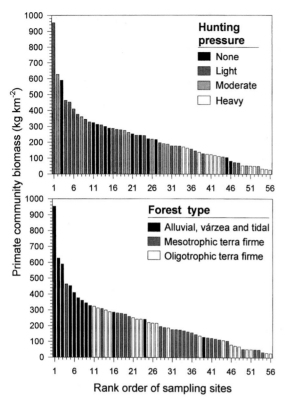

Fig. 15.2. Rank order of crude biomass of all diurnal primates occurring at 56 forest sites of lowland Amazonia and the Guianan Shield, subject to varying degrees of hunting pressure (above) and soil nutrient regimes (below).

degree of hunting pressure alone (Fig. 15.2). For instance, several unhunted sites presented intermediate levels of primate biomass (200–400 kg/km), and two unhunted or lightly hunted sites (Maracá Island and MCSE Reserve) accounted for some of the lowest biomass values ever recorded in the neotropics. However, persistently hunted sites did contain a lower total community biomass (one-way ANOVA, $F_{3,52} = 9.4$, $P < 0.001$), although most of these differences are due to pronounced declines in the biomass of large-bodied species which were most affected by hunting ($F_{3,52} = 27.0$, $P < 0.001$; Table 15.2).

Between-site variation in total primate biomass was, therefore, not merely a function of hunting pressure and is best understood once forest types and the nature of the source fauna are also taken into account. Nutrient-rich Amazonian forests, here represented by floodplain sites,

had the highest primate densities and biomass, even when differences in hunting pressure were taken into account (Fig. 15.3, Table 15.3). Floodplain forests also clearly accounted for the highest densities of large-bodied species, which in unhunted sites contribute with the vast proportion of the overall primate community biomass (Table 15.2). Nine of the ten top-ranking sites in terms of community biomass consisted of floodplain forests (including alluvial, várzea and tidal forests) regardless of the level of hunting pressure to which they had been subjected (Fig. 15.2). Várzea forests of western Brazilian Amazonia, for instance, can sustain an extremely high arboreal vertebrate biomass, and included the highest recorded primate biomass for any neotropical forest (Boa Esperança, Brazil). On the other hand, low-ranking sites were usually represented by oligotrophic terra firme forests, particularly several sites in the Guianan Shield which often sustained a primate biomass under 100 kg/km², even where hunting pressure was light or nonexistent. The three broad classes of forest type were thus a reasonably good indicator of total primate biomass across all sites (Fig. 15.2). In addition, biomass differences between unhunted and hunted areas, and within forest types, clearly decreased from oligotrophic to eutrophic terra firme forests, and were least noticeable within floodplain forests (Fig. 15.3). This pattern suggests different trophic-dependent thresholds in population recovery to a given level of hunting pressure, making nutrient-poor terra firme sites most susceptible (or least resilient) to intensive pulses of primate game harvest.

Once the number of diurnal primate species occurring at any given site was controlled for, there was a significant effect of both forest type and level of hunting pressure on the overall primate biomass (Fig. 15.3, Table 15.3). Surprisingly, however, forest type as defined by broad classes of nutrient input explained a greater proportion of the variation in primate biomass than did hunting pressure. Soil fertility and hunting pressure, however, compensate for one another because primate biomass generally increases with the former and decreases with the latter. Indeed these two factors alone, even at the crude scales used in this study, accounted for over one half of the variation in primate biomass when they were entered simultaneously as independent variables into a multiple regression model ($r^2 = 0.524$, $F_{2,53} = 29.2$, $P < 0.001$). This proportion was clearly greater than that explained by either hunting pressure ($r^2 = 0.249$, $F_{1,54} = 17.9$, $P < 0.001$) or forest type

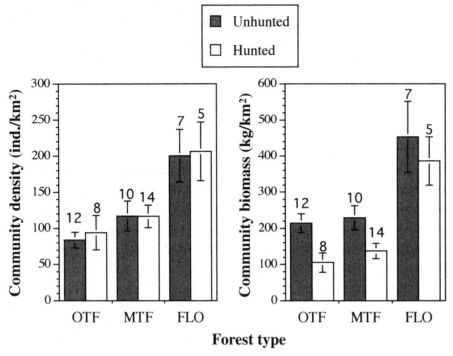

Forest type

Fig. 15.3. Total density and biomass (mean ± SE) of diurnal primates occurring at 56 hunted and unhunted Amazonian sites within oligotrophic terra firme (OTF), eutrophic terra firme (ETF), and floodplain forests (FP). Sites subjected to light or no hunting were combined into 'unhunted'; those subjected to moderate or heavy hunting were combined into 'hunted'. Numbers indicate sample sizes.

($r^2 = 0.213$, $F_{1,54} = 14.6$, $P < 0.001$) when these interacting factors were considered independently. The predictive power of these variables increased even further when considering only the biomass of large-bodied species ($r^2 = 0.719$, $F_{2,53} = 67.6$, $P < 0.001$) which are targeted preferentially by hunters and have the slowest recovery rates.

The same cannot be said about medium- and small-bodied species, the biomass of which was not negatively affected by hunting pressure. Indeed while the proportional contribution of large-bodied species to the total primate biomass at any one site declined sharply from an average of 60% in unhunted sites to 15% in heavily hunted sites, those of small- and medium-bodied species gradually increased (Fig. 15.4). As it might be expected, large-bodied frugivore–folivores, which have lower relative metabolic requirements and harvest the most abundant major food types on offer in neotropical forests, will dominate the community biomass at unhunted sites provided they can remain near carrying capacity. Selective removal of those

species at hunted sites will, therefore, increase the relative abundance of smaller species even if their densities remain unchanged. This resulted in profound shifts in the size structure of primate communities across varying degrees of hunting pressure. For example, the average body weight of primates occurring at different sites crashed from a mean of 2550 g (range = 1790–3420 g, $n = 9$) at unhunted sites to 1020 g (range = 700–1500 g, $n = 13$) at persistently hunted sites (Table 15.2).

Ordination of primate communities

Species dimensionality in a detrended correspondence analysis (DCA) including all sites was largely attributed to the first two axes, out of a cumulative proportion of 52% of the total variation explained by the first four axes (Axis I: 26.5%, II: 13.4%, III: 7.9%, IV: 4.6%). Because the first two axes alone uncovered 40% of the total variance, other dimensions were disregarded in assessing the overall

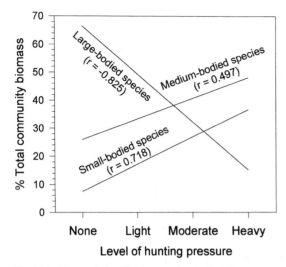

Fig. 15.4. Linear relationships between level of hunting pressure and the percentage contribution of different species size-classes (see text) to the total biomass of diurnal primates at all sites ($n = 56$).

matrix of 56 sampling sites by 16 primate ecospecies (Fig. 15.5). This analysis was, however, clearly affected by inherent differences in the source fauna because it incorporates forest sites right across Amazonia, which includes different species complements at the level of genus (e.g. with the exception of Samuel and Aripuanã, *Callithrix* and *Chiropotes* did not occur in any central-western Amazonian site). Despite these discrepancies, it was still possible to discern three main clusters of forest sites – a western Amazonian terra firme forest cluster; a floodplain forest cluster; and an eastern Amazonian/Guianan Shield cluster – which to a large extent was influenced by habitat differences in ecospecies composition and abundance.

In a subsequent DCA ordination I thus excluded all 15 sites from eastern Amazonia and the Guianan Shield in order to minimize the confounding effects of biogeography on community structure. Considering only the 41 remaining sites of central-western Amazonia, the first two DCA axes accounted for 46% of a total of 57% of the cumulative variance explained by the first four axes (Axis I: 37.1%, Axis II: 8.7%, Axis III: 7.6%, Axis IV: 3.9%). It thus became possible to discern three major clusters of sites diverging primarily along the first two axes: those in hunted terra firme forests; those in unhunted terra firme forests; and those in floodplain/várzea forests (Fig. 15.6).

Environmental loadings on the first and second axes can, therefore, be primarily attributed to site differences in forest types and hunting pressure, respectively, which are associated with reasonably predictable abundance responses in different sets of primate ecospecies.

DISCUSSION

Primates as game vertebrates in neotropical forests

A number of studies have documented the quantitative or qualitative impact of subsistence hunting on neotropical forest primates (Mittermeier, 1987; Bodmer *et al.*, 1988; Ráez-Luna, 1995; Peres, 1990; in press a). It has been estimated that between 3.3 and 8.1 million primates are consumed each year by the rural population of the nine states of Brazilian Amazonia (Peres, in press a). In most cases, hunting for subsistence needs tends to be highly selective towards the large-bodied, prehensile-tailed atelines which weigh 5 kg or more (Peres, 1990; Ayres *et al.*, 1991; Glanz, 1991; Vickers, 1991). However, medium-sized species such as brown capuchins (*Cebus apella*), white-faced capuchins (*C. albifrons*) and bearded sakis (*Chiropotes satanas*), can be harvested on a large scale in many areas (Hill, 1996; C. Peres & H. Nascimento, unpubl. data), particularly where the atelines have already been extirpated or severely reduced in numbers. In contrast, small-bodied species are often bypassed by hunters in favor of alternative prey items that could be killed subsequently during a given hunting foray. The burden of such size-dependent game preferences thus falls heavily upon a few species of frugivore-folivores, which arguably perform critical ecological services such as effective dispersal of large seeds that cannot be ingested by small-bodied species. In the neotropics, this is aggravated by the relative scarcity of primate species weighing >5 kg compared to paleotropical primate faunas (Terborgh & van Schaik, 1987; Peres, 1994a; Fleagle & Reed, 1996), which further increases the extent to which these few species are singled out. This is clearly illustrated by quantitative checklists of game vertebrates harvested by Amazonian forest dwellers from previously undepleted areas in which howlers (*Alouatta*), woolly monkeys (*Lagothrix*), and spider monkeys (*Ateles*) usually rank within the top four most frequently killed species of birds and mammals (Yost & Kelley, 1983; Redford & Robinson, 1987;

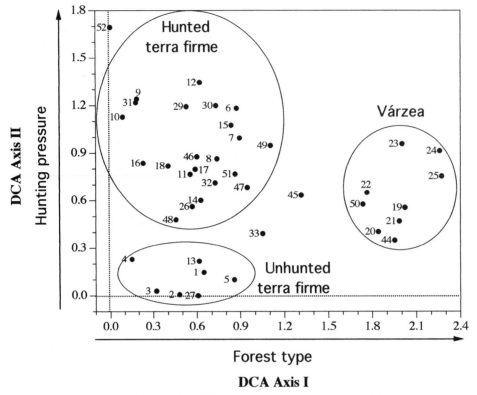

Fig. 15.5. Detrended correspondence analysis (DCA) plot showing the centers of gravity for all 56 sampling sites placed along the first two ordination axes according to the species composition and numeric abundance of diurnal primates. Site groupings distinguished within contour lines include (A) eastern Amazonian and Guianan Shield sites; and (B) terra firme forests of western Amazonia; and (C) floodplain forests. Site numbers correspond to those shown on Fig. 15.1.

Vickers, 1991; Ayres *et al*, 1991; Valenzuela *et al.*, in press; C. Peres & H. Nascimento, unpubl. data). Elsewhere in the neotropics, harvests of currently endangered ateline primates can be seen in the extent to which meat of woolly spider monkeys (*Brachyteles arachnoides*) once fueled the protein needs of entire expeditions by early naturalists venturing inland into the Atlantic forest. This practice survived well into this century as reported by elderly hunters who once routinely shot this increasingly rarer species (Lane, 1990), as contemporary poachers still occasionally do in poorly protected slopes of the Serra do Mar Reserve within the State of São Paulo (F. Campos, pers. comm.).

It is therefore not surprising that population declines in large-bodied species comprise one of the main shifts in primate community structure in overhunted Amazonian forests. Densities of the three ateline genera at unhunted and lightly hunted sites are consistently higher than those at moderately to heavily hunted sites (Peres, 1990; 1991; in press *a,b*). In contrast, small-bodied species (e.g. marmosets, tamarins, squirrel monkeys and titi monkeys) that are usually ignored by hunters were either equally abundant or increased in number at intensively hunted sites. Although this could be interpreted as a classic density-compensation response to competitive release resulting from local extinctions or population declines of large-bodied species, there is virtually no evidence to suggest that this was indeed the case. For instance, the abundance of small-bodied species appeared to depend primarily on environmental gradients independent of hunting pressure, and there is little dietary evidence to suggest that small and large-bodied species actually compete for resources

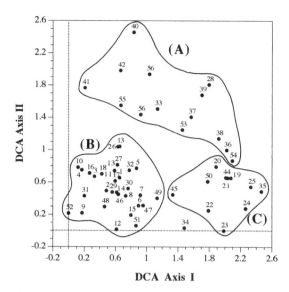

Fig.15.6. DCA plot of 41 western Amazonian sites showing three main site clusters placed along the first two ordination axes, which are largely defined by forest types according to nutrient cycles and degree of hunting pressure. Site numbers correspond to those on Fig. 15.1.

particularly during periods of overall fruit scarcity (Peres, 1994*b*, 1994*c*).

Forest productivity, hunting pressure, and primate abundance

This study clearly shows the pervasive influence of forest types, as defined by broad classes of nutrient availability, on the overall primate abundance once the effects of hunting pressure are controlled for. Indeed even on a crude scale, hunting pressure and forest type when considered simultaneously, can largely predict the overall structure of a large number of primate communities provided they include a comparable pool of species at least in terms of ecological analogs.

Local (α) species diversity had a profound effect on the between-site variation of both censused numbers of individual primates and their combined biomass, even when hunting pressure and forest nutrient status were taken into account. Compared to species-rich communities of western Amazonian terra firme forests, primate density and biomass were markedly reduced in similar forests of eastern Amazonia – where primate α-diversity is consistently

lower (Peres, 1989; Ferrari *et al.*, 1996) – despite the occurrence of considerably more fertile soils in at least some cases (Kayapó Area: C. Peres & C. Baider, unpubl. data). This is theoretically interesting because it suggests that the apparent lack of biomass compensation results from crude partitioning of niche space in species-poor communities along an apparently similar resource spectrum. Alternatively, low primate biomass might also be a function of greater resource seasonality and lower floristic diversity in eastern Amazonia, which are correlated with the east-west gradient of local primate diversity across the region (Ayres, 1986).

The distinction between Amazonian floodplain and terra firme soils is closely tied to the underlying nutrient regime of a given forest (Furch & Klinge, 1989). This has profound consequences for the amount of macronutrients available to plants in different forest types and, in turn, to the overall productivity of vegetative and reproductive plant parts that can be harvested by primary consumers. On average, total primate density and biomass more than doubled from the unflooded forest sites most deprived of external sources of nutrients, to floodplain forests which enjoy the greatest seasonal influxes of nutrient-rich alluvial sediments (Irion, 1978; Junk, 1997). This is particularly the case of the species-rich river basins of western Amazonia where the relatively young soils of Pleistocene and Holocene várzea forests provide the greatest edaphic contrasts with those of terra firme forests. The typically nutrient-poor Ultisols of the terra firme forests considered here roughly match the long folded crystalline rocks of Precambrian origin which form the Guianan and Central-Brazilian (or Guaporé) Shields. These archaic kratons have received no fresh sediment deposits since the Triassic or lower Cretaceous (Putzer, 1984), and are characterized by forests under an extremely tight nutrient budget (Furch, 1984). The consistently low biomass of primates and other non-volant mammals in terra firme forests thus appears to reflect one of the region's large-scale geochemical gradients which is profoundly affected by exposure to fluvial nutrient dynamics (Peres, 1997*a*; 1997*b*; in press *c*).

This can be seen in the strong negative relationship between distance to major white-water rivers and densities of howler monkeys in Amazonia (Peres, 1997*b*). This large-scale geographic pattern is consistent with the importance of soil nutrients to other folivorous primates in mainland Africa (McKey, 1978; Oates *et al.*, 1990), Madagascar

(Ganzhorn, 1992), and south-east Asia (Rodman, 1978; E. Bennett, pers. comm.). Floodplain forests appear to provide a more favorable foliage quality to arboreal folivores, such as howler monkeys, common and royal sloths, hoatzins, and iguanas. These species are some of the faunal hallmarks of Amazonian várzea and tidal forests (Peres, 1989; 1997a), but become strikingly rare in oligotrophic terra firme forests. This is one of the reasons why várzea forests tend to diverge into a cohesive cluster of sites in ordinations of Amazonian primate communities which take into account both species composition and relative abundance (Peres, 1997a, this study).

Clearly, however, approximately 94% of the entire Amazon basin consists of terra firme systems well beyond the influence of large rivers carrying heavy loads of suspended sediments. As a result, only 6% of Amazonian soils have no major limitations of one key nutrient or another to plant photosynthesis (Sanchez, 1981). Accordingly, high levels of primate abundance in the region have to be considered an exception rather than the rule. The paradise-like images of monkey-watching that floodplain sites such as Cocha Cashu (Terborgh, 1983; Janson & Emmons, 1990) and Boa Esperança (Peres, 1997a) may evoke for many a primatologist are therefore highly atypical in this vast region where highly leached soils prevail. Moreover, data presented here suggest that forests under high nutrient status can partially cancel out the detrimental effects of persistent hunting pressure on primate biomass (cf. Peres, in press a,b). By the same token, the effects of persistent hunting can be aggravated by low habitat productivity, which can be graphically illustrated by several cases of rapid population declines and local extinctions driven by overhunting in Amazonian terra firme forests that became suddenly accessible via newly-built roads (Smith, 1976; Peres, 1990; Ayres et al, 1991). In contrast, howler monkeys and other game populations in typical várzea forests can coexist with a moderate harvest regime for hundreds of years (Peres, 1997b).

Herein lies an important implication to game management policy in Amazonia and other major regions of tropical forests which has largely failed to consider the effects of habitat quality on game population and community structure. As shown for primates, the effects of game harvest appear to be invariably mediated by forest types. Terra firme forests of low nutrient status, which comprise the vast proportion of the region, tend to be less resilient to game extraction than equivalent areas of floodplain forest facing a comparable level of hunting pressure. Although more species-rich than flooded forests (Peres, 1997a), primate communities of even entirely undisturbed terra firme forests are deceptively fragile, and can be easily truncated into a more depauperate sub-community once their large-bodied species are wiped out by game hunters. Hunting restrictions in tropical forests should not be based solely on a target-species approach emphasizing the reproductive rates of different game species, but also on a habitat-based approach in which the productivity of different forest types is carefully considered. Unfortunately much additional research and institutional reform are still required before sound game management programs, designed to maintain viable game populations, can be implemented. Until then the full integrity of primate communities, in a region where animal protein is often at a premium, can only be assured within strictly protected nature reserves.

ACKNOWLEDGMENTS

Primate surveys carried out during this study have been funded by the World Wildlife Fund (1987–1988), Wildlife Conservation Society (1991–1995), and the Josephine Bay and Michael Paul Foundations (since 1996). I am grateful to the Brazilian Oil Company (Petrobrás, S.A.) for providing helicopter transportation to several otherwise inaccessible survey sites. A number of people have helped me through the years in the field, but I would particularly like to thank the invaluable assistance of João Pompilho, Raimundo Nonato, Luís Lopes, Cosmes "Toshiba", Edvar Corrêa, Josimar Pereira, Paul Honess, Pedro Develey, and Hilton Nascimento. This manuscript has benefited from comments made by Michael Alvard, Louise Emmons, and an anonymous reviewer.

REFERENCES

Ayres, J. M. (1986). *Uakaris and Amazonian flooded forest.* Ph.D. thesis, University of Cambridge, Cambridge.

Ayres, J. M., Lima, D. M., Martins, E. S. & Barreiros, J. L., (1991). On the track of the road: changes in subsistence hunting in a Brazilian Amazonian village. In *Neotropical Wildlife Use and Conservation*, ed. J. G. Robinson & K.H. Redford, pp. 82–92. Chicago: University of Chicago Press.

Ayres J. M. & Clutton-Brock, T. H. (1992). River boundaries and species range size in Amazonian primates. *American Naturalist*, **140**, 531–7.

Bennett, E., Nyaoi, A. & Sompud, J. (1999). Effects of hunting on wildlife populations in northern Borneo. In *Hunting for Sustainability in Tropical Forests*, ed. J. G. Robinson & E. L. Bennett, pp. 299–318. New York: Columbia University Press.

Bernstein, I. S., Balcean, P., Dresdale, L., Gouzoules, H., Kavanagh, M., Patterson, T. & Neyman-Warner, P. (1976). Differential effects of forest degradation on primate populations. *Primates*, **17**, 401–11.

Bodmer, R. E. (1993). Managing wildlife with local communities: the case of the Reserva Comunal Tamshiyacu-Tahuayo. Paper prepared for the Liz Claiborne Art Ortenberg Foundation Community Based Conservation Workshop, Airlie, Virginia. pp. 1–31.

Bodmer, R. E. (1995). Managing Amazonian wildlife: biological correlates of game choice by detribalized hunters. *Ecological Applications*, **5**, 872–77.

Bodmer, R. E., Fang, T. G. & Ibanez, L. M. (1988). Primates and ungulates: a comparison of susceptibility to hunting. *Primate Conservation*, **9**, 79–83.

Bourlière, F. (1985). Primate communities: their structure and role in tropical ecosystems. *International Journal of Primatology*, **6**, 1–26.

Branch, L. C. (1983). Seasonal and habitat differences in the abundance of primates in the Amazon (Tapajós) National Park, Brazil. *Primates*, **24**, 424–431.

Buckland, S. T., Anderson, D. R., Burnham, K. P. & Laake, J. L. (1993). *Distance Sampling: Estimating Abundance of Biological Populations*. London: Chapman & Hall.

Burnham, K. P., Anderson, D. R. & Laake, J. L. (1980). Estimation of density from line transect sampling of biological populations. *Wildlife Monographs*, **72**, 1–202.

Calouro, A. M. (1995). Caça de subsistência: sustentabilidade e padrões de uso entre seringueiros ribeirinhos e não-ribeirinhos Estado do Acre. MSc thesis, Universidade de Brasília, Brasília.

Cameron, R., Wiltshire, C., Foley, C., Dougherty, N., Aramayo, X. & Rea, L. (1989). Goeldi's monkey and other primates in northern Bolivia. *Primate Conservation*, **10**, 62–70.

Culotta, E. (1995). Many suspects to blame in Madagascar extinctions. *Science*, **268**, 1568–9.

Dewar, R. E. (1984). Recent extinctions in Madagascar: the loss of the subfossil fauna. In *Quaternary Extinctions: A Prehistoric Revolution*, ed. P. S. Martin & R. G. Klein, pp. 574–9. Tucson: University of Arizona Press.

Eisenberg, J. F. (1979). Habitat, economy and society: Some correlations and hypotheses for the neotropical primates. In *Primate Ecology and Human Origins: Ecological Influences on Social Organization*, ed. I. S. Bernstein & E. O. Smith, pp. 215–62. New York: Garland Press.

Emmons, L. H. (1984). Geographic variation in densities and diversities of non-flying mammals in Amazonia. *Biotropica*, **16**, 210–22.

Fa, J. E., Juste, J., del Val, J. P. & Castroviejo, J. (1995). Impact of market hunting on mammal species in Equatorial Guinea. *Conservation Biology*, **9**, 1107–15.

Ferrari, S. F. & Lopes, M. A. (1996). Primate populations in eastern Amazonia. In *Adaptive Radiations in Neotropical Primates*, ed. M. Norconk, pp. 53–68. New York: Plenum Press.

Fleagle, J. G. & Reed, K. E. (1996). Comparing primate communities: a multivariate approach. *Journal of Human Evolution*, **30**, 489–510.

Freese, C. H., Heltne, P. G., Castro, N. R. & Whitesides, G. (1982). Patterns and determinants of monkey densities in Peru and Bolivia, with notes on distributions. *International Journal of Primatology*, **3**, 53–90.

Furch, K. (1984). Water chemistry of the Amazon basin: the distribution of chemical elements among freshwaters. In *The Amazon: Limnology and Lanscape Ecology of a Mighty Tropical River and Its Basin*, ed. H. Sioli, pp. 167–99. Dordrecht: Dr. W. Junk Publ.

Furch, K. & Klinge, H. (1989). Chemical relationships between vegetation, soil, and water in contrasting inundation areas of Amazonia. In *Mineral Nutrients in Tropical Forest and Savanna Ecosystems*, ed. J. Proctor, pp. 189–204. Oxford: Blackwell Scientific.

Ganzhorn, J. U. (1992). Leaf chemistry and the biomass of folivorous primates in tropical forests. *Oecologia*, **91**, 540–7.

Glanz, W. E. (1991). Mammalian densities at protected versus hunted sites in central Panama. In *Neotropical Wildlife Use and Conservation*, ed. J. G. Robinson & K. H. Redford, pp. 163–73. Chicago: Chicago University Press.

Guillotin, M., Dubost, G. & Sabatier, D. (1994). Food choice and food competition among the three major primate species of French Guiana. *Journal of Zoology, London*, **233**, 551–79.

Hill, K. 1996. The Mbracayu Reserve and the Ache of Paraguay. In *Traditional Peoples and Biodiversity Conservation in Large Tropical Landscapes*, ed. K. H. Redford & J. A. Mansour, pp. 159–96. Arlington: The Nature Conservancy.

Irion, G. (1978). Soil infertility in the Amazonian rain forest. *Naturwissenschaften*, **65**, 515–19.

Janson, C. H. & Emmons, L. H. (1990). Ecological structure of the non-flying mammal community at the Cocha Cashu Biological Station, Manu National Park, Peru. In *Four Neotropical Rainforests*, ed. A. Gentry, pp. 314–38. New Haven: Yale University Press.

Johnson, E. G. & Routledge, R. D. (1985). The line transect method: a nonparametric estimator based on shape restrictions. *Biometrics*, **41**, 669–79.

Junk, W. J (ed.) (1997). *The Central Amazon Floodplain: Ecology of a Pulsing System*. Berlin: Springer.

Lane, F. (1990). A hunt for "monos" (*Brachyteles arachnoides*) in the foothills of the Serra de Paranapiacaba, São Paulo, Brazil. *Primate Conservation*, **11**, 23–5.

Lemos de Sá, R. M. (1995). *Effects of the Samuel Hydroelectric Dam on Mammal and Bird Communities in a Heterogeneous Amazonian Lowland Forest*. Ph.D. thesis, University of Florida, Gainesville.

McRae, M. (1997). Road-kill in Cameroon. *Natural History*, **2/97**, 36–75.

Martins, E. S. (1992). *A caça de subsistência de extrativistas na Amazônia: sustentabilidade, biodiversidade e extinção de espécies*. MSc Thesis, University of Brasília, Brasília.

McKey, D. B. (1978). Soils, vegetation, and seed-eating by black colobus monkeys. In *The Ecology of Arboreal Folivores*, ed. G. G. Montgomery, pp. 423–37. Washington, DC: Smithsonian Institution Press.

Mendes-Pontes, A. R. (1994). *Environmental Determinants of Primate Abundance in Maracá Island, Roraima, Brazilian Amazonia*. MPhil Thesis, Cambridge University, Cambridge.

Mitchell, C. L. & Ráez-Luna, E. F. (1991). *El Impacto de la Caza Humana sobre Populaciones de Primates y Aves de Caza en la Reserva de Biosfera de Manu en el Sudeste Peruano*. Unpublished Report to Wildlife Conservation International, New York.

Mittermeier, R. A. (1977). *Distribution, Synecology and Conservation of Surinam Monkeys*. Ph.D. thesis, Harvard University.

Mittermeier, R. A. (1987). Effects of hunting on rain forest primates. In *Primate Conservation in the Tropical Rain Forest*, ed. C. Marsh & R. Mittermeier, pp. 109–48. New York: Alan R. Liss.

Mittermeier, R. A. & van Roosmalen, M. G. M. (1981). Preliminary observations on habitat utilization and diet in eight Surinam monkeys. *Folia Primatologica*, **36**, 1–39.

Muckenhirn, N. A., Mortensen, B. K., Vessey, S., Fraser, C. E. O. & Singh, B. (1975). *Report on a Primate Survey in Guyana*. Pan American Health Organization. Washington, DC

Oates, J. F. (1996). Habitat alteration, hunting and the conservation of folivorous primates in African forests. *Australian Journal of Ecology*, **21**, 1–9.

Oates, J. F., Whitesides, G. H., Davies, A. G., Waterman, P. G., Green, S. M., Dasilva, G. L. & Mole, S. (1990). Determinants of variation in tropical forest primate biomass: new evidence from west Africa. *Ecology*, **71**, 328–43.

Pack, K. S. (1994). *Integration of People and Dynamic Interactions in Tropical Rainforest Conservation*. MPhil Thesis, University of Kent.

Peres, C. A. (1988). Primate community structure in Brazilian Amazonia. *Primate Conservation*, **9**, 83–7.

Peres, C. A. (1989). A survey of a gallery forest primate community, Marajó Island, Pará. *Vida Silvestre Neotropical*, **2**, 32–7.

Peres, C. A. (1990). Effects of hunting on western Amazonian primate communities. *Biological Conservation*, **54**, 47–59.

Peres, C. A. (1991). Humboldt's woolly monkeys decimated by hunting in Amazonia. *Oryx*, **25**, 89–95.

Peres, C. A. (1992). Prey-capture benefits in a mixed-species group of Amazonian tamarins, *Saguinus fuscicollis* and *S. mystax*. *Behavioral Ecology and Sociobiology*, **31**, 339–47.

Peres, C. A. (1993a). Notes on the primates of the Juruá River, western Brazilian Amazonia. *Folia Primatologica*, **61**, 97–103.

Peres, C. A. (1993b). Structure and spatial organization of an Amazonian terra firme forest primate community. *Journal of Tropical Ecology*, **9**, 259–76.

Peres, C. A. (1994a). Which are the largest New World monkeys? *Journal of Human Evolution*, **26**, 245–9.

Peres, C. A. (1994b). Primate responses to phenological changes in an Amazonian terra firme forest. *Biotropica*, **26**, 98–112.

Peres, C. A. (1994c). Plant resource use and partitioning in two tamarin species and woolly monkeys in an Amazonian terra firme forest. In *Current Primatology, Vol. I: Ecology and Evolution*, ed. B. Thierry, J. R. Anderson, J. J. Roeder & N. Herrenschmidt. pp. 57–66. Strasbourg: Université Louis Pasteur.

Peres, C. A. (1997a). Primate community structure at twenty western Amazonian flooded and unflooded forests. *Journal of Tropical Ecology*, **13**, 381–405.

Peres, C. A. (1997b). Effects of habitat quality and hunting pressure on arboreal folivore densities in neotropical forests: a case study of howler monkeys (*Alouatta* spp.). *Folia Primatologica*, **68**, 199–222.

Peres, C. A. (in press *a*). Impact of subsistence hunting on vertebrate community structure in Amazonian forests: a large-scale cross-site comparison. *Conservation Biology*.

Peres, C. A. (1999). Evaluating the impact and sustainability of subsistence hunting at multiple Amazonian forest sites. In *Hunting for Sustainability in Tropical Forests*, eds. J. G. Robinson & E. L. Bennett, pp. 30–52. New York: Columbia University Press.

Peres, C. A. (in press *c*). Nonvolant mammal community structure in different Amazonian forest types. In *Mammals of the Neotropics, Vol 3*,. ed. J. F. Eisenberg & K. H. Redford. Chicago: University of Chicago Press.

Puertas, P. & Bodmer, R. E. (1993). Conservation of a high diversity primate assemblage. *Biodiversity and Conservation*, **2**, 586–93.

Putzer, H. 1984. The geological evolution of the Amazon basin and its mineral resources. In *The Amazon: Limnology and Landscape Ecology of a Mighty Tropical River and Its Basin*, ed. H. Sioli, pp. 15–46. Dordrecht: Dr. W. Junk Publ.

RADAM. (1973–1981). Projeto RADAMBRASIL. Levantamento de recursos naturais. Vol. 1–18. Maps produced by the Departamento Nacional de Produção Mineral, Ministério das Minas e Energia, Rio de Janeiro.

Ráez-Luna, E. F. (1995). Hunting large primates and conservation of the Neotropical rain forests. *Oryx*, **29**, 43–8.

Redford, K. H. (1992). The empty forest. *Bioscience*, **42**, 412–22.

Redford, K. H. & Robinson, J. G. (1987). The game of choice: patterns of Indian and colonist hunting in the neotropics. *American Anthropologist*, **89**, 650–67.

Robinson, J. G. & Redford, K. H. (eds.) (1991). *Neotropical Wildlife Use and Conservation*. Chicago: Chicago University Press.

Robinson, J. G & Bennett, E. L. (eds.) (1999). *Hunting for Sustainability in Tropical Forests*. New York: Columbia University Press.

Rodman, P. S. (1978). Diets, densities, and distributions of Bornean primates. In *The Ecology of Arboreal Folivores*, ed. G. G. Montgomery, pp. 465–480. Washington, DC: Smithsonian Institution Press.

Rylands, A. B. (1983). *The Behavioural Ecology of Three Species of Marmosets and Tamarins (Callitrichidae, Primates) in Brazil*. Ph.D. thesis, University of Cambridge, Cambridge.

Rylands, A. B. & Keuroghlian, A. (1988). Primate populations in continuous forest and forest fragments in central Amazonia. *Acta Amazonica*, **18**, 291–307.

Sanchez, P. A. (1981). Soils of the humid tropics. In *Blowing in the Wind: Deforestation and Long-Term Implications*, pp. 347–410. Dept. of Anthropology, College of William and Mary, Williamsburg, Va.

Smith, N. J. H. (1976). Utilization of game along Brazil's Transamazon highway. *Acta Amazonica*, **6**, 455–66.

Smuts, B. B., Cheney, D. L., Seyfarth, R. M., Wrangham, R. W. & Struhsaker, T. T. (eds.) (1987). *Primate Societies*. Chicago: University of Chicago Press.

Soini, P. (1986). A synecological study of a primate community in the Pacaya-Samiria National Reserve, Peru. *Primate Conservation*, **7**, 63–71.

Terborgh J. (1983). *Five New World Primates: A Study in Comparative Ecology*. Princeton: Princeton University Press.

Terborgh J. & van Schaik, C. P. (1987). Convergence vs. nonconvergence in primate communities. In *Organization of Communities, Past and Present*, ed. J. H. R. Gee & P. S. Giller, pp. 205–26. Oxford: Blackwell Scientific.

ter Braak, C. J. F. (1988). CANOCO – an extension of DECORANA to analyse species-environment relationships. *Vegetatio*, **75**, 159–60.

Valenzuela, P. M., Stallings, J. R., Bolanos, J. R. & Loachamin, R. C. (1996). The sustainability of current hunting practices by the Huaorani. In *Hunting for Sustainability in Tropical Forests*, ed. J. G. Robinson & E. L. Bennett, pp. 54–76. New York: Columbia University Press.

Vickers, W. T. (1991). Hunting yields and game composition over ten years in an Amazon Indian territory. In *Neotropical Wildlife Use and Conservation*, ed. J. G. Robinson & K. H. Redford, pp. 53–81. Chicago: University of Chicago Press.

Yost, J. A. & Kelley, P. M. (1983). Shotguns, blowguns, and spears: the analysis of technological efficiency. In *Adaptive Responses of Native Amazonians*, ed. R. B. Hames & W. T. Vickers, pp. 189–224. London: Academic Press.

16 · Spatial and temporal scales in primate community structure

JOHN G. FLEAGLE, CHARLES H. JANSON AND KAYE E. REED

The chapters in the preceding section have reviewed evidence that the present structure of primate communities is affected by resource abundance (Janson & Chapman, chapter 14, Peres, chapter 15), food quality and digestive strategies (Janson & Chapman), seasonality (Janson & Chapman; Tutin & White, chapter 13), disease (Tutin & White), contemporary hunting (Peres), recent land-use history (Tutin & White), climatic change in the recent past and the Pleistocene (Tutin & White; Eeley & Lawes, chapter 12), and regional species source pools (Eeley & Lawes; Peres). Despite the diversity of these themes, they fall into two broad contrasts which we shall use to review the preceding results: (1) local vs. regional explanations of primate community structure; and (2) equilibrial vs. nonequilibrial views of community structure.

LOCAL VS. REGIONAL DETERMINANTS OF PRIMATE COMMUNITY STRUCTURE

There is a tension between explanations that rely on local versus regional ecological factors to explain local site diversity. Local factors include resource abundance and quality, seasonality, competition and contemporary hunting. Regional analysis focuses on regional source pools of species – a local site cannot have a primate species that is not in the regional source pool. This distinction is essentially parallel to that between alpha- and gamma-diversity in community ecology. Alpha-diversity reflects the number of coexisting species in a single site, whereas gamma-diversity includes additional species that may replace each other in separate areas of similar habitat only because of geographical isolation. Proponents of local factors find support in the fact that primate population densities predictably increase in areas of greater soil fertility (Peres),

high food quality (Janson & Chapman), and low hunting pressure (Peres). By this view, which taxonomic species occur in a site might depend on the regional pool, but how many species occur and what ecological niches they occupy depends on local resource availability and mortality factors, perhaps modified by competition with other primates (see Ganzhorn, chapter 8, this volume) or mammal species (see Emmons, chapter 10, this volume). Enthusiasts of regional factors do not deny that particular primate species within a given site may have their populations limited by food availability and mortality rates. However, they claim that which species are absent from a community, and by extension how many species are present, depends on the regional pool of species to choose from, which is determined by speciation and extinction rates, which in turn depend on latitude, local topography, recent climate change, and phylogenetic or geological history (Eeley & Lawes). Regions with greater species diversity can produce greater local diversity by a modest level of invasion between contiguous areas of distinct evolutionary history.

The strongest argument for local determinants of community structure can be made for primate biomass. Not only should population density (and hence biomass) respond relatively quickly to ecological change, but a number of autecological studies on primates have suggested that food availability limits population size, at least seasonally (Terborgh 1983; Janson, 1984; Gautier-Hion & Michaloud, 1989). Thus, studies that focus primarily on primate biomass have achieved some success relating biomass to food quality or correlates of food production (Janson & Chapman and references therein). Peres (chapter 15, this volume) was able to explain much of the variation in primate biomass among Neotropical sites using local factors of soil fertility (a correlate of food productivity) and hunting pressure.

However, primate biomass in a given site may also be influenced by regional or historical factors. Peres needed to control statistically for variation in species diversity among sites to reveal significant effects of soil fertility and hunting – sites with few species had reduced biomass levels relative to species-rich sites of equal fertility and hunting pressure. Janson & Chapman found that there was a marked difference in folivore and total primate biomass between sites that contained fore-stomach-fermenting versus hind-gut-fermenting folivores. Thus, understanding variation in primate biomass across sites may need to factor in regional species source pools and the evolutionary history of the primate fauna, in addition to site- specific food availability and mortality factors. Evidence for local vs. regional determinants of primate species diversity is inevitably indirect. While it is easy to propose and reject local explanations for the density of a given species in a site, it is much harder to explain why some other species is absent from it. For instance, confusion surrounds the cause of the widespread positive correlation between primate diversity and rainfall between sites (Reed and Fleagle, 1995). Possible local explanations include lack of seasonality, increased productivity, and increased resource diversity (see Kay *et al.*, 1997). Although the pattern for the New World tropics suggests that productivity is the factor most tightly linked to species diversity across sites (Kay *et al.*, 1997), the pattern does not appear to be simple, as Peres documents that species diversity actually decreases on the most fertile Amazonian soils. Eeley & Lawes suggest that the correlation of species diversity with rainfall is not causal but reflects merely a general relationship between regional species diversity and latitude, the causes of which are still hotly debated (Rosenzweig 1995). Conversely, one could argue that regional species diversity may be correlated with local diversity because rainfall patterns tend to occur across broad geographic areas.

As long as isolation by distance or barriers occurs (Ayres & Clutton-Brock, 1992), and species at least on occasion breach former barriers to invade new geographic ranges, regional and local species diversities will tend to be correlated. Distinguishing between these hypotheses may require sophisticated statistical analyses which can control both for phylogenetic history (Brooks & McLennon, 1991;

Purvis, 1995) and geographical spatial autocorrelation (Legendre *et al.*, 1990), not to mention far more complete data sets than currently exist. For instance, the postulated link between primate species diversity and primary productivity could not be tested directly by Kay *et al.* (1997) because both variables were available from so few study sites.

ARE PRIMATE COMMUNITIES AT EQUILIBRIUM, AND DOES IT MATTER?

In parallel to the problem of local vs. regional factors affecting primate communities is the tension between explanations that assume something like ecological equilibrium and those that suggest the primacy of non-equilibrial states. Most local explanations require quasi-equilibrium between population size and resources and mortality, but equilibrium can be assumed also in regional models, assuming a balance of speciation and extinction rates as in island biogeography. However, recent variation in climate has affected extent of rainforest, even over relatively short periods (Tutin & White, chapter 13, this volume; Lovett, 1993). At least in Africa, this variation has played out as contraction vs. expansion of forest vs. savanna biomes (neither of which is unitary). Particular primate species tolerate or prefer distinct ends of this continuum, and so the relative abundance of species almost surely has changed along with habitat types during the period since the end of the last glaciation, and probably earlier as well. Do such changes help to explain current primate species composition, diversity or biomass?

The answer depends on the scale of change of primate populations in response to ecological change. Tutin & White (chapter 13, this volume) claim that primate populations recover slowly, so that communities may be expected to be out of balance for long periods following disturbance. While it is true that most studied primate populations have very low net growth rates, these rates are generally determined in populations suffering little disturbance, as these are the ones preferred by primatologists for long-term study. When in fact primates have been subjected to major ecological change (a yellow-fever epidemic, major provisioning or its withdrawal), in every case we know of, the population has switched to a new level within a decade, suggesting that maximal rates of population change in primates are by no means as low as habitual population growth rates suggest. At these rates, populations of primate species can be expected to be close to their local ecological equilibrium in response to even fairly rapid ecological change at a geological time scale, as long as they do not go extinct.

What about population changes in response to catastrophic fruit-crop failure such as that described by Foster (1982)? At the rate he inferred of one per eight years, is there any hope that primate populations are close to their carrying capacities? Two factors must be kept in mind. First, widespread famines among primates have yet to be documented in most tropical sites outside Central America, despite hundreds of person-years of study. Thus, such famines may be much less frequent in other areas. Second, nearly all studies that have documented fruit phenology or abundance across years have noted marked variation in the species composition and overall levels of fruit available within years and from year to year (see chapter 13 by Tutin & White, and Janson & Chapman, chapter 14). Yet these same studies note little correlation between these resource changes and primate population growth or decline. We would argue that such short-term patterns of resource variation have instead favored in all primate species the ability to use relatively predictable 'worst-case' foods during periods of resource scarcity (Terborgh, 1986). These critical foods may differ between primate species, but need not. The ability of primate species to use these critical foods would dampen any response of the population to short-term fluctuations in resource abundance. Why such buffering appears to be less reliable in Central America remains an interesting question for further research.

What about human-caused habitat disturbance? Logging clearly can affect primate populations, as shown in many published comparisons of logged and unlogged forests. From the point of view of understanding primate community structure, however, these effects presumably act via changes in the availability of resources (but see Cannon et al., 1998). If we had adequate measures of limiting resources for distinct primate species, it might turn out that the modified primate community is in equilibrium with its altered food supply. In the more common case that we have measured only correlates of resource availability (such as habitat type and rainfall: Peres & Janson, this volume), then including strongly-disturbed habitats in comparative analyses will decrease our ability to resolve the underlying determinants of primate community structure, and might even lead to biases if some habitat types are preferred for human use.

The rapid decline and sometimes extinction of particular primate populations in response to hunting should provide a caution about the criteria for selecting study sites and the interpretation of comparative studies. Surely, any primate community subjected to past or current severe hunting pressure is likely to have both a depressed species diversity and biomass, as large species are selectively pursued by human hunters (Peres, chapter 15, this volume). It would seem folly to include such sites in a broad comparative survey to test hypotheses that generally assume equilibrium in response to local resource availability. However, one could extend this argument to say that any analysis of primate community structure that ignores primate mortality factors is necessarily flawed, as the abundance of any population will be a balance determined by both resource input and losses to predation and disease in addition to starvation.

What, if anything, makes human hunting different from the diverse other sources of mortality experienced by primates? First, the mortality rates are often much larger under human hunting than via non-human predation, although hunting by chimpanzees can lead to demonstrable population decline in red colobus (Stanford, 1995), while Amerindians without guns have apparently hunted some primates sustainably for centuries. Second, human settlements are patchy and so hunting tends to be concentrated in the vicinity of villages, producing large impacts in small areas but little effect in others. By contrast, most large predators of primates range extensively throughout large home ranges, and hunting intensity is rarely patchy in space, although sometimes clustered in time (Isbell, 1990). Thus, under non-human predation regimes, most primate species do not go extinct and may show relatively little change in density from one area to another in a single habitat. While epidemics and other sporadic sources of intense primate mortality should be considered and investigated as possible causes of population density, it seems unlikely that these will cause or confound large-scale patterns due to differences in local resource availability or regional patterns of species replacement.

However, if a species goes extinct in an area, there is no simple argument or measure to predict how long it will take it to re-invade, or how quickly the remaining species will take to adjust evolutionarily to its absence. Recent massive extinctions of large lemur species may well help to explain why the relict lemur communities in Madagascar do not conform to ecological and social patterns characteristic of other primate communities, as noted by Ganzhorn in chapter 8 and by Kappeler in chapter 9 in this volume (see also van Schaik & Kappeler, 1996).

CONCLUSION AND PROSPECTS

Clearly, more data are needed to distinguish among the possible causes for site-to-site differences in primate species diversity. In a few cases, it is easy to recognize impoverished primate communities caused by climate change (e.g., eastern African forests: Wasser, 1993). In most areas such historical factors are unknown. Measures of resource availability for distinct primate species are few and usually available for only one or a few sites. Thus, it is hard to test whether local factors can account for differences in primate community composition, beyond the broad latitudinal correlations documented in many chapters of this volume. Proponents of regional-scale control of diversity need to account for known correlations between diversity and local variables such as rainfall or productivity (Kay et al., 1997). Either both scales are important and explain distinct fractions of the variation in local species diversity, or else one scale of explanation must be able to reproduce the results claimed to support the alternative scale (see Kunin, 1998). For explanations of biomass variation, local measures may be adequate, but more work is needed to investigate the degree to which local biomass depends on local species diversity or species composition (cf. Peres, chapter 15, this volume).

REFERENCES

Ayres, J. M. & Clutton-Brock, T.H. (1992). River boundaries and species range size in Amazonian primates. *American Naturalist*, **140**, 531–7.

Brooks, D. R. & McLennon, D. A. (1991). *Phylogeny, Ecology, and Behaviour: A Research Program in Comparative Biology*. Chicago: University of Chicago Press.

Cannon, C. H., Peart, D. R. & Leighton, M. (1998). Tree species diversity in commercially logged Bornean rainforest. *Science*, **281**, 1366–8.

Foster, R. B. (1982). Famine on Barro Colorado Island. In *The Ecology of a Tropical Forest: Seasonal Rhythms and Long-term Changes*, ed. E. G. Leigh, A. S. Rand & D. M. Windsor, pp 201–12. Washington, DC: Smithsonian Institution Press.

Gautier-Hion, A. & Michaloud, G. (1989). Are figs always keystone resources for tropical frugivorous vertebrates? A test in Gabon. *Ecology*, **70**, 1826–33.

Janson, C. H. (1984). Female choice and mating system of the brown capuchin monkey *Cebus apella* (Primates: Cebidae). *Zeitschrift fur Tierpsychologie*, **65**, 177–200.

Isbell, L. A. (1990). Sudden short-term increase in mortality of vervet monkeys (*Cercopithecus aethiops*) due to leopard predation in Amboseli National Park, Kenya. *American Journal of Primatology*, **21**, 41–52.

Kay, R. F., Madden, R. H., Schaik, C. P. van & Higdon, D. (1997). Primate species richness is determined by plant productivity: implications for conservation. *PNAS*, **275**, 13023–27.

Kunin, W. E. (1998). Extrapolating species abundance across spatial scales. *Science*, **281**, 1513–15.

Legendre, P., Oden, N. L., Sokal, R R., Vaudor, A. & Kim J. (1990). Approximate analysis of variance of spatially autocorrelated regional data. *Journal of Classification*, **7**, 53–76.

Lovett, J. C. (1993). Climatic history and forest distribution in eastern Africa. In *Biogeography and Ecology of the Rain Forests of Eastern Africa*, ed. J. C. Lovett & S. K. Wasser, pp. 23–29. Cambridge: Cambridge University Press.

Purvis, A. (1995). A composite estimate of primate phylogeny. *Philosophical Transactions of the Royal Society in London, Series B*, **348**, 405–21.

Reed, K. E. & Fleagle, J. G. (1995). Geographic and climatic control of primate diversity. *PNAS*, **92**, 7874–76.

Rosenzweig, M. L. (1995). *Species Diversity in Space and Time*. Cambridge: Cambridge University Press.

Stanford, C. B. (1995). The influence of chimpanzee predation on group size and anti-predator behaviour in red colobus monkeys. *Animal Behaviour*, **49**, 577–87.

Terborgh, J. W.. (1983). *Five New World Primates*. Princeton: Princeton University Press.

Terborgh, J. W. (1986). Community aspects of frugivory in tropical forests. In *Frugivores and Seed Dispersal*, ed. A. Estrada & T. H. Fleming, pp. 371–84. Dordrecht: D. W. Junk Publ.

van Schaik, C. P. & Kappeler, P. M. (1996). The social systems of gregarious lemurs: lack of convergence with anthropoids due to evolutionary disequilibrium? *Ethology*, **102**, 915–41.

Wasser, S. K. (1993). The socioecology of interspecific associations among the monkeys of the Mwanihana rain forest, Tanzania: a biogeographic perspective. In *Biogeography and Ecology of the Rain Forests of Eastern Africa*, ed. J. C. Lovett & S. K. Wasser, pp. 267-80. Cambridge, Cambridge University Press.

17 • Primate communities in Africa: The consequence of long-term evolution or the artifact of recent hunting?

THOMAS T. STRUHSAKER

INTRODUCTION

Most comparisons of tropical forest communities between specific forests, regions, or larger geographic areas make the tacit assumption that these communities are the result of long-term evolutionary processes that led to ecologically stable systems. For most tropical forests, however, we have no early records (200 years or older) of what was distributed where and, therefore, cannot know if current ecological communities reflect ecologically stable systems that are the result of long-term evolution or whether they are the artifacts of very recent human activities, e.g., hunting, habitat destruction and alteration. In most tropical areas the impact of human activities has greatly escalated in the past 100–200 years and particularly so in the last 50 years because of exponentially increasing human populations. In some tropical areas this impact by humans is so profound and pervasive that it prevents an understanding of the longer-term processes in the evolution of tropical forest communities.

In this chapter I will restrict the analysis of this problem to Africa and concentrate primarily on the issue of hunting by humans. This analysis is complicated by the fact that hunting and forest degradation through logging and shifting cultivation often interact together. I will begin by describing what appear to be fairly well established cases of extinctions within contemporary history. This will be followed by descriptions of less well-known or uncertain extinctions. Finally, I will speculate on likely future extinctions due to hunting by humans and the possible role of chimpanzees in the local extinctions of red colobus populations.

KNOWN EXTINCTIONS

The best documented examples of local extinctions of primates in Africa come from the Bia National Park in Ghana. Within the last 15–20 years three primates have been eliminated from Bia most certainly due to hunting by humans with shotguns. These are Miss Waldron's red colobus (*Procolobus badius waldroni*), the roloway guenon (*Cercopithecus diana roloway*), and white-naped sooty mangabey (*Cercocebus atys lunulatus*) (Struhsaker, 1997, and unpublished report). By 1993 these species were absent even from the old-growth forest of Bia where they had been studied by scientists in the mid to late 1970s. In fact, *P. b. waldroni* appears to be extinct over its entire range of W. Ghana and E. Cote d'Ivoire, while *C. d. roloway* and *C. a. lunulatus* are rapidly on their way to total extinction (Oates *et al.*, submitted; Oates *et al.*, 1996/1997; McGraw, 1998). It is, for example, very likely that *C. d. roloway* has only recently been hunted to extinction in the Kakum National Park of Ghana. Although still persisting in low numbers, the black and white colobus (*Colobus vellerosus*) is at great risk of extinction throughout its range in W. Ghana and E. Cote d'Ivoire.

Preuss's red colobus (*P. b. preussi*) was first collected near Kumba, Cameroon, but by 1966 it was totally unknown by local hunters in that area. It is now restricted to the Korup National Park where it is under serious threat due to overhunting by humans. A single specimen of Elliot's red colobus (*P. b. ellioti*) was collected from the Semliki Forest of W. Uganda in the late 1940's or early 1950's (Haddow, 1952), but by 1970 there were no recent sightings of this species in Semliki (Struhsaker, 1981). More extensive surveys there in 1985–6 also failed to locate this species (Howard, 1991) and it seems fair to conclude that it is now extinct in the Semliki Forest. It does, however, occur further south and west in Zaire (Democratic Republic of Congo).

The mangabey (*Lophocebus aterrimus*) was exterminated from the Wamba area of Zaire (DRC) by human hunting during a ten-year period in the 1980's and 1990's (Dr. T. Kano, pers. comm.; Struhsaker, 1997).

LESS CERTAIN EXTINCTIONS

The Marahoue National Park (1000 km^2) Cote d' Ivoire is dominated by semi- deciduous forest in the transition zone between rainforest and savanna woodland. Records of uncertain origin and quality indicate that this park at one time contained as many as 12 primate species. During a brief survey there in February 1998, our team was able to detect only eight or nine of these species. Three species that may have been eliminated there are the black and white colobus (*C. vellerosus, polykomos,* or *dollmani*), red colobus (*P. b. waldroni* or *badius*), and patas monkeys. Diana monkeys (*C. d. roloway*) may also be extinct there, while the white-naped mangabeys are in perilously low numbers. Equally disturbing is an apparent catastrophic decline in chimpanzee numbers there. Nest count censuses by Marchesi *et al.* (1995) of the chimpanzees in Marahoue indicated that in 1989–1990 this park had the highest density of any area in Cote d'Ivoire, including the Taï National Park. Eight to nine years later, two different surveys indicated that chimpanzees were at very low densities in Marahoue (Barnes, 1997; Struhsaker, 1998). The abundance of empty shotgun cartridges through the forested areas of Marahoue clearly indicated that hunting by humans was the likely factor leading to a major depression of primate numbers and the probable extinction of 2–3 species all within a 20–30 year period.

POSSIBLE EXTINCTIONS

Possible examples of local extinctions are indicated by the very patchy distributions of a number of Africa's forest primates that seem to be independent of the availability of suitable habitat. Perhaps one of the dearest examples of this is the red colobus superspecies (see Figure 1, p. 5, Struhsaker, 1975; Struhsaker, 1998). Although it occurs over the entire breadth of Africa from Senegal to Zanzibar, there are enormous gaps in its distribution even where there is forest that appears ideal for them. Nowhere is this more evident than in W. Uganda. The local extinction of red colobus in the Semliki Forest has already been mentioned. What remains to be explained, however, is the absence of red colobus from all of the other eight forests along the western rift of Uganda except Kibale (Fig. 17.1). In Kibale red colobus reach extremely high densities (Struhsaker, 1975; 1997), but there is no record whatever that red colobus ever existed in these other forests. This is particularly surprising for Itwara, Kalinzu, and Kasyoha-Kitomi which occur at similar altitudes and have a great many of the same tree species as Kibale. These forests appear to me to be good to excellent habitats for red colobus.

The grey-cheeked mangabey (*Lophocebus albigena*) is another species with an extremely limited and patchy distribution in W. Uganda. It occurs in only three (Kibale, Semliki, and Bugoma) of the nine forests along the western rift (Struhsaker, 1981; Howard, 1991) and, like the red colobus, there is no obvious correlation with forest type or altitude. This species occurs further west all the way to Cameroon and in two of Uganda's eastern forests (Mabira and the Sango Bay complex).

Other likely candidates whose present day distribution may be due to local extinctions as the result of hunting by humans include gorillas (huge gap between eastern and western populations and fragmented distribution in west and east), de Brazza's monkey (*Cercopithecus neglectus*), and l'hoesti's monkey (*C. lhoesti*).

The implications of these distributional gaps and possible examples of recent human-induced extinctions are clear. They greatly undermine our confidence in and ability to understand regional and continent-wide comparisons that attempt to generate explanations of primate community structure on the basis of long-term evolutionary and ecological processes. For example, of the nine African forests compared in Chapman *et al.*, this volume, (see Table 1.3, chapter 1) seven are "missing" one or more primate species that one might "expect" them to have given the broad ecological and geographical distributions of these species, i.e., Budongo (no red colobus, mangabeys, de Brazza, or lhoesti), Douala-Edea (no red colobus, *C. torquatus*, or mandrills), Kibale (no de Brazza), Lope (no red colobus), Salonga (no de Brazza), Tiwai and Taï (no *C. nictitans*). Furthermore, although many species are listed for Makokou and Ituri, the fact is that within these enormous areas the distribution of their primate species is far from

Fig. 17.1. Distribution of major blocks of rainforest in Uganda.

uniform and some species are missing from large areas of these forests. In other words, not only could there be human-induced gaps in the distribution of some primate species within these specific forests, but there is a problem of forest size and habitat heterogeneity that makes compaisons with smaller forests, like Tiwai, Kibale, Budongo, Douala-Edea, and Taï, problematic. Are these differences due to recent human activities and/or scaling problems or do they reflect long-term responses to differences in ecology?

DISCUSSION

Why are red colobus so susceptible to hunting by humans?

Hunters throughout the range of red colobus in W. Africa all acknowledge that red colobus are extremely easy to hunt. Not only do they live in large noisy groups, but they seem not to run or hide well enough from hunters once the attack has begun. Hunters claim that they could kill most of an entire group in one day if they wanted to and had enough ammunition. After shooting one, the hunter simply hides behind a tree, waits a few minutes, then steps out and shoots another one and so forth. In sum, red colobus seem poorly adapted to human hunting pressure and unable to modify their behavior accordingly.

General features of susceptibility to human hunting

Primates that are large, conspicuous, and terrestrial seem to be most prone to human hunting. Large bodies (> 4kg) mean a greater return per cartridge spent. Conspicuous animals are easier to locate. Large and noisy groups are most conspicuous, particularly so if they are high in the trees (e.g., diana monkeys, mangabeys, red colobus). Terrestrial primates (e.g., lhoesti, chimps, baboons, drills, mandrills, *Cercocebus torquatus*) are more prone to being caught in wire snares than arboreal species even though the snares may have been set to catch pigs and duikers.

Tribal traditions and hunting of primates

Hunting practices of local tribes will also influence primate communities. In parts of W. Cameroon in the 1960s all primates except chimpanzees were hunted because it was said that chimpanzees are too much like people. Likewise in W. Uganda hunting practices vary from forest to forest depending on the predominant tribe. For example, primates in the Semliki, Ruwenzori, and perhaps Maramagambo forests have been adversely affected by heavy hunting pressure from the Bakonjo and Bamba people. In contrast, the Batoro, who have traditionally predominated in the area around the Kibale forest, do not hunt or eat primates and this may explain in part why Kibale has such a great diversity and high density of primates. Tribal traditions do, however, change with time and we have no idea about hunting practices 100–200 years ago.

Forest area and degree of isolation

Forest size and degree of isolation from humans will also influence the impact of hunting by people, but even low-level, persistent hunting in these areas can have an enormous effect on large-bodied primates with low rates of growth and reproduction. For example, this may explain why red colobus are absent from most of the enormous Dzanga-Sangha Forest Reserve and Park (> 4500 km^2),

Central African Republic and why primate densities are so exceedingly low in the even larger Ituri Forest (> 13 000 km^2), Democratic Republic of Congo (see Chapman *et al.*, chapter 1, this volume; Struhsaker, unpub.). Similar examples can be found in South America, e.g. spider monkeys appear to have been hunted out of the vast tract of forest (> 2000 km^2) along the Rio Tapiche located several hundred km upriver of Iquitos, Department of Loreto, Peru in spite of low-density human populations (Struhsaker *et al.*, 1997) (also see Redford (1992) Peres, chapter 15, this volume, for other examples of empty forests).

How many other forests, for which we have no prior information, have had their primate communities significantly altered by humans within the recent past?

Reduction of primate populations

In addition to extinctions, hunting and forest degradation and fragmentation by humans can also result in significant reductions in primate population densities and in the proportional species composition of the community (see review in Struhsaker, 1997). In fact, these population declines may well represent incipient extinctions. Once a species has been reduced in density, its population may never recover or require decades, if not centuries, to do so, such as apparently occurs with some rainforest primates and tree species after heavy logging (Struhsaker, 1997) or hunting (see Peres, chapter 15, this volume) and with several marine fish populations due to overfishing even after fishing has stopped (Begon *et al.*, 1986). These changes, of course, confound our understanding of the relationship between primate communities, forest type, soil dualities, and other relevant ecological variables. Many more examples may exist than those documented, but, in the absence of early records, we generally have no idea as to how past human activities helped shape extant primate communities. The implications for intercommunity comparisons are obvious. We may be comparing artifacts of recent human activities rather than products of long-term ecological and evolutionary processes.

Selective trends of hunting and forest degradation by humans

The combined effect of hunting and forest degradation and fragmentation (logging and shifting cultivation) is to select for a community of primates dominated by colonizing species. These are species that can be considered to be generalists, doing well in a wide variety of habitats, particularly secondary bush. They also tend to be highly flexible in their behavioral response to humans, i.e., they hide in thick vegetation and remain silent rather than give alarm calls and/or display. Examples of these highly adaptable colonists in Africa include: *Cercopithecus petaurista*, *C. campbelli*, *C. nictitans*, *C. erythrotis*, *C. mona*, *C. ascanius*, *C. mitis*, and *Procolobus verus*.

In contrast, hunting and forest degradation select against the old-growth forest guild of primates, such as red colobus, mangabeys, chimps and *C. lhoesti* (Skorupa, 1986; Struhsaker, 1997). Counterparts in South America include *Ateles* and *Lagothrix* and perhaps *Cacajao*.

The impact of predation by chimpanzees on primate communities

Extremely high rates of predation by chimpanzees on red colobus have been reported from the Gombe Park in Tanzania (20–35% per year) and this may be increasing (e.g., Busse, 1977; Stanford et al., 1994; Stanford, 1995), resulting in smaller groups of red colobus (Stanford, 1998). Immature red colobus are killed proportionally more than their representation in the population. A similar pattern, but perhaps not quite such high levels of predation by chimps on red colobus has also been reported for Taï (Boesch & Boesch, 1989) and Kibale (Drs David Watts & John Mitani, pers. comm.).

Ultimately, this high level of predation could lead to local extinctions of red colobus, particularly in small forests like Gombe (about 70–80 km^2) with relatively small populations of colobus. The reduction in red colobus group size at Gombe by nearly 50% over a 25-year period (Stanford, 1998) is highly indicative of a declining population. As their prey declines, predation by chimps could increase exponentially and drive the colobus to extinction. Has this happened in the past elsewhere? Recall the absence of red colobus from most of the forests in W. Uganda. Although larger than Gombe, these forests are relatively small (86–850 km^2) and all have chimpanzees (Struhsaker, 1981; Howard, 1991). Combined with possible hunting by humans, high levels of predation by chimpanzees may well have reduced, if not driven to extinction, the populations of particularly susceptible primates like

red colobus in some of Africa's forests. The role of chimps in shaping primate communities may become increasingly important as Africa's forests continue to be fragmented and reduced in size by human activities. Combined with hunting by humans, the effect will likely be catastrophic. Counteracting the impact of chimps, however, is the fact that in many parts of Africa hunting pressure on chimps by humans has increased and will likely continue to do so until most chimp populations are extinct.

Who will go next?

Primates most likely to undergo extreme reductions in populations and, ultimately, extinction in the next 20–50 years are those dependent on old-growth forest, that weigh at least 4–5 kg, spend a significant amount of time on the ground, are noisy and conspicuous, and that live in areas with already high or increasing human populations that have a tradition of hunting primates. Likely candidates in Africa include most colobus species, except the olive colobus, drills, and many, if not most, populations of mandrills, chimps, gorillas, mangabeys, and the diana, lhoesti, and preussi guenons. Wright & Jernvall (chapter 18, this volume) have also concluded that large, terrestrial, habitat-specialist primates are particularly prone to extinction, but on the basis of an entirely different approach and perspective from the one in this chapter.

Concluding remark

Any reasonable comparison of primate communities that attempts to understand the similarities and differences from the perspective of long-term ecological and evolutionary processes is potentially flawed because of the impact of current human activities or those in the recent past. This is because hunting and habitat degradation by humans overwhelm most, if not all, other variables that influence the composition of primate communities. When making such broad comparison every possible effort should be made to incorporate information on human impacts (see Tutin & White, chapter 13, this volume). This will be difficult in most cases because of the lack of historical information. Furthermore, atttempts to make inferences about possible human impacts on the fauna will be compounded by the possibiblity of correlations between hunting pressure and ecological variables, such as soil quality, rainfall, and the availability of plants used by humans. For example, those primate communities living in forests with high quality soils, equitable patterns of rainfall, an abundance of fruit or other products harvested by humans, and located near access routes (e.g., navigable rivers) will likely be the ones most severely impacted by people because that is where people will settle and hunt. Even in "pristine" forest with no known history of significant human impact, an understanding of primate and most other ecological relationships will likely be most productive if our paradigm is shifted away from that of equilibrium to non-equilibrium systems.

ACKNOWLEDGMENTS

I thank Dr Charles Janson for his constructive comments on this essay.

REFERENCES

Barnes, R. F. W. (1997). *A brief visit to Marahoue National Park*. Unpublished report to Conservation International.

Begon, M. Harper, J. L. & Townsend, C. R. (1986). *Ecology: Individuals, Populations, and Communities*. Sunderland, Mass: Sinauer Associates.

Boesch, C. & Boesch, H. (1989). Hunting behavior of wild chimpanzees in the Tai National Park. *American Journal of Physical Anthropology*, **78**, 547–73.

Busse, C. D. (1977). Chimpanzee predation as a possible factor in the evolution of red colobus monkey social organization. *Evolution*, **31**, 907–11.

Haddow, A. J. (1952). Field and laboratory studies on an African monkey, *Cercopithecus ascanius schmidti* Matschie. *Proceedings of the Zoological Society, London*, **122**, 297–394.

Howard, P.C. (1991). *Nature Conservation in Uganda's Tropical Forest Reserves*. IUCN, Gland, Switzerland and Cambridge, UK.

McGraw, W. S. (1998). Three monkeys nearing extinction in the forest reserves of eastern Cote d'Ivoire. *ORYX*, **32**, 233–6.

Marchesi, P., Marchesi, N., Fruth, B., & Boesch, C. (1995). Census and distribution of chimpanzees in Cote d'Ivoire. *Primates*, **34**, 591–607.

Oates, J. F., Struhsaker, T. T. & Whitesides, G. H. (1996/1997). Extinction faces Ghana's Red Colobus monkey and other locally endemic subspecies. *Primate Conservation*, **17**, 138–44.

Oates, J. F., Abedi-Larti, M., McGraw, W. S., Struhsaker, T. T. & Whitesides, G. H. (submitted). An impending faunal crisis in West Africa signaled by the extinction of a colobus monkey. *Conservation Biology*.

Redford, K. H. (1992). The empty forest. *BioScience*, **42**, 412–22.

Skorupa, J. P. (1986). Responses of rainforest primates to selective logging in Kibale Forest, Uganda: A summary report. In *Primates: The Road to Self-sustaining Populations*, ed. K. Benirschke, pp. 57–70. New York: Springer-Verlag.

Stanford, C. B. (1995). The influence of chimpanzee predation on group size and anti-predator behaviour in red colobus monkeys. *Animal Behaviour*, **49**, 577–87.

Stanford, C. B. (1998). *Chimpanzee and Red Colobus: The Ecology of Predator-prey Ecology*. Cambridge, MA: Harvard University Press.

Stanford, C. B., Wallis, J., Matama, H. & Goodall, J. (1994). Patterns of predation by chimpanzees on red colobus monkeys in Gombe National Park, Tanzania. 1982–1991. *American Journal of Physical Anthropology*, **94**, 213–28.

Struhsaker, T. T. (1975). *The Red Colobus Monkey*. Chicago: University of Chicago Press.

Struhsaker, T. T. (1981). Forest and primate conservation in East Africa. *African Journal of Ecology*. **19**, 99–114.

Struhsaker, T. T. (1993). *Ghana's forests and primates. Report of field trip to Bia and Kakum National parks and Boabeng-Fiema monkey sanctuary in November 1993*. Unpublished report to Conservation International.

Struhsaker, T. T. (1997). *Ecology of an African Rain Forest: Logging in Kibale and the Conflict between Conservation and Exploitation*. Gainesville, Florida: University Press of Florida.

Struhsaker, T. T. (1998). *A survey of primates and other mammals in Marahoue National Park, Cote d'Ivoire*. Unpublished report to Conservation International.

Struhsaker, T. T., Wiley, H. & Bishop, J. A. (1997). *Conservation Survey of Rio Tapiche, Loreto, Peru*. Unpublished report for Amazon Center for Environmental Education and Research Foundation.

18 • The future of primate communities: A reflection of the present?

PATRICIA C. WRIGHT AND JUKKA JERNVALL

INTRODUCTION

In widely different habitats and among taxonomically diverse groups, recent extinction rates are 100–1000 times their pre-human levels, with regions rich in endemics especially at risk (Wilson, 1988; Pimm *et al.*, 1995). There is a strong recognition that many primate species are in jeopardy of extinction in the near future (Mittermeier, 1988; Rowe, 1996). Half of the 250 primate species are considered to be of conservation concern according to the Primate Specialist Group of the Species Survival Commission of the World Conservation Union (IUCN, 1996). It is an achievement for primate conservationists that we have not lost any species in this millennium, as other groups such as rodents, birds and reptiles have (Mittermeier, 1996). And primatologists should be congratulated that within the last decade more than 15 new primate species have been discovered or rediscovered (Meier *et al.*, 1987; Simons, 1988; Hershkovitz, 1987, 1990; Meier & Albignac, 1991; Mittermeier *et al.*, 1992; Ferrari & Lopes, 1992; Queiroz, 1992; Silva & Noronha, in press). But it is a sobering fact that 96 primate species are in the Critically Endangered or Endangered category and could disappear within the next 100 years (Rowe, 1996).

The inventory data and categories of threat developed by the Conservation Monitoring Center (IUCN) and The United States Endangered Species Act (USESA), (World Conservation Monitoring Centre, 1992; Rowe, 1996) are very useful in targeting primate species with high conservation priority. But the lists of endangered species alone have limited use in estimating causes and consequences of primate extinctions. Indeed, a major limitation of species counting in conservation biology is that ecological measures of biodiversity are not reducible to taxonomic measures

(e.g., Fig. 18.1, Jernvall & Wright, 1998). Therefore focusing only on the pruning of the taxonomic tree, we may cut out all the twigs that bear fruit. Questions that require taking into account the ecological half of biology include: What ecological categories of primates are endangered, and what kind of communities do they live in? Are whole primate communities endangered or individual specializations within the communities endangered? And what have humans to do with primate ecology anyway?

Primates often perform ecological services that are important to the maintenance of tropical habitats. Primates are critical seed dispersers in ecosystems in Africa, Madagascar and South America (Janson, 1983; Gautier-Hion, *et al.*, 1985; Bourlière, 1985; Jordano, 1995; Chapman, 1995; Ganzhorn & Zinner, 1999; Dew & Wright, 1998). Pollination by primates may also be important in South American, African and Malagasy tropical forests (Terborgh, 1983; Sussman & Raven, 1987; Overdorff, 1992; Gautier-Hion & Maisels, 1994; Wright & Martin, 1995). Primates are also food for top predators, especially hawks, eagles and mammalian carnivores (Isbell, 1990; Struhsaker & Leakey, 1990; Boesch, 1991; Goodman *et al.*, 1993; Stanford *et al.* 1994; Wright *et al.*, 1997; Wright, 1998).

Even loss of one or two species out of a primate community may have far-reaching effects on the tropical ecosystem (White *et al.*, 1995). Loss of all the species important in pollination or seed dispersal may lead to extinction cascades in which a whole range of plant species may become extinct (Myers, 1986, 1997; Ingarsson & Lundberg, 1995). In addition, rare plant species competing for pollinators may not reproduce if population levels of pollinators decline, because of the dilution effect which reduces the effective number of visits each plant receives

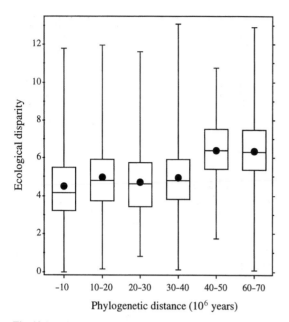

Fig. 18.1. Primate ecological disparity as the function of phylogenetic distance. Note the similar ecological disparities until more than 40-million-years old primate lineages and presence of very disparate (disparity >10) primates among phylogenetically young lineages. To tabulate ecological disparity (as euclidean distances) variables within each ecological category (activity cycle, spatial location, diet, and habitat tabulations) were given values based on their rank order. Body size was included as log-transformed. Variances for each variable were made equal by standardization by standard deviations (for details, see Jernvall & Wright, 1998). For a robust measure of phylogenetic distance, we tabulated, using 10 million year intervals, durations of independent evolutionary history of each primate species from a composite phylogeny (Purvis, 1995). The box shows the upper and lower quartiles and the median as a line across the middle. The black dot is the mean and the lines from the box show the range.

(Straw, 1972; Wright, 1997a). Therefore, in order to maintain the integrity of many tropical forests, it may prove important to preserve areas where ecologically, rather than taxonomically, diverse primate communities are maintained ascertaining that viable populations of all species are conserved.

Among mammals, primates are uniquely well researched. A prodigious amount of data have been collected by primatologists over this past decade, which not only identify

species which have small or vulnerable populations, but also give information on various aspects of primate natural history (Oates, 1996a,b; Eudey, 1995; Mittermeier et al., 1994; Rylands et al., 1997). In this chapter we will review some of the ecological patterns that may result from loss of recently endangered primates species. In particular, we gauge to what degree the loss of current taxa may alter ecological aspects of future primate communities. That is, how grave are the risks of primate extinctions?

PATTERNS OF PRIMATE ENDANGERMENT

It is obvious that several processes function together in primate extinctions. To make a modest start, we examined different ecological patterns of primate risk at the regional level. While regional level generalizations may not apply to all individual conservation cases, regional trends can be useful in identifying the most pertinent ecological correlates of primate endangerment.

We used the endangerment categories from the IUCN-1996 and CITES database for the analysis (IUCN, 1996). Critically endangered (IUCN) and endangered (IUCN, USESA) primates were tabulated as being endangered (E). Vulnerable (IUCN) and threatened (USESA) primates were tabulated as being vulnerable (V). Abundant, non-threatened, and lower-risk categories were tabulated not to go extinct (L).

To make comparisons between endangerment and primate ecology, we grouped primates into various ecological categories that cut across phylogenetic boundaries. We characterized primary resource use of living African, Asian, Malagasy, and South American (including Central American) primates by tabulating their diet, activity pattern, and habitat as discrete specialization types and also their body size. As a robust measure of specialization, dietary and habitat resources were ranked in order of importance. The tabulated primary diets (seven diets) were insects, meat, flowers/nectar, gum, fruits, seeds, and fibrous vegetation. In order to avoid incidental diet items, only the first five ranked primary diets were used in our analysis. The tabulated primary habitats (five habitats) were tropical rain forests, deciduous forests, swamps or flooded forests, scrub or scrub forests, and open habitat (savanna). Similar ranked tabulation was used for activity pattern (diurnal, nocturnal) and spatial location (arboreal, terrestrial). As our categories refer only to primate resource use apart from shape and locomotion, it is readily applicable to other groups (for example, birds and bats would have additional "volant" spatial location). The averaged body weights of males and females were used. The ecological data were largely derived from the compilations in Rowe (1996) and the taxonomic categories follow Wilson & Reeder (1993). (More complete methods can be found in Jernvall & Wright, 1998.)

In order to get a crude approximation of effect of human population densities on primate endangerment, we used World Resource Institute data (WRI, 1994). Then the surface area of the countries that each primate occupied was calculated and pooled and average human population densities were tabulated. The geographic ranges of primates were tabulated from maps in Rowe (1996) and divided into 10^7 hectare-sized categories for a robust measure of range.

Human pressure and geographic range

All four continental regions may lose 29–66% of their extant primate taxa if the predicted extinctions occur, with Madagascar and Asia experiencing the more severe primate extinctions. Africa is in the best position with only 20 endangered out of 68 species (29%). While the continents clearly differ in their primate endangerment, it is a reasonable hypothesis that a common denominator for all the regions is human population growth. For example, 49% of primates that are estimated to be at risk of extinction (vulnerable or greater risk category) live in regions whose human population densities exceed that of the United States (around 0.28 humans per hectare). This is a considerably greater proportion than of the lower extinction risk primates (34%) and therefore high human density alone is a risk factor for primate communities. But as half of the primates at risk of extinction are in regions whose human population densities are lower, other factors are clearly at play. One of them, the geographic range of primate species, is shown in Fig. 18.2. No primate species whose geographic range exceeds the area of India is considered to be at risk of extinction. Thus, small range is another factor that makes primates vulnerable to extinctions, even at low human densities (Fig. 18.2). Small ranges usually mean small population sizes and high vulnerability to catastrophic events and disease, blocks to dispersal, and loss of

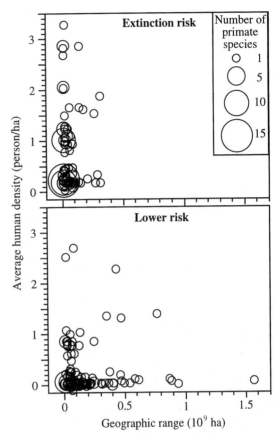

Fig. 18.2. An approximation of the relationship between geographical range of primates and human population densities. Higher proportions of primate species that are at risk of extinction are found in regions with higher human population densities than of primates that are not at risk. Note how no primate species that have broad ecological ranges are at extinction risk.

genetic variability (Lande & Barrowclough, 1987; Dobson & Lyles, 1990; Diamond, 1997). It is noteworthy that the primates with largest geographic ranges are from diverse taxonomic groups (*Chlorocebus aethiops*, *Cebus apella*, *Papio hamadryas anubis*, *Galago senegalensis*) and this diversity may deserve more research attention.

An added risk to primates related to geographic range is political boundaries. Species that are found only in one country are vulnerable to political instability, including the devastation of human warfare which will compromise otherwise successful conservation actions (Diamond, 1997;

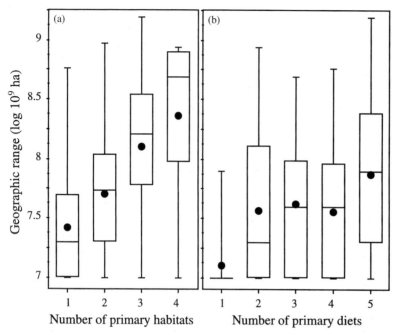

Fig. 18.3. Geographic range of primates as the function of (a) number of primary habitats and (b) number of primary diets. The average range increases linearly with the number of primary habitats but the number of diets appear to not to predict the range. The box shows the upper and lower quartiles and the median as a line across the middle. The black dot is the mean and the lines from the box show the range.

Kramer *et al.*, 1997; Struhsaker, 1997). Endangered species that are restricted to a single country include: *Pan paniscus* (Zaire), *Tarsius syrichta* (Philippines), *Simias concolor* (Mentawai Islands), *Presbytis johnii* (India), *Rhinopithecus roxellana* (China), *Leontopithecus rosalia* (Brazil), and the lemurs (Madagascar). Primate communities found in the coastal rainforests of south-eastern Brazil (Santos *et al.*, 1987; Mittermeier, 1988; Strier, 1990, 1991; Rylands *et al.*, 1997), in Madagascar (Mittermeier, 1988; Mittermeier *et al.*, 1994; Wright, 1997a; Richard & Dewar, in press) and in Vietnam (Nisbett & Ciochon, 1993) are particularly at risk and should continue to receive special attention.

Geographic range and ecology

Small geographic range is known to be linked with vulnerability to extinctions in many organisms (Ceballos & Brown, 1995) but range is also informative for its biological basis (Stotz *et al.*, 1996). "Specialized" animals tend to have small ranges while "generalized" ones have broad ranges (Brown, 1984). While the use of few resources (i.e. specialized) or the use of several resources (i.e., generalized) are both biological specializations, a use of many resources can entail, or require, the use of different habitats. Indeed, primates that use several primary habitats have broader geographic ranges and the relationship between range and habitat number is remarkably linear (Fig. 18.3A). However, while the number of different primary habitats results in the increase in range, the number of different primary diets has a much weaker relationship with range (Fig. 18.3B). Firstly, this suggests that the degree of habitat specialization does not necessarily measure the degree of dietary specializations, a notion that makes the reconstruction of extinct communities difficult from, for example, tooth ecomorphology only. Secondly, higher extinction risk of primates with small geographic ranges help to explain why habitat specialists are at risk but not how dietary specialization affects extinction risks (see Dietary specializations below).

Body size

Within lineages of related mammals, increase in body size is strongly correlated with increased life span and decreased rates of reproductive output (Sacher & Staffeldt, 1974; Terborgh, 1974; Western, 1979; Kiltie, 1984). Large primates have long interbirth intervals and limited lifetime reproductive output and many are endangered, e.g. *Gorilla, Pan, Brachyteles* and *Propithecus* (Fossey, 1983; Goodall, 1986; Strier, 1990; Wright, 1998).

Hunting by humans has made a major impact on many primate communities (Mittermeier, 1987; Peres, 1990, 1993b; Robinson & Redford, 1991; Oates, 1996b). Hunting may eliminate larger species, but have a positive effect on the population densities of smaller species (Peres, chapter 15, this volume). Disproportional endangerment of large body-sized primates can be a sign of hunting pressure and selective decimation of primate communities.

The patterns show that larger primates are endangered in Africa and Madagascar, but there is a little trend in average body size in South America and Asia (Fig. 18.4). While human hunters prefer larger species for food (Alvard, 1993; Alvard et al., 1997; Hill, 1996), at least South American

Indians hunt all sizes of primates with bow and arrow, and only discriminate the large-bodied primates when using guns (Alvard, 1993; Alvard et al., 1997). While anti-poaching and anti-hunting programs are important for all the regions, the selective loss of large primates in Africa and Madagascar can cause irreplaceable changes in certain primate communities.

Circadian rhythm

Body size affects many aspects of an animal's ecology (Fleagle, 1999). In primates, nocturnal species are likely to be smaller and more faunivorous than diurnal ones (Wright, 1989), and diurnal primates are more vulnerable to human hunting. In Africa and Madagascar there is a clear trend that the diurnal species are more endangered than the nocturnal primates (Fig. 18.5). As in Asia the trend is less clear (in South America there is a dearth of nocturnal monkeys), the lower extinction risk of nocturnal primates seem to be part of the same process as the body size trend (Fig. 18.4). However, in Madagascar the impending extinctions threaten almost the whole diurnal primate niche (Fig. 18.5). Only in the cases of massive

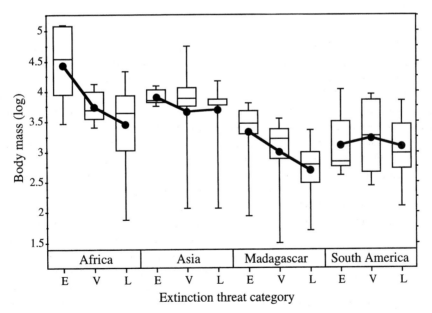

Fig. 18.4. Mean primate body mass comparison for Africa, Asia, Madagascar and South America. Categories are endangered (E), vulnerable (V), and lower risk (L). The box shows the upper and lower quartiles and the median as a line across the middle. The black dot is the mean and the lines from the box show the range.

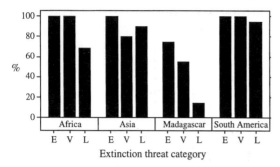

Fig. 18.5. Percentage of diurnal primates in different endangerment categories for Africa, Asia, Madagascar and South America. Categories are endangered (E), vulnerable (V), and lower risk (L).

destruction of habitat in a small geographical range (*Tarsius syrichta*), and/or selective hunting (*Nycticebus pygmaeus*) are nocturnal primates at risk (Heaney, 1993; Tan, 1994). Over three-quarters of the original habitat of these species has been lost (MacKinnon, 1997) and thus nocturnality is no protection against habitat loss.

Dietary specializations

As discussed above, degree of dietary specialization appears to be distinct from degree of habitat specializations in primates (Fig. 18.3). How does the number of primary diets change across endangerment categories? In Asia and Madagascar there is a clear trend in that extinctions are more likely to affect primates with fewer primary diets (Fig. 18.6). African and South American primates have few dietary specialists and it appears that Asian and Malagasy primates may become as generalized. When we look at the data in more detail, primates specializing in fibrous vegetation (mostly folivorous colobines) are disproportionally endangered in Africa and Asia (Fig. 18.7). Indeed, 13 of 37 colobine species (35%) are endangered compared to 15 out of 57 (27%) cercopithecine species (Oates & Davies, 1994). In sympatric colobines, species with the larger, more complex stomach are more endangered (Oates & Davies, 1994).

In contrast, in Madagascar the decline in dietary specialists is projected to happen among frugivores and even specialized insectivory is more endangered in Madagascar than in other regions (Fig. 18.7). Malagasy primates are unique among the continents in that several nocturnal

species are specialized in folivory (i.e., *Lepilemur* and *Avahi*) (Ganzhorn, 1988, 1992, 1995) and thus they may initially "escape" extinction as diurnal lemurs are endangered (Fig. 18.5). Only in South America do primate diets appear to matter less for future survival.

Habitat specializations

The ubiquity of a species usually reflects a wide tolerance of habitat differences (Rosenzweig, 1995). A primate species that has strict habitat requirements, such as living in swamps or along rivers seems to be more at risk if that habitat disappears (Emmons *et al.*, 1983; Emmons, 1984; Peres, 1993a; Barnett & Brandon-Jones, 1997). Other species of the primate community that are using two or three habitat types may survive the loss of a specific habitat or hunting (Emmons *et al.*, 1983; Terborgh, 1992; Peres, 1997). Furthermore, as the use of several habitat types translates to larger geographic range (Fig. 18.3A), habitat generalists are more likely to survive even geographically broader disturbances in ecosystems. Examples of endangered species requiring specialized habitats are *Cacajao melanocephalus* (flooded forests of Central Amazon), *Nasalis larvatus* (peat swamps and mangroves of Borneo), *Allenopithecus nigroviridis* (swamps of Central Zaire and Angola), *Hapalemur griseus aloatrensis* (reed beds of Lake Aloatra, Madagascar).

Interestingly, the proportion of primates specialized to one primary habitat, which is mostly rain forest, is projected to decline in Africa, Asia, and to a lesser degree in South America, but not in Madagascar (Fig. 18.8). Malagasy primates, which have the highest proportion of habitat specialists, seem to be endangered ecologically to such a degree that complete portions of ecosystems are at risk. These parts of ecosystems appear to pertain first to rainforest habitats as the proportional drop of primates specialized to rainforest is projected to be particularly steep in Madagascar (Fig. 18.8).

And all the other factors

No list will be comprehensive on the ecological aspects of endangerment. Geographic range does not measure the abundance or patchiness of species on the landscape. In birds and reptiles, for example, many species are rare either because they are associated with habitat types little

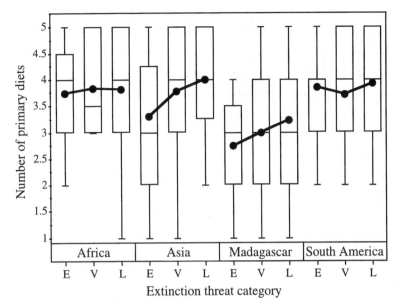

Fig. 18.6. Mean number of primary diets in different endangerment categories for Africa, Asia, Madagascar and South America. Categories are endangered (E), vulnerable (V), and lower risk (L). The box shows the upper and lower quartiles and the median as a line across the middle. The black dot is the mean and the lines from the box show the range.

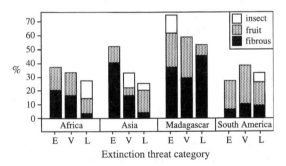

Fig. 18.7. Percentage of different dietary specialists in different endangerment categories for Africa, Asia, Madagascar and South America. The tabulations are on primates that have three or fewer primary diets and the first primary diet is insects (white bars), fruits (gray bars) or fibrous vegetation (black bars). Note that not only can different percentages of dietary specialists be at risk (particularly in Asia and Madagascar) but also different specializations (e.g., folivores in Africa and Asia). Endangerment categories are endangered (E), vulnerable (V), and lower risk (L).

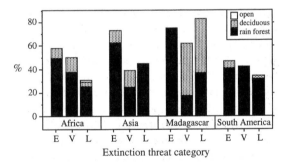

Fig. 18.8. Percentage of different habitat specialists in different endangerment categories for Africa, Asia, Madagascar and South America. The tabulations are on primates that use one primary habitat and the habitat is open (white bars), deciduous forest (gray bars) or rain forest (black bars). Endangerment categories are endangered (E), vulnerable (V), and lower risk (L).

represented in the study area or because they have widely differing levels of abundances between different localities (Schoener, 1987; Thiollay, 1994; Stotz et al., 1996). Inter-locality abundance fluctuations are markedly higher for rare species than for common ones (Terborgh et al., 1990;

Thiollay, 1994). While a general pattern in the distribution of species is that species with wide distributions tend to be locally more abundant than species with narrow distributions (Brown, 1984; Hanski *et al.*, 1993), rare and geographically widespread species do not conform to this pattern. Those primate species that are less abundant and patchily distributed in their geographic range, such as *Callimico goeldii*, *Daubentonia madagascariensis* and *Mandrillus sphinx* are at risk.

Another factor influencing the susceptibility to endangerment is the existence of closely related sympatric species (sibling species) in species-rich rainforest communities. Sibling species are morphologically similar, reproductively isolated, and sympatric (MacArthur & Levins, 1967). Sympatric sibling species have been found in most organisms including birds (Brosset, 1996), plants (Rogstad, 1990), and bats (Kalko *et al.*, 1996) and in some primate communities (Gautier-Hion, 1980; Mate & Colell, 1995; Mate *et al.*, 1995). In primate communities with rich species diversity (12 species or over), some genera (*Cercopithecus, Saguinus, Eulemur, Presbytis*) are represented by two–four species (Gautier-Hion *et al.*, 1985; Overdorff, 1996; Rowe, 1996). If the ecological breadth of each species is partially defined by competition (Rosenzweig, 1995), sibling species may have increased risk of extinction due to competition with congeners. Possible examples of sibling species at risk include *Saguinus imperator* in many sites in the Amazon (Terborgh, 1983; Peres, 1997), and *Cercopithecus lhoesti, Cercopithecus preussi, Hapalemur aureus* and *Callithrix flaviceps* are all rare, and seen in communities with widespread congeners (Gautier-Hion, 1980; Oates *et al.*, 1992; Glander *et al.*, 1989; Wright, 1997*a,b*; Rylands *et al.*, 1997).

Finally, the productivity of an environment can be seen as a regulator of primate biomass and thus indirectly affects the vulnerability of primate diversity. Peres (chapter 15, in this volume) has suggested that the terra firme communities of Brazil have rich species diversity, but lower population densities than those on fluvial and richer soils. In Madagascar a similar pattern of rich species diversity on poorer soils, but higher biomasses on more fertile soils has been documented (Ganzhorn *et al.*, chapter 4, this volume).

THE FUTURE OF PRIMATE COMMUNITIES: A REFLECTION OF THE PRESENT?

Human population growth and the human carrying capacity of the Earth are closely linked with the future existence of forested areas (Ehrlich & Ehrlich, 1981; Cohen, 1995) and therefore, non-human primate communities. Roughly 43% of the Earth's terrestrial vegetated surface has diminished capacity to produce because of direct impacts of human land use (Daily, 1995).

This points to a bleak future if we consider the primates as a whole (presumably including us) (Robinson, 1993; Oates, 1995; Struhsaker, 1997; Kramer *et al.*, 1997). We can predict our primate world in a hundred years with only "weeds and pests" remaining. Rhesus macaques, baboons, bush babies, Hanuman langurs, red howler monkeys, brown capuchins, and mouse lemurs will be the remnants of a once rich and diverse group (Fig. 18.9). However, an immediate danger is that changes in primate communities can begin to convert current ecosystems in ways that make them unpredictably different from the present.

Africa

In Africa, much of the conservation concern has focused on the great apes. This is justifiable as the great apes form a taxonomically unique group among primates. But the loss of the great apes would also have specific changes in many African primate communities as they would eliminate the large body-sized specializations. However, the loss of another group of primates, the colobines, may exert even larger changes in ecosystem processes as the proportion of rainforest folivores declines compared to the present (Oates, 1994). The threat of ecological change is noteworthy because taxonomically a relatively small proportion of African primates are at risk (29%).

Asia

The loss of folivores may have similar impact on ecosystems in Asia as in Africa. Taxonomically, Asia has a much greater proportion of endangered primates including nocturnal species. This suggests that environmental destruction is pronounced in Asia and further loss of habitats will cause extinction of whole communities leaving

Gorilla-Africa

Pygmy Chimpanzee-Africa

Ruffed Lemur-Madagascar

Golden Bamboo Lemur-Madagascar

Fig. 18.9. Endangered species from four geographic areas.
Photos by Russ Mittermeier, Andy Young (gorilla, pygmy
chimpanzee, and orangutan) and Richard Tenaza (Simakobu
monkey).

Woolly Spider Monkey-South America

Golden Lion Tamarin-South America

Orangutan-Asia

Simakobu Monkey-Asia

Fig. 18.9. (*cont.*)

increasingly fewer species with particular ecological properties left.

Madagascar

Malagasy primate communities are unique among the regions in that a large portion of ecological specializations is at risk of extinction. Basically all diurnal primates are endangered making every community under threat. The groups of lemurs that are relatively unaffected are specialized folivores that are nocturnal. The almost total loss of frugivores and species with more generalized habitat use, with larger ranges and, presumably, higher dispersal abilities, all point to an ecosystem collapse. Indeed, Madagascar today may well be a window to what the other regions will be tomorrow. This is despite Asia having a higher percentage of species at risk than Madagascar.

South America

About half of South American primate species may go extinct but the continental-level changes in ecology may initially be less severe. Thus, as a whole, future South American primate communities may resemble the present. However, this means that the loss of half of the taxa eliminates complete communities leaving more or less ecologically equivalent species elsewhere. This could be considered "ecologically random" (Jernvall & Wright, 1998) extinction made possible by the vastness of the Amazon. But even in South America, primates specialized for few primary habitats (i.e., small range) are more at risk and thus the community structure may change as the proportion of habitat generalists increases.

CONCLUSION

It is obvious that the extinction of currently endangered primate species would change the future ecosystems. Conservation efforts focusing on protected areas with endemic species or species with potential to evolve new forms may all fail if the ecosystems that the primate communities are part of collapse. As the ecological patterns in the different regions appear disparate, conservation effort relying on species counting alone may produce different results in different continents. Even in the cases where we can show that the loss of primate biodiversity produces specific eco-

logical changes, the exact changes in ecosystems will be almost impossible to predict. After all, primate communities form only one part of biodiversity. Perhaps a savvy attitude toward primate conservation is to take into account the ecological aspects of primate diversity. At least then we avoid testing our predictions on ecosystem change.

ACKNOWLEDGMENTS

Many thanks to J. Fleagle, C. Janson and K. Reed for the invitation to write this paper. We are grateful to the Wenner Gren Foundation for Anthropological Research for funding the Primate Communities workshop before the International Society of Primatologists Meeting in Madison, Wisconsin. N. Rowe is thanked for the use of his primate database. D. D'Silva assisted with data entry. J. Ganzhorn, J. Fleagle, C. Janson and C. Tan gave excellent advice and comments.

REFERENCES

Alvard, M. (1993). Testing the "ecological noble savage" hypothesis: Interspecific prey choice by Piro hunters of Amazonian Peru. *Human Ecology*, **21**, 355–87.

Alvard, M., Robinson, J., Redford, K. & Kaplan, H. (1997). The sustainability of subsistence hunting in the Neotropics. *Conservation Biology* **11**, 977–82.

Barnett, A. A. & Brandon-Jones, D. (1997). The ecology, biogeography and conservation of the Uakaris, *Cacajao* (Pitheciinae). *Folia Primatologica* **68**, 223–35.

Boesch, C. (1991). The effects of leopard predation on grouping patterns in forest chimpanzees. *Behaviour* **117**, 220–42.

Bourlière, F. (1985). Primate communities: their structure and role in tropical ecosystems. *International Journal of Primatology* **6**, 1–26.

Brosset, A. (1996). Role of the sibling species in the dynamics of the forest bird communities in M'Passa (Northeastern Gabon). In *Long Term Studies of Vertebrate Communities*, ed. M. L. Cody & J. A. Smallwood, pp. 251–89. New York: Academic Press.

Brown, J. H. (1984). On the relationship between abundance and distribution of species. *American Naturalist* **124**, 255–79.

Ceballos, G. & Brown, J. H. (1995). Global patterns of mammalian diversity, endemism, and endangerment. *Conservation Biology*, **9**, 559–68.

Chapman, C. A. (1995). Primate seed dispersal: Coevolution and conservation implications. *Evolutionary Anthropology*, **4**, 74–82.

Cohen, J. E. (1995). Population growth and Earth's human carrying capacity. *Science*, **269**, 341–346.

Daily, G. C. (1995). Restoring value to the world's degraded lands. *Science* **269**, 350–3.

Dew, J. L. & Wright, P. C. (1998). Frugivory and seed dispersal by four species of primates in Madagascar's eastern rainforest. *Biotropica*, **30**, 1–13.

Diamond, J. (1997). *Guns, Germs and Steel*. New York: W.W. Norton & Co.

Dobson, A. P. & Lyles, A. M. (1990). The population dynamics and conservation of primate populations. *Conservation Biology*, **3**, 362–80.

Ehrlich, P. & Ehrlich, A. (1981). *Extinction*. New York: Ballentine Books.

Emmons, L.H. (1984). Geographic variation in densities and diversities of non-flying mammals in Amazonia. *Biotropica*, **16**, 210–22.

Emmons, L.H., Gautier-Hion, A. & Dubost, G. (1983). Community structure of the frugivorous-folivorous forest mammals of Gabon. *Journal of Zoology, London* **199**:209–22.

Eudey, A. (1995). *Asian Primate Action Plan*. Gland: IUCN.

Ferrari, S. F. & Lopes M. A. (1992). A new species of marmoset, genus *Callithrix* Erxleben 1777 (Callitrichidae, Primates) from western Brazilian Amazonia. *Goeldiana Zoologica*, **12**, 1–3.

Fleagle, J.G. (1999). *Primate Adaptation and Evolution*. New York: Academic Press.

Fossey, D. (1983). *Gorillas in the Mist*. Boston: Houghton Mifflin.

Ganzhorn, J. U. (1988). Food partitioning among Malagasy primates. *Oecologia*, **75**, 436–50.

Ganzhorn, J. U. (1992). Leaf chemistry and the biomass of folivorous primates in tropical forests. *Oecologia*, **91**, 540–7.

Ganzhorn, J. U. (1995). Low level forest disturbance effects on primary production, leaf chemistry and lemur populations. *Ecology*, **76**, 2084–96.

Ganzhorn, J. U. & Zinner, D. (1999). Lemurs as essential vectors for dry deciduous forest regeneration in Madagascar. *Conservation Biology*.

Gautier-Hion, A. (1980). Seasonal variation of diet related to species and sex in a community of *Cercopithecus* monkeys. *Journal of Animal Ecology*, **49**, 237–69.

Gautier-Hion, Duplantier, J. M., Quiris, R., Feer, F., Sourd, C, Decous, J. P., Dubost, G., Emmons, L, Erard, C. Hecketweiler, P., Moungazi, A., Roussilhon,

C. & Thiollay, J. M. (1985). Fruit characters as a basis of fruit choice and seed dispersal in a tropical forest vertebrate community. *Oecologia*, **65**, 324–37.

Gautier-Hion, A. & Maisels, F. (1994). Mutualism between leguminous trees and large African monkeys as pollinators. *Behavioral Ecology and Sociobiology*, **34**, 203–10.

Glander, K. E., Wright, P. C., Seigler, D. S., Randrianosolo, V & Randrianosolo, B. (1989). Consumption of cyanogenic bamboo by a newly discovered species of bamboo lemur. *American Journal of Primatology* **19**, 119–24.

Goodall, J. (1986). *The Chimpanzees of Gombe: Patterns of Behavior*. Cambridge, MA: Harvard University Press.

Goodman, S. M., O'Conner, S. & Langrand, O. (1993). A review of predation on lemurs: Implications for the evolution of social behavior in small, nocturnal primates. In *Lemur Social Systems and their Ecological Basis*, ed. P.M. Kappeler & J.U. Ganzhorn, pp. 51–66. New York: Plenum Press.

Hanski, I. (1982). Dynamics of regional distribution: the core and satellite species hypothesis. *Oikos*, **38**, 210–21.

Hanski, I, Kouki, J. & Halkka, A. (1993). Three explanations of the positive relationship between distribution and abundance of species. In *Species Diversity in Ecological Communities: Historical and Geographical Perspectives*, ed. R. Ricklefs & D. Schluter, pp. 108–16. Chicago: University of Chicago Press.

Heaney, L.R. (1993). Biodiversity patterns and the conservation of mammals in the Philippines. *Asia Life Sciences*, **2**, 261–74.

Hershkovitz, P. (1987). The taxonomy of South American sakis, genus *Pithecia* (Cebidae, Platyrrhini): A preliminary report and critical review with the description of a new species and new subspecies. *American Journal of Primatology*, **12**, 387–468.

Hershkovitz, P. (1990). Titis, New World monkeys of the genus *Callicebus* (Cebidae, Platyrrhini): A preliminary taxonomic review. *Fieldiana Zoology New Series*, **55**, 1–109.

Hill, K. (1996). The Mbracayu Reserve and the Ache of Paraguay. In *Traditional Peoples and Biodiversity Conservation in Large Tropical Landscapes*, ed. K. H. Redford & J. A. Mansour, pp. 159–96. Arlington: The Nature Conservancy.

Ingarsson, P. K. & Lundberg, S. (1995). Pollinator functional response and plant population dynamics: pollinators as a limiting resource. *Evolutionary Ecology*, **9**, 421–8.

Isbell, L. A. (1990). Sudden short-term increase in mortality of vervet monkeys (*Cercopithecus aethiops*) due to predation in Amboseli. *American Journal of Physical Anthropology*, **21**, 41–52.

IUCN (1996). *Primate Specialist List of Endangered Species.* Gland: World Monitoring Center.

Janson, C. H. (1983). Adaptation of fruit morphology to dispersal agents in a Neotropical forest. *Science*, **219**, 187–9.

Jernvall, J. & Wright, P. C. (1998). Diversity components of impending primate extinctions. *Proceedings of the National Academy of Sciences USA*, **95** 11279–83.

Jordano, P. (1995). Angiosperm fleshy fruits and seed dispersers: a comparative analysis of adaptation and constraints in plant-animal interactions. *American Naturalist*, **145**, 163–91.

Kalko E. M., Handley, W. O. Jr, & Handley, D. (1996). Organization, diversity, and long-term dynamics of a Neotropical bat community. In *Long Term Studies of Vertebrate Communities*, ed. M. L. Cody & J. A. Smallwood, pp. 503–53. New York: Academic Press.

Kiltie, R. (1984). Gestation in mammals. In *Quarternary Extinctions*, ed. P. S. Martin & R. Klein, pp. 299–314. Tucson: University of Arizona Press.

Kramer, R., van Schaik, C. & Johnson, J. (1997). *Last Stand: Protected Areas and the Defense of Tropical Biodiversity.* New York: Oxford University Press.

Lande, R. & Barrowclough, G. F. (1987). Effective population size, genetic variation and their use in population management. In *Viable Populations for Conservation*, ed. M. E. Soulé, pp. 86–123. Cambridge: Cambridge University Press.

MacArthur, R. H. & Levins, R. (1967). The limiting similarity, convergence and divergence of coexisting species. *American Naturalist*, **101**, 377–85.

MacKinnon, K. (1997). The ecological foundations of biodiversity protection. In *Last Stand: Protected Areas and the Defense of Tropical Biodiversity*, ed. R. Kramer, C. van Schaik & J. Johnson. pp. 36–63. New York: Oxford University Press.

Mate, C. & Colell, M. (1995). Relative abundance of forest cercopithecines in Ariha, Bioko Island, Republic of Equatiorial Guinea. *Folia Primatologica*, **64**, 49–54.

Mate, C., Colell, M. & Escobar, M. (1995) Preliminary observations on the ecology of forest Cercopithecidae in the Lokofe-Ikomaloki region (Ikela, Zaire). *Folia Primatologica*, **64**, 196–200.

Meier, B., Albignac, R., Peyrieras, A, Rumpler, Y. & Wright, P. (1987). A new species of *Hapalemur* (Primates) from southeast Madagascar. *Folia Primatologica*, **48**, 211–15.

Meier, B. & Albignac, A. (1991). Rediscovery of *Allocebus trichotis* Gunther 1875 (Primates) in north east Madagascar. *Folia Primatologica* **56**, 57–63.

Mittermeier, R. A. (1987). Effects of hunting on rain forest primates. In *Primate Conservation in the Tropical Rain Forest*, ed. C. W. Marsh & R. A. Mittermeier, pp. 109–146. New York, Alan R. Liss.

Mittermeier, R. A. (1988). Primate diversity and the tropical forest : case studies from Brazil and Madagascar and the importance of megadiversity countries. In *Biodiversity*, ed. E.O. Wilson & F. M. Peters, pp. 145–54. Washington, DC: National Academy Press.

Mittermeier, R. A. (1996). Primate conservation in the next millenium. *International Society of Primatology Proceedings*. Madison, WI.

Mittermeier, R. A., Schwarz, M. & Ayres, J. M. (1992). A new species of marmoset, genus *Callithrix* Erxleben, 1777 (Callitrichidae, Primates) from the Rio Maues region, state of Amazonas, Central Brazilian Amazonia. *Goeldiana Zoologica*, **14**, 1–17.

Mittermeier, R. A., Tattersall, I., Konstant, W. R., Meyers, D. M. & Mast, R. B. (1994). *Lemurs of Madagascar.* Washington, DC: Conservation International.

Myers, N. (1986). Tropical deforestation and a megaextinction spasm. In *Conservation Biology: the Science of Scarcity and Diversity*, ed. M. E. Soulé, pp. 394–409. Sunderland, MA: Sinaur Associates.

Myers, N. (1997). Mass extinction and evolution. *Science*, **278**, 597–8.

Nisbett, R. A. & Ciochon, R. L. (1993). Primates in northern Vietnam: A review of the ecology and conservation status of extant species, with notes on Pleistocene localities. *International Journal of Primatology*, **14**, 765–95.

Oates, J. F. (1994). The natural history of African colobines. In *Colobine Monkeys: Their Ecology, Behaviour and Evolution*, ed A.G. Davies & J.F. Oates, pp. 75–128. Cambridge: Cambridge University Press.

Oates, J. F. (1995). The dangers of conservation by rural development: A case study from the forests of Nigeria. *Oryx*, **29**, 115–122.

Oates, J. F. (1996a). *African Primates: Status Survey and Conservation Action Plan.* Gland: IUCN.

Oates, J. F. (1996b). Habitat alteration, hunting and the conservation of folivorous primates in African forests. *Australian Journal of Ecology*, **21**, 1–9.

Oates, J. F., Anadu, P. A., Gadsby, E. L. & Lodewijk Were, J. (1992). Sclater's guenon: A rare Nigerian monkey threatened by deforestation. *Research and Exploration*, **8**, 476–91.

Oates, J. F. & Davies, A. G. (1994). Conclusions: the past, present and future of the colobines. In *Colobine Monkeys: Their Ecology, Behaviour and Evolution*, ed. A. G. Davies & J. F. Oates, pp. 347–58. Cambridge: Cambridge University Press.

Overdorff, D. J. (1992). Differential patterns in flower feeding by *Eulemur rufus* and *Eulemur rubriventer* in Madagascar. *American Journal of Primatology*, **28**, 191–203.

Overdorff, D. J. (1996). Ecological correlates to social structure in two lemur species in Madagascar. *American Journal of Physical Anthropology*, **100**, 487–506.

Peres, C. A. (1990). Effects of hunting on western Amazonian primate communities. *Biological Conservation*, **54**, 47–59.

Peres, C. A. (1993*a*) Structure and spatial organization of an Amazonian terra firme forest primate community. *Journal of Tropical Ecology*, **9**, 259–76.

Peres, C. (1993*b*). Notes on the primates of the Jurua River, Western Brazilian Amazonia. *Folia Primatologica*, **61**, 97–103.

Peres, C. A. (1997). Primate community structure at twenty western Amazonian flooded and unflooded forests. *Journal of Tropical Ecology*, **13**, 381–405.

Pimm, S. L., Russell, G. J., Gittleman, J. L. & Brooks, T. M. (1995). The future of biodiversity. *Science*, **269**, 347–50.

Purvis, A. (1995) A composite estimate of primate phylogeny. *Philosophical Transactions of the Royal Society of London*, B, **348**, 405–21.

Queiroz, H. L. (1992). A new species of capuchin monkey, genus *Cebus* Erxleben 1777 (Cebidae, Primates), from eastern Brazilian Amazonia. *Goeldiana Zoologica*, **15**, 1–3.

Richard, A. F. & Dewar, R. (in press). Politics, negotiation, and conservation: a view from Madagascar. In *African Rain Forest Ecology and Conservation*, ed. W. Weber, A. Vedder, H. Simons Morland, L. White, & T. Hart, New Haven: Yale University Press.

Robinson, J. G. (1993). The limits to caring: sustainable living and the loss of biodiversity. *Conservation Biology*, **7**, 20–8.

Robinson, J. G. & Redford, K. H. (1991). *Neotropical Wildlife Use and Conservation*. Chicago: Chicago University Press.

Rogstad, S. H. (1990). The biosystematics and evolution of the *Polyalthia hypoleuca* species complex (Annonaceae) of Malesia. II. Comparative distributional ecology. *Journal of Tropical Ecology*, **6**, 387–408.

Rosenzweig, M. L. (1995). *Species Diversity in Space and Time*. Cambridge: Cambridge University Press.

Rowe, N. (1996). *A Pictorial Guide to the Primates*. East Hampton, NY: Pogonias Press.

Rylands, A. B, Mittermeier, R. A. & Rodriguez-Luna, E. (1997). Conservation of Neotropical primates: Threatened species and an analysis of primate diversity by country and region. *Folia Primatologica*, **68**, 134–60.

Sacher, G. A. & Staffeldt, E. F. (1974). Relation of gestation time to brain weight for placental mammals: Implications for the theory of vertebrate growth. *American Naturalist*, **108**, 593–615.

Santos, I. B., Mittermeier, R. A. & Rylands, A. B. (1987). The distribution and conservation status of primates in southern Bahia, Brazil. *Primate Conservation*, **8**, 126–42.

Schoener, T. W. (1987). Resource partitioning. In *Community Ecology: Patterns and Process*, ed. J. Kikkawa & D.J. Anderson, pp. 91–126. Palo Alto: Blackwell Scientific.

Silva J. S. Jr, & Moronha, M. de A. (in press). On a new species of bare-eared marmoset, genus *Callithrix* Erxleben, 1777, from central Amazonia, Brazil (Primates: Callitrichidae). *Goeldiana Zoologica*, **18**,

Simons, E. L. (1988). A new species of *Propithecus* (Primates) from northeast Madagascar. *Folia Primatologica*, **50**, 143–51.

Stanford, C. B., Wallis, J., Matana, H. & Goodall, J. (1994). Patterns of predation by chimpanzees on red colobus monkeys in Gombe National Park, Tanzania. *American Journal of Physical Anthropology*, **94**, 213–28.

Stotz, D. F., Fitzpatrick, J. W. Parker III, T. A. & Moskovits (1996). *Neotropical Birds Ecology and Conservation*. Chicago: The University of Chicago Press.

Strier, K. (1990). *Faces in the Forest: The Endangered Muriqui Monkeys in Brazil*. New York: Oxford University Press.

Strier, K. (1991). Demography and conservation in an endangered primate, *Brachyteles arachnoides*. *Conservation Biology*, **5**, 214–18.

Straw, R. M. (1972). A Markov model for pollinator constancy and competition. *American Naturalist*, **106**, 597–620.

Struhsaker, T. T. (1997). *Ecology of an African Rain Forest: Logging in Kibale and the Conflict between Conservation and Exploitation*. Gainesville, FL: University Press of Florida.

Struhsaker, T. T. & Leakey, M. (1990). Prey selectivity by crowned hawk eagles on monkeys in the Kibale Forest, Uganda. *Behavioral Ecology and Sociobiology*, **26**, 435–48.

Sussman, R. W. & Raven, P. H. (1987). Pollination by

lemurs and marsupials: an archaic coevolutionary system. *Science*, **200**, 731–6.

Tan, C. L. (1994). Survey of *Nycticebus pygmaeus* in southern Vietnam. IPS Congress, Bali: IPS Handbook and Abstracts: **136**.

Terborgh, J. (1974). Preservation of natural diversity: The problem of extinction-prone species. *BioScience*, **24**, 715–22.

Terborgh, J. (1983). *Five New World Primates*. Princeton: Princeton University Press.

Terborgh J. (1992). Maintenance of diversity in tropical forests. *Biotropica*, **24**, 283–292.

Terborgh, J., Robinson, S. K., Parker, III, T. A., Munn, C. A. & Pierpont, N. (1990). Structure and organization of an Amazonian forest bird community. *Ecological Monographs*, **60**, 213–38.

Thiollay, J. M. (1994). Structure, density and rarity in an Amazonian rainforest bird community. *Journal of Tropical Ecology*, **10**, 449–81.

Western, D. (1979). Size, life-history, and ecology in mammals. *African Journal of Ecology*, **17**, 185–204.

White, F. J., Overdorff, D. J., Balko, E. A. & Wright, P. C. (1995). Distribution of Ruffed Lemurs (*Varecia variegata*) in Ranomafana National Park, Madagascar. *Folia Primatologica*, **64**, 124–31.

Wilson, E. O. (1988). *Biodiversity*. Washington, DC: National Academy of Sciences.

Wilson , D. E. & Reeder, D. M. (1993). *Mammalian Species of the World: A Taxonomic and Geographic Reference*. Washington, DC: Smithsonian Institution Press.

World Conservation Monitoring Centre (1992). *Global Biodiversity*. Cambridge: IUCN.

WRI, IUCN & UNEP (World Resources Institute, World Conservation Union & United Nations Environment Program) (1994). *Global Biodiversity Strategy*. Washington, D.C: IUCN Press.

WRI (World Resources Institute) (1994). *World Resources: 1993–1994*. New York: Oxford University Press.

Wright, P. C. (1989). The nocturnal primate niche in the New World. *Journal of Human Evolution*, **18**, 635–58.

Wright, P. C. (1997a). The future of biodiversity in Madagascar: A view from Ranomafana National Park. In *Natural Change and Human Impact in Madagascar*, ed. S. Goodman & B. Patterson, pp. 381–405. Washington, DC: Smithsonian Institution Press.

Wright, P. C. (1997b). Behavioral and ecological comparisons of Malagasy and Neotropical communities. In *New World Monkeys: Evolution and Ecology*, ed. W.G. Kinzey, pp.127–41. New York: Aldine de Guyter.

Wright, P. C. (1998). Impact of predation risk on the behaviour of *Propithecus diadema edwardsi* in the rain forest of Madagascar. *Behaviour*, **135**, 483–512.

Wright, P. C. & Martin, L. B. (1995). Predation, pollination and torpor in two nocturnal primates: *Cheirogaleus major* and *Microcebus rufus* in the rain forest of Madagascar. In *Creatures of the Dark*, ed. L. Altermann, G. Doyle & M. Izard, pp. 45–60. New York: Plenum Press.

Wright, P. C., Heckscher, S. K. & Dunham, A. E. (1997). Predation on Milne-Edward's sifaka (*Propithecus diadema edwardsi*) by the fossa (*Cryptoprocta ferox*) in the rain forest of Southeastern Madagascar. *Folia Primatologica*, **68**, 34–43.

19 · Concluding remarks

JOHN G. FLEAGLE, CHARLES H. JANSON AND KAYE E. REED

The goal of this volume was to bring together the efforts of scientists with research interests and experience from many parts of the world to provide a comparative perspective of the primate communities or assemblages from different biogeographical regions. The preceding chapters have taken a wide range of approaches to the study of communities including: (1) broad surveys and analyses of the diversity of communities within individual regions; (2) detailed examinations of the factors that underlie community differences at both a microecological level and a macroecological level; (3) comparative study of the relationship between primate diversity and that of other aspects of the fauna and of the flora; and (4) attempts to put extant communities in a temporal perspective through examination of the history of individual and regional faunas and habitats, as well as attempts to predict changes that primate communities are likely to undergo in the coming years if present patterns of extinction continue.

COMMUNITY DIFFERENCES

In general, most authors seemed to find the differences among communities far more impressive than the overall similarities among communities. The chapters by Kappeler, Reed, and Ganzhorn (chapters 9, 7 and 8 respectively) found significant differences in the body size distributions of primates from different regions both at a regional level and for individual communities. Fleagle & Reed's chapter 6 (see also Fleagle & Reed, 1996) documented differences in the ecological space occupied by individual communities of primates of different regions. Recently, Jernvall & Wright (1998) obtained very similar results using the same technique on a different data set for the primates of entire biogeographic regions. Kappeler

(chapter 9, this volume) found differences in the social organization of primates in different areas, following similar results by Kappeler & Heymann (1996).

There are certainly examples of ecological convergence among the primate taxa of different regions. For example, colobine monkeys of Africa and Asia occupy an ecological niche very similar to that of sifakas in Madagascar, and there are small, nocturnal quadrupedal, frugivores and frugivore–insectivores on all continents (see Fleagle & Reed, chapter 6, this volume, Fig. 6.5). Thus, if one compares the ecological space occupied by the faunas of different regions, there are considerable areas of overlap (e.g. Fleagle & Reed, this volume, Fig. 6.8). Nevertheless, it is the distinctive ecological and social features of the primates of each region that particularly call for explanation.

EXPLAINING COMMUNITIES

Compared with documenting differences, understanding the factors underlying these differences is a far more difficult matter. The preceding chapters have identified some major causal factors and at the same time have illuminated the dearth of truly comparable data that are presently available for comparing primate communities both within and between geographical regions. Thus, several authors have documented a positive correlation between primate species diversity at a site and local rainfall (Reed & Fleagle, 1995; see also Janson & Chapman, chapter 14, and Emmons, chapter 10, this volume). In addition, there are suggestions that this relationship is not linear, but that above 2500 mm per year diversity actually declines (Kay *et al.*, 1997). However, different data sets seem to give different curves and several causal relationships have been offered to explain how an increase in rainfall generates increased primate diversity.

Ganzhorn *et al.* (1997) find that increased rainfall is also correlated with increased floral diversity and argue that increased floral diversity should generate more niches for primates. In contrast, Kay *et al.* (1997) have argued that primate diversity parallels plant primary productivity in the New World, although their argument was of necessity indirect as they were able to locate virtually no localities with data on both primate diversity and plant productivity. Obviously a more direct test of the productivity hypothesis is badly needed. In addition, many authors have argued for the importance of soil nutrient status for understanding the relationship between primate diversity and rainfall, but comparative soil data are also limited. Nevertheless, soil differences are the most likely reason to explain differences in habitat and primate community structure among sites within a single rainfall regime, as shown by Peres (1997; chapter 15, this volume).

It might seem impossible to find meaningful correlates of total primate community biomass, given that it is a sum of biomasses of individual species, each of which has distinct ecological requirements. Nevertheless, some patterns stand out and require explanation. In general, most studies have demonstrated that community biomass is inversely related to species diversity and to rainfall. The cause of this relationship is not obvious, as one would expect that, all other things being equal, the more species of primates there are at a site, the greater the biomass would be. Numerous factors have been hypothesized to account for this relationship. Ganzhorn *et al.* (chapter 4, this volume) suggest that high rainfall promotes high plant diversity, which favors high primate diversity but actually reduces the foraging efficiency of each species because their resources are spaced further apart, thus leading to lower biomass. Building on the observation that plant productivity declines above 2500 mm of annual rainfall (Kay *et al.*, 1997), Janson & Chapman (chapter 14, this volume) at first suggest that simple plant productivity could account for the decline in biomass. However, many folivore biomasses reach their peak in seasonally dry forests with 1500–2000 mm of rainfall, even though plant productivity there is less than in moister forests. Seasonality of rainfall is often associated with seasonal leaf drop and flush, which in turn is correlated with low levels of plant defensive chemicals, suggesting that primate folivores are

more sensitive to leaf quality than to absolute leaf production, a conclusion foreshadowed by Oates *et al.* (1990) for African folivores and by Ganzhorn (1992) for Malagasy ones. Clearly, this hypothesis needs to be tested more directly in other communities. In any case, it appears that the relationship of biomass to habitat characteristics may depend on the diet of a primate species.

It should be simpler to describe causal variables relating habitat characteristics to the densities or biomasses of individual species. Several studies of Asian primates have shown a relationship between the biomass of individual primate taxa and the density of particular families or species of trees (see Janson & Chapman, chapter 14, and Gutpa & Chivers, chapter 2, this volume). Similar patterns remain to be documented for primates in other regions, although Ganzhorn (1993) has related differential density of *Lepilemur* and *Avahi* to patterns of plant secondary compounds. Attempts to relate total fruit-eating primate biomass to fruit-fall measures across communities have not been successful, for several reasons (Janson & Chapman, chapter 14, this volume). Because most primatological studies focus on the behavioral variation of one species in a single site, ecological methods are usually focused too narrowly and are too qualitative to provide useful numerical production data for comparison between communities. Total annual production may turn out not to be the relevant predictor of primate biomass because seasonal changes are common in tropical areas, and many primates may suffer critical periods of resource scarcity when they eat fallback foods that otherwise constitute a minor fraction of the diet (e.g., Chapman *et al.*, chapter 1, this volume). We urge future primatologists to make an effort to define more clearly for their study species if there exists a critical time of year for survival and if so when that is and what foods are used then. An added problem is that variation in plant reproduction between years appears to be nearly universal in the moist tropics, and so any meaningful averages of fruit production for a study site will surely require multiple years of data collection. While these difficulties are not trivial, they should only spur us to refine and extend our measures of resource production relevant to primates. For a deeper understanding of the causal variables, we urge field researchers to take soil and leaf samples to allow more rigorous testing of the hypotheses raised in this book.

PRIMATES IN CONTEXT

Primate communities do not evolve in an ecological void, but interact in many ways with the plants and other animals present. Thus, Ganzhorn (chapter 8, this volume) concludes that some of the peculiarities of the Asian primate community first pointed out by Reed & Fleagle (1995) may be due to certain ecological niches, particularly for smaller species, having been preempted by other mammals. Similarly, the great diversity of primate forms found in Madagascar undoubtedly owes its existence to the lack of competing species there. In Africa and South America, however, Ganzhorn (chapter 8, this volume) and Emmons (chapter 10, this volume) find that the diversity of primates parallels that of other mammal species, suggesting that they respond in similar ways to ecological variables. This inference is challenged by Eeley & Lawes (chapter 12, this volume), whose analysis of biogeographic patterns places the emphasis on regional factors controlling species diversity: range size, geographic barriers, and broad-scale habitat features. In this interpretation, the parallelism between primate and other mammalian taxa in species diversity merely shows that they respond similarly to these regional factors, although this is not clearly the case outside Africa and the New World. It remains a challenge for future workers to disentangle the local vs. regional effects on species diversity and specialization. It would be especially helpful if it were possible to specify the critical resources each species requires and then check whether these resources are present in the areas where the species is not. Such studies would have obvious conservation value as well.

PAST, PRESENT, AND FUTURE

Certainly no potential explanation for community differences has received more attention in recent years than phylogeny (DiFiori & Rendall, 1994; Rendall & DiFiori, 1995; Kappeler & Heymann, 1996; Fleagle & Reed, 1996). It is often unclear what different researchers mean when they attribute behavioral and ecological differences or similarities to phylogeny, or how phylogeny affects the ecological diversity of species or communities. In many cases, phylogeny seems to be the catch-all residual explanation for differences that can not be accounted for by the measured environmental factors. Several of the chapters in this volume have directly addressed the role of phylogeny or the temporal aspects of phylogeny in explaining the patterns of ecological similarities and differences found in extant communities. Fleagle & Reed (chapter 6, this volume; see also Wright & Jernvall, chapter 18, this volume; Jernvall & Wright, 1998) showed that ecological similarity is correlated with recency of estimated common ancestry (and hence phylogenetic proximity), but that correlation is not very high, leaving a large amount of ecological similarity that is independent of phylogeny. Kappeler (chapter 9, this volume) suggests that the apparent phylogenetic differences in primate social systems in different regions may occur because life history traits are more conservative than other aspects of behavior. However, others (e.g. Alberch, 1990) have argued that many life history traits are more variable at the species level and subject to homoplasy. To date, no one has done a broad empirical study of the distribution of various behavioral and ecological characteristics of living primate species to determine which aspects of behavior and ecology are more or less subject to homoplasy than others, and why.

Phylogeny is just one aspect of what is perhaps the most difficult factor to evaluate in understanding primate communities—time. The primate communities and the environments they inhabit both have histories. Contributions to this volume have documented both habitat and faunal changes at various time scales ranging from changes due to logging, deforestation, and hunting in recent decades (e.g. Chapman et al., chapter 1, and Struhsaker, chapter 17, this volume) to climatic induced habitat changes during recent millennia (Tutin & White, chapter 13, this volume) to major biogeographical and global climate changes during the past 60 million years (Fleagle & Reed, chapter 6, this volume). At the greatest scale, there are strikingly regular effects of time on the adaptive diversity of communities. Fleagle & Reed (chapter 6, this volume) showed that despite considerable scatter, there is a remarkably constant rate at which ecological diversity evolved in different biogeographical areas, with no evidence of a saturation effect. Thus diversity increases with time and communities made up of taxa that have relatively old divergences from one another are more diverse than communities made up of taxa with relatively recent common ancestors. Their analysis also indicates that the temporal patterning of ancient biogeographical events influences the pattern of current ecological diversity within communities. Regional faunas, and the communities within them, that result from a few adaptive explosions tightly clustered in time show a lower correlation between divergence times and ecological distance between taxa than do regional faunas and communities which have been formed by a long sequence of phylogenetic and biogeographical events. However, these patterns are based on reconstruction of the history of extant taxa, and necessarily provide a very limited view of past radiations. Attempts to reconstruct paleo-communities are fraught with difficulties of sampling biases as well as the hazards of reconstructing behavior and ecology from fragmentary remains, but nevertheless need to be pursued.

For more recent time periods, such as the last several thousand years, we have more detailed climatic records (Tutin & White, chapter 13, this volume) and in some cases good fossil remains (Godfrey *et al.*, 1997). These show profound changes in many aspects of ecology and species compositions. Perhaps the most extensive studies of middle-range influences on primate community structure have been the few attempts to relate species distributions and speciation patterns to Pleistocene refugia (e.g. Kinzey, 1997). The biogeography of individual species distributions and the composition of assemblages are intimately related (Eeley & Lawes, chapter 12, this volume). By and large, biogeographical studies of primate distributions in conjunction with available information on recent climate change and the distribution of other plant and animal taxa are an area of research that is largely untapped, but offers great potential.

It has become obvious in recent years that there are few if any recent primate habitats that have not been affected to some degree by human activities in the form of selective logging, deforestation, or removal or introduction of species. A critical issue raised by Struhsaker in his chapter is whether such disturbance precludes any meaningful comparisons of extant communities to determine the mechanisms generating similarities and differences. Certainly this remains an open question that deserves careful thought and carefully planned study. However, there are many reasons to believe that while human activities are a factor worth studying themselves (see Peres, chapter 15, this volume; Wilson & Johns, 1982; Cannon *et al.*, 1998), they should not preclude our abilities to understand the mechanisms underlying community structure.

First of all, it has become increasingly clear that the concept of a static pristine rainforest is more a platonic ideal than a reality (Chazdon, 1998; Whitmore, 1998). Virtually all forests and their faunas are constantly undergoing change due to climatic factors, natural disasters, disease and successional processes (e.g. Tutin & White, chapter 13, this volume). Plant and animal communities are dynamic phenomena. Stasis is rare. In many cases human disturbances can be shown to parallel changes such as tree falls that are normally occurring in the absence of humans. In some cases the difference may be in rate only. When the human factors such as hunting or tree removal can actually be quantified, this can provide a valuable test of the effects of individual factors on individual species and community structure (e.g. Peres, chapter 15, this volume; Wilson & Johns, 1982; Johns, 1992).

Secondly, as the chapters of this book demonstrate, despite many gaps in our knowledge, there are many consistent patterns in the structure of communities within regions and in the processes that seem to be controlling community structure that cut across differences in the types of recent disturbances that individual communities may have received. This result suggests that many aspects of community structure may be robust to minor insults (e.g. Cannon *et al.*, 1998). Obviously selection of samples requires some caution and clear exceptions deserve careful examination. Nevertheless, the presence of past disturbance should not be an *a priori* justification for not studying primate communities. On the contrary as the chapters by Struhsaker (chapter 17, this volume) and Wright and Jernvall (chapter 18, this volume) both emphasize, the effects of human activities on the world habitats and primates are only going to increase. It is critical that we make every effort to understand how these activities, as well as other factors, affect the communities we have today if we are to have any success in preserving today's communities for the future.

REFERENCES

Alberch, P. (190). Natural selection and developmental constraints: external vs. internal determinants of order in nature. In *Primate Life History and Evolution*, ed. C. Jean DeRousseau, pp. 15–36. New York: John Wiley and Sons.

Cannon, C.H., Peart, D.R. & Leighton, M. (1998). Tree species diversity in commercially logged Bornean rainforest. *Science*, **281**, 1366–8.

Chazdon, R.L. (1998). Ecology – Tropical forests – Log 'em or leave 'em? *Science*, **281**, 1295–6.

Difiore, A. & Rendall, D. (1994). Evolution of social organization – a reappraisal for primates by using phylogenetic methods. *Proceedings of the National Academy of Sciences USA*, **91**, 9941–5.

Fleagle, J.G. & Reed, K.E. (1996). Comparing primate communities, a multivariate approach. *Journal of Human Evolution* **30**, 489–510.

Ganzhorn, J.U. (1992). Leaf chemistry and the biomass of folivorous primates in tropical forests – test of a hypothesis. *Oecologia*, **91**, 540–7.

Ganzhorn, J.U. (1993). Flexibility and constraint of *Lepilemur* ecology. In *Lemur Social Systems and their Ecological Basis*, ed. P.M. Kappeler & J.U. Ganzhorn, pp. 153–65. New York: Plenum Press.

Ganzhorn, J.U., Malcomber, S., Andrianantoanina, O. & Goodman, S.M. (1997). Habitat characteristics and lemur species richness in Madagascar. *Biotropica*, **29**, 331–43.

Godfrey, L.R., Jungers, W.L., Reed, K.E., Simons, E.L. & Chatrath, P.S. (1997). Subfossil lemurs: Inferences about past and present primate communities in Madagascar. In *Natural Change and Human Impact in Madagascar*, ed. S.M. Goodman & B.D. Patterson, pp. 218–56. Washington, DC: Smithsonian Institution.

Jernvall, J. & Wright, P.C. (1998). Diversity components of impending primate extinctions. *Proceedings of the National Academy of Sciences USA*, **95**, 11279–83.

Johns, A.D. (1992). Vertebrate responses to selective logging – implications for the design of logging systems. *Philosophical Transactions of the Royal Society B*, **335**, 437–42.

Kappeler, P.M. & Heymann, E.W. (1996). Nonconvergence in the evolution of primate life history and socio-ecology. *Biological Journal of the Linnean Society*, **59**, 297–326.

Kay, R.F., Madden, R.H., van Schaik, C.P. & Higdon, D. (1997). Primate species richness is determined by plant productivity: Implications for conservation. *Proceedings of the National Academy of Sciences USA*, **94**, 13023–7.

Kinzey, W.G. (1997). Synopsis of New World Primates (16 Genera). In *New World Primates, Ecology, Evolution and Behavior*, ed. W.G. Kinzey, pp. 169–324. New York: Aldine de Gruyter.

Oates, J.F., Whitesides, G.H., Davies, A.G., Waterman, P.G., Green, S.M., Dasilva, G.L. & Mole, S. (1990). Determinants of variation in tropical forest primate biomass – new evidence from West Africa. *Ecology*, **71**, 328–43.

Peres, C.A. (1997). Effects of habitat quality and hunting pressure on arboreal folivore densities in neotropical forests: A case study of howler monkeys (*Alouatta* spp.). *Folia Primatologica*, **68**, 199–222.

Reed, K.E. & Fleagle, J.G. (1995). Geographic and climatic control of primate diversity. *Proceedings of the National Academy of Sciences USA*, **92**, 7874–6.

Rendall, D. & Di Fiore, A. (1995). The road less traveled: phylogenetic perspectives in primatology. *Evolutionary Anthropology*, **4**, 43–52.

Whitmore, T.C. (1998). *An Introduction to Tropical Rain Forests*, 2nd edn. New York: Oxford University Press, 296 pp.

Wilson, W.L. & Johns, A.D. (1982). Diversity and abundance of selected animal species in undisturbed forest, selectively logged forest and plantations in East Kalimantan, Indonesia. *Biology and Conservation*, **24**, 205–18.

Systematic index

Subject index